Ecological Research Monographs

Series editor
Yoh Iwasa

More information about this series at http://www.springer.com/series/8852

Jotaro Urabe • Tohru Nakashizuka
Editors

Ecological Impacts of Tsunamis on Coastal Ecosystems

Lessons from the Great East Japan Earthquake

Springer

Editors
Jotaro Urabe
Graduate School of Life Sciences
Tohoku University
Sendai, Japan

Tohru Nakashizuka
Graduate School of Life Sciences
Tohoku University
Sendai, Japan

ISSN 2191-0707 ISSN 2191-0715 (electronic)
Ecological Research Monographs
ISBN 978-4-431-56446-1 ISBN 978-4-431-56448-5 (eBook)
DOI 10.1007/978-4-431-56448-5

Library of Congress Control Number: 2016955041

Cover illustration: *Left*: Matsukawa-ura Lagoon in Fukushima Prefecture (photo by Takao Suzuki); *right*: a coastal pine forest in Iwate Prefecture (photo by Tohru Nakashizuka), damaged and destroyed, respectively, by the tsunamis caused by the Great East Japan Earthquake in March 2011.

Printed on acid-free paper

This Springer imprint is published by Springer Nature
The registered company is Springer Japan KK
The registered company address is: Chiyoda First Bldg. East, 3-8-1 Nishi-Kanda, Chiyoda-ku, Tokyo 101-0065, Japan

Preface

Disturbance is a key concept in understanding various ecological issues such as biodiversity, community assemblages, ecosystem stability and resilience, and nature conservation. Earthquakes and resulting tsunamis are extreme examples of such disturbance. Because these events can result in rare but intensive disturbance of large areas, monitoring and examination of organismal response to these events at species, population, community and landscape levels provide invaluable opportunity to uncover the ecological significance of large-scale disturbances.

In our society, earthquake and tsunami disasters have the potential to cause great loss to human lives and properties. Therefore, society has made extensive efforts to reduce the risks of such natural disasters. However, knowledge of biological and ecological responses regarding such extreme disturbances is also essential and has led to the recent development of the Ecosystem based Disaster Risk Reduction (EcoDRR). Ecosystems routinely provide us with important ecosystem services. Therefore, a harmonious balance needs to be considered between ecosystem properties and man-made infrastructure to reduce disturbance risk to ecosystems.

The Great East Japan Earthquake on 11 March 2011, registering 9.0 on the Moment Magnitude Scale (Mw), caused large tsunamis that struck the Pacific coastline of eastern Japan (Tohoku), creating intense high impact disturbances to various ecosystems along the coastline. In advance of the disaster, the coastal ecosystems in Tohoku area had been frequently visited and studied by animal and plant ecologists, in part due to the great natural value of the area. Immediately after the Great East Japan Earthquake it was difficult to access and examine coastal ecosystems in detail because infrastructures were heavily damaged in the region. However, as soon as possible following the disturbance, a number of ecologists initiated studies to examine and evaluate the impacts and implications of the Great East Japan Earthquake tsunamis to these coastal ecosystems in the Tohoku area. This book is a compilation of edited articles from these studies. Most authors of these articles have previously studied the coastal ecosystems in Tohoku area prior to this earthquake and thus were highly qualified to describe the changes in the ecosystems as a result of the tsunamis.

This book begins by presenting an outline of the Great East Japan Earthquake disaster (Chap. 1). The book is then divided into two parts; Part I reports on 14 aquatic studies (Chaps. 2, 3, 4, 5, 6, 7, 8, 9, 10, 11, 12, 13, 14, and 15) and Part II is composed of eight terrestrial studies (Chaps. 16, 17, 18, 19, 20, 21, 22, and 23).

The Great East Japan Earthquake affected various aquatic ecosystems not only because of the tsunamis but also as a result of land subsidence. In Chaps. 2, 3, 4, and 5, responses of sessile animal populations and rocky shore communities to these disturbances are described. In Chap. 6, successional changes in a subtidal benthic community after the tsunami disturbances are analyzed. Among the coastal ecosystems, tidal flats were probably the most vulnerable to disturbances by the tsunamis and land subsidence. In the following six chapters (Chaps. 7, 8, 9, 10, and 11), ecological impacts of these disturbances to the tidal flat communities are reported. These studies uncover how and why the community response differed depending on the tidal flats. In Chaps. 12, 13, and 14, direct and indirect effects of disturbances by the earthquake on molluscan populations are described. These studies serve to understand species-specific differences in vulnerability to the tsunami disturbances. In addition to natural ecosystems, the earthquake and resulting tsunamis significantly damaged coastal infrastructures. Accordingly, harmful substances such as fuels were released from some storages facilities in the coastal areas of Tohoku region. In Chap. 15, the ecological consequences of these fuel spills and the subsequent conflagration are reported.

The Great East Japan Earthquake also seriously affected terrestrial vegetation along the Tohoku Pacific coastal area. In Chapter 16, a remote sensing study of the impacts of the tsunamis on a wide range of coastal ecosystems are reported. Chapters 17 and 18 give accounts of the damages and recovery of coastal sand dune communities, in particular the role of buried seeds in the recovery process as seen in Chap. 18. In Chapter 19, a comprehensive evaluation of the tsunami impacts on coastal flora is presented. In addition, damages to the coastal pine forests by the tsunamis are analyzed in detail in Chap. 20. The tsunamis also created wetlands behind sand dunes. Chapter 21 reports on unexpected disturbance outcomes that lead to a revival of certain endangered species. In Chapter 22, the effect of the tsunamis on forest vegetation and its recovery are reported. And, in the final chapter (Chapter 23), the comprehensive monitoring system about the impact and recovery of ecosystems conducted by Ministry of Environment Japan, is introduced.

Since these are natural events, earthquakes and tsunamis will inevitably recur sometime and somewhere in the future. However, such large and rare disturbances are limited, and thus so is ecological knowledge about these disturbances. We believe that the studies in this book are useful not only to deepen our understanding of these disturbances in ecology but also to seek a way for practical applications of EcoDRR to coastal ecosystems.

We acknowledge Drs. Yoh Iwasa, Tetsukazu Yahara and Takashi Saitoh for their recommendations and encouragements for publishing this book. Finally, we would like to dedicate this book to all the peoples who fell victim and suffered greatly due to the Great East Japan Earthquake and subsequent tsunamis.

Sendai, Japan

Jotaro Urabe
Tohru Nakashizuka

Contents

Chapter 1
Outline of the Great East Japan Earthquake

Naoaki Shimada

Abstract An earthquake occurred at 2:46 P.M. on March 11, 2011, off the Pacific coast of Tohoku. It measured 9.0 Mw on the moment magnitude scale and was the largest recorded earthquake in known Japanese history. Extremely tall tsunami waves were observed along with this earthquake. Ogatsu-cho in Ishinomaki City had the highest level of tsunami flooding at 34.94 m and Omoe in Miyako City had the highest tsunami run-up at 38.38 m. This earthquake brought about enormous damages including 19,418 casualties, 2592 missing persons, and 121,809 completely destroyed buildings. It was decided to call this disaster arising from the earthquake the "Great East Japan Earthquake" at the Cabinet Meeting on April 1, 2011.

Keywords Great East Japan Earthquake • Tsunami • State of damage

1.1 The Earthquake

The 2011 off the Pacific coast of Tohoku Earthquake occurred at 2:46 P.M. on March 11, 2011. It measured 9.0 Mw on the moment magnitude scale and was the largest recorded earthquake in Japanese history. The hypocenter was located at 38°06.2′ north latitude, 142°51.6′ east longitude, and a depth of 24 km. The hypocenter lay on the Sanriku coast (approximately 130 km east-southeast of the Oshika Peninsula in Miyagi Prefecture) (Japan Meteorological Agency 2012b).

By the standards of the Japan Meteorological Agency, the largest measure on the seismic scale, a seven, was observed in Kurihara City in Miyagi Prefecture. In 37 municipalities of the four prefectures of Miyagi, Fukushima, Ibaraki, and Tochigi, a reading above six on the seismic scale was observed. The locations that were observed to have a reading on the seismic scale of six or greater (from Yahaba-cho in Iwate Prefecture to Narita City in Chiba Prefecture) extended over approximately

N. Shimada (✉)
Faculty of Policy Studies, Iwate Prefectural University,
152-52 Sugo, Takizawa, Iwate 020-0693, Japan
e-mail: naoaki@iwate-pu.ac.jp

J. Urabe, T. Nakashizuka (eds.), *Ecological Impacts of Tsunamis on Coastal Ecosystems*, Ecological Research Monographs,
DOI 10.1007/978-4-431-56448-5_1

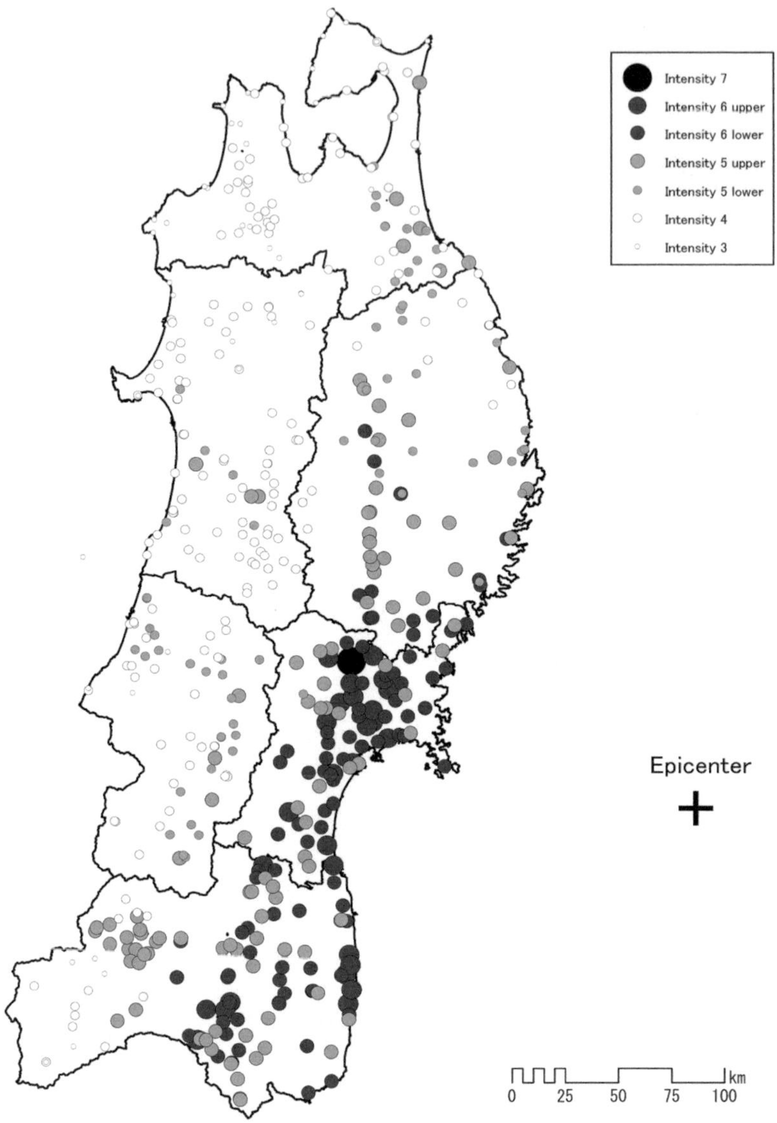

Fig. 1.1 Seismic scale in the Tohoku Area for the 2011 off the Pacific coast of Tohoku Earthquake (Figure created using material from a publication by the Japan Meteorological Agency (2012a))

430 km, suggesting that the strong vibrations extended over a wide area. Strong vibrations were observed mainly in Miyagi Prefecture but also from the center of Fukushima to the eastern part of the prefecture and in southern Iwate Prefecture (Fig. 1.1). The vibrations of the earthquake were observed throughout all of Japan with the exception of Miyazaki Prefecture and Okinawa Prefecture (Japan Meteorological Agency 2012a).

The area around the Oshika Peninsula in Miyagi Prefecture was moved horizontally 5.4 m toward east-southeast by the earthquake. The area between Kamaishi in Iwate Prefecture and Natori in Miyagi Prefecture was moved horizontally more than 3 m and crustal change was observed over a wide area ("Summary of the Amount of Change by the Coastal Observation Authority": http://www.gsi.go.jp/common/000136983.pdf, checked on 5/7/2016). In general it was a movement toward the hypocenter in all locations ("Electronic Reference Points for Crustal Change Received from Continuous GPS Observation" (Geospatial Information Authority of Japan): http://www.gsi.go.jp/chibankansi/chikakukansi40005.html, checked on 5/7/2016).

In the upward and downward movement of the crust, a significant subsidence was observed on the Pacific coast of Tohoku. The area around the Oshika Peninsula in Miyagi Prefecture had the greatest subsidence with a crustal subsidence of approximately 1.1 m. From Yamada in Iwate Prefecture to Iwaki in Fukushima Prefecture, there was a crustal subsidence of more than 0.5 m (Fig. 1.2a). Crustal changes were also observed over a wide area from Hokkaido to the Kinki region ("Electronic Reference Points for Crustal Change Received from Continuous GPS Observation" (Geospatial Information Authority of Japan): http://www.gsi.go.jp/chibankansi/chikakukansi40005.html, checked on 5/7/2016).

1.2 The Tsunamis

Extremely tall tsunami waves were observed along with this earthquake. The tsunami gauges set up by Japan Meteorological Agency and others on the Pacific coast of Tohoku were unable to collect accurate data due to power outages, leakage, and damage from the tsunami (Japan Meteorological Agency 2012b). According to Haraguchi and Iwamatsu (2011a, b) who carried out a field survey of traces of the tsunamis, such as those left behind in buildings, through the 2011 Tohoku Earthquake Tsunami Joint Survey Group, Oguchi-cho in Ishinomaki City had the highest level of tsunami flooding at 34.94 m, and Omoe in Miyako City had the highest tsunami run-up at 38.38 m. The totaled results for each municipality are shown in Fig. 1.2b. This shows that the locations with tsunami flood height above 20 m and with exceptionally large tsunamis were developed ria coast areas of Sanriku. Large tsunamis and tsunami flooding over 10 m in height were observed across a broad area from Hachinohe City in Aomori Prefecture to Souma City in Fukushima Prefecture. Locations where tsunami flooding greater than 5 m in height was observed extended across approximately 580 km in a straight line from Misawa City in Aomori Prefecture at the northernmost point to Asahi City in Chiba Prefecture at the southernmost point (Haraguchi and Iwamatsu, 2011a, b). Tsunamis above 1 m in height were observed along the Pacific coast from Hokkaido to Kagoshima and in the Bonin Islands. Tsunamis were also observed along the Pacific coast of various other countries. A 2.47 m tsunami was observed in Crescent City in the United States and a 2.45 m tsunami was observed in Arica, Chile. There were also observations of tsunamis over 2 m in height in many areas (Japan Meteorological Agency 2012b).

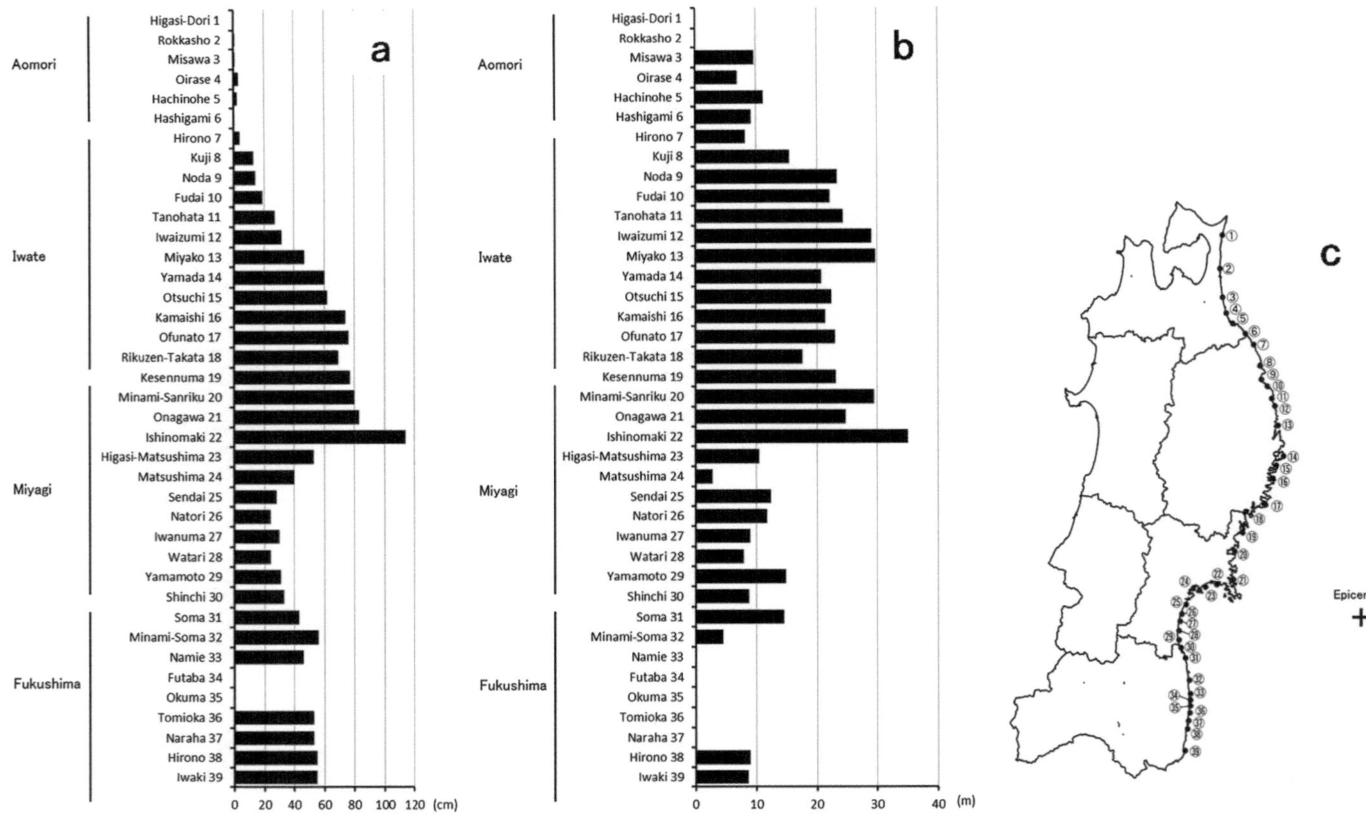

Fig. 1.2 Earth subsidence amount (**a**), Height of Tsunami Flooding (**b**), and Municipal observations (**c**) on the Pacific Coast of the Tohoku Region. The highest value for each municipality is displayed. (*a*) consists of information published by the Geospatial Information Authority of Japan (http://www.gsi.go.jp/BOUSAI/h23_tohoku.html) and (*b*) consists of information from Haraguchi and Iwamatsu (2011a, b)

Table 1.1 Extent of tsunami flooding by prefecture and land use in the 2011 off the Pacific coast of Tohoku earthquake

Flooded area (km^2)						Ratio of flood area to non-flooded area (%)			
	Building and arterial traffic sites	Other	Arable land, forests, golf courses	Rivers, lakes, marshes, beaches, etc.	Total	Building and arterial traffic sites	Other	Arable land, forests, golf courses	Rivers, lakes, marshes, beaches, etc.
Aomori	3	5	5	11	24	11	21	21	47
Iwate	21	9	18	10	58	37	15	31	17
Miyagi	74	27	183	43	327	23	8	56	13
Fukushima	15	10	67	19	112	13	9	60	17
Ibaraki	3	5	3	11	23	16	23	14	46
Chiba	4	1	7	6	17	15	8	40	37
Total	120	58	283	100	561	21	10	50	18

Created using materials from http://www.gsi.go.jp/BOUSAI/h23_tohoku.html (the Geospatial Information Authority of Japan)

There was flooding damage due to the tsunamis arising from this earthquake over a broad area extending from Aomori to Chiba Prefectures (Table 1.1). Large parts of coastal areas of the Miyagi and Fukushima Prefectures were flooded, but the area flooded was not so broad in Iwate Prefecture where there were high flood levels. It is thought that this is because the ria coast in the southern part of Iwate Prefecture and the sea cliffs in the northern part are developed while the plains areas are not developed.

The ratio for the extent of tsunami flooding by land use for each prefecture is shown in Table 1.1. Arable land, forests, and golf courses had the most tsunami flooding between all of the regions at 50 %. Next were areas used for buildings and roads at 21 %. The rate of flooding for arable land, forests, and golf courses in Miyagi and Fukushima prefectures was at 56 % and 60 %, respectively, while the rate of flooding was larger in areas used for buildings and roads for Iwate Prefecture at 37 %. It is thought that this difference is due to the concentration of urban areas along the narrow coastal plains in Iwate Prefecture because of the ria coast and sea cliffs, while in the Miyagi and Fukushima prefectures the coastal plains are broad and the arable land is distributed widely.

1.3 State of Damage

This earthquake brought about enormous damages including 19,418 casualties, 2592 missing persons, 6220 injured, 121,809 completely destroyed buildings, 278,496 half destroyed buildings, and 744,190 partially damaged buildings (current as of March 1, 2016) (Table 1.2). Around 470,000 evacuees occurred directly

Table 1.2 Human and building losses from the 2011 off the Pacific coast of Tohoku earthquake

	Deceased	Missing	Injured	Total destruction	Half destroyed	Partially destroyed
Aomori	3	1	110	308	701	1005
Iwate	5132	1124	211	19,597	6571	18,959
Miyagi	10,549	1239	4145	82,999	155,129	224,195
Fukushima	3626	225	183	15,169	78,960	141,454
Other	108	3	1571	3736	37,135	358,577
Total	19,418	2592	6220	121,809	278,496	744,190

Created from materials from the Fire and Disaster Management Agency (2016) (http://www.fdma.go.jp/bn/280308_%E7%AC%AC153%E5%A0%B1.pdf)

Table 1.3 Amount of monetary losses in agriculture, forestry, and fisheries

	Agriculture (billion yen)	Forestry (billion yen)	Fisheries (billion yen)	Total damage (billion yen)
Iwate	68.8	29.6	564.9	663.3
Miyagi	550.5	55.1	689.6	1295.20
Fukushima	239.5	49.5	92.4	381.4

Created from information from the Tohoku Regional Agricultural Administration Office, Ministry of Agriculture, Forestry and Fisheries of Japan (2016)
http://www.maff.go.jp/tohoku/osirase/higai_taisaku/hukkou/pdf/fukkou160309.pdf

following the earthquake disaster. Even in February 2016, 5 years after the disaster, there were still around 174,000 people living as evacuees, such as those residing in temporary housing ("The Present State of and Efforts Towards Reconstruction" (Reconstruction Agency): http://www.reconstruction.go.jp/topics/main-cat7/sub-cat7-2/20160304-4_sankousiryou1-1.pdf, check on 5/7/2016).

The damages to the agriculture, forestry, and fisheries industries totaled to 1.12 trillion yen for the agricultural and forestry industries and 1.26 trillion yen for the fisheries industry, a total of approximately 2.38 trillion yen ("Efforts to Support the Reconstruction of the Agriculture, Forestry, and Fishery Industries Following the Tohoku Earthquake" (Ministry of Agriculture, Forestry and Fisheries of Japan): http://www.maff.go.jp/j/kanbo/kihyo02/fukkou/pdf/1603_zen.pdf, checked on 5/7/2016). The conditions in Iwate, Miyagi, and Fukushima Prefectures, where damage was particularly significant, are collected in Table 1.3. Fisheries suffered the largest monetary loss in Iwate. In Miyagi the agricultural and fishery industries suffered around the same monetary losses, while agriculture suffered the largest loss in Fukushima ("Efforts of the Tohoku Regional Agricultural Administration Office with the Restoration of Agriculture and Agricultural Communities", Tohoku Regional Agricultural Administration Office, Ministry of Agriculture, Forestry and Fisheries of Japan: http://www.maff.go.jp/tohoku/osirase/higai_taisaku/hukkou/pdf/fukkou160309.pdf, checked on 5/7/2016).

The affected agricultural areas amounted to roughly 20,000 ha in these three prefectures with around 70 % of the area being in Miyagi Prefecture (Table 1.4). Among

Table 1.4 Conditions in agricultural land restoration

	Affected agricultural area	Area reopened for agriculture by March 2016	Area with instructions for evacuation	Agricultural land put to another use	Agricultural land restoration rate (%)
Iwate	730	490		60	67.1
Miyagi	14,340	12,660		630	88.3
Fukushima	5460	1820	2120	580	33.3
Total	20,530	14,970	2120	1270	72.9

Created from information from the Tohoku Regional Agricultural Administration Office, Ministry of Agriculture, Forestry and Fisheries of Japan (2016)
http://www.maff.go.jp/tohoku/osirase/higai_taisaku/hukkou/pdf/fukkou160309.pdf

Table 1.5 Weight of catches, value of catches, and comparisons with before the earthquake disaster

	February 2015–January 2016		Comparison to before the earthquake disaster	
	Weight of catch (tons)	Value of catch (billion yen)	Weight of catch	Value of catch
Iwate	93,000	16	67 %	82 %
Miyagi	245,000	58	79 %	98 %
Fukushima	6000	1	59 %	36 %
Total	345,000	74.3	74 %	93 %

Created from information from the Fisheries Agency of Japan (2016)
http://www.jfa.maff.go.jp/j/yosan/23/pdf/1603kadaigenjou.pdf

these agricultural lands, 73 % in total has been restored. However, in Fukushima Prefecture, only 33 % of agricultural land has been restored ("Efforts to Support the Reconstruction of the Agriculture, Forestry, and Fishery Industries Following the Tohoku Earthquake" (Tohoku Regional Agricultural Administration Office, Ministry of Agriculture, Forestry and Fisheries of Japan): http://www.maff.go.jp/tohoku/osirase/higai_taisaku/hukkou/pdf/fukkou160309.pdf, checked on 5/7/2016).

In the fishing industry, approximately 29,000 fishing boats and 319 fishing harbors were damaged. The weight of catches and the value of catches also fell significantly (Table 1.5). The objective for restoring fishing boats was 20,000 boats and approximately 18,000 boats, or 91 %, have been restored. Among fishing harbors damaged, 311 have been restored until 2016 so that they are able to land boats. The weight of the catch has recovered 74 % compared to before the disaster and the value of the catch has recovered 93 %, but the situation remains still poor in Fukushima Prefecture (Table 1.5) ("The Current Situation and Issues with Restoring the Fisheries Industry" (Fisheries Agency of Japan): http://www.jfa.maff.go.jp/j/yosan/23/pdf/1603kadaigenjou.pdf, checked on 5/7/2016).

It was decided to call this disaster arising from the earthquake the "Great East Japan Earthquake" at the Cabinet Meeting on April 1, 2011.

References

Haraguchi T, Iwamatsu A (2011a) Detailed maps of the impacts of the 2011 Japan Tsunami Vol. 1: Aomori, Iwate and Miyagi prefectures. Koko-Shoin Publishers, Tokyo, in Japanese with English abstract

Haraguchi T, Iwamatsu A (2011b) Detailed maps of the impacts of the 2011 Japan Tsunami Vol. 2: Fukushima, Ibaraki and Chiba prefectures. Koko-Shoin Publishers, Tokyo, in Japanese with English abstract

Japan Meteorological Agency (2012a) Monthly report on earthquakes and volcanoes in Japan December 2012: 177–187 (in Japanese)

Japan Meteorological Agency (2012b) Report on The 2011 off the Pacific coast of Tohoku Earthquake. Technical Report of the Japan Meteorological Agency. 133:479pp. (in Japanese with English abstract)

Part I
Aquatic Ecosystems

Chapter 2
Normality of Succession of an Intertidal Community After the Great East Japan Earthquake

Ryusuke Kado and Nobuyoshi Nanba

Abstract On March 11, 2011, the Great East Japan Earthquake occurred with an epicenter off the Tohoku area in northern Honshu, Japan. By the Tohoku earthquake tsunami and following ground subsidence, coastal organisms seem to have experienced various physiological and biological effects. We have continuously examined influence of the earthquake to the intertidal organisms at Sakihama in Iwate Prefecture for 4 years since August 2011 by using video camera footage. The surveyed site between E.H.W.S. and E.L.W.S. that has about a 180 cm tidal range experienced a subsidence of 130 cm due to the earthquake. The first colonizers on the newly submerged substrate were barnacles *Semibalanus cariosus*. The hard uneven substrate formed by the settlement of the barnacle facilitated the recruitment of the green alga *Ulva intestinalis* first and then the mussel *Mytilus galloprovincialis* followed. This multilayered community structure increased in species richness for 2 years. However, the species richness subsequently decreased with increasing abundance of the sea urchin *Strongylocentrotus nudus* which was the causative species in rocky shore denudation in this area before the earthquake. This biological succession was almost the same as that of a previous experiment from 2004 to 2006 which was carried out to examine the ecological roles of *S. cariosus* using test plates installed at the same location. The main difference between the two biological successions during the two experimental periods was the occurrence of a new alien barnacle *Perforatus* (*Balanus*) *perforatus* after the earthquake. According to the monthly monitoring data of the sedentary organisms on the test plates set on the jetty wall in Sakihama, this barnacle was found for the first time in 2012 and has been settling on them every summer. Just after the earthquake, many barges and tugboats came from various localities in Japan for reconstruction activities along the Tohoku coast. One barge, registered in Niigata, a city on the Japan seacoast where *P. perforatus* has become common, was anchored near the research site in 2013 and had this barnacle on its hull and fenders. Since the settlement of the alien

R. Kado (✉) • N. Nanba
School of Marine Biosciences, Kitasato University,
1-15-1, Kitasato, Minami-ku, Sagamihara, Kanagawa 252-0373, Japan
e-mail: rukado@kitasato-u.ac.jp

J. Urabe, T. Nakashizuka (eds.), *Ecological Impacts of Tsunamis on Coastal Ecosystems*, Ecological Research Monographs,
DOI 10.1007/978-4-431-56448-5_2

barnacle occurred after the earthquake, it was likely that these boats played a role in the dispersion of *P. perforatus* into this area. This possibility implies that the earthquake affected on the intertidal communities not only directly via the tsunamis and subsidence but also indirectly via anthropogenic activities.

Keywords Biological succession • Intertidal community • The Great East Japan Earthquake • Tsunami • Alien species

2.1 Introduction

The littoral zone is a transitional ecotone formed in between tides. It is 1.7–1.8 m in height along the northeast coast of Honshu Island, Japan, and the vertical zonation formed by several major dominant organisms has been shown to be dependent on environmental gradients (Raffaelli and Hawkins 1996). *Semibalanus cariosus*, a barnacle widely distributing in northern parts of the North Pacific, is one of the dominant species forming a clear zonal distribution. This species is called a thatched barnacle as it has many radiating ribs on its shell (Newman and Abbott 1980) and forms a hard substratum for other organisms such as the green alga *Ulva (Enteromorpha) intestinalis*, red alga *Analipus japonicus*, and mussels *Mytilus* spp. in the littoral zone and *Undaria pinnatifida* and *Laminaria religiosa* in the upper sublittoral zone (personal observation). Such a multilayered structure based on *S. cariosus* has provided interstitial spaces for small crustaceans, mollusks, and annelids and has contributed to the increase in biodiversity of the littoral and upper sublittoral zones (Kado, unpublished data). Mussels in particular could not recruit in the intertidal zone on flat concrete jetty walls without the presence of the shells of the thatched barnacle (Kado, unpublished data).

However, *S. cariosus* had a mass mortality during the period from 1998 to 1999 (for unknown reasons), and most of them had dropped off from the jetty walls along the Sanriku coast (Kado, unpublished data). As a result, the community on the jetty walls was simultaneously lost with the exfoliated barnacles. Accordingly, only several species dominated in the littoral zone on the jetty walls of the Sanriku coast before 2011. For example, the surf barnacle, *Chthamalus challengeri*, occurred abundantly in the upper littoral; the oyster, *Crassostrea gigas*, was abundant in the middle littoral; and the serpulid, *Hydroides ezoensis*, dominated in the zone just below *C. gigas*.

A previous study showed that at a jetty wall of the Sanriku coast, *S. cariosus* recruited every year since 1989 on test plates that were set at 40 cm below the mean sea level (M.S.L.) along the jetty wall, although the recruited individuals varied according to years (Kado 2003a). However, juveniles of the barnacle have been unable to establish a clear zonation pattern since 2000 on the jetty wall. The low barnacle abundance on the jetty wall may have been due to predation by the sea urchin *Strongylocentrotus nudus*, which has increased recently along the Sanriku coast (Fujiwara 2008). *S. nudus* consumed not only barnacles but also other sedentary organisms like algae, mussels, and other organisms before they are able to grow large enough to mature. Therefore, this species was often regarded as a causative species of rocky shore denudation in this area (Kado 2006).

On March 11 in 2011, the Great East Japan Earthquake occurred along northeast coast of Honshu in Japan and caused large tsunamis and ground subsidence. In this study, to understand the communities grown on jetty walls on the Sanriku coast, we first briefly described results of the experiment that was carried out before the earthquake on the jetty wall at Sakihama on the Sanriku coast. Based on the results, I hypothesized that biological interactions such as predation by sea urchins were principal factors regulating structures of the intertidal communities on the jetty walls. We examined the hypothesis by comparing some results of an experiment before the earthquake and continual observation of the community developed on a jetty wall after the earthquake. In addition, we examined the possibility that an invasion of a barnacle species that was presumably introduced by anthropogenic activities after the earthquake has played a pivotal role in determining the community structure on the jetty walls on the Sanriku coast in recent years.

2.2 An Experiment Before the Great East Japan Earthquake

To examine the ecological role and pattern of formation of littoral communities, we conducted an experiment on a wall of the Second Sakihama Jetty, Iwate Prefecture, by using plastic test plates as an artificial substrate from 2004 to 2006 (Fig. 2.1). In this experiment, we prepared three treatments each composed of three PVC test plates (25 × 25 cm) (Fig. 2.2); sets A and B were plates with *S. cariosus* settled on their surface. The *S. cariosus* were allowed to settle on the plates which were set in the intertidal at Sakihama in April in 2004 and grew to around 10 mm in diameter

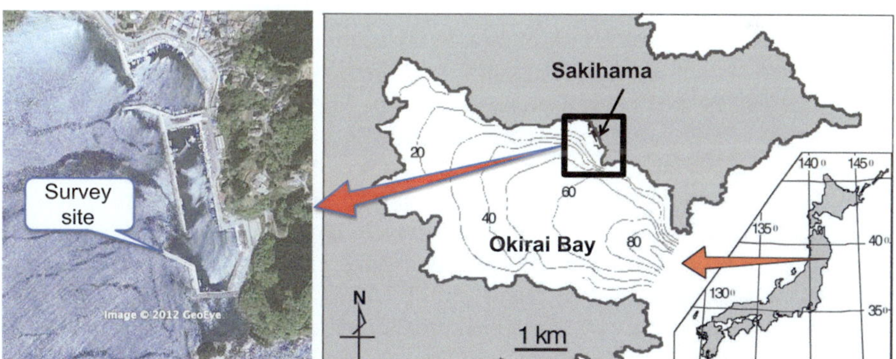

Fig. 2.1 Maps of Okirai Bay, Iwate Prefecture, showing the Sakihama area and the Second Jetty where the present study was performed

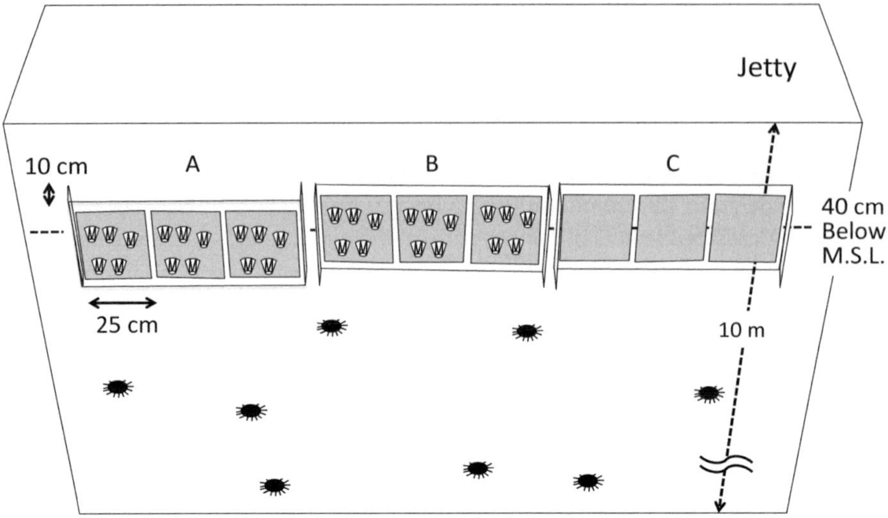

Fig. 2.2 Diagram showing the schematic arrangement of the test plates set on the wall of the jetty for assessing the roles of the barnacle *Semibalanus cariosus* and sea urchin *Strongylocentrotus nudus* in community assemblage processes. Three test plates (25 cm^2 on each side) were set inside three stainless frames. Plates A and B with the barnacles already settled on the plates before this experiment were placed 10 cm apart from the wall (Plate A) using the stainless frame guard and placed directly on the wall (Plates B), respectively. Plates C without barnacles on the plates was placed directly on the wall

when the experiment started. The third set, set C, was used as a control without any sessile animals on the plates. Set A was placed 10 cm apart from the jetty wall surface to decrease chance of predation by sea urchins, and set B and set C were placed directly on the wall where sea urchins, *S. nudus*, were distributed. All of these plates were installed at 40 cm below MSL during June of 2004 (Fig. 2.2) when the experiment was initiated. Then, the number of species and total weight including the plate itself were investigated every 2 months over 2 years.

An outline of the results is as follows. In the autumn, many mussels *M. galloprovincialis* had settled only on the plates with the barnacles and not on the plates without barnacles, i.e., control plates (set C). In May 2005, i.e., 11 months later from the start of experiment, small numbers of surf barnacles and green and red macroalgae had colonized on the set C (Fig. 2.3C). Contrastingly, set A and B plates showed a clear two-layered structure with many grown mussels that covered over the barnacles. The barnacles lived and even grew to large individuals underneath the upper layer of mussels. Green and red macroalgae grew moderately on the set A plates but were limited on set B plates (Fig. 2.3A, B). Between the space with barnacles and the mussel canopy, some interstitial organisms such as small snails were found on the plates of all treatments.

In May 2006, 2 years after the start of the experiment, a few *C. challengeri* and *C. gigas* and small numbers of green and red algae were found on the set C plates. On the set A plates, mussels had continually grown on the barnacles which were still

Fig. 2.3 Succession of the community assemblages on the test plates manipulated with the three different conditions: (**A**) plates started with barnacle *Semibalanus cariosus* and placed them 10 cm apart from the wall that inhibited access of sea urchins; (**B**, **C**) plates started with or without *S. cariosus*, respectively, and were placed directly on the wall enabling access of sea urchins

alive but grew little, and green and red algae flourished on the mussels. On the set B plates, number of the mussels and barnacles decreased and bare spaces appeared with a few oysters and green algae. In December 2006, 30 months after the start of experiment, communities on the set C and B plates showed almost the same structure as that found on the natural jetty walls where only *C. challengeri* and *C. gigas* dominated. On the A set plates, however, a number of large mussels dominated externally; the reddish barnacle *Megabalanus rosa* settled on the mussel shells and *S. cariosus* attached underneath. In addition, a number of small snails, annelids, flatworms, and amphipods were found interstitially on the A set plates.

These differences in communities between the A and B set plates and the A and C set plates showed that the sea urchin *S. nudus* and barnacle *S. cariosus* are key species in the littoral community. Excess density of the former resulted in a "sea urchin barren," and recruitment of *S. cariosus* contributed to the formation of a littoral community with a high biodiversity. Inside the community, the growth of *S. cariosus* was inhibited by overgrowth of the mussel *M. galloprovincialis*. However, the barnacle seemed to benefit from the cover of the mussels because they were apparently protected from predation by sea urchins and predatory gastropods. On the other hand, *M. galloprovincialis* received an advantage from the barnacle since the barnacles form a substrate for the settlement of this mussel although they had a risk to be preyed on by sea urchins. Thus, there may be a trade-off effect between these two species. Based on these observations. We hypothesized that as far as sea urchins were abundant, the community would be continually dominated by oysters and serpulids and would not return to the previous status dominated by *S. cariosus* (Fig. 2.3C).

2.3 Succession of Intertidal Community After the Great East Japan Earthquake

2.3.1 Methods

The post-tsunami research was performed at the former experimental site mentioned above. The experiment began on July 2, 2011, using an underwater camera (WTW-WA7000H, Wireless Tsukamoto Co., Ltd.) which was fixed on the top of a stainless steel frame-shaped pyramid (Fig. 2.4). The camera, focused on the vertical wall surface, was brought down slowly using a rope attached to a frame from the uppermost part of the jetty to the bottom. Video images were recorded on a portable hard disk (PoliceNote 2400S, Sun-Mechatronics) with a monitor screen. Depth and density of organisms were measured with a measure scale tape that was lowered from the top of the jetty and was displayed on each image. The top of the jetty was 32 cm higher than M.S.L. at that time and about 11 m above the seabed. Surveys were carried every 2 months during the first year and four times in the following years. In addition, to compare the communities developed on the jetty wall between before and after the earthquake, monthly monitoring on the abundance of sedentary

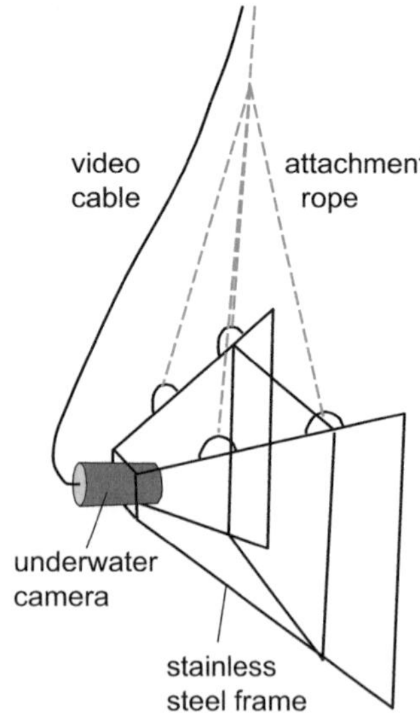

Fig. 2.4 A diagram showing the underwater camera system used in this study to observe the littoral and sublittoral communities on the jetty wall

organisms on two PVC test plates (25 × 25 cm) has been investigated. The test plates were set at 40 cm below M.S.L., kept for a month on the Sakihama jetty wall, and then changed each month to new ones.

2.3.2 Results and Discussion

2.3.2.1 Pioneer Species into the New Habitats

At the jetty where this research was made in July 2011, the extent of the subsidence was 1.3 m (Fig. 2.5b), which was relatively more than other adjacent land areas, due to difference in substrata hardness and that the site stands out about 200 m apart from the land (Fig. 2.1). The observed scale of the subsidence means that almost two thirds of the sessile littoral organisms in this site sank into the subtidal zone.

The first settlers were the barnacle *S. cariosus.* They were abundant and completely covered the bare area from the mid-intertidal to about 6 m below M.S. L. not only on the wall but also on rocky shores around the bay (Fig. 2.5e). They survived with a high rate except loss due to detachment by their own overcrowded settlement. According to our previous study (Kado et al. 2002), this barnacle releases its nauplii when the surface chlorophyll *a* concentration exceeds 1 mg/m^3 at the beginning of the spring bloom. The nauplii settled on the shore 3–5 weeks later when the chlorophyll a concentration remained at approximately the same level (Kado et al. 2002). Although the chlorophyll *a* concentrations were unknown during March to June 2011, it was most likely that the concentration had been kept over 1 mg/m^3 at least for more than a month after the earthquake. This high recruitment of *S. cariosus* right after the earthquake might have been attributed to two reasons. One of the reasons may be the decrease in filter feeders. Tsunamis caused by the Great East Japan Earthquake in 2011 washed away a number of organisms including cultured species such as oysters *C. gigas*, scallop *Mizuhopecten yessoensis*, and sea squirt *Halocynthia roretzi* and the fouling organisms such as barnacles *M. rosa*, mussels *M. galloprovincialis*, and other sedentary animals that settled on the culture rafts, buoys, and ropes. All of these filter feeders fed on phytoplankton. Accordingly, nauplii of *S. cariosus* could have sufficiently fed on phytoplankton during the algal spring bloom and maintained a high survival rate throughout the larval stages. The heavy settlement of this barnacle was, however, not the first case in this area. Indeed, the same phenomenon had been observed in May 2004. However, at that time, the abundance of the sea urchin, *Strongylocentrotus nudus*, reached more than ten individuals/m^2 and almost all of the settled *S. cariosus* had been consumed by the sea urchin within the next 3 months (Kado personal observations). This high sea urchin abundance had been continued until February in 2011. However, the sea urchin abundance decreased due to the tsunami. Kawamura et al. (2014) also reported that abundance of sea decreased in Otsuchi Bay, Iwate Prefecture, by the backwash of tsunamis caused by the earthquake. Thus, the second reason of high settlement rate of *S. cariosus* right after the tsunamis was likely due to low sea urchin abundance (<1 individual/m^2) (Fig. 2.6).

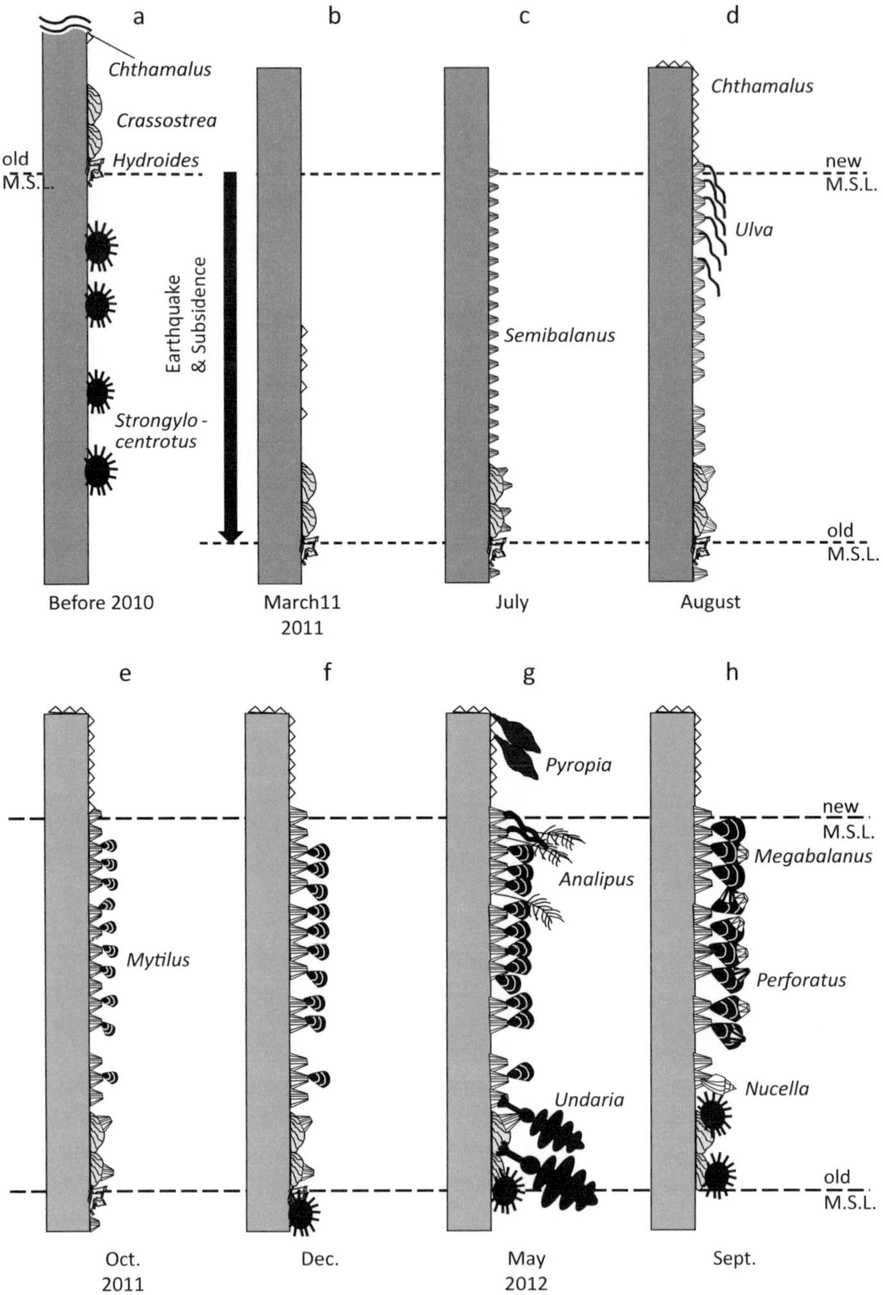

Fig. 2.5 Diagrams showing successional changes in the intertidal communities on the wall of the Second Jetty in Sakihama before and 1 year after the Great East Japan Earthquake. Details are described in the text

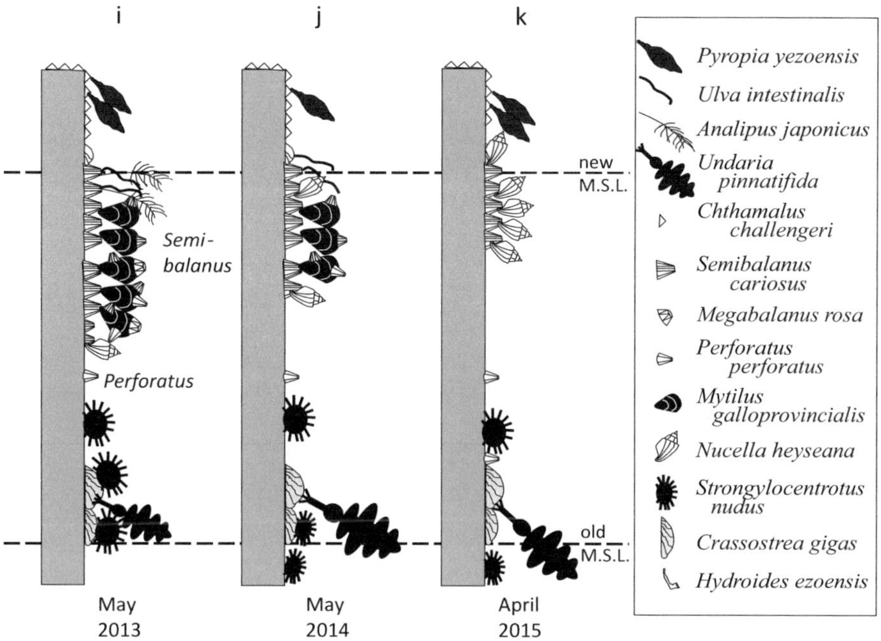

Fig. 2.5 (continued)

2.3.2.2 Formation of Littoral Communities

2011

Soon after the heavy settlement of *S. cariosus*, several organisms were initiated to form a new community over the barnacles as substrates. For example, the green alga *Ulva* (*Enteromorpha*) *intestinalis* initially settled on the shells of *S. cariosus* growing over the mid-littoral in August (Fig. 2.5d). During the same period, surf barnacle *C. challengeri* recruited in the high littoral zone with a high density. Then, the mussel *M. galloprovincialis* followed to settle on the shells of the barnacles from the mid-littoral to the upper sublittoral zone from summer to fall in 2011 (Fig. 2.5e). During the summer of 2011, oysters also settled over the shells of barnacles in the mid-littoral zone. Oysters *C. gigas* and serpulids *H. ezoensis* which had colonized in the mid-littoral zone before the tsunamis were taken to the sublittoral zone due to the subsidence of substrate jetty by the earthquake. However, these animals lived there even after the earthquake.

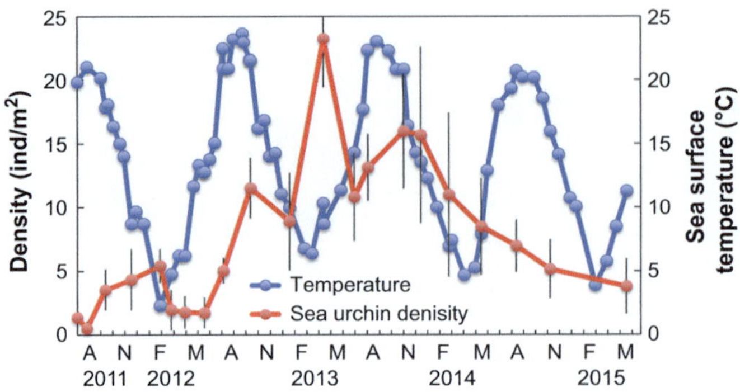

Fig. 2.6 Changes in average abundance of sea urchin *Strongylocentrotus nudus* (*red line*) on the wall of the Second Jetty in Sakihama and surface seawater temperature (*blue line*) in front of the wall at Sakihama fishing port in Iwate Prefecture. *Error bars* indicate standard deviation

2012

In winter 2012, *S. cariosus* grew to more than 10 mm in shell diameter (Fig. 2.5f). Abundance of the sea urchin had increased to about five individuals/m^2 by the summer of 2012 (Fig. 2.6), and the abundance of *S. cariosus* in the sublittoral zone decreased to about 3×10^3 individuals/m^2. *M. galloprovincialis* in the mid- to lower littoral zones grew larger, resulting in forming a black mussel bed (about 15×10^2 individuals/m^2) over the barnacles.

In spring 2012, the brown alga *Undaria pinnatifida* and red alga *Analipus japonicus* grew on the barnacles in the sublittoral zone and on the mussels in the mid-littoral zone, respectively (Fig. 2.5g). *S. cariosus* usually settled in spring but had a poor recruitment in this year. In the summer and fall of this year, *C. challengeri*, *S. cariosus*, and *M. galloprovincialis* that were sessile animals were abundantly found in the upper, the middle to the lower, and the lower intertidal zones, respectively (Fig. 2.5h). The mussels anchored by the byssus matrix on the barnacles. Such a community structure was found before 1998 (Kado personal observation) and accorded well with those in the experiment done before the earthquake in which predation pressure of sea urchin was artificially decreased (see above). This concordance implies that biological succession observed in the littoral zone after the earthquake was not an ecologically rare event.

However, we found some rare phenomena. In the summer 2012, the reddish barnacle *Megabalanus rosa* settled on the mussel shells (Fig. 2.5h). This barnacle was common in this area on the culture buoys and ropes before the earthquake but was not observed to occur on the shells of mussels. This rare event was also observed in the experiment done before the earthquake in which predation pressure of sea urchin was artificially decreased (see above). In addition to this, a new alien barnacle *Perforatus perforatus* appeared and settled on the mussel shells.

P. perforatus is a barnacle distributed in a range from the English Channel to the Mediterranean (Herbert et al. 2003) and is reported to have been introduced along the east coast of Korea in 2004 (Choi et al. 2013). In Japan, this barnacle was noted in 2012 for the first time at Toga Bay in Akita Prefecture, northern Honshu along the Japan Sea (Nogata et al. 2015). According to Nogata et al. (2015), this alien barnacle settled on culture buoys and ropes abundantly. In some cases, more than two thirds of the settled barnacles on these culture buoys and ropes were *P. perforatus*. A field survey along the coast of the Japan Sea revealed that this barnacle settled on piers and tetrapods with a large abundance and on shells of various mollusks such as turbo shell *Turbo cornutus*, rock shell *Thais clavigera*, and oyster *Crassostrea nippona* (Kado, unpublished data). According to the monthly monitoring data on sedentary organisms on the Sakihama jetty wall, *P. perforatus* appeared for the first time on the plates in September 2012 (Kado unpublished data).

2013

From late 2012 to 2013, abundance of the sea urchin on the jetty wall further increased and reached to be more than ten individuals/m^2 (Fig. 2.6). As a result, barnacles that settled below the lower littoral zone almost disappeared, and the ecological situation returned to a sea urchin-dominated barren that was similar to that before the earthquake.

In February 2013 when the seawater temperature was 1.5–2.0 °C higher than that of an average year, only a small number of macroalgae recruited (Fig. 2.5i). Although *S. cariosus* recruited on the mussels, abundance of the mussels decreased from 10.7×10^2 individuals/m^2 in June 2012 to 1.3×10^2 individuals/m^2 in April 2013. In the summer of 2013, the reddish barnacle *M. rosa* that had settled in the previous year on the mussels almost disappeared with the decrease in abundance of mussels. *P. perforatus* that had settled the previous summer survived and overwintered with a low mortality. They reached to about 10 mm in shell diameter in the summer of 2013. In addition, their newly settled individuals were observed on the mussels and on the walls.

2014

In the spring of 2014, we found *S. cariosus* individuals only in the tidal level higher than 60 cm M.S.L. In this zone, sea urchins could not remain for a long time. The *S. cariosus* individuals grew to about 20 mm in size. Some individuals of this species were fed on even in this zone by predatory gastropods such as *Nucella heyseana* and *Ceratostoma burnetti* (Fig. 2.5j). A number of recruited individuals of this barnacle were observed in the spring of 2014 as in 2011. However, *S. cariosus* individuals that had settled below the lower littoral zone disappeared before the late autumn. During this period, abundance of the mussels also decreased, and their recruitment was very limited probably due to the lack of *S. cariosus* as an appropriate substrate.

2015

In the spring of 2015, abundances of *S. cariosus* and *M. galloprovincialis* were highly limited in the tide level between 15 and 30 cm below M.S.L. In contrast, carnivorous rock snails accumulated at the wall of the jetty in this experimental site (Fig. 2.5k). Besides, few oysters *Crassostrea gigas* settled on the jetty wall after the earthquake. Even when the number of barnacle individuals was reduced by the sea urchins in 2015, the oyster never dominated in the mid-littoral zone.

2.4 Conclusion

A variety of benthic organisms occurred on the jetty wall of Sakihama at the Sanriku coast after the ground subsidence caused by the Great East Japan Earthquake in March 2011. The increase in abundance of some species such as *S. cariosus* was attributable to two reasons. One reason was reduction of the sea urchin abundance due to the tsunamis. The other reason was the temporal decrease in abundance of filter feeders that enabled a rapid colonization of *S. cariosus* which, in turn, provided a hard uneven substrate for other sedentary settlers. Accordingly, the community with diverse organisms was developed on the jetty wall in 2012, a year after the earthquake. However, the community with various organisms did not last for a long time. In 2013, with an increasing abundance of the common sea urchin *S. nudus*, the community on the jetty wall has returned to be a poor species composition. In the sublittoral zone on the wall, the community with a very poor biodiversity like "rocky shore denudation" was commonly observed from the third to fourth years after the Great East Japan Earthquake. Thus, the community structure in the intertidal zone of Sakihama area has been gradually approaching that found before the earthquake. In addition to the common sea urchin, increase in the number of gastropods has accelerated this succession by predating on the barnacles and mussels left on the jetty wall.

We found an alien barnacle *P. perforatus* for the first time in 2012 in Sakihama area. This barnacle survived in the following winter seasons and colonized successfully at the mid-littoral to upper sublittoral zone, although the mean sea surface temperature (SST) in winter in this area (6.8 °C in Kamaishi near Sakihama) was much lower than that (8.2 °C in Brighton in the eastern English Channel) which is the most northeastern native locality in Europe (Herbert et al. 2003; World Sea Temperatures a, b). One may suppose that the appearance of *P. perforatus* in Sakihama area was a result of natural range expansion by the larval dispersion from Japan Sea to the Pacific Ocean by means of the Tsugaru Warm Current. However, although we examined their distribution range along the coasts of northeast Japan, we did not find this species at least in coastal areas of the Pacific side of Aomori Prefecture (Kado unpublished data). This implies that there is a large gap in their distribution range between coasts of Japan seaside and Pacific Ocean seaside. Rather, *P. perforatus* may have been introduced by anthropogenic activities. It is reported that plastic flotsam provided a means of range expansion of various sessile organisms in North Wales (Rees and Southward 2009). Considering that plastic debris with Korean alphabet is occasionally found close to the Pacific coast in Iwate Prefecture, floating objects such as aquaculture buoys may act as a vector for this barnacle from the Japan Sea side to the Pacific side. However, it has never reported that this species had settled on such plastic debris to the present. Thus, another human-mediated introduction of *P. perforatus* seems to be much more probable. Many barges and tugboats came into Iwate Prefecture from various localities in Japan from Hokkaido in the north to Kagoshima in the south for reconstruction of harbors and shore areas that were damaged heavily by the earthquake and tsunami in 2011. Furthermore, in 2013, we found several individual *P. perforatus* (about 10 mm diameter) on the hull and fenders of a barge that was registered in Niigata, a city faced to Japan Sea, and anchored at Ofunato near Sakihama (Kado personal observation). These results may indicate that the barges and tugboats with home ports on the Japan Sea may have aided the range expansion of this barnacle. Although the possibility should be examined by genetic analysis like as the case of *Balanus glandula* (Geller et al. 2008; Kado 2003b; Kado and Nanba 2006), we need to consider that these boats may have already accelerated further expansion of this barnacle in various localities in Japan when they terminate their mission and return to the home harbors where *Balanus glandula* were not distributed.

Acknowledgements We express our appreciation to the following graduate and undergraduate students for their assistance to this research: H. Hamaguchi, T. Kimura, A. Sakai, H. Ozasa, F. Yoshida, S. Nagano, T. Tsuji, Y. Abe, and H. Arai. We also express our thanks to the two editors and Dr. C. Norman who checks English text and gave some valuable comments. This research was supported by Grant-in-Aid for Scientific Research (C) (Grant No. 16580276, 24580279) and Tohoku Ecosystem-Associated Marine Sciences (TEAMS) of the Ministry of Education, Culture, Sports, Science and Technology Japan.

References

Choi KH, Choi HW, Kim IH, Hong JS (2013) Predicting the invasion pathway of *Balanus perforatus* in Korean seawaters. Ocean Polar Res 35:63–68

Fujiwara T (2008) 3.4 Sanriku coast/ Iwate Prefecture, chapter 3. Sea urchin barren in Japan. In: Fujita D, Machiguchi Y, Kuwahara H (eds) Recovery from urchin barrens – ecology, fishery and utilization of sea urchins. Seizando-Shoten, Tokyo, pp 39–43 (in Japanese)

Geller J, Sotka EE, Kado R, Palumbi SR, Schwindt E (2008) Sources of invasions of a northeastern Pacific acorn barnacle, *Balanus glandula* Darwin 1854 in Japan and Argentina. Mar Ecol Prog Ser 358:211–218

Herbert RJH, Hawkins SJ, Sheader M, Southward AJ (2003) Range expansion and reproduction of the barnacle *Balanus perforatus* in the eastern English Channel. J Mar Biol Ass UK 83:73–82

Kado R (2003a) Recruitment of thatched barnacle *Semibalanus cariosus* and concerning environmental factors. Sessile Org 20:63–68 (in Japanese)

Kado R (2003b) Invasion of Japanese shores by the NE Pacific barnacle *Balanus glandula* and its ecological impacts. Mar Ecol Prog Ser 249:199–206

Kado R (2006) Life cycle of barnacles and ecology of its early stage – how they live and how to reach appropriate site for settlement. In: Sessile Organisms Society of Japan (ed) Newest studies of Cirripedia – the relationship between barnacles and human. Koseisha Kouseikakku, Tokyo, pp 93–111 (in Japanese)

Kado R, Hayakawa Y, Hayashizaki K, Nanba N, Ogawa H, Okano K (2002) Spatial distribution and abundance of barnacle larvae in Okkirai Bay, northeast Honshu, Japan, a case study of *Semibalanus cariosus* (Pallas). Fish Sci 68(1):405–408

Kado R, Nanba N (2006) *Balanus glandula*: a new alien species from the west coast of North America, established on the northeast coast of Japan. In: Koike F, Clout MN, Kawamichi M, De Poorter M, Iwatsuki K (eds) Assessment and control of biological invasion risks, Shoukadoh Book Sellers, /World Conservation Union. IUCNa9, Kyoto/Gland, pp 210–211

Kawamura T, Tamaki H, Hayakawa J, Won NI, Muraoka D, Kurita Y (2014) Changes in abalone and sea urchin populations in rocky reef ecosystems on the Sanriku coast damaged by the massive tsunami and other environmental changes associated with the Great East Japan Earthquake in 2011. Glob Environ Res 18:47–56

Newman WA, Abbott DP (1980) Cirripedia: the barnacles. In: Morris RH, Abbott DP, Haderlie EC (eds) Intertidal invertebrates of California. Stanford University Press, Stanford, pp 504–535

Nogata Y, Yoshimura E, Sato K, Kado R, Okano K (2015) The first find of invasive alien barnacle *Perforatus perforatus* in Japan and the development of a detection method using real-time PCR. Sessile Org 32:1–6 (in Japanese with English abstract)

Raffaelli D, Hawkins SJ (1996) Intertidal ecology. Kluwer Academic Publishers, Dordrecht/Boston/London

Rees EIS, Southward AJ (2009) Plastic flotsam as an agent for dispersal of *Perforatus perforatus* (Cirripedia: Balanidae). Mar Biodivers Rec 2:1–3

World Sea Temperatures (a) http://www.seatemperature.org/asia/japan/kamaishi-october.htm

World Sea Temperatures (b) http://www.seatemperature.org/europe/united-kingdom/brighton-october.htm

Chapter 3
Rocky Intertidal Zonation: Impacts and Recovery from the Great East Japan Earthquake

Takashi Noda, Aiko Iwasaki, and Keiichi Fukaya

Abstract We assessed the course and status of the recovery of rocky intertidal zonation after the Great East Japan Earthquake by conducting a census of the vertical distribution of 11 dominant macrobenthos (7 sessile and 4 mobile species) around the mid-tidal elevation at 23 sites along the Sanriku Coast, 150–160 km north-northwest of the epicenter of the earthquake. Sites were observed from 2011 to 2013, and the vertical distributions of each species were compared with those from the pre-earthquake period. Our results show that the earthquake considerably altered rocky intertidal zonation, mainly through coseismic subsidence rather than by the subsequent tsunami. By 28 months after the earthquake, the zonation of late-successional sessile taxa had not recovered, suggesting that the rocky intertidal community will experience a long delay before recovering from the influence of the earthquake. The dynamics of rocky intertidal zonation after the earthquake and accompanying subsidence includes two unique features: a delayed negative impact and an occasional increase in population sizes of several taxa. Neither of which has been reported following earthquakes of similar magnitude with accompanying uplift, in which there were mass mortalities of zone-forming species within 1 year after the event, preceding downward shifts in their zonations.

Keywords Benthos • Distribution • Disturbance • Earthquake • Rocky intertidal • Seaweed • Subsidence • Succession • Tsunami • Zonation

---

T. Noda (✉) • A. Iwasaki
Faculty of Environmental Earth Science, Hokkaido University, Sapporo, Hokkaido, Japan
e-mail: noda@ees.hokudai.ac.jp

K. Fukaya
The Institute of Statistical Mathematics, Tachikawa, Tokyo, Japan

J. Urabe, T. Nakashizuka (eds.), *Ecological Impacts of Tsunamis on Coastal Ecosystems*, Ecological Research Monographs,
DOI 10.1007/978-4-431-56448-5_3

3.1 Introduction

The Great East Japan Earthquake (Mw 9.0) caused a large tsunami, with run-up heights exceeding 30 m and subsidence of 35–70 cm throughout the Tohoku region (Lay and Kanamori 2011; Tajima et al. 2013). There is no doubt that the tsunami had great potential to affect a variety of marine benthos. This is because tsunami waves can transport benthos away from their original habitats (Castilla 1988; Castilla et al. 2010; Whanpetch et al. 2010; Lomovasky et al. 2011; Takami et al. 2013) and often severely damage benthic habitats, such as seagrass beds (Whanpetch et al. 2010) and coral reefs (Chavanich et al. 2005; Campbell et al. 2007), by removing and redepositing coastal substrata at large spatial scales (Lomovasky et al. 2011). In contrast to the widespread impact of a tsunami on various organisms living in diverse habitats, a subsidence of several tens centimeters should affect relatively few species living in specific habitats, such as the intertidal rocky shore. In particular, sessile species living in this habitat, whose distributions are often restricted to within a narrow vertical range of several tens of centimeters (Stephenson and Stephenson 1972), might be quite sensitive to land-level changes caused by an earthquake. Indeed, previous studies have demonstrated that earthquake-related uplifts caused mass mortalities of various sessile species and consequently altered rocky intertidal communities at regional scales (Haven 1972; Johansen 1972; Bodin and Klinger 1986; Castilla 1988; Castilla et al. 2010).

In rocky intertidal habitats, zonation (Fig. 3.1) is the most general spatial distribution pattern of organisms (Stephenson and Stephenson 1972). A mega-earthquake would immediately alter the zonation of each species through the direct effect of the tsunami and subsidence; thereafter, there could be additional changes in zonation via modification of population processes after the earthquake, such as mortality and recruitment. In addition, the immediate impacts of the earthquake and subsequent changes in zonation should vary among species. Although Castilla (1988) described the impact and recovery of rocky intertidal zonation from the uplift associated with a megaquake for a single sessile species, there are no studies yet concerning the impact and recovery from an earthquake and associated subsidence except for several qualitative evaluations of changes in vertical position of zones (Haven 1972; Johansen 1972).

Long-term community census data from before the event provide a unique opportunity to evaluate the community impacts and recovery from the rare disturbance event (Parker and Wiens 2005). For 8 years before the 2011 Tohoku earthquake, we regularly carried out censuses of the distribution and abundance of rocky intertidal organisms at 23 sites on five shores close to the epicenter (Okuda et al. 2004; Nakaoka et al. 2006; Fukaya et al. 2010; Munroe and Noda 2010). To assess the impacts and recovery from the earthquake, we have been continuing the surveys using the same methods at the same places after the earthquake, so as to compare the post-earthquake results with the data collected before the event. To our knowledge this is a unique example of research on the zonation dynamics of rocky intertidal organisms encompassing periods before and after a megaquake. Here we report

Fig. 3.1 Photograph of rocky intertidal zonation in a study site (Myojin). Horizontal bands with different color are dominated by different sessile organisms

the course and status of recovery of the vertical distribution of 11 dominant rocky intertidal macrobenthos (seven sessile and four mobile species) in the mid-shore of 23 sites located 150–160 km north-northwest of the epicenter of the 2011 Tohoku earthquake. We compare the vertical gradients of abundance of each species 4, 16, and 28 months after the megaquake with those obtained before the earthquake (2003–2010).

3.2 Materials and Methods

3.2.1 Census Design

Rocky intertidal zonation was monitored at 23 sites along five shores located 150–160 km north-northwest of the epicenter (38°06′12.0″N, 142°51′36.0″E) of the 2011 Tohoku earthquake (Fig. 3.2). For each shore, four or five sites were haphazardly chosen from semi-exposed locations. At each site a permanent plot (50 cm horizontally by 100 cm vertically, centered at mid-tide) was marked with plastic anchors drilled into roughly vertical rock delimiting the study area. The vertical extent of the permanent plots covered 72.4 % of the tidal range. In March 2011,

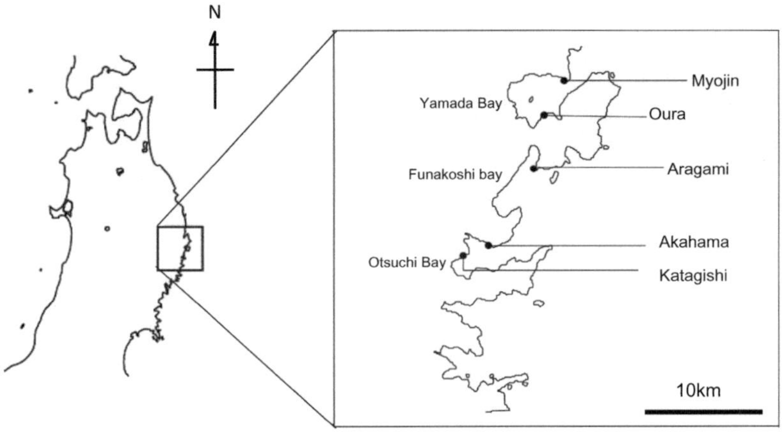

Fig. 3.2 Map of study sites. Five rocky shores were chosen for the census of intertidal organisms on the Sanriku Coast along the Pacific coast of Japan (Reproduced from Noda et al. 2015)

the census area experienced subsidence of 50 cm due to the Tohoku earthquake. We therefore extended the permanent plot at each site (100 cm vertically by 50 cm horizontally) so that its new low edge shared the top edge of each previous permanent plot.

Permanent plots were monitored at low tide in July, from 2003 to 2010 and from 2011 to 2013. At each census, the abundance of each mobile species was estimated in each 50 cm wide by 10 cm high band of the 10 (before the earthquake) or 20 quadrats (after the earthquake), representing vertical sections of each permanent plot. In addition, the coverage by each sessile species was estimated by a point sampling method, counting the occurrence of each focal species at 20 points on grid with 5-cm intervals both vertically and horizontally superimposed on each plot.

3.2.2 Data Analysis

We used the data accumulated from all permanent plots to obtain vertical distributions as mean abundances at each elevational position of the quadrats. We compared the vertical zonation of species having unimodal vertical distributions using before/after comparisons in quadrats located at the same height. To obtain reliable estimates of zonation of each species for this comparison, we excluded rare species, that is, those with mean values of total abundance in each year at the permanent plots before the earthquake of fewer than 25 for sessile species and fewer than 100 for mobile species.

3.3 Results

3.3.1 Four Months After the Earthquake

All sessile organisms shifted downward in zonation according to the subsidence in 2011 (Fig. 3.3). Two of the seven sessile taxa showed clear changes in abundance immediately after the earthquake: mussels, including *Septifer virgatus* and *Mytilus* spp., decreased in abundance, whereas the barnacle *Semibalanus cariosus* drastically increased. The other sessile organisms, the barnacle *Chthamalus challengeri*, Pacific oyster *Crassostrea gigas*, perennial brown alga *Analipus japonicus*, and crustose red alga *Hildenbrandia rubra*, did not show clear changes in abundance immediately after the earthquake.

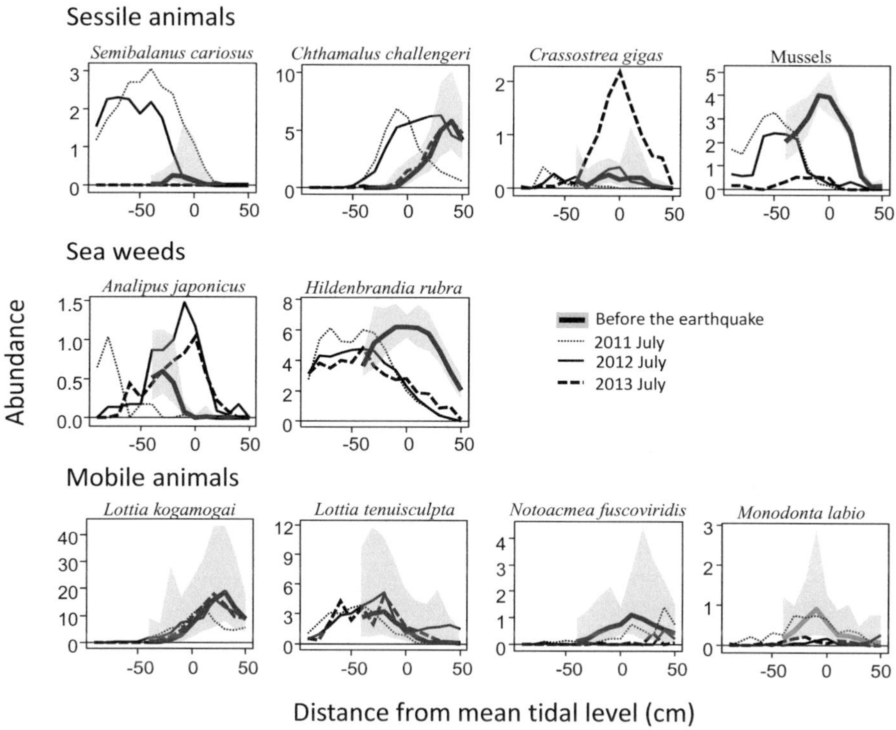

Fig. 3.3 Course and status of recovery of vertical distribution of rocky intertidal macrobenthos within 3 years after the megaquake along the coast located 150–160 km north-northwest of the epicenter of the 2011 Tohoku earthquake. Tidal level on the y-axis represents elevational position of the quadrat (see the subsection Census design in the Materials and Methods section for details). Coverage and abundance at each tidal level were calculated using data accumulated from all plots at each census. Coverage and abundance before earthquake at each tidal level indicates mean ± 1.96 standard deviation of data obtained from each census of pre-earthquake period (2003–2010) (Reproduced from Noda et al. 2015)

None of the mobile species showed a clear change in their vertical distribution immediately after the earthquake (Fig. 3.3). Two limpets, *Lottia tenuisculpta* and *Notoacmea fuscoviridis*, and the herbivorous trochid *Monodonta labio* showed no clear change in abundance. The limpet *Lottia kogamogai* seemed to slightly decrease in abundance immediately after the earthquake.

3.3.2 Sixteen Months After the Earthquake

From 4 to 16 months after the earthquake, changes in zonation varied qualitatively among sessile taxa. The mode zonal elevations of *C. challengeri*, *C. gigas*, and *A. japonicus* shifted upward (Fig. 3.3). *Chthamalus challengeri* increased in abundance and expanded its zonation vertically upward without any upward shift in the lower range of its zonation. *Analipus japonicus* also increased in abundance. *Crassostrea gigas* did not show a clear change in abundance but did recover to its normal vertical position of zonation. The abundances of these latter two species apparently increased compared to before the earthquake. Taxa that did not show a vertical shift in their zonation, including *S. cariosus*, *S. virgatus*, *Mytilus* spp., and *H. rubra*, seem to have slightly decreased in abundance.

From 4 to 16 months after the earthquake, changes in abundance varied qualitatively among mobile species (Fig. 3.3). *Monodonta labio* and *N. fuscoviridis* decreased in abundance. *Lottia tenuisculpta* did not show a clear change in abundance. *Lottia kogamogai* seemed to increase slightly and recover its zonation from before the earthquake.

3.3.3 Twenty-Eight Months After the Earthquake

From 16 to 28 months after the earthquake, changes in zonation varied qualitatively among sessile taxa (Fig. 3.3). *Chthamalus challengeri* decreased in abundance in the lower half of its zonation and consequently in 2013 overlapped with its zonation from before the earthquake. Although mussels also shifted their zonation upward, they did not show any increase in their normal zonation positions. Therefore, the mussel species did not recover their vertical ranges and their abundances continuously decreased until 28 months after the earthquake. Other sessile taxa did not show clear changes in their vertical zonation. Among these, *S. cariosus*, which had drastically increased in its abundance right after the earthquake, drastically decreased, with the result that its abundance in 2013 was clearly lower than before the earthquake. *Crassostrea gigas* drastically increased in number and became clearly more abundant than before the earthquake. *Analipus japonicus* seemed to decrease slightly in number but was still more abundant than before earthquake. *Hildenbrandia rubra* did not show any clear changes in abundance, and its vertical range and abundance did not recover to their states before the earthquake.

Among the mobile animals, *M. labio* and *N. fuscoviridis*, which had decreased from 4 to 16 months after the earthquake (Fig. 3.3), did not increase from 16 to 28 months after the earthquake, and therefore they were less abundant 28 months after the earthquake than before the event. During the same period, there were no clear changes in the abundance of *L. kogamogai* or *L. tenuisculpta*, and their levels were similar to those before the earthquake.

3.4 Discussion

The most unexpected response of the rocky intertidal community to the Great East Japan Earthquake in 2011was that four of the seven sessile taxa became temporarily more abundant than before the earthquake. Except for *S. cariosus*, this increase seems to have been caused by heavy recruitment occurring in free space, which had increased within their normal vertical position because the proportion of rock surface unoccupied by any sessile organisms increased with elevation in the rocky intertidal zone of our study area (Noda et al. 2015). In addition, these species exhibit higher colonizing ability compared with competitive dominants such as mussels (Paine and Levin 1981; Wootton 1993) and *H. rubra* (Wennberg 1992). Interspecific differences in the timing of their increase in numbers, as seen between *C. challengeri* and *C. gigas*, might be related to differences in their growth patterns; the former exhibits faster growth that stops within the first year of benthic life (Miyamoto et al. 1999), whereas the latter grows for several years after settlement (Brown and Hartwick 1988). The increase in population size of *S. cariosus* was likely mainly caused by the heavy settlement among the adults, responding to cues associated with conspecific adults (Burke 1986). There are several supports that the heavy settlement might be caused by the huge larval supply in 2011: the recruitment density per area available for larvae was considerably higher in 2011 than in the years before the earthquake, although the amount of free space was not (T. Noda, personal observation).

While it is not surprising that the earthquake and associated tsunami and widespread subsidence resulted in obvious population decreases in about half of the rocky intertidal organisms, from an ecological perspective, it is surprising that not all of the populations decreased immediately after the earthquake, but rather with a delay of several years. This finding suggests that the negative impacts may have been mainly caused by the coseismic subsidence rather than by the tsunami. It is plausible that the transport below a normal vertical range by subsidence did not cause immediate death for sessile organisms (Haven 1972; Johansen 1972), but a more gradual death through competition with sessile taxa (Connell 1961, 1972) or by predation (Paine 1974; Underwood and Denley 1984). In contrast, there is no plausible mechanistic explanation for the delayed decline of mobile organisms.

Another implication comes from the failed recovery of two sessile taxa—the mussels and the perennial crustose red alga *H. rubra*—all of which often play an important role in community organization in rocky intertidal communities by providing secondary substrata (Suchanek 1986; Lohse 1993; Miyamoto and Noda 2004) or through competitive displacement (Paine 1966; Dayton 1971; Menge 1976; Wennberg 1992; Miyamoto and Noda 2004). This suggests that the rocky intertidal community will experience a long delay in recovering from the influence of the earthquake. The species that did not recover are late-successional sessile species, which grow slowly and have weak colonization ability (Paine and Levin 1981; Wennberg 1992; Wootton 1993). Indeed, even in a one-square-meter experimental disturbance patch for which there is an external supply of larvae and propagules to contribute to their populations' recovery, it takes decades for these species to recover to their coverage before the disturbance (T. Noda, personal observation). Therefore, it is reasonable that more time will be required for the recovery of these species after the earthquake, because there may not be a ready external supply of larvae and propagules from surrounding populations, as the coseismic subsidence occurred at a large spatial scale encompassing entire metapopulations of those species (Kinlan and Gaines 2003).

3.5 Conclusions

The Great East Japan Earthquake considerably altered rocky intertidal zonation along the Sanriku Coast 150–160 km north-northwest of the epicenter of the earthquake, mainly as a result of coseismic subsidence rather than the tsunami. At 28 months after the earthquake, the zonation of late-successional sessile taxa had not recovered, suggesting that the rocky intertidal community will experience a long delay in recovering from the effects of the earthquake. The dynamics of rocky intertidal zonation after the earthquake and accompanying subsidence includes two unique features: a delayed negative impact and the occasional increase in populations of several taxa. Neither of these effects has been reported from similar magnitude megaquakes with accompanying uplift, in which mass mortality of zone-forming species occurred within 1 year of the event, preceding downward shifts in their zonation (Haven 1972; Johansen 1972; Bodin and Klinger 1986; Castilla 1988; Castilla and Oliva 1990; Castilla et al. 2010).

Acknowledgments We thank Dr. T. Okuda, Dr. M. Tsujino, T. Hagino, M. Iida, R. Sahara, M. Ohira, and Y. Kanamori for field assistance. For field and laboratory facilities, we are grateful to the staff and students of the International Coastal Research Center of the Atmosphere and Ocean Research Institute, The University of Tokyo. This study was made possible by the generous support and encouragement of local fisherman and the fishery office of the Fisherman's Cooperative Associations in Iwate Prefecture, Japan. This research was supported by the Cooperative Program of the Atmosphere and Ocean Research Institute, The University of Tokyo (Nos. 108 [2006], 104 [2007], 107 [2008], 101 [2009], and 103 [2010] to T.N.), and was partly supported by Japan Society for the Promotion of Science KAKENHI Grants (Nos. 20570012, 24570012, and 15K07208 to T.N.) and the Tohoku Ecosystem-Associated Marine Sciences (TEAMS) project. The funders had no role in study design, data collection and analysis, decision to publish, or preparation of the manuscript.

References

Bodin P, Klinger T (1986) Coastal uplift and mortality of intertidal organisms caused by the September 1985 Mexico earthquakes. Science 233:1071–1073

Brown JR, Hartwick EB (1988) Influences of temperature, salinity and available food upon suspended culture of the Pacific oyster. Crassostrea gigas. I. Absolute and allometric growth. Aquaculture 70:231–251

Burke RD (1986) Pheromones and the gregarious settlement of marine invertebrate larvae. Bull Mar Sci 39:323–331

Campbell SJ, Pratchett MS, Anggoro AW, Ardiwijaya RL, Fadli N, Herdiana Y, Kartawijaya T, Mahyiddin D, Mukminin A, Pardede ST, Rudi E, Siregar AM, Baird AH (2007) Disturbance to coral reefs in Aceh, Northern Sumatra: impacts of the Sumatra-Andaman tsunami and pre-tsunami degradation. Atoll Res Bull 544:55–78

Castilla JC (1988) Earthquake-caused coastal uplift and its effects on rocky intertidal kelp communities. Science 242:440–443

Castilla JC, Oliva D (1990) Ecological consequences of coseismic uplift on the intertidal kelp belts of *Lessonia nigrescens* in central Chile. Estuar Coast Shelf Sci 31:45–56

Castilla JC, Manríquez PH, Camaño A (2010) Effects of rocky shore coseismic uplift and the 2010 Chilean mega-earthquake on intertidal biomarker species. Mar Ecol Prog Ser 418:17–23

Chavanich S, Siripong A, Sojisuporn P, Menasveta P (2005) Impact of tsunami on the seafloor and corals in Thailand. Coral Reefs 24:535

Connell JH (1961) The influence of intra-specific competition and other factors on the distribution of the barnacle *Chthamalus stellatus*. Ecology 42:710–723

Connell JH (1972) Community interactions on marine rocky intertidal shores. Annu Rev Ecol Syst 3:169–192

Dayton PK (1971) Competition, disturbance, and community organization: the provision and subsequent utilization of space in a rocky intertidal community. Ecol Monogr 41:351–389

Fukaya K, Okuda T, Nakaoka M, Hori M, Noda T (2010) Seasonality in the strength and spatial scale of processes determining intertidal barnacle population growth. J Anim Ecol 79:1270–1279

Haven HW (1972) Effects of land-level changes on intertidal invertebrates, with discussion of postearthquake ecological succession. In: Krauskopf KB (ed) The great Alaska earthquake of 1964. National Academy of Sciences, Washington, DC, pp 82–126

Johansen HW (1972) Effects of elevation change on benthic algae in Prince William Sound. In: Krauskopf KB (ed) The great Alaska earthquake of 1964. National Academy of Sciences, Washington, DC, pp 35–68

Kinlan BP, Gaines SD (2003) Propagule dispersal in marine and terrestrial environments: a community perspective. Ecology 84:2007–2020

Lay T, Kanamori H (2011) Insights from the great 2011 Japan earthquake. Phys Today 64:33–39

Lohse DP (1993) The importance of secondary substratum in a rocky intertidal community. J Exp Mar Biol Ecol 166:1–17

Lomovasky BJ, Firstater FN, Salazar AG, Mendo J, Iribarne OO (2011) Macro benthic community assemblage before and after the 2007 tsunami and earthquake at Paracas Bay, Peru. J Sea Res 65:205–212

Menge BA (1976) Organization of the New England rocky intertidal community: role of predation, competition, and environmental heterogeneity. Ecol Monogr 46:355–393

Miyamoto Y, Noda T (2004) Mussel effects on competitive inferior species: competitive exclusion to facilitation. Mar Ecol Prog Ser 276:293–298

Miyamoto Y, Noda T, Nakao S (1999) Zonation of two barnacle species not determined by competition. J Mar Biol Assoc UK 79:621–628

Munroe DM, Noda T (2010) Physical and biological factors contributing to changes in the relative importance of recruitment to population dynamics. Mar Ecol Prog Ser 421:151–162

Nakaoka M, Ito N, Yamamoto T, Okuda T, Noda T (2006) Similarity of rocky intertidal assemblages along the Pacific coast of Japan: effects of spatial scales and geographic distance. Ecol Res 21:425–435

Noda T, Iwasaki A, Fukaya K (2015) Recovery of rocky intertidal zonation: 3 yr after the 2011 Tohoku Earthquake. J Mar Biol Ass UK DOI:10.1017/S002531541500212X

Okuda T, Noda T, Yamamoto T, Ito N, Nakaoka M (2004) Latitudinal gradient of species diversity: multi-scale variability in rocky intertidal sessile assemblages along the North Western Pacific coast. Popul Ecol 46:159–170

Paine RT (1966) Food web complexity and species diversity. Am Nat 100:65–75

Paine RT (1974) Intertidal community structure: experimental studies on the relationship between a dominant competitor and its principal predator. Oecologia 15:93–120

Paine RT, Levin SA (1981) Intertidal landscapes: disturbance and the dynamics of pattern. Ecol Monogr 51:145–178

Parker KR, Wiens JA (2005) Assessing recovery following environmental accidents: environmental variation, ecological assumptions, and strategies. Ecol Appl 15:2037–2051

Stephenson TA, Stephenson A (1972) Life between tidemarks on rocky shore. Freeman, San Francisco

Suchanek TH (1986) Mussels and their role in structuring rocky shore communities. In: Moore PG, Seed R (eds) The ecology of rocky coasts. Columbia University Press, New York, pp 70–96

Tajima F, Mori J, Kennett BLM (2013) A review of the 2011 Tohoku-Oki earthquake (Mw 9.0): large-scale rupture across heterogeneous plate coupling. Tectonophysics 586:15–34

Takami H, Won N-I, Kawamura T (2013) Impacts of the 2011 mega-earthquake and tsunami on abalone *Haliotis discus hannai* and sea urchin *Strongylocentrotus nudus* populations at Oshika Peninsula, Miyagi, Japan. Fish Oceanogr 22:113–120

Underwood AJ, Denley EJ (1984) Paradigms, explanations and generalizations in models for the structure of intertidal communities on rocky shore. In: Strong DR Jr, Simberloff D, Abele LG, Thistle AB (eds) Ecological communities: conceptual issues and the evidence. Princeton University Press, Princeton, pp 151–180

Wennberg T (1992) Colonization and succession of macroalgae on a breakwater in Laholm Bay, a eutrophicated brackish water area (SW SWEDEN). Acta Phytogeogr Suec 78:65–77

Whanpetch N, Nakaoka M, Mukai H, Suzuki T, Nojima S, Kawai T, Arythaka C (2010) Temporal changes in benthic communities of seagrass beds impacted by a tsunami in the Andaman Sea, Thailand. Estuar Coast Shelf Sci 87:246–252

Wootton JT (1993) Size-dependent competition: effects on the dynamics vs. the endpoint of mussel bed succession. Ecology 74:195–206

Chapter 4
Quantitative Evaluation of the Impact of the Great East Japan Earthquake and Tsunami on the Rocky Intertidal Community

Aiko Iwasaki, Keiichi Fukaya, and Takashi Noda

Abstract It has been widely assumed that tsunamis associated with mega-earthquakes severely damage coastal ecosystems. In no previous studies, however, have the community-level impacts of tsunamis been credibly quantified. We evaluated the impacts of the tsunami following the Great East Japan Earthquake (the 2011 Tohoku earthquake tsunami) on species abundances in a rocky intertidal metacommunity encompassing 30 km of shoreline located 150–160 km north-northwest of the epicenter of the earthquake. For this purpose, we generated an indicator of the tsunami impact—the change in mean abundances between before and after the tsunami, standardized to population variability before the earthquake—for each of five sessile and six mobile species. We then used the averaged values of this indicator as a quantitative measure of tsunami impacts for the entire community and for the sessile and mobile subgroups. The tsunami had an overall significant negative impact on the rocky intertidal community at the regional scale, although the negative impact seemed smaller than those of similar large-scale disturbances on rocky intertidal communities. There was a significant negative impact of the tsunami on mobile species, but not on sessiles. The results suggest that mobiles on rocky shores are more vulnerable than sessiles to tsunamis. At a regional scale, the rocky intertidal community suffered minor damage from the 2011 Tohoku earthquake tsunami.

Keywords Tsunami • Community impact • Population growth • Effect size • Benthic community • Rocky intertidal • Sessile organisms • Mobile mollusca

---

A. Iwasaki (✉) • T. Noda
Faculty of Environmental Earth Science, Hokkaido University, Sapporo, Hokkaido, Japan
e-mail: aiwasaki@ees.hokudai.ac.jp

K. Fukaya
The Institute of Statistical Mathematics, Tachikawa, Tokyo, Japan

J. Urabe, T. Nakashizuka (eds.), *Ecological Impacts of Tsunamis on Coastal Ecosystems*, Ecological Research Monographs,
DOI 10.1007/978-4-431-56448-5_4

4.1 Introduction

It has been widely assumed that tsunamis associated with mega-earthquakes severely damage coastal ecosystems (Lomovasky et al. 2011; Rife 2014) because of the tremendous strength of the wave forces and their vast spatial coverage, encompassing entire metapopulations of most coastal organisms. Indeed, many previous reports have documented that tsunamis cause mass mortality of coastal marine populations (e.g., Lomovasky et al. 2011; Jaramillo et al. 2012; Seike et al. 2013). Most of previous studies that reported the community-level effect of tsunamis are not necessarily sufficient for evaluating the direct impacts of tsunami's disturbance on coastal communities for several reasons. First, most of these studies estimated damage by using only one or a few species (Campbell et al. 2007; Takami et al. 2013). In such cases the stochastic effect of selecting a species for sampling, which strengthens the influence of an individual species' properties on the estimated susceptibility to tsunamis, can become too large to obtain a general picture of community-level impacts of the tsunami. Second, most of these studies have estimated the impact of tsunamis on populations at relatively small spatial scales, i.e., at one location or a few sites (Whanpetch et al. 2010; Lomovasky et al. 2011; Kanaya et al. 2015). In such cases, the stochastic effect of selecting a site for sampling, which strengthens the influence of local coastal topography on the wave impact of the tsunami (Wijetunge 2006; Mori et al. 2011), can become too large to obtain a general picture of regional-scale impacts of the tsunami. Third, most studies have estimated tsunami damage on the basis of (1) changes in abundance based on a comparison between data from two sampling times (before and after the impact) or (2) comparison between abundance data from impact and control sites, obtained after the tsunami in a snapshot sampling (e.g., Paterson et al. 2012; Urabe et al. 2013). In the former case, any population difference detected between before and after the event will inevitably include the effects of environmental and demographic stochasticity as well as the tsunami effects. In the latter case, spatial differences in abundance will reflect not only the tsunami effects but also the spatial variation of environmental factors. Ideally, tsunami damage on a focal community should therefore be evaluated on the basis of its influence on many species randomly selected from the focal community. In addition, the tsunami impact on populations of the selected species should be estimated over a large spatial scale encompassing the entire metapopulation. Furthermore, changes in abundances between before and after the tsunami, as an indicator of the tsunami impact on each species, should be standardized to population variability under normal conditions for each species. To our knowledge, however, no previous studies have met these conditions.

The Great East Japan Earthquake occurred on 11 March 2011, which triggered a mega-tsunami (the 2011 Tohoku earthquake tsunami) with a run-up height that reached 40 m (Mori et al. 2011, 2012). Although mega-tsunamis are rare events with a return period of multiple decades at a regional scale (Dominey-

Howes 2007; Usami 2013), they occur almost every year somewhere in the world (Lander et al. 2003; Scheffers and Kelletat 2003). Mega-tsunamis should have negative impacts on rocky intertidal communities, as expected for other coastal communities; however, the following two basic questions remain unanswered with regard to these impacts. The first question, which has not yet been precisely answered for all coastal communities, is whether tsunamis significantly decrease species population abundances in the rocky intertidal community. The second question is whether there are any differences in the magnitude of tsunami impacts between sessile and mobile species, two major components of the community. Mobile species are generally predicted to be more vulnerable and to be more easily dislodged by hydrodynamic forces, than sessile species (Denny 1985, 1993, 1995).

For 8 years before the 2011 Tohoku earthquake tsunami, we regularly performed community censuses targeting all macrobenthic species at 23 sites along over 30 km of shoreline located 150–160 km north-northwest of the earthquake epicenter (Okuda et al. 2004; Nakaoka et al. 2006; Fukaya et al. 2010; Munroe and Noda 2010). The biological data obtained in these censuses met all conditions necessary for accurate evaluation of the ecological impacts of the tsunami, as mentioned above. Our purpose here was to evaluate the impacts of the 2011 Tohoku earthquake tsunami on species abundances in the rocky intertidal metacommunity in our census area. Although this earthquake triggered not only tsunamis but also subsidence, this study focuses on effects of tsunamis alone by excluding those of subsidence from overall effects of earthquake. We address the following question: did the tsunami significantly reduce the species abundances of the community as a whole, or of two subgroups, sessiles and mobiles? For this purpose, we evaluated an indicator of tsunami impact, namely, the change in mean abundances between before and after the tsunami (as standardized against population variability before the tsunami), for each of five sessile and six mobile species. From this we obtained, as averaged values, quantitative measures of the 2011 Tohoku earthquake tsunami's impact at community-wide level and on the two subgroups. These results allow us to discuss the actual impact of the tsunami on the rocky intertidal community at a regional scale.

4.2 Materials and Methods

4.2.1 Study Area

Our study sites were located along 30 km of shoreline on the Sanriku Coast (39°N, 142°E; see Fig. 3.2 in Chap. 3). Our study sites suffered effects from the large tsunamis, with a local height of 8–15 m, and from coseismic subsidence (Haraguchi and Iwamatsu 2011). The tidal range at our study sites, defined as the vertical extent between mean high water and mean low water of spring tide, was approximately 140 cm. Commonly found at our study sites were various rocky intertidal organisms, including sessiles such as mussels, barnacles, and perennial algae, and mobile invertebrates such as herbivorous gastropods (Nakaoka et al. 2006). These species showed zonation, as evidenced by a vertical distribution with a peak abundance at an intermediate position within a vertical range of several tens of centimeters. Judging from dispersal distances (Kinlan and Gaines 2003), the horizontal extent of our study sites, encompassing a 30-km shoreline, covered an entire range of metapopulations of most rocky intertidal organisms.

4.2.2 Census Design

Abundances of rocky intertidal organisms were monitored at 23 sites from 5 shores along the Sanriku Coast. Distances between neighboring shores ranged from 4.1 to 25.7 km (see Fig. 3.2 in Chap. 3, for details). At each shore, four or five sites were haphazardly chosen from among semi-exposed locations. In July 2002 at each site, a permanent plot was marked with plastic anchors drilled into roughly vertical rock delimiting a study area that was 50 cm horizontally by 100 cm vertically, centered at the mean tidal level. Distances between neighboring plots ranged from 7.8 to 172.5 m (mean ± SD, 49.4±48.3 m). The vertical extent of the permanent plots covered 72.4 % of the tidal range. In March 2011, the census area experienced subsidence of around 50 cm due to the Great East Japan earthquake. In July 2011, we therefore vertically extended the permanent plot at each site (an additional 100 cm high by 50 cm wide), with the lower border of the new part of the plot sharing the top border of each original permanent plot, so that the vertical observation range was 200 cm.

Permanent plots were monitored at low tide in July from 2003 to 2011. At each census, the abundance of each mobile species was estimated in each 50 cm wide by 10 cm high band of the 10 (before tsunamis) or 20 (after tsunamis) quadrats, representing vertical sections of each permanent plot. In addition, the coverage of each sessile species was estimated by a point sampling method, counting the occurrence of the focal species at 20 points on a grid superimposed on each plot with 5-cm intervals both vertically and horizontally.

To estimate the vertical subsidence at each shore, we measured the vertical deviations of seawater level from the height of anchors drilled at the midpoint of each permanent plot in July 2011, May 2012, and July 2012, assuming a measurement precision of 5 cm based on our minimum sampling resolution of the vertical distribution of the sessile species. We then used the measured deviations and tide level data from the Japan Meteorological Agency to estimate the vertical subsidence as 50 cm at four shores (Myojin, Oura, Aragami, and Katagishi) and 60 cm at Akahama.

4.2.3 Data Analysis

4.2.3.1 Estimation of Species-Level Tsunami Impacts

The Great East Japan Earthquake caused subsidence of around 50 cm as well as subsequent tsunamis in the study area; therefore, the abundances and vertical profiles of rocky intertidal organisms could have been altered not only by the tsunamis but also by the subsidence. The subsidence would have resulted in an immediate 50- or 60-cm downward shift in the distribution of all sessile organisms without changing their abundances. Therefore, our estimated changes caused in the abundances of sessiles by the tsunami needed to be based on a comparison of abundances as measured within the borders of the plots established in 2002. In contrast, all mobile organisms would have been able to return to their original elevational ranges right after the subsidence, because the vertical range of subsidence was narrower than the daily tidal range. Therefore, our estimated changes caused in the abundances of mobiles by the tsunami needed to be based on a comparison of abundances measured within the same elevational range (between 50 or 60 cm above and 50 or 60 cm below mean tidal level) within the same plot.

To estimate the impact of the tsunami on the abundance of each species, we first calculated the annual population growth rates for sessile and mobile organisms. For the years before the tsunami (2003–2010), the annual population growth rates r_t (from year t–1 to year t) were calculated as the log ratio of densities (Royama 1992):

$$r_t = \log N_t - \log N_{t-1} \tag{4.1}$$

where N_t is the mean abundance per plot in year t, obtained as the mean coverage or the mean density in permanent plots of sessile and mobile species, respectively. For growth between 2010 and the year of the tsunami (2011), the population growth rates of sessile species were obtained by subtracting the logarithm of mean coverage per plot in 2010 from the logarithm of mean coverage in 2011 measured within the range of the plots established in 2002; the population growth rates for mobile species were obtained by subtracting the logarithm of mean abundance per plot in 2010 from the logarithm of mean density in the same elevational range (between 50 cm above and 50 cm below mean tidal level) in 2011.

By using annual population growth rates r_t, we then obtained the no assumption effect size of single-subject data (Busk and Serlin 1992; Olive and Smith 2005; Petursdottir et al. 2009; Hurwitz et al. 2015), ES_i, representing the impact of the tsunami on species i, as calculated by dividing the difference between each species' baseline and intervention-phase means by the standard deviation of the baseline (Glass 1976; Busk and Serlin 1992; Busse et al. 1996):

$$ES_i = \frac{r^i_{2010-2011} - \bar{r}^i_{\text{before}}}{SD(r^i_{\text{before}})} \tag{4.2}$$

Table 4.1 Species used for estimating the tsunami impact

Species	Taxonomic group	Trophic group
Sessile		
Analipus japonicus	Phaeophyceae	Primary producer
Hildenbrandia rubra	Florideae	Primary producer
Crassostrea gigas	Bivalvia	Suspension feeder
Chthamalus challengeri	Crustacea	Suspension feeder
Mytilidae spp.	Bivalvia	Suspension feeder
Mobile		
Cellana grata	Gastropoda	Herbivore
Lottia kogamogai	Gastropoda	Herbivore
Lottia lindbergi	Gastropoda	Herbivore
Lottia tenuisculpta	Gastropoda	Herbivore
Monodonta confusa	Gastropoda	Herbivore
Notoacmea fuscoviridis	Gastropoda	Herbivore

Here, $r^i_{2010-2011}$ and $\overline{r}^i_{\text{before}}$ represent the population growth rate from 2010 to 2011 and the mean annual population growth rate in the pre-tsunami period for species *i*, respectively, and $SD(r^i_{\text{before}})$ denotes the standard deviation of the annual population growth rate in the pre-tsunami period for species *i*.

4.2.3.2 Estimation of Community-Level Tsunami Impacts

We evaluated the impact of the tsunamis on the whole community and on two subgroups: sessiles and mobiles. We calculated the mean effect size ($\overline{ES}$) for each group by simply averaging the effect size for species (ES_i) that were included in each group (Glass 1976; Busk and Serlin 1992; Hurwitz et al. 2015):

$$\overline{ES} = \frac{\sum_{n}^{i=1} ES_i}{n} \tag{4.3}$$

Here, *n* denotes the number of species included in each group. The 95 % confidence intervals for mean effect sizes ($\overline{ES}$) were generated by bootstrapping, corrected for bias in unequal distribution around the original value by using the bias-corrected and accelerated (BCa) method (Busk and Serlin 1992; Kirby and Gerlanc 2013).

Our dataset contained 75 sessile and 38 mobile species. Among them, we selected five sessile and six mobile species (Table 4.1) for calculating ES_i to ensure the accuracy of $\overline{ES}$ the community-level effect of tsunamis alone, using five criteria. First, we selected species for which the abundance and distribution seemed unaffected by the subsidence. For sessile species, we obtained an ES_i based on a comparison of abundances within the ranges of plots established in 2002. However, for some sessiles, the vertical distribution might have changed between the time of the tsunami and the post-tsunami survey (4 months after the earthquake), through mortality caused by transport below their normal vertical range by subsidence or through recruitment into the normal vertical range following subsidence. Such species therefore needed to be eliminated from estimates of direct tsunami impacts. To detect such species, we visually determined whether the vertical profiles of their abundances within the range of plots established in 2002 clearly changed from 2010 to 2011 for each species. This was because such changes in vertical profiles would have been quite sensitive to mortality or recruitment, which in turn would have been strongly dependent on the elevation after the subsidence. For mobile species we obtained an ES_i based on a comparison of abundances measured within the same elevational range (between 50 cm above and 50 cm below mean tidal level) in the same plot. However, the vertical distribution of some of these species might have changed after the subsidence, because the vertical distributions of some sessile organisms, which provide food or microhabitat for mobiles, might have been changed by the subsidence. Therefore, such species needed also to be eliminated from estimates of the tsunami impacts. To detect such species, we visually determined, for each species, whether the vertical profile of abundance within the same elevational range (between 50 cm above and 50 cm below mean tidal level) clearly changed from 2010 to 2011.

Second, we eliminated those species for which abundance increased after the subsidence, because the tsunami could not have directly increased abundance—it could only have decreased it. We applied a paired t-test to each species to determine whether its abundance increased from 2010 to 2011 within the same original 2002 plot (in the case of sessile species) or in the same elevational range (in the case of mobile species).

Third, we selected species that had been observed in all years. This was because inclusion of any years without observations would have reduced the sample size used to estimate the mean and standard deviation of population growth rates during the pre-tsunami period and thus would have reduced the accuracy of the estimated tsunami impacts.

Fourth, we selected species for which the pre-tsunami mean total coverage value was at least 0.007 (for sessile species) or for which the number of individuals each year in permanent plots was more than 30 (for mobile species). This was because estimates of population growth rates in species with small population sizes would have included an observation error too large to obtain a general picture of the population-level impact of the tsunami.

Fifth, we selected those species for which the mode of the vertical distribution of abundance was located within the vertical range of the permanent plot. This is because the population dynamics of rocky intertidal organisms often vary across their elevational range (e.g., Fukaya et al. 2014); therefore, any estimated tsunami impact near the limits of this range might not reflect impacts across the entire elevational range. To select appropriate species, we first obtained the mean number of individuals (for mobile species) or the mean coverage (for sessile species) for each of the ten 10×50 cm quadrats covering the vertical range of each plot. These mean values were calculated by using data from all plots in all years before the tsunami. We then visually judged for each species whether or not the vertical abundance distribution profile was unimodal.

4.3 Results and Discussion

The 2011 Tohoku earthquake tsunami had a significant negative impact on the rocky intertidal community at the regional scale (Fig. 4.1), with a decline in abundance of 11 species by 45.25 % on average (Table 4.2). Unfortunately, this decrease cannot be simply compared with population declines caused by other disturbance events at similar large spatial scales, such as oil spills or massive ice scours, because the ecological impacts of such events have rarely been quantified and have not yet been reported for the community level at regional scales (Moyse and Nelson-Smith 1964). However, several reports indicate that such large-scale disturbances cause catastrophic damage to many species on intertidal rocky shores. For example, the *Torrey Canyon* oil spill incident in 1967 resulted in the disappearance of almost all limpets, carnivorous snails, calcareous algae, and kelp from the rocky intertidal

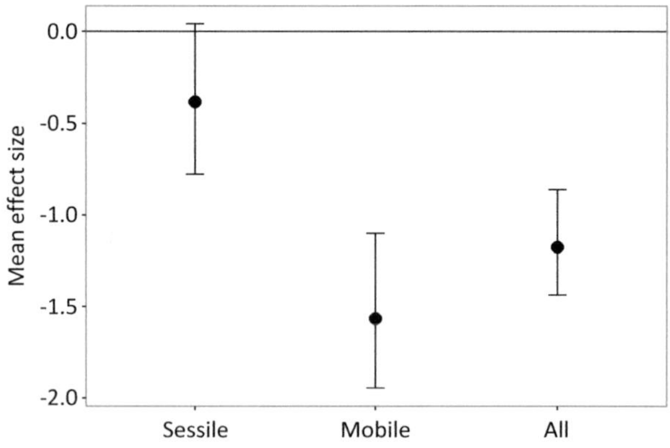

Fig. 4.1 Mean effect size of tsunami impacts on the abundance of all species combined and on two subgroups in the rocky intertidal community. Vertical lines denote bias-corrected 95 % confidence intervals, which were generated by bootstrapping procedures (10,000 iterations). The zero line indicates no effect, and mean effects are considered significant when the 95 % confidence interval does not cross the zero line

Table 4.2 Mean population growth rates for each group in the year of the tsunami

Group	Population growth rate Mean ± SE ($\log_{10}$)
All	−0.26 ± 0.14
Sessile	0.04 ± 0.05
Mobile	−0.45 ± 0.23

shore along tens of kilometers of shoreline around Cornwall, England (Southward and Southward 1978). Furthermore, a massive ice scour of the rocky intertidal seashore in Canada in 1987 removed almost all barnacles and mussels and the macroalga *Fucus* sp. from more than 50 km of exposed rocky shoreline (McCook and Chapman 1991, 1997). These reports on an oil spill and a massive ice scour impacts and this study involve a similar length of the time lag between the event and post-event survey (i.e., 2–4 months), which may cause the subequal underestimation of the direct negative impact through the population recovery for these studies. Thus the impact of these large-scale disturbances on rocky intertidal communities seems larger than the negative impact of the tsunami caused by the megaquake.

The tsunami did not have a significant negative impact on sessiles, but it did on mobiles (Fig. 4.1). This suggests that mobiles are more vulnerable to tsunamis than sessiles on rocky shores. Indeed, in the Alaska earthquake of 1964, populations of mobiles (limpets *Acmaea persona* and *Acmaea digitalis*) declined along rocky intertidal shores after the resultant tsunami, whereas the species composition and zonation of rocky intertidal sessile assemblages did not clearly change except in the area of heavy uplift (Haven 1971). Following the Philippines earthquake tsunamis in 2004, there were no reported changes in abundance of any rocky intertidal sessile species, but there were decreases in the abundance of the intertidal littorinid snails *Littoraria undulata* and *Echinolittorina feejeensis* after the tsunami (Sanpanich et al. 2006). In addition, the 2011 Tohoku earthquake tsunami caused decreases in populations of the abalone *Haliotis discus* by about 50 %, whereas populations of the kelp *Eisenia bicyclis* did not clearly decrease on subtidal rocky shores (Takami et al. 2013).

The decrease in populations of mobiles caused by the 2011 Tohoku earthquake tsunami could have affected the abundances of their prey species and consequently could alter the community structure via indirect interactions for decades, because community structure on the intertidal rocky shore is often strongly influenced by top-down effects (Lubchenco and Gaines 1981; Menge 2000). Indeed, previous studies have documented that disturbance events with spatial scales similar to that of the tsunami reduce the populations of mobile invertebrate consumers and consequently alter the rocky intertidal community structure for decades. The decrease in populations of the limpet *Patella vulgate* caused by the *Torrey Canyon* oil spill incident triggered unusual growth of the green algae *Enteromorpha* and *Ulva*, preventing recovery of barnacle populations for more than 10 years (Southward and Southward 1978). The *Exxon Valdez* oil spill reduced populations of molluscan grazers (limpets and periwinkles) and carnivorous gastropods (whelks). The declines in populations of these consumers triggered blooms of ephemeral green algae and an opportunistic barnacle, *Chthamalus dalli*, which inhibited the recovery of *Fucus* sp., lengthening the recovery process on rocky shorelines in Alaska for decades (Peterson et al. 2003). In addition, a massive ice scour event in Canada in 1987 also reduced numbers of the molluscan grazers *Littorina obtusata* and *Littorina rudis* and the predatory whelk *Nucella lapillus*. Their declines facilitated the recruitment of ephemeral green algae and mussels, consequently inhibiting the establishment of the perennial seaweed *Fucus vesiculosus*—which was a dominant sessile species before the massive ice scour event—for almost a decade (McCook and Chapman 1991, 1997).

4.4 Conclusions

Although it has been widely assumed that tsunamis can severely damage coastal ecosystems, our study strongly suggested that the rocky intertidal community at a regional scale suffered significant but only relatively minor direct damage from the tsunami caused by the Great East Japan Earthquake in 2011 than other catastrophic events such as oil spill (Southward and Southward 1978) and massive ice scour (McCook and Chapman 1991, 1997). This gap between actual and assumed impacts asserts the primacy of exact quantification in ecological consequences of disturbance. The approach used here should be applicable to quantification of the direct influence of a variety of disturbance events on various communities.

Acknowledgments We are grateful to T. Hagino, Dr. T. Okuda, Dr. M. Tsujino, M. Iida, R. Sahara, M. Sakaguchi, M. Ohira, and Y. Kanamori for field assistance. We also thank Dr. Y. Nishimura for helpful comments regarding tsunamis. For field and laboratory facilities, we are grateful to the staff and students of the International Coastal Research Center of the Atmosphere and Ocean Research Institute, The University of Tokyo. This study was made possible by the generous support and encouragement of local fishers and the fishery office of the Fisherman's Cooperative Associations in Iwate. This research was supported by the Cooperative Program of the Atmosphere and Ocean Research Institute, The University of Tokyo (Nos. 108, 104, 107, 101, and 103 in 2006, 2007, 2008, 2009, and 2010, respectively, to TN), and was supported partly by Japan Society for the Promotion of Science (JSPS) KAKENHI Grants (Nos. 20570012, 24570012, and 15 K07208 to TN) and the Tohoku Ecosystem-Associated Marine Sciences (TEAMS) project. The funders had no role in study design, data collection and analysis, decision to publish, or preparation of the manuscript.

References

Busk P, Serlin R (1992) Meta-analysis for single-case research. In: Kratochwill T, Levin J (eds) Single-case research design and analysis. Erlbaum, Hillsdale, pp 187–212

Busse RT, Kratochwill TR, Elliott SN (1996) Meta-analysis for single-case consultation outcomes: applications to research and practice. J Sch Psychol 33(4):269–285

Campbell SJ, Pratchett MS, Anggoro AW et al (2007) Disturbance to coral reefs in Aceh, Northern Sumatra: impacts of the Sumatra-Andaman tsunami and pre-tsunami degradation. Atoll Res Bull 544:55–78

Denny MW (1985) Wave forces on intertidal organisms: a case study. Limnol Oceanogr 30(6):1171–1187

Denny MW (1993) Disturbance, natural selection and the prediction of maximal wave-induced forces. Contemp Math 141:65–90

Denny M (1995) Predicting physical disturbance: mechanistic approaches to the study of survivorship on wave-swept shores. Ecol Monogr 65(4):371–418

Dominey-Howes D (2007) Geological and historical records of tsunami in Australia. Mar Geol 239(1):99–123

Fukaya K, Okuda T, Nakaoka M et al (2010) Seasonality in the strength and spatial scale of processes determining intertidal barnacle population growth. J Anim Ecol 79:1270–1279

Fukaya K, Okuda T, Nakaoka M et al (2014) Effects of spatial structure of population size on the population dynamics of barnacles across their elevational range. J Anim Ecol 83(6):1334–1343

Glass GV (1976) Primary, secondary, and meta-analysis of research. Educ Res 5:3–8

Haraguchi T, Iwamatsu A (2011) Detailed maps of the impacts of the 2011 Japan Tsunami, vol 1, Aomori, Iwate and Miyagi prefectures. Kokin-Shoin Publishers, Tokyo (in Japanese)

Haven SB (1971) Effects of land level changes on intertidal invertebrates with discussion of post-earthquake ecological succession. In: The Great Alaska Earthquake of 1964: Biology. National Academy of Science, Washington, DC, pp. 82–106

Hurwitz JT, Kratochwill TR, Serlin RC (2015) Size and consistency of problem-solving consultation outcomes: an empirical analysis. J Sch Psychol 53(2):161–178

Jaramillo E, Dugan JE, Hubbard DM, Melnick D, Manzano M et al (2012) Ecological implications of extreme events: footprints of the 2010 earthquake along the Chilean coast. PLoS One 7(5):e35348. doi:10.1371/journal.pone.0035348

Kanaya G, Suzuki T, Kikuchi E (2015) Impacts of the 2011 tsunami on sediment characteristics and macrozoobenthic assemblages in a shallow eutrophic lagoon, Sendai Bay, Japan. PLoS One 10(8):e0135125. doi:10.1371/journal.pone.0135125

Kinlan BP, Gaines SD (2003) Propagule dispersal in marine and terrestrial environments: a community perspective. Ecology 84(8):2007–2020

Kirby KN, Gerlanc D (2013) BootES: an R package for bootstrap confidence intervals on effect sizes. Behav Res Methods 45(4):905–927

Lander JF, Whiteside LS, Lockridge PA (2003) Two decades of global tsunamis. Sci Tsunami Haz 21(1):3–82

Lomovasky BJ, Firstater FN, Salazar AG et al (2011) Macro benthic community assemblage before and after the 2007 tsunami and earthquake at Paracas Bay, Peru. J Sea Res 65(2):205–212

Lubchenco JL, Gaines SD (1981) A unified approach to marine plant-herbivore interactions. I. Populations and communities. Annu Rev Ecol Syst 12:405–437

McCook LJ, Chapman ARO (1991) Community succession following massive ice-scour on an exposed rocky shore: effects of Fucus canopy algae and of mussels during late succession. J Exp Mar Biol Ecol 154(2):137–169

McCook LJ, Chapman ARO (1997) Patterns and variations in natural succession following massive ice-scour of a rocky intertidal seashore. J Exp Mar Biol Ecol 214(1):121–147

Menge BA (2000) Top-down and bottom-up community regulation in marine rocky intertidal habitats. J Exp Mar Biol Ecol 250(1):257–289

Mori N, Takahashi T, Yasuda T, Yanagisawa H (2011) Survey of 2011 Tohoku earthquake tsunami inundation and run-up. Geophys Res Lett 38(L00G14):6 pp, doi:10.1029/2011GL049210

Mori N, Takahashi T, The 2011 Tohoku Earthquake Tsunami Joint Survey Group (2012) Nationwide post event survey and analysis of the 2011 Tohoku earthquake tsunami. Coast Eng J 54(01):1250001

Moyse J, Nelson-Smith A (1964) Effects of the severe cold of 1962-63 upon shore animals in South Wales. J Anim Ecol 33:1–31

Munroe DM, Noda T (2010) Physical and biological factors contributing to changes in the relative importance of recruitment to population dynamics. Mar Ecol Prog Ser 421:151–162

Nakaoka M, Ito N, Yamamoto T, Okuda T, Noda T (2006) Similarity of rocky intertidal assemblages along the Pacific coast of Japan: effects of spatial scales and geographic distance. Ecol Res 21:425–435

Noda T, Iwasaki A, Fukaya K (2016) Rocky intertidal barnacle population dynamics: impacts and recovery from the Great East Japan Earthquake. In: Urabe J, Nakashizuka T (eds) Ecological impacts of tsunamis on coastal ecosystems: lessons from the Great East Japan Earthquake, Ecological research monographs, Springer, pp 47–58

Okuda T, Noda T, Yamamoto T, Ito N, Nakaoka M (2004) Latitudinal gradient of species diversity: multi-scale variability in rocky intertidal sessile assemblages along the North Western Pacific coast. Popul Ecol 46:159–170

Olive ML, Smith BW (2005) Effect size calculations and single subject designs. Educ Psychol 25:313–324

Paterson GL, Kendall MA, Aryuthaka C et al (2012) Tsunami impacts in the marine environment: review and results from studies in Thailand. In: Gower D et al (eds) Biotic evolution and environmental change in Southeast Asia. Cambridge University Press, Cambridge, pp 246–269

Peterson CH, Rice SD, Short JW, Esler D, Bodkin JL, Ballachey BE, Irons DB (2003) Long-term ecosystem response to the Exxon Valdez oil spill. Science 302:2082–2086

Petursdottir AL, McMaster K, McComas JJ, Bradfield T et al (2009) Brief experimental analysis of early reading interventions. J Sch Psychol 47(4):215–243

Rife GS (2014) Impacts of tsunami events on ecosystem services provided by benthic macroinvertebrate assemblages of marine coastal zones. In: Kontar YA, Santiago-Fandiño V, Takahashi T (eds) Tsunami events and lessons learned. Springer, Dordrecht, pp 147–159

Royama T (1992) Analytical population dynamics. Chapman & Hall, London

Sanpanich K, Wells FE, Chitramvong Y (2006) Effects of the 26 December 2004 tsunami on littorinid molluscs near Phuket, Thailand. J Molluscan Stud 72(3):311–313

Scheffers A, Kelletat D (2003) Sedimentologic and geomorphologic tsunami imprints worldwide—a review. Earth Sci Rev 63(1):83–92

Seike K, Shirai K, Kogure Y (2013) Disturbance of shallow marine soft-bottom environments and megabenthos assemblages by a huge tsunami induced by the 2011 M9.0 Tohoku-Oki earthquake. PLoS One 8(6):e65417. doi:10.1371/journal.pone.0065417

Southward AJ, Southward EC (1978) Recolonization of rocky shores in Cornwall after use of toxic dispersants to clean up the Torrey Canyon spill. J Fish Board Can 35(5):682–706

Takami H, Won N, Kawamura T (2013) Impacts of the 2011 mega-earthquake and tsunami on abalone Haliotis discus hannai and sea urchin *Strongylocentrotus nudus* populations at Oshika Peninsula, Miyagi, Japan. Fish Oceanogr 22(2):113–120

Urabe J, Suzuki T, Nishita T, Makino W (2013) Immediate ecological impacts of the 2011 Tohoku earthquake tsunami on intertidal flat communities. PLoS One 8(5):e62779. doi:10.1371/journal.pone.0062779

Usami T (2013) Materials for comprehensive list of destructive earthquakes in Japan. 599–2012, Bull. Earthq. Res. Inst., University of Tokyo Press, Tokyo

Whanpetch N, Nakaoka M, Mukai H, Suzuki T, Nojima S, Kawai T, Aryuthaka C (2010) Temporal changes in benthic communities of seagrass beds impacted by a tsunami in the Andaman Sea, Thailand. Estuar Coast Shelf Sci 87(2):246–252

Wijetunge JJ (2006) Tsunami on 26 December 2004: spatial distribution of tsunami height and the extent of inundation in Sri Lanka. Sci Tsunami Haz 24(3):225–239

Chapter 5
Rocky Intertidal Barnacle Population Dynamics: Impacts and Recovery from the Great East Japan Earthquake

Takashi Noda, Masayuki Sakaguchi, Aiko Iwasaki, and Keiichi Fukaya

Abstract The abundances and distributions of rocky intertidal sessile organisms were considered to be changed immediately as a direct result of either the tsunami or the subsidence and then further altered by changes in population processes such as mortality and recruitment. We investigated the influence of the Great East Japan Earthquake on the population dynamics of a rocky intertidal barnacle, *Chthamalus challengeri*, especially focusing on changes in vertical distribution, as well as alteration of recruitment and its causes and consequences. Light after the earthquake, *C. challengeri* zonation shifted vertically downward 50 cm, but the abundance did not clearly changed, suggesting that the barnacle did not suffer immediate mortality from either the tsunami or the subsidence. After the earthquake, the zone of *C. challengeri* expanded upward, and the species' abundance increased over the entire study area until 2012. Then the abundance below and around the lower limit of the species' normal vertical range decreased, and consequently, the zonation returned almost to the state before the earthquake by July 2013. During the study period, the population dynamics of this barnacle were driven in complex ways by interactions among larval supply, density-dependent intraspecific interactions, and interspecific competition, which were affected by factors including free-space availability for larval setting and the densities of conspecifics and competing species, of all which varied with elevation and time after the earthquake. One important finding is that the earthquake altered larval supply at the metapopulation level by changing the benthic population size.

Keywords Barnacle • *Chthamalus challengeri* • Density dependence • Earthquake • Larval supply • Population dynamics • Rocky intertidal • Sessile organisms • Subsidence • Tsunami

---

T. Noda (✉) • M. Sakaguchi • A. Iwasaki
Faculty of Environmental Earth Science, Hokkaido University, Sapporo, Hokkaido, Japan
e-mail: noda@ees.hokudai.ac.jp

K. Fukaya
The Institute of Statistical Mathematics, Tachikawa, Tokyo, Japan

J. Urabe, T. Nakashizuka (eds.), *Ecological Impacts of Tsunamis on Coastal Ecosystems*, Ecological Research Monographs,
DOI 10.1007/978-4-431-56448-5_5

5.1 Introduction

Sessile organisms such as mussels, barnacles, and seaweeds are a major component of rocky intertidal ecosystems. They play important roles in community organization by serving as a food resource for other macrobenthos (Paine 1966; Hori and Noda 2001) or altering the habitat structure (Suchanek 1986) or microenvironment (Bertness and Leonard 1997; Bertness et al. 1999) of other organisms. They also can competitively exclude other organisms from primary substrata (Connell 1961; Paine 1966; Dayton 1971; Miyamoto and Noda 2004) or provide secondary substrata for attachment of other organisms (Lohse 1993; Noda et al. 2003; Miyamoto and Noda 2004).

The Great East Japan Earthquake (Mw 9.0) caused large tsunami waves and coseismic subsidence of several tens of centimeters throughout the entire Tohoku region of Japan (Lay and Kanamori 2011; Tajima et al. 2013). These catastrophic events had great potential to affect the distribution and abundance of rocky intertidal sessile organisms. This is because extremely strong wave action and scouring by floating debris associated with a tsunami can dislodge or scrape these organisms from the substrata to which they are attached. In addition, tsunamis can redeposit coastal sediments and abrade or bury and smother sessile organisms. Furthermore, subsidence may have a substantial impact because it can shift a substrate with attached organisms to a deeper position below the lower limit of their normal range, which is often restricted to within a narrow vertical range of several tens of centimeters (Stephenson and Stephenson 1972). Indeed, previous studies have demonstrated that uplifts resulted in mass mortality of various sessile species and consequently altered rocky intertidal communities at regional scales (Haven 1972; Johansen 1972; Bodin and Klinger 1986; Castilla 1988; Castilla et al. 2010). However, there are no reports concerning the influence of subsidence on population dynamics of rocky intertidal sessile organisms except for several qualitative evaluations of population changes after subsidence (Haven 1972; Johansen 1972).

Acorn barnacles are ideal for evaluating the influence of the Great East Japan Earthquake on population dynamics of rocky intertidal sessile species. Their life histories are well understood, and it is easy to quantify the distribution and abundance of both the benthic population and recruitment. All rocky intertidal sessile animals including barnacles exhibit complex life histories that include pelagic larval and benthic phases linked by recruitment (Todd 1998). Therefore, the transition between these two phases can be crucial to understanding the population dynamics of these organisms (Thorson 1950). The structure and size of benthic populations are often governed by larval recruitment variability at the local scale (Gaines and Roughgarden 1985), whereas at broader spatial scales encompassing an entire metapopulation, larval recruitment should depend on the stock size via regional reproductive output (Hughes et al. 2000). At local and intermediate scales, the recruitment variability is strongly influenced by passive transport processes that depend on coastal topography and hydrodynamic features of coastal currents (reviewed by Shanks 1995). A simultaneous evaluation of abundance, distribution, and larval recruitment obtained from multiple sites should therefore greatly benefit for progressing our understanding of the population dynamics of these organisms.

Before the Great East Japan Earthquake, we regularly carried out censuses targeting all rocky intertidal macrobenthic species and their recruitment. These censuses covered 8 years at 25 plots on five shores close to the epicenter of the earthquake (Okuda et al. 2004; Nakaoka et al. 2006; Fukaya et al. 2010; Munroe and Noda 2010). The data from these censuses show that the barnacle *Chthamalus challengeri* (Fig. 5.1) is the dominant sessile animal in the mid to high intertidal zone in this region (Nakaoka et al. 2006; Fukaya et al. 2010). This species has the highest seasonal recruitment from July to October (Iwaki 1975), and individuals live several years (Miyamoto et al. 1999). The population dynamics of *C. challengeri* in this region are driven by seasonally variable density-dependent and density-independent processes (Fukaya et al. 2010) with considerable spatial variability in both abundance and recruitment intensity (Munroe and Noda 2010). This barnacle is consumed by various invertebrates, mainly whelks (Hori and Noda 2001), and competes with other sessile organisms, such as other barnacles (*Balanus glandula* and *Semibalanus cariosus*), mussels, and seaweeds (Miyamoto et al. 1999; Chiba and Noda 2000; Kado 2003).

To assess the impacts of the earthquake on the barnacle population using before/after comparisons, we have been carrying out surveys after the earthquake using the same methods at the same places as before the earthquake. To our knowledge, this is the only example of a study of population dynamics of a rocky intertidal sessile organism encompassing periods before and after a megaquake. Here, we show the influence of the Great East Japan Earthquake on the population dynamics of the rocky intertidal barnacle *C. challengeri*, especially focusing on changes in vertical distribution, as well as the alteration of recruitment and its causes and consequences.

Fig. 5.1 Barnacle *Chthamalus challengeri*, the dominant sessile organisms in the mid to high rocky intertidal zone along the Pacific coast of Japan

5.2 Materials and Methods

5.2.1 Census Design

The censuses were conducted along five shores located 150–160 km north-northwest of the epicenter of the Great East Japan Earthquake (Fig. 5.2; 38°06′12.0″N, 142°51′36.0″E). Detailed descriptions of the study sites and biogeographic features of the area are shown in previous works (Okuda et al. 2004; Nakaoka et al. 2006; Fukaya et al. 2010).

Within each shore, four or five sites were haphazardly chosen in semi-exposed locations, with distances between neighboring sites ranging from 7.8 to 209 m. At each site, adjacent recruitment and control plots separated by several tens of centimeters were marked with permanent anchors drilled into roughly vertical rock. The vertical heights of the plots were 100 cm, and the vertical midpoint of each plot corresponded to the mean tidal level. The horizontal widths of the recruitment and control plots were 50 cm and 30 cm, respectively.

In July 2011, 4 months after the coseismic subsidence, we extended the control plots vertically (100 cm higher and 50 cm wide), with the lower border of the new part of the plot sharing the top border of each original control plot, so that the vertical observation range was 200 cm. In April 2012, we also reestablished recruitment plots (100 cm vertically by 30 cm horizontally) at the same location, centered mid-tide.

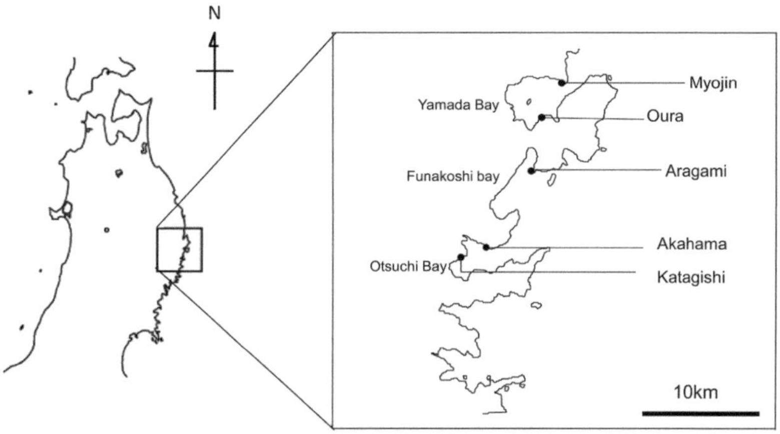

Fig. 5.2 Map of study sites. Five rocky shores were chosen for the census of intertidal organisms along the Pacific coast of Japan (Reproduced from Noda et al. 2015)

5.2.2 *Data Collection*

Recruitment plots were cleared of all surface organisms each April and July during low-tide events by burning and then scraping the rock surface with a wire brush. Recruitment plots were observed each July and October, first from 2004 to 2007 (before the earthquake) and then in 2012 and 2013. Recruitment plots were sampled approximately 90 days after clearing to assess the magnitude of recruitment during the major recruitment period for *C. challengeri* each year. Sampling of recruitment plots involved photographing 5 × 5 cm quadrats. Forty replicate quadrat photographs taken within each plot were used to estimate the recruitment density in each plot each year. The annual recruitment density of *C. challengeri* was calculated by adding the number of recruits from both seasons.

Control plots were observed each April, July, and October from 2004 to 2007 (before the earthquake) and then in 2012 and 2013. *Chthamalus challengeri* coverage was estimated by using a grid overlain on the plot with observation points at evenly spaced intervals (5 cm vertically and 5 cm horizontally, 200 grid points total). The presence or absence of *C. challengeri* at each grid point was used to estimate total coverage of this species in each plot for each census.

5.3 Results and Discussion

5.3.1 *Immediate Changes in Distribution and Abundance*

The Great East Japan Earthquake caused large tsunami waves (run-up height several to 30 m) and subsidence of 50 cm throughout the study area (Lay and Kanamori 2011; Tajima et al. 2013). We expected these catastrophic events to immediately alter not only the elevational position of *C. challengeri* in the rocky intertidal zonation but also its abundance throughout its entire range of distribution. Although *C. challengeri* zonation shifted vertically downward 50 cm after the earthquake (Fig. 5.3), its abundance in July 2011, 4 months after the earthquake, was not significantly different from that before the earthquake (Iwasaki et al. 2016 in this volume). This suggests that the barnacle did not suffer immediate mortality from either the tsunami or the subsidence.

5.3.2 *Changes in Passive Transport of Larvae*

The abundances and distributions of rocky intertidal sessile organisms were considered to be changed immediately as a direct result of the effects of the earthquake and then further altered by changes in population processes such as mortality and recruitment. For example, the collapse of coastal cliffs and landslides caused by the

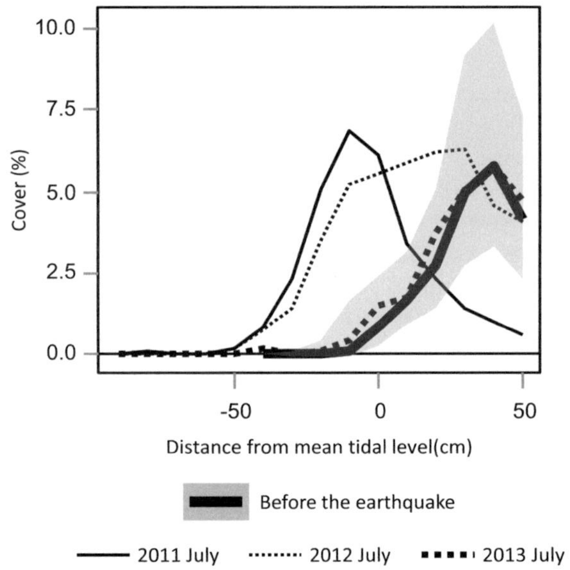

Fig. 5.3 Course and status of recovery of vertical distribution of *Chthamalus challengeri* within 3 years after the megaquake along the coast located 150–160 km north-northwest of the epicenter of the Great East Japan Earthquake. Tidal level on the y-axis represents elevational position of the quadrat (see the subsection "Census Design" in the "Materials and Methods" section for details) so that a smaller number corresponds to the upper side of the plot. Coverage and abundance at each tidal level were calculated using the data accumulated from all plots at each census. Coverage before earthquake at each tidal level indicates mean ± 1.96 standard deviation of data obtained from each census of pre-earthquake period (2003–2010) (Reproduced from Noda et al. 2015)

earthquake could change passive larval transport processes via modification of coastal topography. To examine such a possibility, we examined the spatial consistency of annual recruitment density at each site within the same shore. Specifically, it was evaluated by comparing Spearman's rank correlation coefficients (r_s) among three combinations of years: (1) among the 4 years before the earthquake, (2) between 2012 and each year before the earthquake, and (3) between 2013 and each year before the earthquake. There were no clear differences between the r_s values in the three combinations of years (mean ± s.e. (1) 0.57 ± 0.07; (2) 0.51 ± 0.07; (3) 0.50 ± 0.05), suggesting that the horizontal pattern of larval recruitment after the earthquake did not differ from that before the earthquake (Noda et al. in press).

5.3.3 Population Changes Following the Earthquake

The immediate change in vertical position associated with coseismic deformation likely results in subsequent changes in zonation and abundance of rocky intertidal sessile organisms (Haven 1972; Johansen 1972; Castilla 1988). From July 2011 to

Fig. 5.4 Massive recruitment of *Chthamalus challengeri* observed at Myojin, July 2011

July 2012, the zone of *C. challengeri* expanded upward (Fig. 5.3), and the mean coverage at plots significantly increased from 7.5 % to 12.8 % (paired t test: t = 2.842, df = 22, $p = 0.009$), suggesting that the species' abundance increased over the entire study area. A reasonable hypothesis for explaining this increase is that larvae that had been released during the settlement season in 2011 settled on free space within their normal range, which was notably abundant during this period compared to before the earthquake (Noda et al. in press). There are several pieces of evidence supporting this hypothesis. Although the barnacle coverage in post-earthquake control plots was extremely low just after the earthquake (July 2011), it reached at an exceptionally high level in April 2012, before the recruitment season for *C. challengeri* (Noda et al. in press). In addition, photographs taken at several shores in July 2011 show that there had been massive recruitment of *C. challengeri* in the 2011 recruitment season (Fig. 5.4).

During 2011–2012, the population dynamics of this barnacle species was mainly driven by density-independent processes at two contrasting spatial scales (Noda et al. in press). At a broader spatial scale encompassing an entire metapopulation, the increase in the size of the barnacle metapopulation resulted in an increased larval supply throughout the study area (Noda et al. in press), whereas at the plot scale, local population growth was almost linearly dependent on larval supply (Fig. 5.5: the regression slop did not significantly differ from 1 in 2012). This latter observation might be explained by the fact that free-space availability was notably high from July 2011 to July 2012 compared to before the earthquake (Noda et al. in press). Recruitment becomes increasingly important to population dynamics as more free space becomes available (Todd 1998).

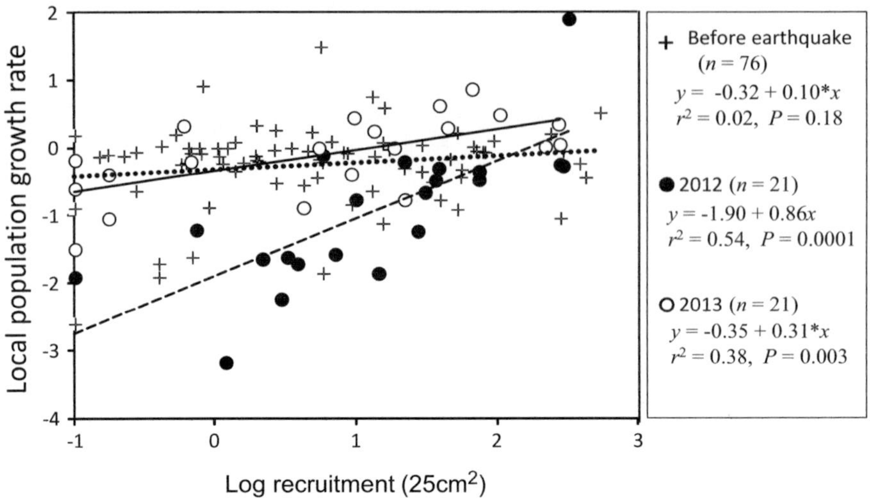

Fig. 5.5 Regression of the population growth rate (from April to December) in each plot against the annual recruitment density of *Chthamalus challengeri*. Estimated regression equations and fitted lines are shown. The *dotted, dashed, and solid lines* represent the years before the earthquake (2004–2007), 2012, and 2013, respectively. The *asterisks* next to the regression coefficients indicate that the values differ from 1 at a significance level of $P < 0.05$ (Reproduced from Noda et al. in press)

5.3.4 Recovery of Zonation and Abundance

Although the distribution and abundance of rocky intertidal sessile organisms can be altered by direct and subsequent effects of coseismic deformation, they are generally expected to return to their pre-earthquake status after a certain time. From July 2012 to July 2013, the abundance of *C. challengeri* below and around the lower limit of the species' normal vertical range decreased, and consequently, the zonation returned almost to the state before the earthquake by July 2013 (Fig. 5.3). This change in the lower half of the zonation was mainly caused by competition with other sessile organisms rather than by predation. This is evident in the observations that the decrease in coverage was greater in plots where the coverage of other sessile species was high, whereas it did not depend on the abundance of whelks, which are the main predators of this barnacle (Table 5.1). Indeed, a superior competitor, the Pacific oyster *Crassostrea gigas*, which can competitively displace *C. challengeri* (Hoshiai 1958), had drastically increased in coverage during 2012–2013 (Noda et al. Chap. 3 in this volume).

During 2012–2013, coverage within the range of zonation before the earthquake was almost unchanged (Fig. 5.3). In this vertical range, the local population growth is still dependent on larval supply, while density-dependent processes resume as before the earthquake (Noda et al. in press, Fig. 5.5), probably due to a decrease in available free space for setting larvae and an increase in numbers of benthic individuals (Noda et al. in press).

Table 5.1 Result of multiple regression analysis predicting annual change in coverage of *Chthamalus challengeri* during 2011 and 2012 and during 2012 and 2013 ($n = 184$, $R^2 = 0.49$). In this analysis, control plots and post-earthquake plots were divided into four subquadrats (50 cm vertically by 50 cm horizontally) and plot with the barnacle coverage at time t is 100 % or 0 % or with missing data were excluded from the analysis

Independent variables	Standard partial regression coefficients
Years (2012 or 2013)	–0.40*
Elevation	–0.39*
Elevation^2	–0.03
Own cover in time t	–0.62*
Cover of other sessiles in time t	–0.24***
Density of consumers in time t	0.21
Elevation × own cover	0.16
Elevetion^2 × own cover	0.08
Elevation × density of consumers	0.04
Elevation^2 × density of consumers	0.43

*$P < 0.05$, **$P < 0.01$, ***$P < 0.001$

5.4 Conclusions

Within 3 years of the 2011 Tohoku earthquake, which caused a large tsunami and subsidence of 50 cm over the entire study area, the abundance and zonation of the rocky intertidal barnacle *C. challengeri* dynamically changed but subsequently recovered to their status before the earthquake. One implication of this study is that the dynamics of zonation and abundance for this barnacle 1–2 years after the earthquake were driven in a complex way by the interaction of various processes, such as larval supply and intraspecific and interspecific competitions, which were influenced by factors such as free-space availability for setting larvae and densities of conspecifics and competing species, all of which varied with elevation and time after the earthquake. This finding leads to a significant suggestion: differences in the timing and relative importance of complex factors as described above may have caused large interspecific variability in zonation dynamics after the earthquake (Noda et al. 2015, Noda et al. 2016). In this case, the dynamics of individual species cannot be predicted from ecological knowledge obtained from pre-earthquake communities, such as the factors limiting upper and lower zonation limits, nor can their susceptibility to subsidence and tsunamis. In other words, the dynamics of zonation after large earthquakes and associated coseismic subsidence and tsunamis will be highly context dependent and therefore highly unpredictable.

Another implication arises from the fact that the Great East Japan Earthquake altered larval supply at the metapopulation level by changing the benthic population size. This finding leads to an important prediction. While such large-scale positive feedback for the stock–recruitment relationship resulted in population increases for the barnacle, a negative spiral was also plausible; that is, following an earthquake a reduction in a benthic metapopulation could cause a decrease in larval supply at the metapopulation level. Indeed, numbers of several sessile species have been continuously decreasing over the entire study area since the earthquake (Noda et al. 2015, Noda et al. 2016). Populations of those species will experience a long delay in recovering from the influence of the earthquake. In addition, if there is the potential for either the Allee effect, such as an inverse density dependence of fertilization success, which is often reported for broadcast spawners (Peterson and Summerson 1992; Levitan and Petersen 1995), or alternative stable community states (Petraitis and Latham 1999; Petraitis and Dudgeon 2004), then the damaged populations will most likely not recover to their former state.

Acknowledgments We thank Dr. T. Okuda, Dr. M. Tsujino, T. Hagino, M. Iida, R. Sahara, M. Ohira, and Y. Kanamori for field assistance. For field and laboratory facilities, we are grateful to the staff and students of the International Coastal Research Center of the Atmosphere and Ocean Research Institute, University of Tokyo. This study was made possible by the generous support and encouragement of local fisherman and the fishery office of the Fisherman's Cooperative Associations in Iwate Prefecture, Japan. This research was supported by the cooperative program of the Atmosphere and Ocean Research Institute, University of Tokyo (Nos. 108 [2006], 104 [2007], 107 [2008], 101 [2009], and 103 [2010] to T.N.), and was partly supported by the Japan Society for the Promotion of Science KAKENHI Grants (Nos. 20570012, 24570012 and 15K07208 to T.N.) and the Tohoku Ecosystem-Associated Marine Sciences (TEAMS) project. The funders had no role in study design, data collection and analysis, decision to publish, or preparation of the manuscript.

References

Bertness MD, Leonard GH (1997) The role of positive interactions in communities: lessons from intertidal habitats. Ecology 78:1976–1989

Bertness MD, Leonard GH, Levin JM, Schmidt PR, Ingraham AO (1999) Testing the relative contribution of positive and negative interactions in rocky intertidal communities. Ecology 80:2711–2726

Bodin P, Klinger T (1986) Coastal uplift and mortality of intertidal organisms caused by the September 1985 Mexico Earthquakes. Science 233:1071–1073

Castilla JC (1988) Earthquake-caused coastal uplift and its effects on rocky intertidal kelp communities. Science 242:440–443

Castilla JC, Manríquez PH, Camaño A (2010) Effects of rocky shore coseismic uplift and the 2010 Chilean mega-earthquake on intertidal biomarker species. Mar Ecol Prog Ser 418:17–23

Chiba S, Noda T (2000) Factors maintaining topography-related mosaic of barnacle and mussel on a rocky shore. J Mar Biol Assoc UK 80:617–622

Connell JH (1961) The influence of intra-specific competition and other factors on the distribution of the barnacle *Chthamalus stellatus*. Ecology 42:710–723

Dayton PK (1971) Competition, disturbance, and community organization: the provision and subsequent utilization of space in a rocky intertidal community. Ecol Monogr 41:351–389

Fukaya K, Okuda T, Nakaoka M, Hori M, Noda T (2010) Seasonality in the strength and spatial scale of processes determining intertidal barnacle population growth. J Anim Ecol 79:1270–1279

Gaines SD, Roughgarden J (1985) Larval settlement rate: a leading determinant of structure in an ecological community of the marine intertidal zone. Proc Natl Acad Sci U S A 82:3707–3711

Haven HW (1972) Effects of land-level changes on intertidal invertebrates, with discussion of postearthquake ecological succession. In: Krauskopf KB (ed) The Great Alaska Earthquake of 1964. National Academy of Sciences, Washington, DC, pp 82–126

Hori M, Noda T (2001) Spatio-temporal variation of avian foraging in the rocky intertidal food web. J Anim Ecol 70:122–137

Hoshiai T (1958) Synecological study on intertidal communities I. The zonation of intertidal animal community with special reference to the interspecific relation. Bull Mar Biol Stn Asamushi 9:27–33

Hughes TP, Baird AH, Dinsdale EA, Moltschaniwskyj NA, Pratchett MS, Tanner JE, Willis BL (2000) Supply-side ecology works both ways: the link between benthic adults, fecundity and larval recruits. Ecology 81:2241–2249

Iwaki T (1975) Breeding and settlement of *Chthamalus challengeri* Hoek on the southern coast of Hokkaido. Bull Fac Fish Hokkaido Univ 26:1–10

Iwasaki A, Fukaya K, Noda T (2016) Quantitative evaluation of the impact of the Great East Japan Earthquake and Tsunami on the rocky intertidal community. In: Urabe J, Nakashizuka T (eds) Ecological impacts of tsunamis on coastal ecosystems: lessons from the Great East Japan Earthquake, Ecological research monographs, Springer, pp 35–46

Johansen HW (1972) Effects of elevation change on benthic algae in Prince William Sound. In: Krauskopf KB (ed) The Great Alaska Earthquake of 1964. National Academy of Sciences, Washington, DC, pp 35–68

Kado R (2003) Invasion of Japanese shores by the NE Pacific barnacle *Balanus glandula* and its ecological and biogeographical impact. Mar Ecol Prog Ser 249:199–206

Lay T, Kanamori H (2011) Insights from the great 2011 Japan earthquake. Phys Today 64:33–39

Levitan D, Petersen CW (1995) Sperm limitation in marine organisms. Trends Ecol Evol 10:228–231

Lohse DP (1993) The importance of secondary substratum in a rocky intertidal community. J Exp Mar Biol Ecol 166:1–17

Miyamoto Y, Noda T (2004) Mussel effects on competitive inferior species: competitive exclusion to facilitation. Mar Ecol Prog Ser 276:293–298

Miyamoto Y, Noda T, Nakao S (1999) Zonation of two barnacle species not determined by competition. J Mar Biol Assoc UK 79:621–628

Munroe DM, Noda T (2010) Physical and biological factors contributing to changes in the relative importance of recruitment to population dynamics. Mar Ecol Prog Ser 421:151–162

Nakaoka M, Ito N, Yamamoto T, Okuda T, Noda T (2006) Similarity of rocky intertidal assemblages along the Pacific coast of Japan: effects of spatial scales and geographic distance. Ecol Res 21:425–435

Noda T, Minamiura N, Miyamoto S (2003) Seasonal changes in an intertidal annual algal assemblage in the Northern Japan: the role of preemption and grazing on algal replacement. Ecol Res 18:695–709

Noda T, Iwasaki T, Fukaya K (2015) Recovery of rocky intertidal zonation: two years after the 2011 Great East Japan Earthquake. J Mar Biol Assoc UK. doi:10.1017/S002531541500212X

Noda T, Sakaguchi M, Iwasaki A, Fukaya K (in press) Influence of the 2011 Tohoku Earthquake on population dynamics of a rocky intertidal barnacle: cause and consequence of alternation in larval recruitment. Coast Mar Sci

Noda T, Iwasaki A, Fukaya K (2016) Rocky intertidal zonation: impacts and recovery from the Great East Japan Earthquake. In: Urabe J, Nakashizuka T (eds) Ecological impacts of tsunamis on coastal ecosystems: lessons from the Great East Japan Earthquake, Ecological research monographs, Springer, pp 25–34

Okuda T, Noda T, Yamamoto T, Ito N, Nakaoka M (2004) Latitudinal gradient of species diversity: multi-scale variability in rocky intertidal sessile assemblages along the North Western Pacific coast. Popul Ecol 46:159–170

Paine RT (1966) Food web complexity and species diversity. Am Nat 100:65–75

Peterson CH, Summerson HC (1992) Basin-scale coherence of population dynamics of an exploited marine invertebrate, the bay scallop: implications of recruitment limitation. Mar Ecol Prog Ser 90:257–272

Petraitis PS, Dudgeon SR (2004) Do alternate stable community states exist in the Gulf of Maine rocky intertidal rocky intertidal zone? Comment. Ecology 85:1160–1165

Petraitis PS, Latham RE (1999) The importance of scale in testing the origins of alternative community states. Ecology 80:429–442

Shanks AL (1995) Mechanisms of cross-shelf dispersal of larval invertebrates and fish. In: McEdward L (ed) Ecology of marine invertebrate larvae. CRC Press, Boca Raton, pp 323–368

Stephenson TA, Stephenson A (1972) Life between tidemarks on rocky shore. Freeman, San Francisco

Suchanek TH (1986) Mussels and their role in structuring rocky shore communities. In: Moore PG, Seed R (eds) The ecology of rocky coasts. Columbia University Press, New York, pp 70–96

Tajima F, Mori J, Kennett BLM (2013) A review of the 2011 Tohoku-Oki earthquake (Mw 9.0): large-scale rupture across heterogeneous plate coupling. Tectonophysics 586:15–34

Thorson G (1950) Reproductive and larval ecology of marine bottom invertebrates. Biol Rev Camb Philos Soc 25:1–45

Todd CD (1998) Larval supply and recruitment of benthic invertebrates: do larvae always disperse as much as we believe? Hydrobiologia 375(376):1–21

Chapter 6
Ecological Impacts of the Great East Japan Earthquake and Tsunami and the Following Succession on the Subtidal Macrobenthic Community in Onagawa Bay, Northeastern Japan, with Special Reference to the Dominant Taxon, Polychaetes

Hirokazu Abe, Genki Kobayashi, and Waka Sato-Okoshi

Abstract The Great East Japan Earthquake of March 2011 caused huge tsunami waves. In the present study, field surveys were conducted in the innermost part of Onagawa Bay, from July 2007 to December 2013 to examine whether the benthic community was damaged severely by disturbances caused by the tsunami. Before the earthquake, the benthic community was relatively stable in the species composition and dominated by cirratulid and magelonid polychaetes. The density, biomass, and diversity of polychaetes decreased after the earthquake and tsunami, and the macrobenthic community structure fluctuated for 2 years after the disturbance. In this fluctuation period, spionid and capitellid polychaetes and opportunistic bivalves occurred abundantly, and the initial succession was driven by larval recruitment. In June 2013, the community entered a new state, which is different from the pre-

---

H. Abe (✉)
Graduate School of Agricultural Science, Tohoku University,
Amamiya 1-1, Tsutsumidori, Aoba-ku, Sendai 981-8555, Japan

Current address: Faculty of Science, Toho University,
Miyama 2-2-1, Funabashi, Chiba 274-8510, Japan
e-mail: hirokazuabe@affrc.go.jp; abehiro13@gmail.com

G. Kobayashi
Graduate School of Agricultural Science, Tohoku University,
Amamiya 1-1, Tsutsumidori, Aoba-ku, Sendai 981-8555, Japan

Current address: Graduate School of Frontier Sciences, The University of Tokyo,
Chiba 277-8564, Japan

W. Sato-Okoshi
Graduate School of Agricultural Science, Tohoku University,
Amamiya 1-1, Tsutsumidori, Aoba-ku, Sendai 981-8555, Japan
e-mail: wsokoshi@tohoku.ac.jp

J. Urabe, T. Nakashizuka (eds.), *Ecological Impacts of Tsunamis on Coastal Ecosystems*, Ecological Research Monographs,
DOI 10.1007/978-4-431-56448-5_6

quake community, and is dominated by maldanid polychaetes. Although both density and biomass of polychaetes gradually increased after the period of disturbance, density recovered to the pre-tsunami level, while biomass did not. Ecological impacts due to chemical pollution, derived from tsunami debris, were observed in addition to the physical disturbance of the tsunami. The additive effects by these drivers resulted in the initial succession similar to patterns of biological recolonization following large-scale hypoxia. The polychaete community has shown nonlinear responses during the recovery process, and the macrobenthic community in Onagawa Bay is still in the process of succession though 3 years have passed after the tsunami disturbance.

Keywords Tsunami • Polychaete • Community structure • Succession • Disturbance • Great East Japan Earthquake

6.1 Introduction

Physical disturbances play a significant role in ecosystem dynamics. Disturbances, such as earthquakes and tsunamis, can have drastic effects on marine organisms and the marine ecosystem, especially the benthic community. These large-scale disturbances cause biotic and abiotic changes within ecosystems. Recolonization after the large-scale disturbance is in general a lengthy process, because immigration from nearby undisturbed communities is often difficult. According to the review by Duarte et al. (2013), (1) estuarine and coastal ecosystems usually show partial recovery and do not necessarily reach full of their pre-disturbance status; (2) recovery trajectories may exhibit hysteresis, due to the presence of buffers which slow down recovery and the difference in thresholds during the degradation and recovery phases; (3) recovery depends likely on the type, magnitude, frequency, and timing of disturbances; and (4) the connectivity of ecosystems is probably essential for prompt recovery, because nearby communities must supply propagules and colonizers. Additionally, the order in which species will colonize a disturbed site may depend on the scale, frequency, and timing of the disturbance, because they each have different life history traits (Sousa 1984). This means that the trajectory of community succession will also vary with the type of disturbance (Yoshino et al. 2014). This makes it difficult to assess the impact of a disturbance and the following recovery process. Examination of the community level structural changes caused by large-scale disturbances, and the following recolonization process, would provide meaningful insights into the mechanisms underlying community structuring (Thibault and Brown 2008). However, the impacts by earthquakes and tsunamis on marine ecosystems are not well understood, because these types of disturbances are rare and unpredictable.

A megathrust earthquake that measured 9.0 on the moment magnitude scale occurred off the Pacific coast of northeastern Japan on March 11, 2011 (herein referred to as the Great East Japan Earthquake), as documented by a number of studies (Ide et al. 2011; Ozawa et al. 2011; Sato et al. 2011; Simons et al. 2011). Huge tsunami waves generated by the earthquake struck ca. 2000-km stretch of the

Pacific coastline of eastern Japan (Goto et al. 2011, 2012; Hinata et al. 2011; Kawai et al. 2011; Mori et al. 2011, 2012; The 2011 Tohoku Earthquake Tsunami Joint Survey Group 2011). Studies looking at the ecological impacts of the 2011 Great East Japan Earthquake and tsunami on intertidal soft bottom (Miura et al. 2012; Urabe et al. 2013; Kanaya et al. 2014), subtidal soft bottom (Seike et al. 2013), subtidal rocky shore (Takami et al. 2013), and deep-sea ecosystems (Kitahashi et al. 2014) were conducted. These studies suggest that the impacts differed among areas, habitats, and species. Other than these studies, the impacts of earthquakes and tsunamis on the marine benthic community have been conducted, including those evaluating the impacts of the 1989 Loma Prieta earthquake (Okey 1997), the 2004 Indian Ocean earthquake and tsunami (Hunsigi et al. 2006; Chavanich et al. 2008; Fujioka et al. 2008; Whanpetch et al. 2010), the 2007 Peru earthquake and tsunami (Lomovasky et al. 2011), and the 2010 Chilean earthquake (Jaramillo et al. 2012). However, all of these studies did not examine long-term or yearly scale changes in marine benthic communities, which are essential for evaluating the community dynamics after the disturbance.

Onagawa Bay, located on the Pacific coast of northeastern Japan, was severely affected by the Great East Japan Earthquake and tsunami. The tsunami reached inundation heights of 20 m and destroyed the central area of Onagawa Town. It may have damaged the benthic community in Onagawa Bay. Moreover, oil contamination of the bottom sediment by spilled heavy oil was observed in the bay after the tsunami. Thus, it is important to examine the structural changes of the macrobenthic community after the tsunami and during the recovery process in order to understand the effects of tsunami and the mechanism by which marine ecosystems fluctuate.

The sea bottom of Onagawa Bay was mainly comprised of muddy sediment (Sasaki et al. 1995), and the macrobenthic community was dominated by polychaetes before the earthquake and tsunami. Polychaetes are one of the most diverse and abundant invertebrate groups in soft bottom habitats worldwide, and they play an important role in the functioning of benthic communities, such as in marine food webs and material recycling (Hutchings 1998). Polychaetes are known to be the group most tolerant to environmental stresses such as organic enrichment, heavy metal pollution, and low oxygen concentrations in the macrobenthic community and are the pioneers of recolonization after disturbance events (Dean 2008; Steckbauer et al. 2011). Thus, polychaetes are good indicators of environmental disturbance, marine biodiversity, and community patterns in benthic invertebrate assemblages (Giangrande et al. 2005; Dean 2008). Additionally, infaunal macrobenthos usually have limited mobility and are tightly associated with physical and chemical condition of sea bottom sediments in contrast to epifaunal macrobenthos. The adults of the former taxa have limited abilities to immigrate to open patches for recolonization following a disturbance. Considering these characteristics, the successional dynamics of infaunal polychaete communities is a useful indicator for evaluating the impacts of disturbances in the benthic environment.

The aim of the present study is to evaluate the ecological impacts of the Great East Japan Earthquake and tsunami on the macrobenthic community in Onagawa Bay and to examine the response of the community to a large-scale disturbance, especially focusing on benthic polychaete communities.

6.2 Materials and Methods

6.2.1 Study Site

Onagawa Bay is located on the base of Oshika Peninsula in southernmost part of Sanriku ria coast, northeast mainland of Japan (Fig. 6.1). Its coastline is generally rocky, and the sea bottom of the bay is mainly comprised of muddy sediment (Sasaki et al. 1995). There is no significant inflow of freshwater into the bay except small river pours. The bay opens to the Pacific Ocean by about 3.8 km bay mouth. The innermost part of the bay was separated from the main basin by two long breakwaters jutted out from both sides with about 150 m opening mouth, forming a semi-enclosed embayment. The breakwaters collapsed due to the tsunami and had not reconstructed during the study period. Several heavy oil tanks placed around the bay also collapsed by the tsunami. Large volume of heavy oil spilled into the bay caused oil contamination of the bottom sediment after the tsunami.

6.2.2 Samplings

Samplings were conducted at a station (38°26.14′N, 141°27.83′E; 21–23 m depth) located in the innermost part of Onagawa Bay, from July 2007 to December 2013 (Fig. 6.1). Sampling was temporally ceased from March to October 2011, just after the earthquake, because sampling boats were unavailable and the sampling devices were lost. One to four replicate samples were collected once a month using a Smith-McIntyre grab (collected area: 0.05 m^2) from July 2007 to February 2011 and an Ekman-Birge grab (collected area: 0.02 m^2) from November 2011 to November 2013. The samples were sieved through a 1 mm mesh, and all organisms remaining on the sieve were fixed with 10 % neutralized formalin. Each organism was sorted, identified, and counted under a stereomicroscope. The wet weight biomass of the

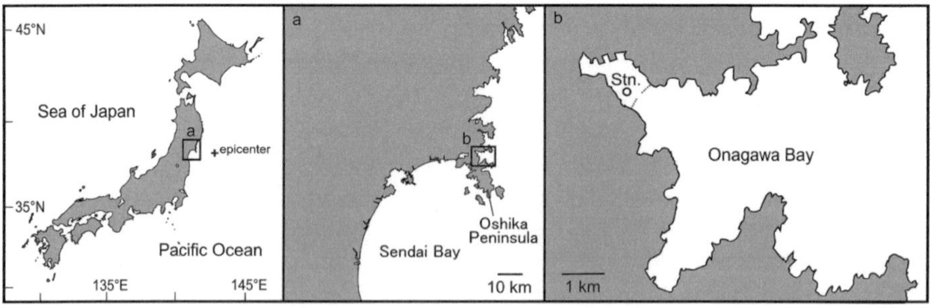

Fig. 6.1 Maps showing the location of the sampling station. Open circle indicates a sampling station. Broken lines in map b indicate sites where breakwaters were located before the earthquake. The breakwaters collapsed due to the tsunami (Modified after Abe et al. 2015)

macrobenthos was measured using an electric balance (GR-120, A&D Co. Ltd., Tokyo, Japan). Benthic species with a weight of more than 1 g such as holothurians, asteroids, bivalves, decapods, and spatangoids were excluded from data analysis because rarely collected large specimens spoil the accuracy in estimating the biomass. In the polychaete species, large specimens of glycerid, pectinariid, and sabellid species and maldanid species *Metasychis gotoi* were excluded from biomass data analysis during the study period before the tsunami.

6.2.3 Environmental Variables

Water temperature, salinity, and dissolved oxygen (DO) concentration for the bottom layer in the water column (about 20 m depth) were determined using a CTD RINKO-Profiler (JFE Advantech Co., Ltd., Kobe, Japan) and HQd digital water quality meters (HQ40d. HACH Co. Ltd., Loveland, CO, USA), during the sampling period. These environmental factors were determined also from July to October 2011, after the tsunami, when the sample collection was not conducted. Mud temperatures were measured using a mercury thermometer. The data for grain size distribution, concentration of normal-hexane extracts, and total sulfide near the sampling station, in the innermost part of Onagawa Bay, were obtained from annual reports published by Miyagi Prefecture and Tohoku Electric Power Company (2009, 2010, 2011, 2012, 2013, 2014, 2015) and were used to construct figures. To determine the changes in the chemical composition of the bottom sediment between before and after the tsunami, Welch *t*-test was carried out using the data of chemical oxygen demand (COD), concentration of polychlorinated biphenyl (PCB), total sulfide (acid-volatile sulfide: AVS), total mercury, alkyl mercury, organic phosphorus, copper, zinc, cadmium, and normal-hexane extracts of bottom sediments, obtained from annual reports by Miyagi Prefecture and Tohoku Electric Power Company (2009, 2010, 2011, 2012, 2013, 2014, 2015).

6.2.4 Data Analyses

Taxonomic diversity of benthic polychaete communities was quantified by computing the Shannon-Wiener index (H′). Cluster analysis was conducted based on the group average linkage method using the Bray-Curtis dissimilarity matrix calculated from the density of each polychaete family. To visualize the impact of the earthquake and tsunami on the benthic polychaete community, two-dimensional nonmetric multidimensional scaling (nMDS: Anderson and Willis 2003) based on the Bray-Curtis dissimilarity index was used. A similarity percentage analysis (SIMPER: Clarke 1993) with 9,999 permutations was conducted to identify a family's contributions to the differences among the clusters detected in the dendrogram. One-way analysis of similarity (ANOSIM: Clarke 1993) with 9,999 permutations was used to test for differences in community structure between the group of each cluster, and post hoc pairwise comparisons were performed using one-way ANOSIM. Mean density, mean biomass, family richness (number of families), and the Shannon-Wiener diversity index at the family level for benthic polychaetes were calculated at each chronological group. To examine whether these parameters differed depending upon the chronological groups defined by cluster analysis and nMDS plot, nonparametric multiple comparison by the Steel-Dwass test was conducted. ANOSIM was conducted using the software PRIMER v. 6 (Clarke and Gorley 2006), and the other analyses were conducted with R3.1.3 statistical software using the Vegan (Okansen et al. 2015) and MASS (Venables and Ripley 2002) packages.

6.3 Results

6.3.1 Temporal Fluctuation of Density and Biomass in the Macrobenthic Community

During the study period, species from total 21 taxonomic groups were identified. These taxonomic groups were Polychaeta, Bivalvia, Gastropoda, Decapoda, Amphipoda, Cumacea, Sipuncula, Nemertea, Asteroidea, Ophiuroidea, Holothuroidea, Pennatulacea, Actiniaria, Isopoda, Scaphopoda, Pycnogonida, Stomatopoda, Turbellaria, Spatangoida, Ascidiacea, and Porifera. Before the tsunami, the density and biomass of macrobenthos varied from 390 to 2,360 ind (individuals) m^{-2} and from 10.0 to 93.1 g m^{-2} (mean±SD: 1,151±509 ind m^{-2} and 33.2±18.5 g m^{-2}, N=41) (Fig. 6.2). The density and biomass of macrobenthos decreased to 100 ind m^{-2} and 1.1 g m^{-2}, respectively, and both recorded the lowest value in November 2011. After that, the density fluctuated temporally in 2012 (1,287±1,117 ind m^{-2}, N=12), and returned to the pre-disturbance level in 2013 (1,392±551 ind m^{-2}, N=11). On the other hand, the biomass did not return to the pre-disturbance level both in 2012 (14.0±12.4 g m^{-2}, N=11) and 2013 (22.8±14.8 g m^{-2}, N=9). Polychaetes were the most dominant taxonomic group in terms of both density and biomass in almost all months, except the period from May to September 2012. In May 2012, there was an exceptional increase of a minute gastropod *Voorwindia paludinoides*, and small bivalves were the most dominant group from July to September 2012. Among the bivalves, a small opportunistic bivalve *Theora lubrica* was the most common, followed by *Raetellops pulchellus*.

6.3.2 Temporal Fluctuation of Density and Biomass in the Benthic Polychaete Community

During the study period, a total of 32 polychaete families were identified: Spionidae, Cirratulidae, Magelonidae, Capitellidae, Maldanidae, Nephtyidae, Lumbrineridae, Ampharetidae, Glyceridae, Paraonidae, Terebellidae, Hesionidae, Sabellidae,

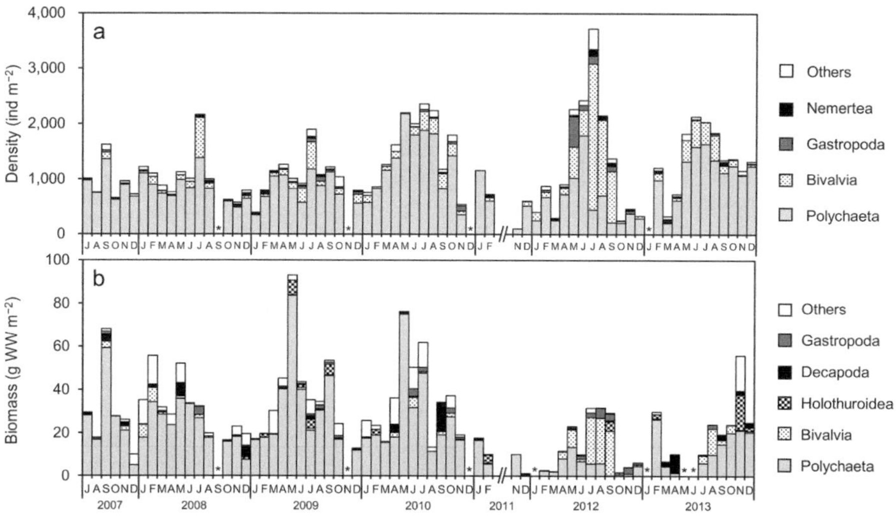

Fig. 6.2 Composition of macrobenthos density (**a**) and biomass (**b**) at a station in innermost part of Onagawa Bay from July 2007 to December 2013. Asterisks (*) and double slashes (//) indicate no data and the period when the sampling could not be conducted after the earthquake, respectively

Pilargidae, Nereididae, Syllidae, Oweniidae, Dorvilleidae, Aphroditidae, Pectinariidae, Cossuridae, Chaetopteridae, Sigalionidae, Polynoidae, Phyllodocidae, Apistobranchidae, Trichobranchidae, Goniadidae, Orbiniidae, Flabelligeridae, Opheliidae, and Eunicidae. Before the tsunami, during the period from July 2007 to February 2011, the families Cirratulidae and Magelonidae were dominant in terms of density (Fig. 6.3), comprising 30.9 % and 21.5 % of the benthic polychaetes, respectively. The family Cirratulidae was comprised of a number of species belonging to several genera such as *Aphelochaeta*, *Chaetozone*, *Cirriformia*, *Cirratulus*, and *Tharyx*, and the family Magelonidae was comprised of one species, *Magelona japonica*. The density and composition of the benthic polychaete community had little seasonal variation and was relatively stable before the tsunami. However, the density and diversity of polychaetes drastically decreased after the earthquake, and the lowest density was recorded in November 2011. Between November 2011 and April 2013, the density and composition of benthic polychaetes fluctuated and cirratulid and magelonid polychaetes almost disappeared during this period. Spionid polychaetes were the most dominant during the period from December 2011 to May 2012, just after the earthquake. Capitellid polychaetes *Capitella* sp. temporarily appeared at high density (1,475 ind m^{-2}) in June 2012. In May 2013, maldanid polychaetes gradually increased the density and became the most dominant family in the benthic polychaete community. During this period, the density of polychaetes almost returned to pre-earthquake levels. However, the community structure was completely different from that before the tsunami. The dominant feeding guilds

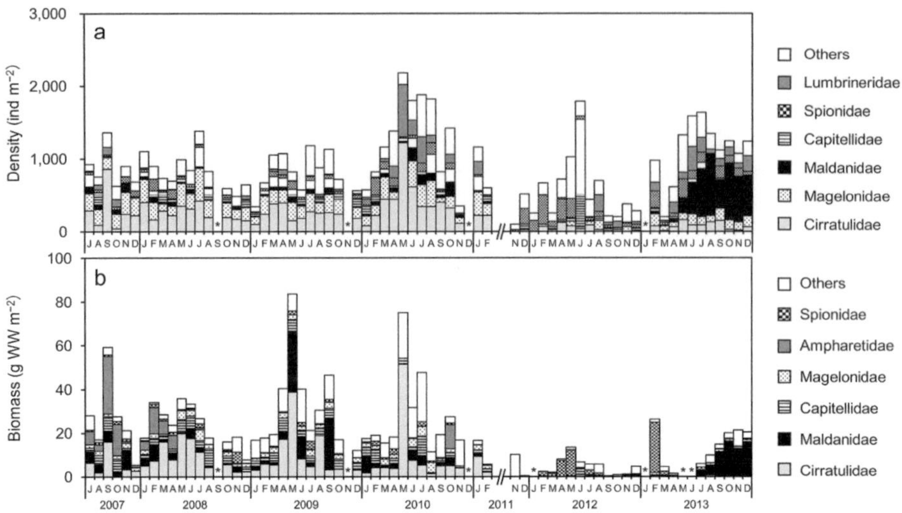

Fig. 6.3 Density (**a**) and biomass (**b**) of each family of benthic polychaete at a station in innermost part of Onagawa Bay from July 2007 to December 2013. Asterisks (*) and double slashes (//) indicate no data and the period when the sampling could not be conducted after the earthquake, respectively

shifted from surface deposit feeder (cirratulid and magelonid polychaete) to subsurface deposit feeder (maldanid polychaete) between before and after the tsunami.

At least five species of maldanid polychaetes were present before the earthquake. However, no maldanid polychaetes were collected during the period from November 2011 to August 2012, just after the earthquake (Fig. 6.4a). However, the density of maldanid polychaetes suddenly increased from May 2013, with two main species, *Clymenella collaris* and *Praxillella pacifica*. Although *C. collaris* was uncommon before the earthquake, it became a dominant species following the earthquake, and *Maldane* spp., which were collected before the earthquake, were not collected after the earthquake.

In spionid polychaetes, a total of 11 species were found. These were *Pseudopolydora achaeta*, *Paraprionospio coora*, *Laonice* sp., *Prionospio membranacea*, *Prionospio ehlersi*, *Prionospio sexoculata*, *Prionospio aucklandica*, *Rhynchospio glutaea*, *Spiophanes bombyx*, *Spiophanes kroyeri*, and *Scolelepis* sp. The dominant species of spionid polychaetes before the earthquake were *Sp. kroyeri* and *Ps. achaeta* (Fig. 6.4b). However, after the earthquake, species composition of spionids changed and densities of *Pa. coora*, *Laonice* sp., and *Prionospio* spp. increased. It is noteworthy that *Laonice* sp. was found frequently after the tsunami but rarely collected before the tsunami.

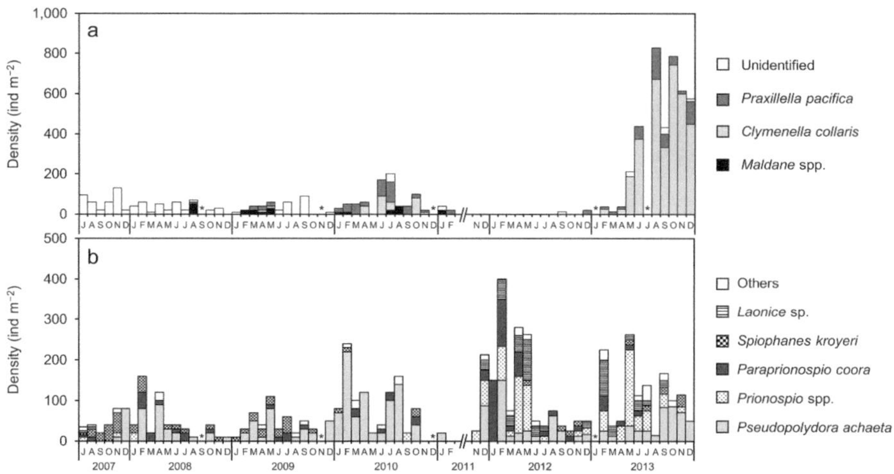

Fig. 6.4 Species composition of the density of maldanid (**a**) and spionid (**b**) polychaete at a station in innermost part of Onagawa Bay from July 2007 to December 2013. Asterisks (*) and double slashes (//) indicate no data and the period when the sampling could not be conducted after the earthquake, respectively (Modified after Abe et al. 2015)

6.3.3 Changes of Benthic Polychaete Community Structures

Results of cluster analysis identified eight clusters with 60% dissimilarity (Fig. 6.5a). Cluster I includes the months in the period before the earthquake, from July 2007 to February 2011, except September 2007 and July 2010. July 2012 was the only month after the earthquake included in cluster I. The community structures in September 2007 and July 2010 were classified as cluster II. The period from November 2011 to May 2013 was classified as clusters III, IV, V, VI, and VII. Cluster VIII was defined as the period from June to December 2013. ANOSIM showed that the clusters were significantly different (Global $R=0.91$, $P<0.01$) and the post hoc pairwise comparisons showed that cluster I was significantly different from clusters III, IV, VI, and VIII (R=0.99, 0.87, 0.93, and 0.88, respectively, and $P<0.01$ in all the cases). SIMPER analysis showed that each cluster was branched due to the contribution of cirratulid, spionid, capitellid, lumbrinerid, cossurid, and maldanid polychaetes (Table 6.1).

Mean densities of major polychaete families within each cluster are shown in Table 6.2. Cluster I and II were dominated by the families Cirratulidae and Magelonidae, and cluster III and IV were dominated by the family Spionidae. Cluster V was dominated by capitellid polychaetes, and clusters VI, VII, and VIII were dominated by the families Lumbrineridae, Cossuridae, and Maldanidae, respectively.

Each cluster was arranged separately in the nMDS ordination (Fig. 6.5b). The arrangement of nMDS plots showed that the months could be divided into four chronological groups. Cluster I (excluding July 2012) and cluster II comprise a

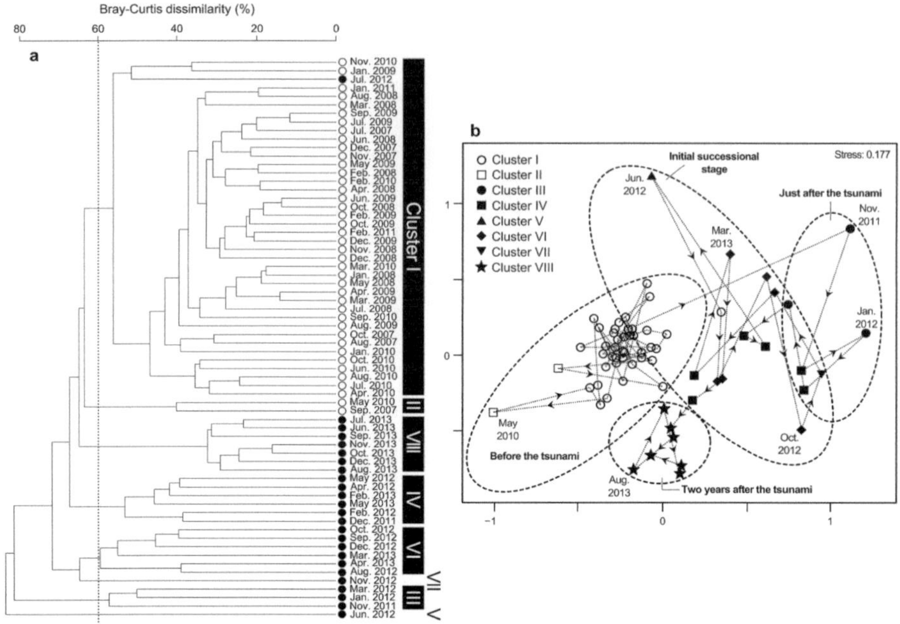

Fig. 6.5 Hierarchical cluster dendrogram (**a**) and 2-dimensional nMDS ordinal configuration (**b**) according benthic polychaete communities, using group average linkage by Bray-Curtis dissimilarities calculated on the density data of polychaete families in innermost part of Onagawa Bay. **a** Cluster analysis depicting eight assemblages at 60 % dissimilarity level, and which named cluster I to VIII. Open and closed circles indicate the polychaete communities before and after the tsunami, respectively. **b** Polychaete communities belonging to each cluster are represented by different symbols. Before and after the tsunami are represented by open and closed symbols, respectively. The dotted line and arrows indicate the trajectory of the community succession along the time. The dashed circles represent each chronological group; that is, (1) before the tsunami (samples collected in July 2007–February 2011), (2) just after the tsunami (November 2011–March 2012), (3) initial successional stage (April 2012–May 2013), and (4) two years after the tsunami (June–December 2013) (Modified after Abe et al. 2015)

group defined as (1) before the tsunami. December 2011 and February 2012, which were included in cluster IV and cluster III, construct a group defined as (2) just after the tsunami. July 2012, remaining cluster IV, and clusters V, VI, and VII comprise a group defined as (3) initial successional stage. Cluster VIII was defined as (4) 2 years after the tsunami. The community structures of benthic polychaetes just after the tsunami were distant and segregated, on the opposite side, from the community structures before the tsunami in the nMDS plot. In the initial successional stage, the community structure fluctuated and was dispersive. Although the community structure 2 years after the tsunami seemed similar to the structure before the tsunami, there were no overlaps in sites in nMDS plots between these two groups.

There were significant differences in mean density, mean biomass, family richness, or the Shannon-Wiener index between before and just after the tsunami (Steel-Dwass test, all $P<0.05$) (Fig. 6.6). Biomass significantly differed between before

Table 6.1 The results of SIMPER analysis. Polychaete families which have more than 6 % contribution to the differences in community structure between each cluster were listed with mean dissimilarity between communities

	Cluster I		Cluster II		Cluster III		Cluster IV		Cluster V	Cluster VI	Cluster VII
Cluster II	Mean dissimilarity = 63.5										
	Cirratulidae	47.3									
	Lumbrineridae	20.5									
	Magelonidae	7.9									
Cluster III	Mean dissimilarity = 83.3		Mean dissimilarity = 93.5								
	Cirratulidae	26.4	Cirratulidae	56.5							
	Magelonidae	23.0	Lumbrineridae	19.6							
	Capitellidae	8.9									
	Spionidae	7.3									
Cluster IV	Mean dissimilarity = 66.6		Mean dissimilarity = 82.1		Mean dissimilarity = 71.5						
	Spionidae	20.0	Cirratulidae	44.5	Spionidae	27.3					
	Cirratulidae	14.9	Lumbrineridae	15.8	Dorvilleidae	10.6					
	Magelonidae	13.2	Spionidae	12.2	Cirratulidae	9.7					
	Dorvilleidae	7.3			Cossuridae	8.7					
					Magelonidae	7.5					
Cluster V	Mean dissimilarity = 82.1		Mean dissimilarity = 92.4		Mean dissimilarity = 88.3		Mean dissimilarity = 81.2				
	Capitellidae	63.8	Capitellidae	44.9	Capitellidae	83.3	Capitellidae	67.0			
	Cirratulidae	9.4	Cirratulidae	30.2			Spionidae	10.5			

(continued)

Table 6.1 (continued)

	Cluster I		Cluster II		Cluster III		Cluster IV		Cluster V		Cluster VI		Cluster VII
Cluster VI	Mean dissimilarity = 70.3		Mean dissimilarity = 82.1		Mean dissimilarity = 70.8		Mean dissimilarity = 69.7		Mean dissimilarity = 85.9				
	Cirratulidae	26.5	Cirratulidae	59.0	Lumbrineridae	21.0	Spionidae	28.9	Capitellidae	79.1			
	Magelonidae	18.2	Lumbrineridae	17.4	Spionidae	13.7	Dorvilleidae	9.5					
	Lumbrineridae	9.1			Glyceridae	13.4	Lumbrineridae	9.3					
	Capitellidae	8.9			Magelonidae	12.8	Cirratulidae	8.2					
					Nephtyidae	10.4	Cossuridae	7.7					
Cluster VII	Mean dissimilarity = 78.7		Mean dissimilarity = 92.0		Mean dissimilarity = 80.5		Mean dissimilarity = 68.7		Mean dissimilarity = 89.5		Mean dissimilarity = 64.8		
	Cirratulidae	24.6	Cirratulidae	52.8	Cossuridae	34.8	Spionidae	27.5	Capitellidae	76.0	Cossuridae	34.7	
	Cossuridae	17.3	Lumbrineridae	17.7	Glyceridae	14.2	Cossuridae	12.1			Lumbrineridae	14.3	
	Magelonidae	14.0			Magelonidae	13.8	Dorvilleidae	9.5			Magelonidae	6.6	
	Capitellidae	8.9			Spionidae	10.4	Cirratulidae	9.2			Nephtyidae	6.3	
Cluster VIII	Mean dissimilarity = 64.3		Mean dissimilarity = 75.5		Mean dissimilarity = 86.6		Mean dissimilarity = 68.1		Mean dissimilarity = 88.4		Mean dissimilarity = 72.4		Mean dissimilarity = 80.9
	Maldanidae	41.0	Cirratulidae	41.0	Maldanidae	47.3	Maldanidae	39.8	Capitellidae	52.9	Maldanidae	50.6	Maldanidae
	Cirratulidae	12.1	Maldanidae	26.0	Lumbrineridae	11.3	Spionidae	12.0	Maldanidae	22.4	Lumbrineridae	8.4	Cossuridae
	Lumbrineridae	9.1	Lumbrineridae	12.6	Magelonidae	10.1	Lumbrineridae	8.4			Magelonidae	7.0	Lumbrine-ridae
	Magelonidae	6.5											

Table 6.2 Mean densities (ind m^{-2}) and standard deviation (SD) of each polychaete family which appeared abundantly in each cluster were listed by descending order. N is the number of months which included in each cluster

Cluster I		**Cluster II**		**Cluster III**		**Cluster IV**	
Family	**Mean density (SD)**	**Family**	**Mean density (SD)**	**Family**	**Mean density (SD)**	**Family**	**Mean density (SD)**
Cirratulidae	255 (121)	Cirratulidae	1040 (255)	Spionidae	83 (63)	Spionidae	274 (67)
Magelonidae	209 (90)	Lumbrineridae	410 (438)	Glyceridae	67 (29)	Dorvilleidae	88 (144)
Capitellidae	92 (68)	Magelonidae	80 (113)	Capitellidae	13 (22)	Cirratulidae	83 (55)
Spionidae	58 (51)	Paraonidae	40 (57)	Cirratulidae	8 (14)	Magelonidae	62 (58)
Maldanidae	49 (47)	Ampharetidae	40 (0)	Nephtyidae	13 (22)	Cossuridae	58 (54)
Lumbrineridae	45 (73)	Capitellidae	20 (28)	Nereididae	8 (14)	Sabellidae	45 (60)
Paraonidae	31 (26)	Spionidae	20 (0)	Sabellidae	4 (7)	Glyceridae	42 (28)
Glyceridae	24 (19)	Maldanidae	20 (0)	Cossuridae	4 (7)	Capitellidae	40 (25)
Apistobranchidae	22 (66)	Glyceridae	20 (0)			Lumbrineridae	39 (54)
Terebellidae	21 (26)	Pectinariidae	20 (28)			Nephtyidae	39 (42)
Nephtyidae	18 (19)	Nephtyidae	10 (14)			Maldanidae	38 (75)
Ampharetidae	18 (21)	Pilargidae	10 (14)			Pilargidae	10 (26)
	(*N*=40)		(*N*=2)		(*N*=3)		(*N*=6)
Cluster V		**Cluster VI**		**Cluster VII**		**Cluster VIII**	
Family	**Mean density (SD)**	**Family**	**Mean density (SD)**	**Family**	**Mean density (SD)**	**Family**	**Mean density (SD)**
Capitellidae	1,475	Lumbrineridae	93 (80)	Cossuridae	163	Maldanidae	613 (162)
Spionidae	50	Magelonidae	53 (41)	Magelonidae	63	Lumbrineridae	154 (96)
Sabellidae	50	Nephtyidae	50 (38)	Spionidae	50	Magelonidae	130 (28)
Cirratulidae	38	Spionidae	46 (17)	Nephtyidae	25	Spionidae	99 (52)

(continued)

Table 6.2 (continued)

Glyceridae	38	Glyceridae	22 (24)	Syllidae	25	Cirratulidae	80 (49)
Dorvilleidae	38	Capitellidae	20 (23)	Sigalionidae	25	Pilargidae	46 (28)
Magelonidae	25	Cirratulidae	19 (23)	Lumbrineridae	13	Ampharetidae	31 (34)
Goniadidae	25	Sabellidae	15 (30)	Sabellidae	13	Capitellidae	26 (32)
Nephtyidae	13	Maldanidae	10 (15)			Nephtyidae	28 (25)
Terebellidae	13	Polynoidae	10 (15)			Cossuridae	25 (40)
Cossuridae	13	Cossuridae	8 (20)			Glyceridae	17 (14)
Nereididae	13	Nereididae	6 (7)			Trichobranchidae	17 (28)
	(N=1)		(N=6)		(N=1)		(N=7)

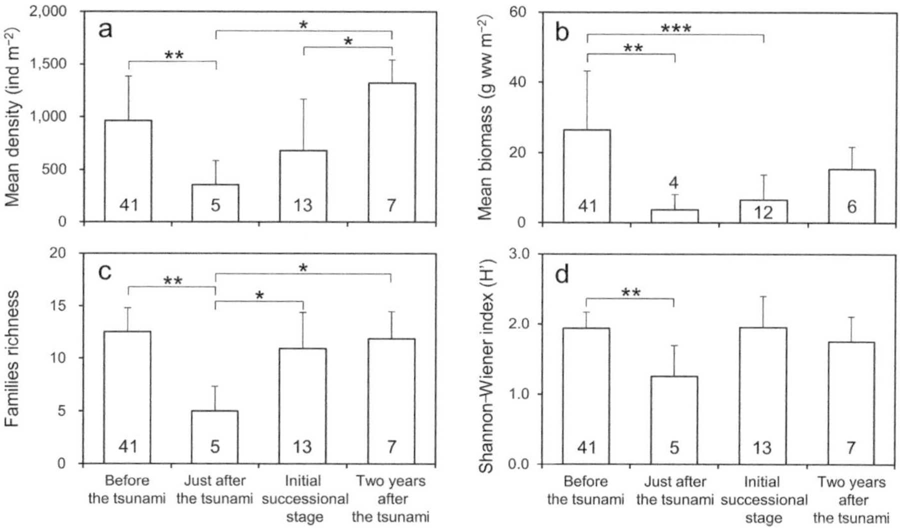

Fig. 6.6 Successional patterns of mean density (**a**), mean biomass (**b**), families richness (**c**), and Shannon-Wiener index (**d**) along with the chronological groups defined in Fig. 6.5b. The numbers in the columns indicate the number of months which included in each chronological group. The results of Steel-Dwass test are denoted by $^{*}P<0.05$, $^{**}P<0.01$, and $^{***}P<0.001$. Bars indicate standard deviations

the tsunami and the initial successional stage (Steel-Dwass test, $P<0.05$). The biomass has been increasing gradually but has still not fully recovered 2 years after the tsunami, although there was no significant difference between before the tsunami and 2 years after the tsunami (Steel-Dwass test, $P>0.05$). The community structure before the tsunami was characterized by high biomass and was mainly comprised of cirratulid and magelonid polychaetes (Fig. 6.6, Table 6.3). The community structure observed just after the tsunami was characterized by low density, low biomass, low family richness, and a low diversity index (Fig. 6.6). Spionid polychaetes dominated in this period (Table 6.3). Capitellid and spionid polychaetes were the major taxa observed in the next successional stage, and maldanid polychaetes were dominant 2 years after the tsunami (Table 6.3).

6.3.4 Environmental Factors

Seasonal variations in the water temperature, salinity, and DO concentration in the bottom layer of the water column and sediment temperature at the sampling station in Onagawa Bay are shown in Fig. 6.7. Water and sediment temperatures ranged from 4.4 to 21.7 °C and from 5.9 to 22.3 °C, respectively. Salinity ranged from 32.6 to 34.7, and dissolved oxygen concentration ranged from 3.0 to 11.8 mg L^{-1}. DO concentration was relatively high in winter and spring and low in summer and

Table 6.3 Mean densities (ind m^{-2}) and standard deviation (SD) of each polychaete family which appeared abundantly in each chronological group defined in Fig. 6.5b were listed by descending order. N is the number of months which included in each chronological group

Before the tsunami		Just after the tsunami		Initial successional stage		Two years after the tsunami	
Family	Mean density (SD)	Family	Mean density (SD)	Family	Mean density (SD)	Family	Mean density (SD)
Cirratulidae	297 (210)	Spionidae	173 (146)	Capitellidae	136 (403)	Maldanidae	613 (162)
Magelonidae	207 (92)	Cossuridae	52 (66)	Spionidae	111 (103)	Lumbrineridae	154 (96)
Capitellidae	90 (68)	Glyceridae	46 (35)	Lumbrineridae	63 (69)	Magelonidae	130 (28)
Lumbrineridae	64 (127)	Capitellidae	23 (34)	Magelonidae	59 (45)	Spionidae	99 (52)
Spionidae	56 (51)	Nephtyidae	20 (29)	Cirratulidae	52 (51)	Cirratulidae	80 (49)
Maldanidae	49 (46)	Cirratulidae	18 (29)	Nephtyidae	42 (36)	Pilargidae	46 (28)
Paraonidae	32 (27)	Magelonidae	12 (16)	Dorvilleidae	46 (102)	Ampharetidae	31 (34)
Glyceridae	23 (18)	Nereididae	5 (11)	Glyceridae	33 (27)	Nephtyidae	28 (25)
Apistobranchidae	21 (65)	Dorvilleidae	3 (7)	Sabellidae	32 (45)	Capitellidae	26 (32)
Terebellidae	20 (26)	Sabellidae	3 (6)	Cossuridae	28 (45)	Cossuridae	25 (40)
Ampharetidae	19 (21)	Sigalionidae	3 (6)	Maldanidae	22 (52)	Glyceridae	17 (14)
Nephtyidae	18 (19)			Polynoidae	8 (12)	Trichobranchidae	17 (28)
	(*N*=41)		(*N*=5)		(*N*=13)		(*N*=7)

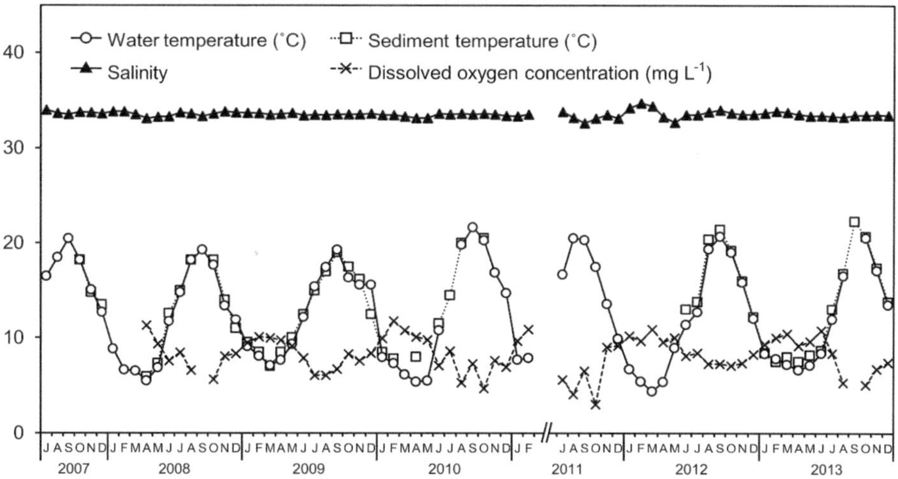

Fig. 6.7 Temporal variations of temperature, salinity, and dissolved oxygen concentration at bottom layer (20 m depths) in the water column and sediment temperature at a station in innermost part of Onagawa Bay

autumn. There were notable decreases in DO concentration in 2011 after the tsunami, and the lowest value was recorded in October 2011.

6.4 Discussion

This study clearly showed that Great East Japan Earthquake and tsunami affected largely the macrobenthic community in Onagawa Bay, both in the short term and long term. Just after the earthquake, the density of macrobenthos significantly decreased and almost all macrobenthic species disappeared (Fig. 6.2). Only spionid, glycerid, and nereidid polychaetes were observed in November 2011, when the first field survey was conducted after the earthquake (Fig. 6.3). The infaunal organisms in tidal flats emerged from the bottom sediment, which was liquefied by the shock of the earthquake (Okoshi 2015). The same liquefaction phenomenon likely occurred in the subtidal zone, and almost all of the macrobenthic organisms would have been swept away, along with the sediments, by the tsunami. Since the macrobenthic organisms collected just after the tsunami were not newly settled individuals, based on their size, they were considered to be the survivors of this large-scale disturbance.

The density and biomass of macrobenthos gradually increased, and the community structure fluctuated after disturbance by the tsunami, until May 2013 (Figs. 6.2, 6.3, and 6.5b). Starting in June 2013, the macrobenthic community transitioned to a stage with low fluctuation in the community structure that was different from what existed before the tsunami (Figs. 6.3 and 6.5). Although macrobenthos density returned to the pre-tsunami level, the biomass did not (Fig. 6.6). The families Cirratulidae, Maldanidae, and Ampharetidae, which include relatively large species, were the main contributors to the total biomass of the macrobenthic community before the earthquake. However, the number of relatively large species decreased, and the community was mainly comprised of relatively small polychaete species after the earthquake. Although family richness and the Shannon-Wiener diversity index both decreased immediately following the tsunami, they returned to the levels observed before the tsunami during the initial successional stage (Fig. 6.6). It is important to note that the present study evaluated polychaete diversity at family level, suggesting that diversity of polychaetes in Onagawa Bay was much larger at the species level compared with the results shown here.

The benthic organisms in coastal areas are typically reestablished by larval recruitment and succession following a disturbance (Pearson and Rosenberg 1978). Therefore, the resilience of a benthic community depends upon species composition and the life cycles, reproduction periods, and larval dispersal patterns of the various species (Levin 1984; Pechenik 1999; Zajac and Whitlatch 1982a; Zajac et al. 1998). Spionid polychaetes, *Paraprionospio coora*, *Prionospio* spp., and *Laonice* sp.; capitellid polychaetes, *Capitella* sp.; and the bivalve species, *Theora lubrica*, were the most dominant macrobenthos in the period shortly after the tsunami and during the initial successional stage after the tsunami (Figs. 6.2 and 6.3, Table 6.3). These macrobenthic groups are well known as opportunistic benthic taxa and have high intrinsic rate of natural increase and environmental tolerance abilities (Pearson and Rosenberg 1978; Bellan et al. 1988; Noji and Noji 1991). Planktonic larvae of spionid and capitellid polychaetes were common in Onagawa Bay (Abe et al. 2011) and were found in the bay even after the tsunami (Abe 2013). Additionally, *T. lubrica* has a long planktonic larval period (Imabayashi and Iwatani 1988). These macrobenthic organisms with planktonic development and long-distance dispersal patterns may be able to take advantage of large-scale disturbances, by rapidly recolonizing uninhabited areas after the tsunami. On the other hand, benthic polychaetes belonging to families such as Cirratulidae, Maldanidae, Paraonidae, and Ampharetidae, which have either no planktonic larval phase, or a relatively short planktonic larval phase (Wilson 1991; McHugh and Tunnicliffe 1994; Giangrande 1997; Petersen 1999), seem to be more affected by large-scale disturbance. Although the density of benthic polychaetes belonging to these families was relatively high before the earthquake (Fig. 6.3), their larvae were rarely observed at the same sampling station (Abe et al. 2011; Abe 2013). The fact supports the assumption that these polychaetes have either no planktonic larval phase or a relatively short planktonic larval phase in Onagawa Bay.

The recovery process of macrobenthic communities after the disturbance also depends on the spatial and temporal scale of disturbance (Karakassis et al. 1999; Lardicci et al. 2001; Munari et al. 2003). The dominance of opportunistic species during the initial succession stage, following the tsunami, was similar to the pattern of biological recolonization observed following a long-term severe hypoxia event (Steckbauer et al. 2011). The tsunami, which swept up almost all of the macroben-

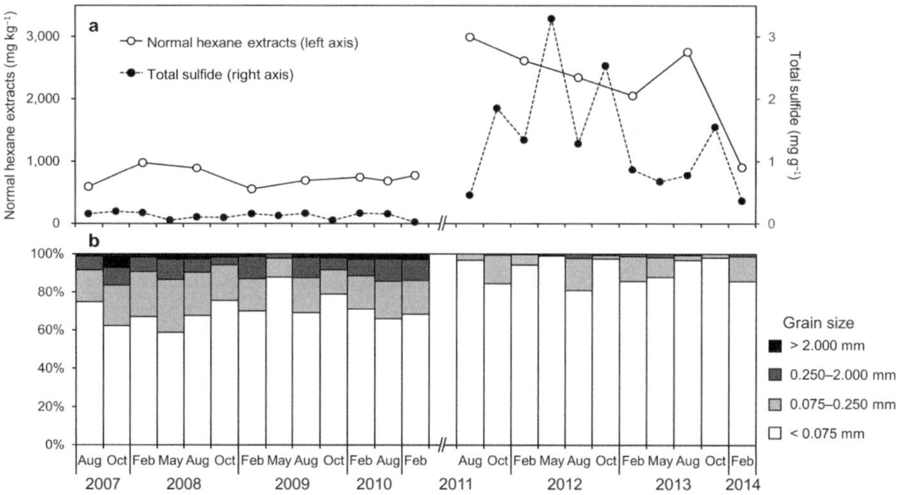

Fig. 6.8 Temporal variation of concentration of normal-hexane extracts and total sulfide (**a**) and grain size distribution of bottom sediments (**b**) near sampling station in innermost part of Onagawa Bay from August 2007 to February 2014 (The data were obtained from Miyagi Prefecture and Tohoku Electric Power Company (2009, 2010, 2011, 2012, 2013, 2014, 2015))

thic taxa, had an impact comparable to long-term severe hypoxia, in that both remove all animal taxa. Recovery after severe hypoxia is usually dominated by relatively small, fast-growing, short-lived opportunistic species, such as capitellid and spionid polychaetes, in its initial phase (Diaz and Rosenberg 1995; Pearson and Rosenberg 1978; Wildsmith et al. 2009). These species only play a minor role in the community prior to hypoxia. After a hypoxia event, polychaetes are the pioneers of recolonization, followed by mollusks. This sequence corresponds to each group's tolerance of chemical pollution and low oxygen concentrations (Steckbauer et al. 2011). Polychaetes were the pioneers in the present study, and the increase of the polychaete species, *Capitella* sp., and *Paraprionospio coora*, common indicators of pollution (Dean 2008; Kodama et al. 2012), was observed in the initial successional stage. Following polychaete recolonization, another species indicative of pollution, the bivalve species *Theora lubrica* (Tamai 1996 as *Theora fragilis*; Hayward et al. 1999; Saito 2006) was observed, during the period from July to September 2012.

Changes in the chemical composition of the bottom sediment were observed in the innermost part of Onagawa Bay after the tsunami. Chemical oxygen demand (COD), concentration of polychlorinated biphenyl (PCB), total sulfide, total mercury, alkyl mercury, organic phosphorus, copper, zinc, cadmium, and normal-hexane extracts of bottom sediments were measured before and after the tsunami (Miyagi Prefecture and Tohoku Electric Power Company 2009, 2010, 2011, 2012, 2013, 2014, 2015). The concentrations of total sulfide, normal-hexane extracts, and cadmium of bottom sediments significantly increased after the tsunami (Welch *t*-test, all $P<0.01$). The concentration of normal-hexane extracts, which is the primary indicator of oil contamination, was the most dramatically increased after the tsunami (Fig. 6.8a). There were several heavy oil tanks near the sampling station before the tsunami. The tanks collapsed because of the tsunami and a large volume of heavy oil spilled into the bay. The increase in normal-hexane extracts may be due to the spilled heavy oil. After the tsunami, the smell of oil was detected in the collected sediments, and oil was observed floating up from the bottom sediments when sieving. In addition to this, sedimental sulfide, which is the primary indicator of hypoxia (Sasaki and Arai 1997), increased by up to 3.3 mg g^{-1}, after the tsunami (Fig. 6.8a). These results suggest that the inner most part of Onagawa Bay was exposed to chemical pollution after the earthquake. Notable decrease in DO concentration was observed in 2011, after the tsunami (Fig. 6.7). The natural (earthquake and tsunami) and anthropogenic (chemical pollution) disturbances in Onagawa Bay after the tsunami may result in a pattern of biological recolonization similar to those observed following large-scale hypoxia events. The argument implies that, in addition to physical disturbance by the tsunami, chemical pollution played a role in recolonization of the macrobenthic community after the tsunami.

Oil contamination explains the presence of toxic hydrocarbons, such as PAHs, in sediment (Kanaya et al. 2016). This contamination will cause acute damage and long-term effects on the benthic community (Peterson et al. 2003). Dissolved sulfide is also known as a toxicant for aquatic organisms (Bagarinao 1992). Therefore, the slow recovery of the benthic polychaete community after the tsunami may be due to the contamination of the sediment. This is supported by the absence of macrobenthic animals that are highly vulnerable to pollution, such as crustaceans (Bagarinao 1992) and *Clymenella collaris* (Kodama et al. 2012). These organisms were almost absent in 2011 and 2012 (Figs. 6.2 and 6.4). However, crustaceans reappeared and *C. collaris* was dominant in 2013. Normal-hexane extracts and sulfide gradually decreased, and they reached levels of approximately 900 mg kg^{-1} and 0.4 mg g^{-1} in February 2014, respectively (Fig. 6.8a). The changes in sediment concentrations of harmful chemicals and the differential tolerances of each macrobenthic species to those poor conditions are the main factors affecting the succession of the macrobenthic community following the tsunami in Onagawa Bay.

Two years after the tsunami, the macrobenthic community in Onagawa Bay has transitioned into a stage which was temporally stable and significantly different from the pre-earthquake community when the maldanid polychaetes *Clymenella collaris* and *Praxillella pacifica* dominated (Figs. 6.3 and 6.4). These maldanid species sometimes appear in abundance in the muddy bottom sediments of Japanese waters (Ishikawa 1989; Hayashi et al. 1992; Lee and Paik 1995; Kodama et al. 2012). The tube construction and feeding activities of maldanid polychaetes can modify the community structure of a seabed assemblage with "subduction" (Levin et al. 1997) and "spongy" effects (Mach et al. 2012) on the sediments. Maldanid polychaetes are sometimes a dominant species in benthic communities. When their density increases, the firmness of the sediment decreases (Mach et al. 2012), and the maldanid polychaetes could be regarded as ecosystem engineers (Meadows et al. 2012). Thus, it is likely that their environmental modification could contribute to the continuance of a community structure with dominance of maldanid polychaetes. However, the life cycle, reproduction, developmental mode, and physiological features of the maldanid polychaetes abundantly found in the present study are not known. More information about the specific biology of these worms is needed in order to understand why the maldanid species were dominated after the tsunami. Additionally, it has been reported that a stress-tolerant maldanid polychaete, *Clymenella torquata*, removed oil from contaminated sediment by reworking and irrigating the sediments (Koerting-Walker and Buck 1989). Considering the fact, it is possible that congeneric *C. collaris* is removing oil from the bottom sediments in Onagawa Bay.

In marine benthic ecology, sediment granulometry is the most important factor in determining the characteristics of benthic communities (Sanders 1958; Gray 1974; Rhoads 1974; Snelgrove and Butman 1994; Snelgrove et al. 1997; Wijsman et al. 1999). Sediment granulometry was measured in the innermost part of Onagawa Bay before the tsunami, and the composition of silt clay ranged from 58.7 to 87.7 % (Fig. 6.8b; Sasaki et al. 1995). However, after the tsunami, it increased to 80.8–98.9 % (Fig. 6.8b), indicating that the tsunami altered the composition of the bottom sediments. This increase in the silt-clay component may affect the constitutive alteration of the macrobenthic community after the earthquake. Suspension feeders are generally confined to sandy bottom sediments, while subsurface deposit feeders favor muddy bottom sediments (Rhoads and Young 1970; Sanders 1958; Snelgrove and Butman 1994; Wijsman et al. 1999). Therefore, this granulometric change may be one of the factors that induced the shift in dominant feeding guilds before (surface deposit feeder: Cirratulidae and Magelonidae) and after the disturbance (subsurface deposit feeder: Maldanidae). Additionally, the origin of the bottom surface sediments deposited after the tsunami (likely derived from offshore marine and/or terrestrial areas) could potentially affect which feeding guild is the dominant.

In this study, ecological impacts due to natural disturbances as well as anthropogenic disturbances were observed in Onagawa Bay, and a combination of physical, chemical, and biological factors affected the recovery processes of the benthic community after the earthquake and tsunami. Benthic communities tend to respond rapidly to environmental changes (Zajac and Whitlatch 1982b). On the other hand, although recovery time is variable, the recovery process after a severe disturbance is generally slower (Karakassis et al. 1999; Lardicci et al. 2001; Munari et al. 2003), and can take decades to complete (Diaz and Rosenberg 2008; Wildsmith et al. 2009; Veríssimo et al. 2012). Cirratulid and magelonid polychaetes, which were dominant before the tsunami, have gradually increased in 2013, but the macrobenthic community in Onagawa Bay may still be in the process of succession, almost 3 years after the disturbance. Further study is thus needed to assess whether or not the community of the macrobenthos in Onagawa Bay returns to the pre-tsunami state.

Acknowledgments A special note of thanks to Captain Toyokazu Hiratsuka, and the staff of the Field Science Center, Graduate School of Agriculture Science, Tohoku University, for the help provided while collecting the samples in Onagawa Bay. We express our sincere appreciation to Dr. Leslie Harris of Natural History Museum Los Angeles County and Naoto Jimi of Hokkaido University for her and his identification of maldanid and cirratulid polychaetes, respectively. We also express our sincere appreciation to Drs. Kazunori Hasegawa of National Museum of Nature and Science and Kenji Okoshi of Toho University for their identification of mollusk shell. We are grateful to the members of the Laboratory of Biological Oceanography, Graduate School of Agricultural Science, Tohoku University, for their cooperation and assistance during field observations. We also thank Dr. Wataru Teramoto for his support during community analysis. This study was partially funded by a research grant from the Research Institute of Marine Invertebrates Foundation awarded to Hirokazu Abe and from the Ministry of Education, Science, Sports and Culture of Japan to Waka Sato-Okoshi.

References

Abe H (2013) Taxonomy and dynamics of planktonic larvae of spionid polychaetes in Onagawa Bay, northeastern Japan. Dissertation, Tohoku University

Abe H, Sato-Okoshi W, Endo Y (2011) Seasonal changes of planktonic polychaete larvae and chlorophyll *a* concentration in Onagawa Bay, northeastern Japan. Ital J Zool 78:255–266

Abe H, Kobayashi G, Sato-Okoshi W (2015) Impacts of the 2011tsunami on the subtidal polychaete assemblage and the following recolonization in Onagawa Bay, Northeastern Japan. Mar Environ Res 112:86–95

Anderson MJ, Willis TJ (2003) Canonical analysis of principal coordinates: a useful method of constrained ordination for ecology. Ecology 84:511–525

Bagarinao T (1992) Sulfide as an environmental factor and toxicant: tolerance and adaptations in aquatic organisms. Aquat Toxicol 24:21–62

Bellan G, Desrosiers G, Willsie A (1988) Use of an annelid pollution index for monitoring a moderately polluted littoral zone. Mar Pollut Bull 19:662–665

Chavanich S, Viyakarn V, Sojisuporn P, Siripong A, Menasveta P (2008) Patterns of coral damage associated with the 2004 Indian Ocean tsunami at Mu Ko Similan Marine National Park, Thailand. J Nat Hist 42:177–187

Clarke KR (1993) Non-parametric multivariate analyses of changes in community structure. Aust J Ecol 18:117–143

Clarke KR, Gorley RN (2006) PRIMER v6: user manual/tutorial. PRIMER-E Ltd., Plymouth, p 190

Dean H (2008) The use of polychaetes (Annelida) as indicator species of marine pollution: a review. Rev Biol Trop 56:11–38

Diaz RJ, Rosenberg R (1995) Marine benthic hypoxia: a review of its ecological effects and the behavioural responses of benthic macrofauna. Oceanogr Mar Biol Ann Rev 33:245–303

Diaz RJ, Rosenberg R (2008) Spreading dead zones and consequences for marine ecosystems. Science 321:926–929

Duarte CM, Borja A, Carstensen J, Elliott M, Karawuse-Jensen D, Marbà N (2013) Paradigms in the recovery of estuarine and coastal ecosystems. Estuar Coasts. doi:10.1007/s12237-013-9750-9

Fujioka Y, Tabuchi R, Hirata Y, Yoneda R, Patanaponpaiboon P, Poungparn S, Shibuno T, Ohba H (2008) Disturbance and recovery of mangrove forests and macrobenthic communities in Andaman Sea, Thailand following the Indian Ocean tsunami. In: Proceedings of the 11th international coral reef symposium, Ft. Lauderdale, Florida, 7–11 July 2008

Giangrande A (1997) Polychaete reproductive patterns, life cycles and life histories: an overview. Oceanogr Mar Biol Annu Rev 35:323–386

Giangrande A, Licciano M, Musco L (2005) Polychaetes as environmental indicators revisited. Mar Pollut Bull 50:1153–1162

Goto K, Chagué-Goff C, Fujino S, Goff J, Jaffe B, Nishimura Y, Richmond B, Sugawara D, Szczuciński W, Tappin DR, Witter RC, Yulianto E (2011) New insights of tsunami hazard from the 2011 Tohoku-oki event. Mar Geol 290:46–50

Goto K, Chagué-Goff C, Goff J, Jaffe B (2012) The future of tsunami research following the 2011 Tohoku-oki event. Sed Geol 282:1–13

Gray JS (1974) Animal–sediment relationships. Oceanogr Mar Biol Annu Rev 12:223–261

Hayashi I, Iguchi Y, Yoshida T, Takahashi T (1992) Macrobenthic assemblages of Nanao Bay, Sea of Japan. Bull Coast Oceanogr 30:91–107 (in Japanese)

Hayward BW, Morley MS, Hayward JJ, Stephenson AB, Blom WM, Hayward KA, Grenfell HR (1999) Monitoring studies of the benthic ecology of Waitemata harbour, New Zealand. Rec Auckland Mus 36:95–117

Hinata H, Fujii S, Furukawa K, Kataoka T, Miyata M, Kobayashi T, Mizutani M, Kokai T, Kanatsu N (2011) Propagating tsunami wave and subsequent resonant response signals detected by HF radar in the Kii Channel, Japan. Estuar Coast Shelf Sci 95:268–273

Hunsigi G, Yekkeli N, Perumal L, Thippannavar M (2006) Effects of 2004 tsunami on marine ecosystems–a perspective from the concept of disturbance. Curr Sci 90:772–773

Hutchings P (1998) Biodiversity and functioning of polychaetes in benthic sediments. Biodivers Conserv 7:1133–1145

Ide S, Baltay A, Beroza GC (2011) Shallow dynamic overshoot and energetic deep rupture in the 2011 Mw 9.0 Tohoku-Oki Earthquake. Science 332:1426–1429

Imabayashi H, Iwatani T (1988) Growth and distribution in planktonic larvae of the small bivalve *Theora lubrica* (preliminary report). Benthos Res 33:17–24 (in Japanese with English abstract)

Ishikawa K (1989) Relationship between bottom characteristics and benthic organisms in the shallow water of Oppa Bay, Miyagi. Mar Biol 102:265–273

Jaramillo E, Dugan JE, Hubbard DM, Melnick D, Manzano M, Duarte C, Campos C, Sanchez R (2012) Ecological implications of extreme events: Footprints of the 2010 earthquake along the Chilean coast. PLoS ONE 7, e35348

Kanaya G, Maki H, Suzuki T, Sato-Okoshi W, Kikuchi E (2014) Tsunami-induced changes in a Shallow Brackish Lagoon Ecosystem (Gamo Lagoon) in Sendai Bay, Japan. Global Environ Res 18:35–46

Kanaya G, Maki H, Chiba F, Miura K, Fukuchi S, Sasaki H, Nishimura O (2016) Impacts of fuel spills caused by the 2011 Japanese tsunami on the subtidal soft-bottom communities of a semi-enclosed bay located on the Sanriku coast. In: Urabe J, Nakashizuka T (Eds). Ecological impacts of tsunamis on coastal ecosystems: lessens from the Great East Japan Earthquake. Springer, pp 223–252

Karakassis I, Hatziyanni E, Tsapakis M, Plaiti W (1999) Benthic recovery following cessation of fish farming: a series of successes and catastrophes. Mar Ecol Prog Ser 184:205–218

Kawai H, Satoh M, Kawaguchi K, Seki K (2011) The 2011 off Pacific coast of Tohoku earthquake tsunami observed by GPS buoys. J Jpn Soc Civ Eng Ser B2(67):1291–1295

Kitahashi T, Jenkins RG, Nomaki H, Shimanaga M, Fujikura K, Kojima S (2014) Effect of the 2011 Tohoku Earthquake on deep-sea meiofaunal assemblages inhabiting the landward slope of the Japan Trench. Mar Geol 358:128–137

Kodama K, Lee JH, Oyama M, Shiraishi H, Horiguchi T (2012) Disturbance of benthic macrofauna in relation to hypoxia and organic enrichment in a eutrophic coastal bay. Mar Environ Res 76:80–89

Koerting-Walker C, Buck JD (1989) The effect of bacteria and bioturbation by *Clymenella torquata* on oil removal from sediment. Water Air Soil Pollut 43:413–424

Lardicci C, Como S, Corti S, Rossi F (2001) Recovery of the macrozoobenthic community after severe dystrophic crises in a Mediterranean coastal lagoon (Orbetello, Italy). Mar Pollut Bull 42:202–214

Lee SW, Paik EI (1995) Density composition and feeding guild of the dominant polychaetous community in shallow muddy bottom in Tomioka Bay, Amakusa, Japan. J Korean Fish Soc 28:793–804

Levin LA (1984) Life history and dispersal patterns in a dense infaunal polychaete assemblage: community structure and response to disturbance. Ecology 65:1185–1200

Levin L, Blair N, DeMaster D, Plaia G, Fornes W, Martin C, Thomas C (1997) Rapid subduction of organic matter by maldanid polychaetes on the North Carolina slope. J Mar Res 55:595–611

Lomovasky BJ, Firstater FN, Salazar AG, Mendo J, Iribarne OO (2011) Macro benthic community assemblage before and after the 2007 tsunami and earthquake at Paracas Bay, Peru. J Sea Res 65:205–212

Mach ME, Levings CD, McDonald PS, Chan KMA (2012) An Atlantic infaunal engineer is established in the Northeast Pacific: *Clymenella torquata* (Polychaeta: Maldanidae) on the British Columbia and Washington Coasts. Biol Invasions 14:503–507

McHugh D, Tunnicliffe V (1994) Ecology and reproductive biology of the hydrothermal vent polychaete *Amphisamytha galapagensis* (Ampharetidae). Mar Ecol Prog Ser 106:111–120

Meadows PS, Meadows A, Murray JMH (2012) Biological modifiers of marine benthic seascapes: their role as ecosystem engineers. Geomorphology 157–158:31–48

Miura O, Sasaki Y, Chiba S (2012) Destruction of populations of *Batillaria attramentaria* (Caenogastropoda: Batillariidae) by tsunami waves of the 2011 Tohoku earthquake. J Molluscan Stud 78:377–380

Miyagi Prefecture, Tohoku Electric Power Company (2009) Measurements of warm drainage water of the Onagawa Nuclear Power Plant for fiscal 2007. Miyagi Prefecture, p 261 (in Japanese)

Miyagi Prefecture, Tohoku Electric Power Company (2010) Measurements of warm drainage water of the Onagawa Nuclear Power Plant for fiscal 2008. Miyagi Prefecture, p 261 (in Japanese)

Miyagi Prefecture, Tohoku Electric Power Company (2011) Measurements of warm drainage water of the Onagawa Nuclear Power Plant for fiscal 2009. Miyagi Prefecture, p 261 (in Japanese)

Miyagi Prefecture, Tohoku Electric Power Company (2012) Measurements of warm drainage water of the Onagawa Nuclear Power Plant for fiscal 2010. Miyagi Prefecture, p 261 (in Japanese)

Miyagi Prefecture, Tohoku Electric Power Company (2013) Measurements of warm drainage water of the Onagawa Nuclear Power Plant for fiscal 2011. Miyagi Prefecture, p 261 (in Japanese)

Miyagi Prefecture, Tohoku Electric Power Company (2014) Measurements of warm drainage water of the Onagawa Nuclear Power Plant for fiscal 2012. Miyagi Prefecture, p 263 (in Japanese)

Miyagi Prefecture, Tohoku Electric Power Company (2015) Measurements of warm drainage water of the Onagawa Nuclear Power Plant for fiscal 2013. Miyagi Prefecture, p 269 (in Japanese)

Mori N, Takahashi T, Yasuda T, Yanagisawa H (2011) Survey of 2011 Tohoku earthquake tsunami inundation and run-up. Geophys Res Lett 38:L00G14

Mori N, Takahashi T, The 2011 Tohoku Earthquake Tsunami Joint Survey Group (2012) Nationwide post event survey and analysis of the 2011 Tohoku Earthquake Tsunami. Coast Eng J 54:1250001

Munari C, Modugno S, Ghion F, Castaldelli G, Fano EA, Rossi R, Mistri M (2003) Recovery of the macrobenthic community in the Valli di Comacchio, northern Adriatic Sea, Italy. Oceanol Acta 26:67–75

Noji CI-M, Noji TT (1991) Tube lawns of spionid polychaetes and their significance for recolonization of disturbed benthic substrates. Meeresforschung 33:235–246

Okey TA (1997) Sediment flushing observations, earthquake slumping, and benthic community changes in Monterey Canyon head. Cont Shelf Res 17:877–897

Okoshi K (2015) Impact of repeating massive earthquakes on intertidal mollusk community in Japan. In: Ceccaldi H-J, Hénocque Y, Koike Y, Komatsu T, Stora G, Tusseau-Vuillemin M-H (eds) Marine productivity: perturbations and resilience of socio-ecosystems. Springer, Cham, pp 55–62

Oksanen J, Blanchet FG, Kindt R, Legendre P, Minchin PR, O'Hara RB, Simpson GL, Solymos P, Stevens MHH, Wagner H (2015) Package 'vegan'. Community ecology package. R Packag Version 2:2–1. http://cran.r-project.org/web/packages/vegan/index.html. Accessed 17 Apr 2015

Ozawa S, Nishimura T, Suito H, Kobayashi T, Tobita M, Imakiire T (2011) Coseismic and post-seismic slip of the 2011 magnitude-9 Tohoku-Oki earthquake. Nature 475:373–376

Pearson TR, Rosenberg R (1978) Macrobenthic succession in relation to organic enrichment and pollution of the marine environment. Oceanogr Mar Biol Annu Rev 16:229–311

Pechenik JA (1999) On the advantages and disadvantages of larval stages in benthic marine invertebrate life cycles. Mar Ecol Prog Ser 177:269–297

Petersen ME (1999) Reproduction and development in Cirratulidae (Annelida: Polychaeta). Hydrobiologia 402:107–128

Peterson CH, Rice SD, Short JW, Esler D, Bodkin JL, Ballachey BE, Irons DB (2003) Long-term ecosystem response to the Exxon Valdez oil spill. Science 302:2082–2086

Rhoads DC (1974) Organisms–sediment relation on the muddy sea floor. Oceanogr Mar Biol Annu Rev 12:263–300

Rhoads DC, Young DK (1970) The influence of deposit-feeding organisms on sediment stability and community trophic structure. J Mar Res 28:150–178

Saito H (2006) Population dynamics of a pollution indicator bivalve *Theora lubrica*. Bull Fish Res Agents 16:29–95 (in Japanese)

Sanders HL (1958) Benthic studies in Buzzards Bay. I. Animal– sediment relations. Limnol Oceanogr 3:245–258

Sasaki K, Arai N (1997) Sulfide concentration in bottom sediments of Onagawa Bay, Miyagi. Bull Jpn Soc Fish Oceanogr 61:162–167, in Japanese with English abstract

Sasaki K, Aoyagi K, Arai N (1995) Bottom sediments of Onagawa bay, Miyagi. Bull Jpn Soc Fish Oceanogr 59:241–249 (in Japanese with English abstract)

Sato M, Ishikawa T, Ujihara N, Yoshida S, Fujita M, Mochizuki M, Asada A (2011) Displacement above the hypocenter of the 2011 Tohoku-Oki Earthquake. Science 332:1395

Seike K, Shirai K, Kogure Y (2013) Disturbance of shallow marine soft-bottom environments and megabenthos assemblages by a huge tsunami induced by the 2011 M9. 0 Tohoku-Oki Earthquake. PLoS ONE 8, e65417

Simons M, Minson SE, Sladen A, Ortega F, Jiang J, Owen SE, Meng L, Ampuero J-P, Wei S, Chu R, Helmberger DV, Kanamori H, Hetland E, Moore AW, Webb FH (2011) The 2011 magnitude 9.0 Tohoku-Oki Earthquake: mosaicking the megathrust from seconds to centuries. Science 332:1421–1425

Snelgrove PVR, Butman CA (1994) Animal–sediment relationships revisited: cause versus effect. Oceanogr Mar Biol Annu Rev 32:111–117

Snelgrove PVR, Blackburn TH, Hutchings PA, Alongi DM, Grassle JF, Hummel H, King G, Koike I, Lamshead PJD, Ramsing NB, Solis-Weiss V (1997) The importance of marine sediment biodiversity in ecosystem processes. Ambio 26:578–583

Sousa WP (1984) The role of disturbance in natural community. Ann Rev Ecol Syst 15:353–391

Steckbauer A, Duarte CM, Carstensen J, Vaquer-Sunyer R, Conley DJ (2011) Ecosystem impacts of hypoxia: thresholds of hypoxia and pathways to recovery. Environ Res Lett 6:025003. doi:10.1088/1748-9326/6/2/025003

Takami H, Won NI, Kawamura T (2013) Impacts of the 2011 mega-earthquake and tsunami on abalone *Haliotis discus hannai* and sea urchin *Strongylocentrotus nudus* populations at Oshika Peninsula, Miyagi, Japan. Fish Oceanogr 22:113–120

Tamai K (1996) Temporal tolerance of larval *Theora fragilis* (Bivalvia: Semelidae) to hypoxic conditions. Fish Sci 62:996–997

The 2011 Tohoku Earthquake Tsunami Joint Survey Group (2011) Nationwide field survey of the 2011 off the Pacific coast of Tohoku earthquake tsunami. J Jpn Soc Civ Eng Ser B2(67):63–66

Thibault KM, Brown JH (2008) Impact of an extreme climatic event on community assembly. Proc Natl Acad Sci Unit States Am 105:3410–3415

Urabe J, Suzuki T, Nishita T, Makino W (2013) Immediate ecological impacts of the 2011 Tohoku Earthquake Tsunami on intertidal flat communities. PLoS One 8, e62779

Venables WN, Ripley BD (2002) Modern applied statistics with S, 4th edn. Springer, New York

Veríssimo H, Neto JM, Teixeira H, Franco JN, Fath BD, Marques JC, Patrício J (2012) Ability of benthic indicators to assess ecological quality in estuaries following management. Ecol Indic 19:130–143

Whanpetch N, Nakaoka M, Mukai H, Suzuki T, Nojima S, Kawai T, Aryuthaka C (2010) Temporal changes in benthic communities of seagrass beds impacted by a tsunami in the Andaman Sea, Thailand. Estuar Coast Shelf Sci 87:246–252

Wijsman JWM, Herman PMJ, Marian-Traian G (1999) Spatial distribution in sediment characteristics and benthic activity on the northwestern Black Sea shelf. Mar Ecol Prog Ser 181:25–39

Wildsmith MD, Rose TH, Potter IC, Warwick RM, Clarke KR, Valesini FJ (2009) Changes in the benthic macroinvertebrate fauna of a large microtidal estuary following extreme modifications aimed at reducing eutrophication. Mar Pollut Bull 58:1250–1262

Wilson WH (1991) Sexual reproductive modes in polychaetes: classification and diversity. Bull Mar Sci 48:500–516

Yoshino K, Katano T, Ito Y, Hamada T, Fujii N, Hayami Y (2014) Community assembly by limited recruitment in a hypoxia-stressed soft bottom: A case study of macrobenthos in Ariake Bay. Plankton Benthos Res 9:57–66

Zajac RN, Whitlatch RB (1982a) Responses of estuarine infauna to disturbance. I. Spatial and temporal variation of initial recolonization. Mar Ecol Prog Ser 10:1–14

Zajac RN, Whitlatch RB (1982b) Responses of estuarine infauna to disturbance. II. Spatial and temporal variation of succession. Mar Ecol Prog Ser 10:15–27

Zajac RN, Whitlatch RB, Thrush SF (1998) Recolonization and succession in soft-sediment infaunal communities: the spatial scale of controlling factors. Hydrobiologia 375(376):227–240

Chapter 7
Ecological Consequences of the Tsunamis Caused by the Great East Japan Earthquake and Subsequent Disturbance Events in a Shallow Brackish Lagoon in Sendai Bay, Japan

Gen Kanaya, Takao Suzuki, Keiichi Kanou, Tomohiko Kondoh, Waka Sato-Okoshi, and Eisuke Kikuchi

Abstract The impacts of the tsunami caused by the Great East Japan Earthquake and subsequent disturbance events were examined in Gamo Lagoon, a shallow brackish lagoon in Sendai Bay, Japan. A major tsunami with a height of 7.2 m struck the lagoon on 11 March 2011, followed by a 2-month estuary closing and a typhoon-induced flood within a year. These events induced drastic changes in the salinity and sediment characteristics as well as a washout of vegetation. After the tsunami, the plant community was characterized chiefly by the proliferation of "pioneer species." The annual helophyte *Suaeda maritima* became dominant in the bare high-tide zone created through the disappearance of the reed marsh. Sand dune vegetation (SDV) had recovered only in patchily after 4 years, and the community was characterized by the invasive species *Cakile edentula*. The density and species richness of macrozoobenthos changed after each disturbance event (i.e., the tsunami, estuary closing, and typhoon). Macrozoobenthic density recovered within 6 months after the tsunami and typhoon, chiefly due to the density overshoots of several opportunistic taxa. Though tsunami- and typhoon-induced changes in macro-

G. Kanaya (✉)
National Institute for Environmental Studies, 16-2 Onogawa, Tsukuba 305-8506, Japan
e-mail: gen@nies.go.jp

T. Suzuki • K. Kanou
Graduate School of Life Science, Tohoku University, Aoba 6-3, Aramaki, Aoba-ku, Sendai 980-8578, Japan

T. Kondoh • W. Sato-Okoshi
Graduate School of Agricultural Science, Tohoku University, Amamiya 1-1, Tsutsumidori, Aoba-ku, Sendai 981-8555, Japan

E. Kikuchi
EEC, Miyagi University of Education, Aoba 149, Aramak, Aoba-ku, Sendai 980-0845, Japan

J. Urabe, T. Nakashizuka (eds.), *Ecological Impacts of Tsunamis on Coastal Ecosystems*, Ecological Research Monographs,
DOI 10.1007/978-4-431-56448-5_7

zoobenthic community structure were significant, by 3 years post-tsunami, the community had begun to recover toward pre-tsunami conditions. These results emphasize both the vulnerability and resiliency of estuarine biotic communities against large pulsed disturbance events.

Keywords Brackish lagoon • Community succession • Pulsed disturbance • Macrozoobenthos • Sand dune vegetation (SDV) • Salt marsh

7.1 Introduction

Estuarine ecosystems consist of several types of habitats including sand dunes, salt marshes, and tidal flats. These environments are the main transitions between land and sea and provide essential ecosystem functions such as nutrient cycling, decomposition, biological productivity, and biodiversity (Levin et al. 2001). However, an estuary often experiences natural and anthropogenic disturbance events that induce drastic changes in habitat structure and environmental conditions (e.g., Ritter et al. 2005; Cardoso et al. 2008; Middleton 2009). Due to the harsh environmental conditions, dominant species in estuarine habitats are potentially adaptive to stressors such as salinity fluctuation, desiccation, and physical disturbance (e.g., Dauer 1984; Lissner and Schierup 1997; Yokoya et al. 1999; Tessier et al. 2002; Ritter et al. 2005). Furthermore, they often have opportunistic life history traits (e.g., frequent reproduction, rapid population growth, and short time to maturity), allowing them to reestablish populations rapidly (e.g., Grassle and Grassle 1974; Ritter et al. 2005; Kanaya et al. 2011).

A major pulsed disturbance such as a tsunami, flood, or hurricane can devastate aquatic ecosystems (e.g., Cardoso et al. 2008; Middleton 2009; Jaramillo et al. 2012). Above all, massive tsunamis are some of the most intensive disturbance events affecting estuarine coastal habitats. They modify estuarine habitat structure through extraordinary physical disturbances such as scouring, deposition of tsunami debris, and loss of vegetation on a broad spatial scale (e.g., Hayasaka et al. 2012; Jaramillo et al. 2012; Szczuciński et al. 2012; Seike et al. 2013). Generally, it takes several or more years to recover the biotic community in an aquatic ecosystem after a defaunation event (e.g., Niemi et al. 1990; Beukema et al. 1999; Cardoso et al. 2008; Middleton 2009). Borja et al. (2010) reported that full recovery of coastal marine and estuarine ecosystems can take a minimum of 15–25 years from the viewpoint of community composition. As huge pulsed disturbance events are intense but rare, their ecological consequences have not been fully examined in estuarine coastal ecosystems.

On 11 March 2011, huge tsunami waves (up to 20 m high) caused by the Great East Japan Earthquake struck the Pacific coast of northeastern Japan. The tsunami induced extraordinary physical disturbances in brackish lagoons, salt marshes, and sand dunes along the Sendai Bay (e.g., Kanaya et al. 2012; Szczuciński et al. 2012; Urabe et al. 2013). For example, Gamo Lagoon, a shallow eutrophic lagoon, was intensively disturbed by

7.2 m high tsunami waves, which led to drastic changes in topography, loss of vegetation, and depression of macrozoobenthic species diversity (Kanaya et al. 2012, 2014, 2015). Furthermore, the lagoon experienced additional disturbance events in 2011 and 2012: a 2-month estuary closing, a typhoon-induced flood, and excavation works (see Tanaka et al. 2014). These disturbance events may have seriously affected the habitat structure and associated biotic community to varying degrees of intensity.

In this study, ecological impacts of the disturbance events to Gamo Lagoon were assessed using the pre- and post-tsunami data sets. Changes in the lagoon topography and areas of vegetation, including reed marsh, sand dune vegetation (SDV), and macroalgae, were assessed by GPS-based mapping. Historical data sets of macrozoobenthos were used to assess the relative impacts of each disturbance event. The aims of this study were to reveal (1) the relative impacts of the pulsed disturbance events in 2011 and 2012 on the lagoon ecosystem and (2) the taxa- or habitat-specific recovery potential of SDV, salt marsh, and macrozoobenthos after a disturbance event. We specifically focused on the vulnerability and resiliency of estuarine biotic communities, which are frequently subjected to pulsed disturbance events.

7.2 Study Site and the Disturbance Events

Gamo Lagoon (0.11 km^2; mean water depth: 0.8 m) is located on the north side of the Nanakita River Estuary (38°15′10.8″ N, 141°00′36.0″ E) in Sendai Bay, Japan (Fig. 7.1a). The lagoon is separated from the estuary by a stone levee with three water gates and from Sendai Bay by a 150 m wide sandbar. Salinity fluctuated tidally from < 10 to >25 (Kanaya et al. 2014). The sediment was sandy in the vicinity of the lagoon mouth (silt-clay <10 %) and muddy (silt-clay >50 %) and highly reduced in the inner portion (Kanaya et al. 2014). Reed marsh, dominated by *Phragmites australis*, developed in the lagoon edge, and SDV consisting of seashore plants including *Calystegia soldanella*, *Lathyrus japonicus*, *Ischaemum anthephoroides*, *Carex kobomugi*, and several other taxa developed on the sandbar (Council for the Promotion of Nature Restoration in Gamo Tidal Flat 2008). The macroalgae *Gracilaria vermiculophylla* and *Ulva prolifera* grew densely in inner subtidal areas (coverage, 50–100 %; Kanaya et al. 2014).

On 11 March 2011, huge tsunami waves (7.2 m in height) heavily disturbed the lagoon. The sandbar was devastated completely, and most of the muddy sediment was flushed away (Kanaya et al. 2012, 2015). In June 2011, the sandbar was reformed through the deposition of drifting sand (see Fig. 7.1b). From August to September 2011, the lagoon became dominated by freshwater due to the estuary closing. On 22 September 2011, a typhoon-induced flood destroyed the sandbar, resulting in the formation of a new river mouth at the central lagoon (Tanaka et al. 2014). The new opening was closed by drifting sea sand in March 2012, shortly after the excavation of the Nanakita River mouth at its original location. In September 2012, a new tidal channel that opened to the river mouth was created by a storm surge. However, it was refilled by deposited sea sand in October 2013.

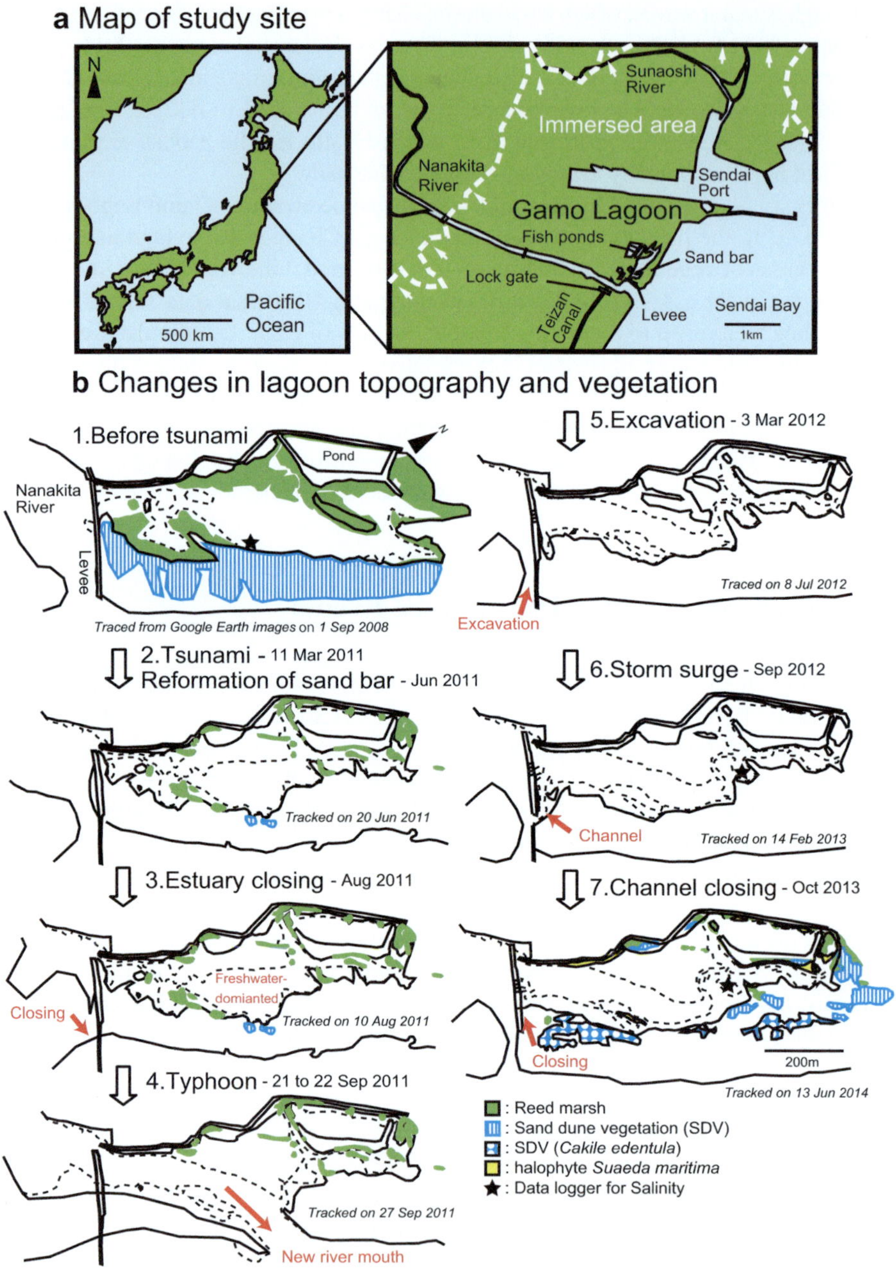

Fig. 7.1 Location of Gamo Lagoon (**a**) and changes in lagoon topography from 2011 to 2014 (**b**). Reed marsh, sand dune vegetation (SDV), and patches of the invasive species *C. edentula* and the annual halophyte *S. maritima* are shown (The 2008 vegetation map was modified from Kanaya et al. 2012. The location of the salinity data logger is indicated by a *black star*)

7.3 Methods

7.3.1 Lagoon Topography, Vegetation, and Macroalgae

Shorelines were tracked in the field using a handy GPS (GPS map 60CSx, Garmin) during spring low and high tides on 20 June 2011, 27 September 2011, 18 July 2012, 14 February 2013, and 13 June 2014. Topographical changes in the Nanakita River mouth during the estuary closing were assessed from an aerial photograph (Kahoku Shimpo Newspaper, 16 August 2011). Pre-tsunami topographies were traced from satellite images obtained between 1 September 2008 and 4 April 2010 (Google ™ Earth, http://www.google.com/earth/index.html).

Areas of reed marsh and SDV were tracked in the field on 21 June 2011, 27 September 2011, and 13 June 2014. Patches of the invasive sand dune plant *Cakile edentula* (Fig. 7.2e) that was introduced from the USA to Northern Japan during 1990s (Hokkaido 2010, http://bluelist.ies.hro.or.jp/db/detail.php?k=08&cd=118) and the annual halophyte *Suaeda maritima* (Fig. 7.2f) were tracked on 13 June 2014. A pre-tsunami vegetation map was obtained from the Council for the Promotion of Nature Restoration in Gamo Tidal Flat (2008). Macroalgal coverage (%) was determined on 12 July 2014 by measuring the coverage in 2×2 m areas at 45 stations (see Fig. 7.3a). Macroalgal coverage in 2011, 2012, and 2013 was obtained from Kanaya et al. (2014).

7.3.2 Salinity and Sediment Characteristics

The salinity of the lagoon water was measured during ebb tide on each sampling occasion using conductivity meters (CM-21P or CM-31P, TOA-DKK). Salinity was also measured continuously at 30-min intervals using a data logger (600 OMS, YSI Nanotech Japan), positioned in the central lagoon (Fig. 7.1b) from 30 July 2012 to 15 April 2014. Pre-tsunami salinity data were obtained from Kurihara et al. (2000).

The silt-clay contents and redox potential (Eh) of sediments were measured at 12–20 stations (see Fig. 7.3d) on 17 occasions from July 2002 to June 2013. In 2011 and 2012, sampling was conducted after each disturbance event (Table 7.1). In the field, Eh was measured at a 5 cm depth using a handheld oxidation-reduction potential (ORP) meter (RM-12P, RM-20P, RM-30P, TOA DKK, and TRX-90, TOKO). Surface sediment (1–3 cm deep) was sampled and dried at 70 °C for 48 h, and the silt-clay content (<0.063 mm mesh) was determined using a wet sieving method.

7.3.3 Macrozoobenthos

The macrozoobenthos was quantitatively sampled concurrently with the sediment sampling, with an additional sampling on 2 September 2011. At each station, 176.7 cm^2 sediment was collected to a 30 cm depth using a polyvinyl chloride (PVC) core

Fig. 7.2 Photos of sand dune vegetation (SDV) in August 2009 (**a**), June 2011 (**b**), and June 2014 (**c**). Dead shells (mainly *N. japonica*) and marsh rhizomes appeared on the scoured sediment surface after the typhoon (**d**). The invasive SDV species *C. edentula* (**e**) and the annual halophyte *S. maritima* (**f**) are also shown

sampler (φ5 × 80 cm long or φ15 × 50 cm long). In 2002 and 2003, a pooled total of 235.6 cm^2 sediment was collected to a 30-cm depth at each station for macrozoobenthos sampling. The sediment was sieved through a 1 mm mesh sieve, and the retained fraction was fixed in 5 % neutralized formalin. In the laboratory, animals were sorted and identified under a dissecting stereomicroscope.

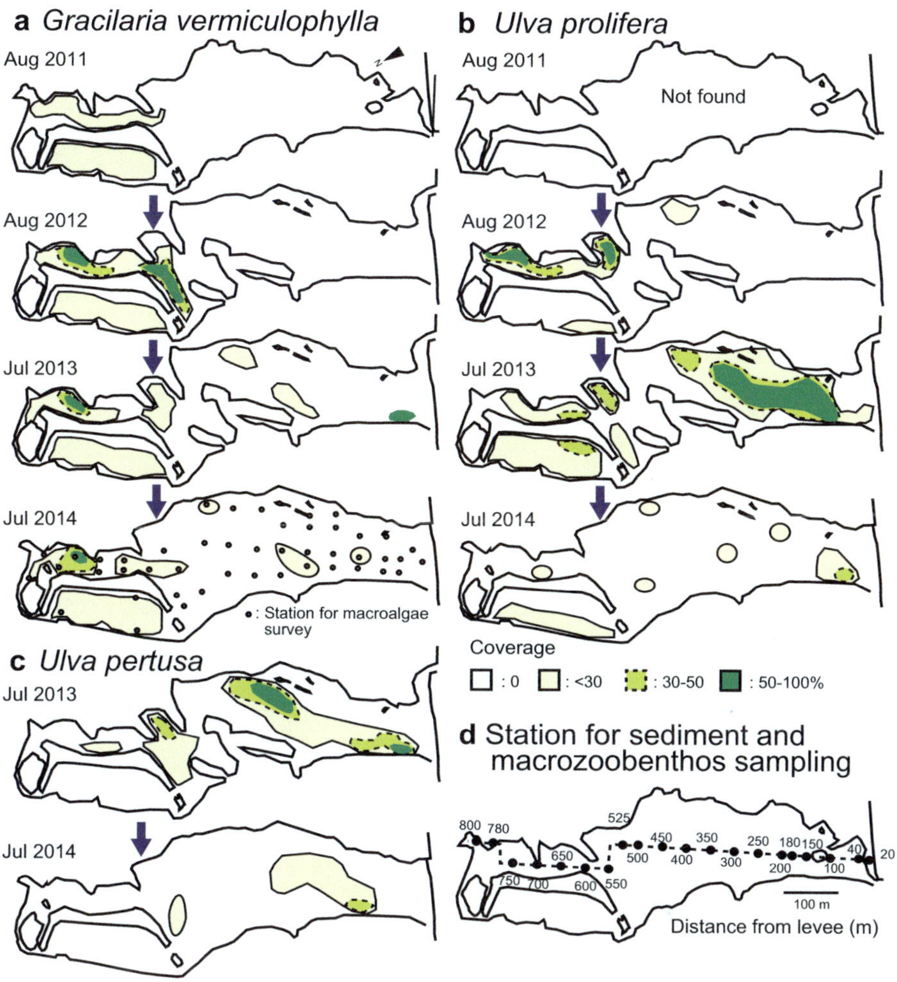

Fig. 7.3 Coverage (%) of the dominant macroalgae *G. vermiculophylla* (**a**), *U. prolifera* (**b**), and *U. pertusa* (**c**). *Circles* in (**a**) indicate survey stations (Data obtained from 2011 to 2013 are from Kanaya et al. 2014. The stations of sediment and macrozoobenthos sampling are shown in (**d**))

7.3.4 Data Analyses

Nonmetric multidimensional scaling (nMDS) was used to assess temporal changes in macrozoobenthic community structure, using data sets from 17 sampling occasions from July 2002 to June 2013. Bray-Curtis similarity was calculated from the lagoonal mean densities of all macrozoobenthic taxa. The square roots of macrozoobenthic densities were used for analyses. Analyses were performed using PRIMER ver. 6.2 (Clarke and Gorley 2006).

Table 7.1 Silt-clay contents (%) and Eh (mV) of sediment samples collected in Gamo Lagoon

			Sampling stations[a]	
Sampling occasion	Silt-clay (%)	Eh (mV)	n	Area (m)
Before tsunami: Jul 2002–Jun 2008				
Jul, Aug, Sep, Oct 2002	22 (18)	18 (175)	48[b]	175–690
Mar, Jun 2003	23 (16)	44 (182)	24[b]	175–690
Jun 2004	n.d.	103 (214)	20	20–800
Jun 2005	n.d.	86 (200)	20	20–800
Jul 2007	n.d.	43 (197)	19	20–800
Jun 2008	29 (16)	97 (204)	19[c]	20–800
Tsunami: 11 Mar 2011				
Jul 2011	5 (6)	296 (105)	20	20–800
Estuary closing: beginning of Aug 2011				
2 Sep 2011	n.d.	n.d.		50–500
Typhoon: 21–22 Sep 2011				
28 Sep 2011	7 (12)	202 (167)	12	200–750
24 Oct 2011	n.d.	259 (119)	13	300–800
Excavation: 3 Mar 2012				
13 Mar 2012	6 (6)	324 (80)	16	150–800
Jun 2012	8 (11)	191 (87)	17	100–800
Storm surge (channelization): September 2012				
Jun 2013	5 (7)	245 (91)	16	100–780
Channel closing: October 2013				

Data are presented as the lagoonal mean (standard deviation). The numbers of stations (*n*) and sampled areas (distance from the levee) are also shown

[a]Sampling stations were located at 20–50 m intervals along a transect line from the levee to inner lagoon (see Fig. 7.3d)

[b]$n = 12$ for each sampling month

[c]$n = 12$ for silt-clay

n.d. no data

7.4 Results

7.4.1 Topography and Vegetation

The tsunami altered the lagoon topography through intensive scouring and deposition of drifting sea sand (see Fig. 7.1b), which led to the removal of salt marsh and SDV (see Fig. 7.2). In August 2011, the Nanakita River mouth was closed completely by deposited sea sand. On 21–22 September 2011, reed marsh, SDV, and bottom sediment near the floodway were washed away by a typhoon-induced flood. SDV and reed marsh recovered little during 2012–2013, while the annual halophyte *S. maritima* gradually increased its patch size in the bare high-tide zone. In 2014, SDV on the sandbar had partly recovered and was dominated by the invasive seashore plant *C. edentula*.

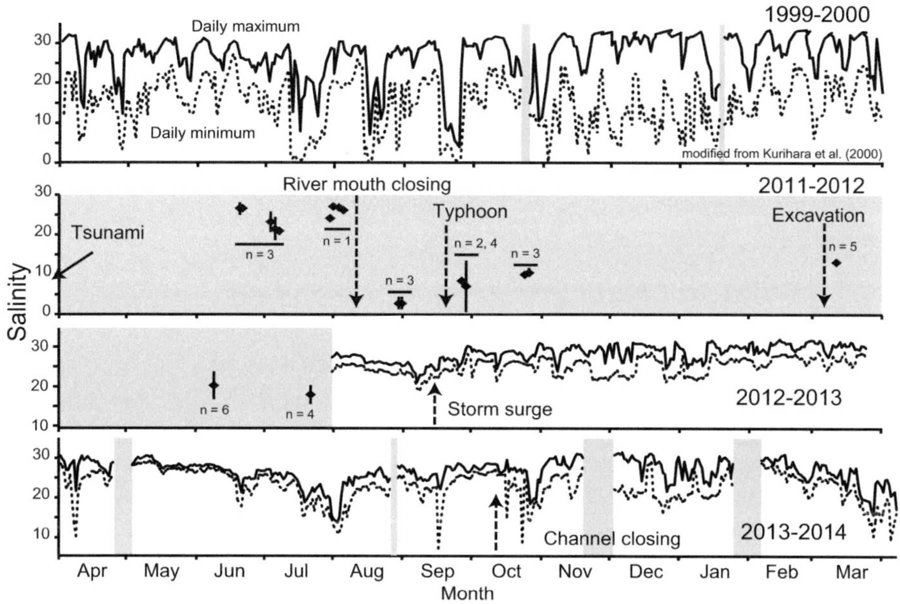

Fig. 7.4 Seasonal changes in lagoon salinity. *Gray*-colored areas indicate that logger data were not available. *Black diamonds* indicate salinities measured using a handheld electrical conductivity (EC) meter during ebb tides. *Bars* indicate standard deviations (Data from 1999 to 2000 are from Kurihara et al. 2000)

The sizes of macroalgal patches varied drastically over the years. Before the tsunami, *G. vermiculophylla* and *U. prolifera* grew densely in most of the inner subtidal areas (Kanaya et al. 2014). However, in 2011, only few patches of *G. vermiculophylla* remained. In 2012–2013, patch sizes of *G. vermiculophylla* and *U. prolifera* gradually recovered from the innermost to the central portion of the lagoon. In 2013, *Ulva pertusa* proliferated densely, although this species occurred rarely before the tsunami. The two *Ulva* species declined sharply in 2014, while *G. vermiculophylla* retained its patch sizes.

7.4.2 Salinity

Daily maximum and minimum salinity values fluctuated widely during the sampling period (Fig. 7.4). Pre-tsunami salinity measurements often fell below 15. From June to August 2011, salinity ranged from 20 to 30, while it fell below 5 during the estuary closing from August to September 2011. After the typhoon on 21–22 September 2011, the salinity was more fluctuant: It was higher than 25 during high tides but fell below 10 during ebb tides (Kanaya unpublished data) since the lagoon became a river course. After the excavation works by Miyagi Prefecture in March 2012, salinity in the lagoon became more stable. When the lagoon became dominated by seawater after channelization by the storm surge in September 2012, daily minimum salinity of the lagoon water was higher than 20. When the channel was refilled in October 2013, however, the salinity decreased and sometimes fell below 15.

7.4.3 *Sediment Characteristics: Silt-Clay and Eh*

Before the tsunami, the sediment was muddy and highly reduced in the inner lagoon (silt-clay: 20–60 %; Eh <–100 mV, Fig. 7.5). However, most of the muddy sediment was washed away by the tsunami, and the sediment became sandy and more oxidized throughout the lagoon (silt-clay: < 10 %; Eh: 0–+400 mV; see data from July 2011). The typhoon in September 2011 also intensively scoured the sediment in the area from ~200 to 525 m inward from the levee. During 2012–2013, the sediment was still sandy and well oxidized (silt-clay: < 5 %; Eh: +100 to +450 mV), except for several muddy stations in the inner lagoon.

7.4.4 *Macrozoobenthos*

Before the tsunami, the density of total macrozoobenthos ranged from ~0 to >10,000 individuals (ind.) m^{-2} (Fig. 7.6), with relatively higher densities in the vicinity of the levee. After the tsunami, the density sharply increased, to as high as 25,000 ind. m^{-2} in July 2011 at most stations, followed by an overall decline after the estuary closing and typhoon. Within 6 months after the typhoon, the macrozoobenthos recovered their densities throughout the lagoon (5000 to >20,000 ind. m^{-2}), including typhoon-disturbed area in the southern portion (100 to 450 m north from the levee; Fig. 7.6).

The lagoonal average density of macrozoobenthos (Fig. 7.7a) increased sharply after the tsunami (9292 ind. m^{-2} in July 2011) compared with the densities in 2007 and 2008 (3313 and 4193 ind. m^{-2}, respectively), followed by a steep decline during the estuary closing and typhoon (2531 and 683 ind. m^{-2} in 2 and 28 September 2011, respectively). Subsequently the density gradually recovered to 9360 ind. m^{-2} within 10 months after the typhoon. In 2013, the macrozoobenthos density (5502 ind. m^{-2}) approached to 2007–2008 levels (3313–4193 ind. m^{-2}).

Taxa richness also exhibited a clear temporal change (Fig. 7.7b). Richness was depressed by the tsunami and estuary closing (10–18 taxa) compared with 2007 and 2008 levels (21–23 taxa) and reached its minimum after the typhoon (9 taxa). It gradually recovered from 13 to 30 taxa over the following 10 months and reached to 40 taxa in June 2013. This high number of taxa was due chiefly to the occurrence of stenohaline marine species in the samples, such as the polychaetes *Ceratonereis erythraeensis*, *Scalibregma inflatum*, and *Scoletoma nipponica* and the sea cucumber *Patinapta ooplax*.

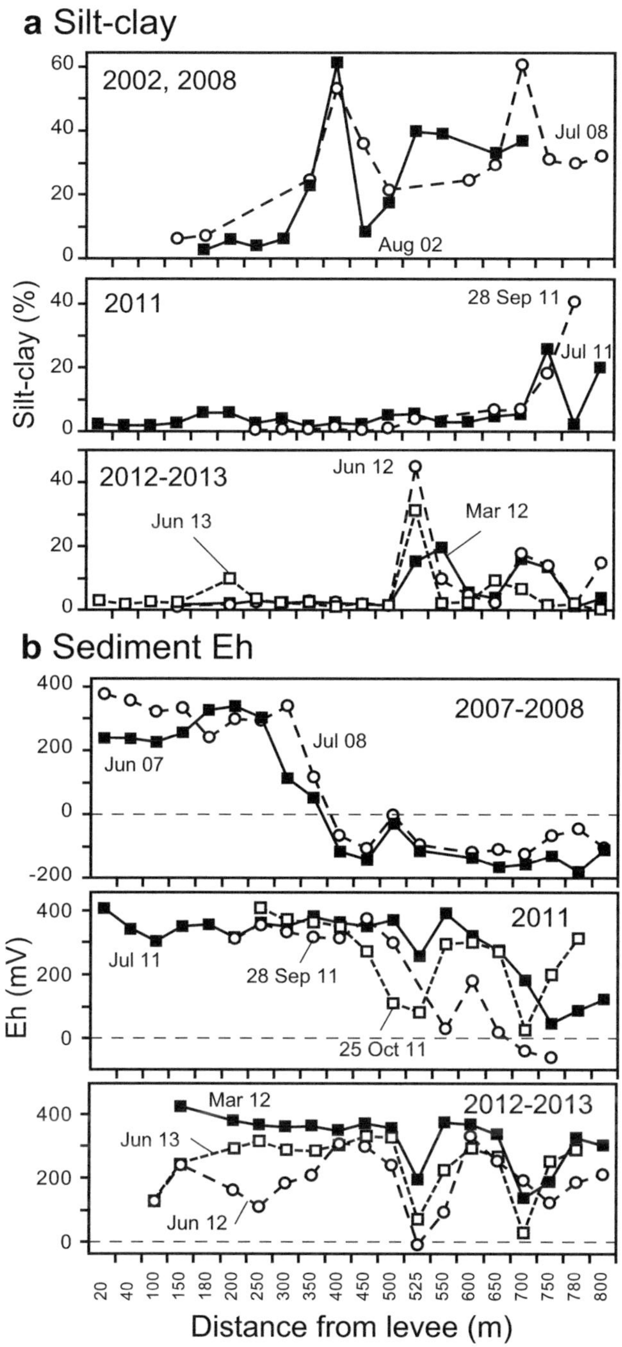

Fig. 7.5 Spatiotemporal changes in silt-clay contents (**a**) and sediment Eh (**b**) at 10–20 stations (see Fig. 7.3d)

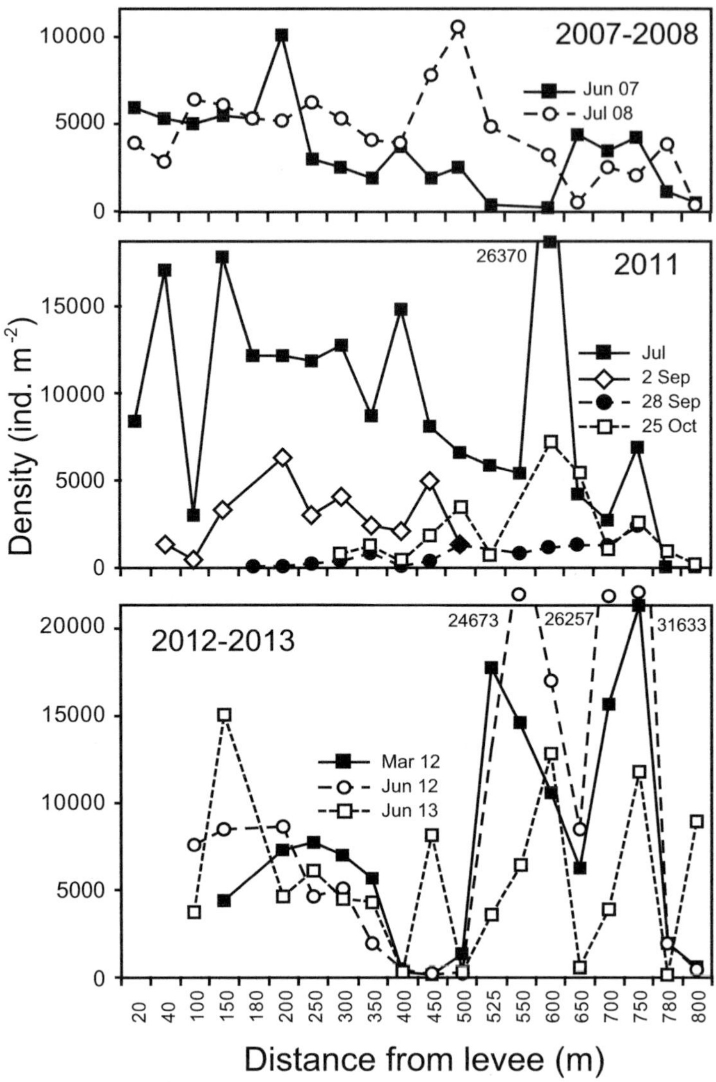

Fig. 7.6 Spatiotemporal changes in total macrozoobenthos density at 10–20 stations (see Fig. 7.3d)

After the tsunami, the bivalves largely disappeared (Fig. 7.7c), and the community was predominated by polychaetes and amphipods. During the estuary closing and typhoon, polychaetes and amphipods also declined sharply. Recolonization by polychaetes and amphipods occurred during March–June 2012, 6–9 months after the typhoon. In 2012, the bivalves recovered their population size through the recruitment of *Nuttallia japonica*, *Ruditapes philippinarum*, and *Musculus senhousia*. In 2013, the composition of polychaetes, amphipods, and bivalves was similar to pre-tsunami conditions.

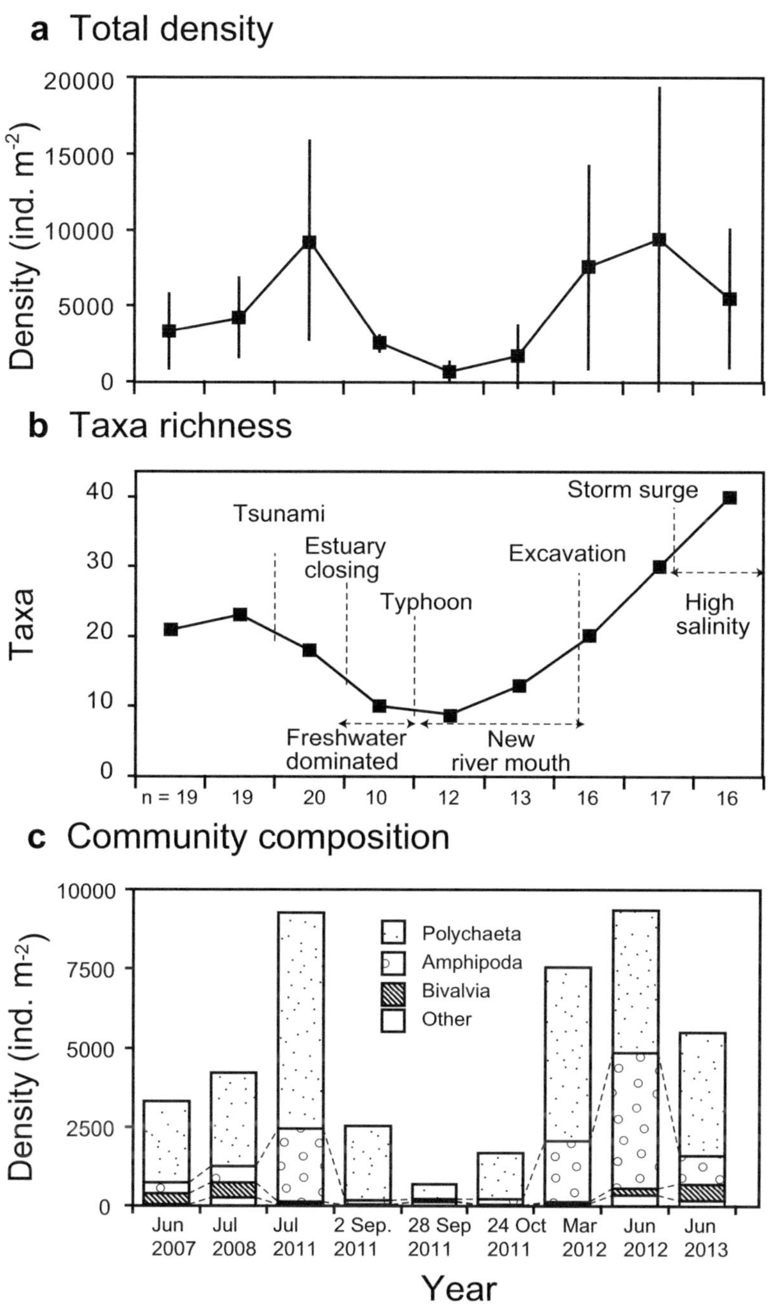

Fig. 7.7 Temporal changes in lagoonal mean density (**a**), taxa richness (**b**), and community composition (**c**) of macrozoobenthos. Values represent the total (**a**), or average (**b**, **c**), of 10–20 stations. *Bars* indicate standard deviations

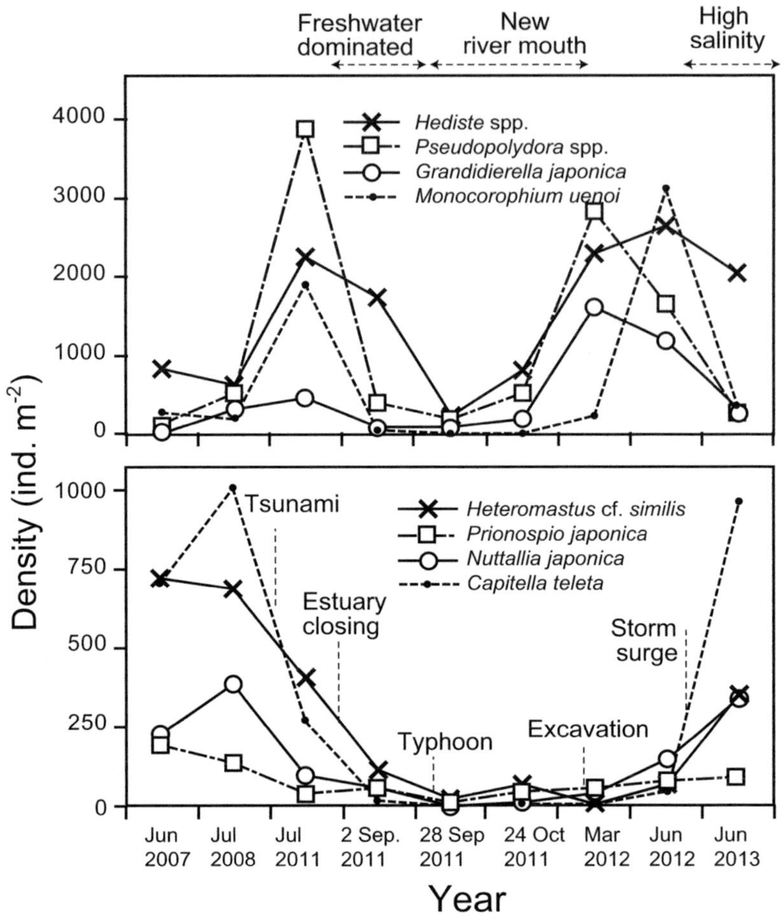

Fig. 7.8 Temporal changes in densities of dominant macrozoobenthos in Gamo Lagoon. Values are the lagoonal averages from 10 to 20 stations. For sample numbers (n), please see Fig. 7.7a. Upper and lower panels: 1st–4th and 5th–8th dominant taxa, respectively

7.4.5 Numerical Responses of Dominant Macrozoobenthic Taxa

The dominant taxa exhibited species-specific responses to the disturbance events (Fig. 7.8). Of the dominant taxa, four taxa, including the polychaetes *Hediste* spp. and *Pseudopolydora* spp. and the amphipods *Grandidierella japonica* and *Monocorophium uenoi*, exhibited rapid population recoveries (6–9 months) after the disturbance events. In contrast, population recovery took longer (>1 year) for the polychaetes *Heteromastus* cf. *similis*, *Prionospio japonica*, and *Capitella teleta* and the bivalve *N. japonica*.

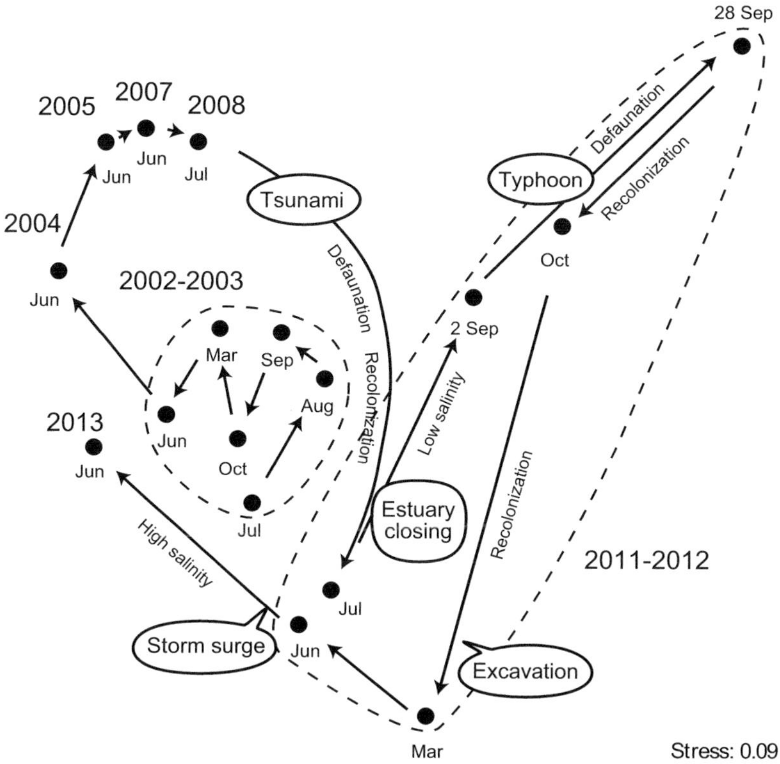

Fig. 7.9 A **nMDS plot** showing changes in macrozoobenthic community structure from 2002 to 2013. Bray-Curtis similarity was calculated from the square root of the lagoonal mean density. Each sample was indicated with a *black circle*

7.4.6 Changes in Taxonomic Composition of Macrozoobenthos

The nMDS plot clearly demonstrated the ecological impacts of each disturbance event on the macrozoobenthic community structure (Fig. 7.9). Intra- and interannual variations from 2002 to 2008 were much smaller than those during the post-tsunami period. The effect sizes of the tsunami, estuary closing, and typhoon were extraordinary but comparable with one another. The community structure recovered within 9 months after the typhoon, to a structure similar to that of July 2011. From 2012 to 2013, the community structure recovered to that of the pre-tsunami period.

7.5 Discussion

During the study period, disturbance events altered the habitat structure and associated biota to different degrees of intensity. The tsunami and typhoon resulted in drastic changes in topography, sediment characteristics, and vegetation areas of

Gamo Lagoon on larger spatial scales. After the tsunami, the species composition and population density of macrozoobenthos fluctuated for years and differed from those of the pre-tsunami period. These results suggest that the community was not yet achieved to a steady state, even after 3 years. Furthermore, salt marsh and SDV were characterized by the pioneer species, suggesting that the communities were still at an early succession stage.

The most significant consequences of the tsunami and typhoon were the intensive physical disturbances. Scouring and/or deposition of sediment resulted in changes in sediment granulometry and redox conditions in the lagoon. Before the tsunami, sediment H_2S accumulation caused mass mortalities of macrozoobenthos in muddy-reduced areas during the warmer months (Kanaya 2014). However, most of the mud was washed away by the tsunami, and the sediment became sandy and more oxidized throughout the lagoon (Kanaya et al. 2015). This may have facilitated the colonization of several polychaete and amphipod species in the habitat after the tsunami. Therefore, the replacement of the highly reduced mud with drifting sea sand was, in this instance, a positive consequence of the huge pulsed disturbances in this lagoon.

Another consequence of the disturbance events was the fluctuation of lagoon salinity. Although estuarine macrozoobenthos and plant species can cope with salinity changes, extreme changes within a short time can have sublethal effects (e.g., Yamamuro 1996; Lissner and Schierup 1997; Yokoya et al. 1999; Yamochi 2013). In fact, the salinity depression (<5) that occurred during the estuary closing in August to September 2011 caused a sharp decline in the species richness and total density of macrozoobenthos due to the mortality of polyhaline species such as *C. teleta*, *H.* cf. *similis*, and *Pseudopolydora* spp. (referred to Yamamuro 1996; Kanaya et al. 2011). Only the nereididae polychaete *Hediste* spp. maintained its population size during the estuary closing, which is consistent with its high tolerance for salinity fluctuation (Inamori and Kurihara 1979). On the other hand, in 2012, high salinity seemed to support the survival of stenohaline marine species such as the macrogreen algae *U. pertusa* (Yamochi 2013) and several macrozoobenthic taxa. The disappearance of *U. pertusa* and *U. prolifera* in 2014 was concurrent with decreasing lagoon salinity. These results suggest that the salinity regime controlled the population sizes of the opportunistic macroalgae in the lagoon. In contrast, no apparent change was observed in the patch size of the macrored algae *G. vermiculophylla*. This is possibly due to its greater tolerance to salinity fluctuations as shown in Yokoya et al. (1999).

In Gamo Lagoon, the reed marsh was an essential habitat for birds and marsh-associated macrozoobenthos such as the endangered gastropod *Cerithidea rhizophorarum* (Council for the Promotion of Nature Restoration in Gamo Tidal Flat 2008; Kanaya et al. 2012). However, the tsunami and typhoon washed away most of the reed marsh in this lagoon. For the recovery of *P. australis* patches in the area, the increased salinity after the tsunami disaster was a potential threat. Warren et al. (2002) observed that the coverage and frequency of *P. australis* correlated negatively with soil salinity; the species disappeared abruptly when the salinity was >26. Lissner and Schierup (1997) demonstrated that juveniles of *P. australis* barely survived at a salinity of 22.5, which was similar to the salinity in Gamo Lagoon after the storm surge (daily minimum >20). In addition, fishponds adjacent to the lagoon ceased pumping up groundwater (Kanaya et al. 2014). This would have also reduced the supply of freshwater to the lagoon.

The salt marsh and sand dune plant communities were still at an early stage of recolonization 4 years after the tsunami. The bare high-tide zone created by the loss of reed marsh was densely vegetated by the annual halophyte *S. maritima*, although it was rarely observed before the tsunami (Council for the Promotion of Nature Restoration in Gamo Tidal Flat 2008). Generally, such annual plants dominate the pioneer stages of strandline ephemeral populations, while they become less abundant at later succession stages (Watkinson and Davy 1985). Tessier et al. (2002) experimentally demonstrated that *S. maritima* is less competitive but more adaptive to frequently disturbed habitats. This suggests that the tsunami-induced gaps in the marsh area allowed proliferation of this less competitive species.

The invasive sand dune plant *C. edentula* proliferated on the sandbar after the tsunami. Hayasaka et al. (2012) reported that the species occurred on heavily disturbed sandy beaches along the Sanriku region in August 2011. Watkinson and Davy (1985) suggested that *C. edentula* is a pioneer species that particularly dominates strandline, and dry dune habitats were forced to undergo frequent disturbances. This species was not recorded on the sand dune before the tsunami disaster (Council for the Promotion of Nature Restoration in Gamo Tidal Flat 2008). It appears that *C. edentula* immigrated to the habitat after the tsunami event.

The present results indicate that the devegetated zones in the marsh and SDV provided habitats for the pioneer plant species, *S. maritima* and *C. edentula*. Even 4 years after the tsunami, the most long-lived plant taxa that dominated during the pre-tsunami period had not recovered their population sizes. Even if the environmental conditions are maintained properly, a longer period of time (e.g., 5 to >20 years) may be needed for complete recovery of the salt marsh, SDV, and associated biota in the area (see Tessier et al. 2002; Warren et al. 2002; Borja et al. 2010).

In Gamo Lagoon, overshoots of population densities were also observed for several opportunistic species in the macrozoobenthic community. For example, four dominant taxa (*Hediste* spp., *Pseudopolydora* spp., *M. uenoi*, and *G. japonica*) recovered their population sizes within 6–9 months after the tsunami and typhoon. These species all seemed to have opportunistic life history traits, allowing them to recover their population sizes within a shorter period of time (Aikins and Kikuchi 2002; Kikuchi and Yasuda 2006; Kanaya et al. 2011). Other dominant taxa, including *H.* cf. *similis*, *P. japonica*, and *C. teleta*, also have opportunistic or multivoltine life histories (Grassle and Grassle 1974; Kanaya et al. 2011). These observations demonstrate that the macrozoobenthic community in the lagoon consisted chiefly of short-lived polychaetes and amphipods with high population turnover. Most of the taxa also dominated the community before the tsunami (data in 2005–2008; Kanaya et al. 2015). Therefore, they would potentially be more adaptive to estuarine soft-bottom habitats forced to undergo frequent disturbances.

Population recovery of bivalves (e.g., *N. japonica* and *R. philippinarum*) took more than 1.5 years after the disturbance events. This may reflect their limited periods of reproduction. For example, Miyawaki and Sekiguchi (1999) demonstrated that recruitment of spat of *N. japonica* (reported as *Nuttallia olivacea*) and *R. philippinarum* occurred once or twice per year. In the summer of 2011, we observed the recruitment of spat of *N. japonica*; however, most disappeared during the estuary closing in August to September 2011. In the summer of 2012, clam spat were recruited again, and the population size recovered successfully during the following summer. These observations suggest that bivalves with relatively low population turnover are more vulnerable to pulsed disturbances including tsunami and typhoon and require much longer periods for population recovery in the future.

Borja et al. (2010) mentioned that the full recovery of estuarine ecosystems can take a minimum of 15–25 years after an intensive degradation, from the view of biotic composition. At our study site, however, the macrozoobenthic community structure returned to pre-tsunami conditions within 3 years after the tsunami. As discussed above, this is likely due to the rapid population recovery of the numerically dominant taxa in the community, although several long-lived taxa have not yet recolonized (e.g., the bivalve *Macoma contabulata*, gastropods *Hima festiva* and *C. rhizophorarum*, hermit crab *Pagurus minutus*, and mud shrimp *Upogebia yokoyai*). The dominant taxa are common residents in Japanese brackish waters (Yamamuro 1996; Miyawaki and Sekiguchi 1999; Aikins and Kikuchi 2002; Kanaya et al. 2011). They would primarily be adaptive to estuarine habitats that experience frequent disturbances, since they have high population turnovers and/or high ecophysiological capacities toward stressors (e.g., Grassle and Grassle 1974; Dauer 1984; Ritter et al. 2005). These findings emphasize the high resiliency of estuarine macrozoobenthic communities toward pulsed disturbances.

The present data revealed the impacts of the 2011 tsunami and following disturbance events in the Gamo Lagoon ecosystem. At present, the habitat structure and environmental conditions have not attained at a steady state ecosystem condition. The salinity regime in the lagoon is not stable, and restoration works in the area such as reconstruction of seawalls are currently ongoing (see Kanaya et al. 2014). The environmental changes by such restoration works in the lagoon ecosystem may induce changes in the associated biotic community and alter its ecosystem functioning (e.g., Levin et al. 2001). Therefore, long-term monitoring and adaptive management are necessary to determine the relevant solutions for future restoration and conservation of the lagoon ecosystem.

Acknowledgments We would like to thank the students in our laboratories at Tohoku University, as well as K. Kinoshita, H. Maki, H. Misui, Y. Nakamura, A. Oishi, and the staff at the Marine Biological Research Institute of Japan Co., Ltd. for their help in the field and in the laboratory. We also thank Dr. J. Urabe and Dr. S. Shikano for allowing us to use their laboratory facilities. In 2007 and 2008, part of the survey was conducted together with the Miyagi Prefecture monitoring program. The research was partly supported by NIES and Tohoku Ecosystem-Associated Marine Sciences (TEAMS).

References

Aikins S, Kikuchi E (2002) Grazing pressure by amphipods on microalgae in Gamo Lagoon, Japan. Mar Ecol Prog Ser 245:171–179

Beukema JJ, Flach EC, Dekker R, Starink M (1999) A long-term study of the recovery of the macrozoobenthos on large defaunated plots on a tidal flat in the Wadden Sea. J Sea Res 42:235–254

Borja Á, Dauer DM, Elliott M, Simenstad CA (2010) Medium- and long-term recovery of estuarine and coastal ecosystems: patterns, rates and restoration effectiveness. Estuar Coasts 33:1249–1260

Cardoso PG, Raffaelli D, Lillebø AI, Verdelhos T, Pardal MA (2008) The impact of extreme flooding events and anthropogenic stressors on the macrobenthic communities' dynamics. Estuar Coast Shelf Sci 76:553–565

Clarke KR, Gorley RN (2006) PRIMER v6: user manual/tutorial. PRIMER-E, Plymouth

Council for the promotion of nature restoration in Gamo tidal flat (2008) Total plan for nature restoration for Gamo tidalflat. Miyagi Prefecture, p 52 (in Japanese)

Dauer DM (1984) High resilience to disturbance of an estuarine polychaete community. Bull Mar Sci 34:170–174

Grassle JF, Grassle JP (1974) Opportunistic life histories and genetic systems in marine benthic polychaetes. J Mar Res 32:253–284

Hayasaka D, Shimada N, Konno H, Sudayama H, Kawanishi M, Uchida T, Goka K (2012) Floristic variation of beach vegetation caused by the 2011 Tohoku-oki tsunami in northern Tohoku, Japan. Ecol Eng 44:227–232

Hokkaido (2010) List of invasive species in Hokkaido–Hokkaido blue list 2010–. Hokkaido Prefecture. Available: http://bluelist.hokkaido-ies.go.jp/. Accessed on 5 Mar 2015

Inamori Y, Kurihara Y (1979) Analysis of the environmental factors affecting the life of the brackish polychaete, *Neanthes japonica* (Izuka), 1: the effects of the environmental factors on survival and growth. Bull Mar Biol Stn Asamushi Tohoku Univ 16:87–100

Jaramillo E, Dugan JE, Hubbard DM, Melnick D, Manzano M, Duarte C, Campos C, Sanchez R (2012) Ecological implications of extreme events: footprints of the 2010 earthquake along the Chilean coast. PLoS ONE 7:e35348

Kanaya G (2014) Recolonization of macrozoobenthos on defaunated sediments in a hypertrophic brackish lagoon: effects of sulfide removal and sediment grain size. Mar Environ Res 95:81–88

Kanaya G, Suzuki T, Kikuchi E (2011) Spatio-temporal variations in macrozoobenthic assemblage structures in a river-affected lagoon (Idoura Lagoon, Sendai Bay, Japan): influences of freshwater inflow. Estuar Coast Shelf Sci 92:169–179

Kanaya G, Suzuki T, Maki H, Nakamura Y, Miyajima Y, Kikuchi E (2012) Effects of the 2011 tsunami on the topography, vegetation, and macrobenthic fauna in Gamo Lagoon, Japan. Jpn J Benthol 67:20–32

Kanaya G, Maki H, Suzuki T, Sato-Okoshi W, Kikuchi E (2014) Tsunami-induced changes in a shallow brackish lagoon ecosystem (Gamo Lagoon) in Sendai Bay, Japan. Glob Environ Res 18:35–46

Kanaya G, Suzuki T, Kikuchi E (2015) Impacts of the 2011 tsunami on sediment characteristics and macrozoobenthic assemblages in a shallow eutrophic lagoon, Sendai Bay, Japan. PLoS ONE 10:e0135125

Kikuchi E, Yasuda K (2006) Comparison of the life cycles of two sympatric estuarine polychaetes, *Hediste diadroma* and *H. atoka* (Polychaeta: Nereididae), in the Nanakita River estuary, northeastern Japan. Limnology 7:103–115

Kurihara Y, Kikuchi E, Uehara T, Suzuki T (2000) Report of fundamental investigation of environmental conservation measure for Gamo tidal flat in 1999. Miyagi Prefecture, p 37

Levin LA, Boesch DF, Covich A, Dahm C, Erséus C, Ewel KC, Kneib RT, Moldenke A, Palmer MA, Snelgrove P, Strayer D, Weslawski JM (2001) The function of marine critical transition zones and the importance of sediment biodiversity. Ecosystems 4:430–451

Lissner J, Schierup H-H (1997) Effects of salinity on the growth of *Phragmites australis*. Aquat Bot 55:247–260

Middleton BA (2009) Regeneration of coastal marsh vegetation impacted by hurricanes Katrina and Rita. Wetlands 29:54–65

Miyawaki D, Sekiguchi H (1999) Interannual variation of bivalve populations on temperate tidal flats. Fish Sci 65:817–829

Niemi GJ, DeVore P, Detenbeck N, Taylor D, Lima A, Pastor J, Yount JD, Naiman RJ (1990) Overview of case studies on recovery of aquatic systems from disturbance. Environ Manag 14:571–587

Ritter C, Montagna PA, Applebaum S (2005) Short-term succession dynamics of macrobenthos in a salinity-stressed estuary. J Exp Mar Biol Ecol 323:57–69

Seike K, Shirai K, Kogure Y (2013) Disturbance of shallow marine soft-bottom environments and megabenthos assemblages by a huge tsunami induced by the 2011 M9.0 Tohoku-Oki earthquake. PLoS ONE 8:e65417

Szczuciński W, Kokociński M, Rzeszewski M, Chagué-Goff C, Cachão M, Goto K, Sugawara D (2012) Sediment sources and sedimentation processes of 2011 Tohoku-oki tsunami deposits on the Sendai Plain, Japan – insights from diatoms, nannoliths and grain size distribution. Sediment Geol 282:40–56

Tanaka H, Adityawan MB, Mano A (2014) Morphological changes at the Nanakita River mouth after the Great East Japan Tsunami of 2011. Coast Eng 86:14–26

Tessier M, Gloaguen JC, Bouchard V (2002) The role of spatio-temporal heterogeneity in the establishment and maintenance of *Suaeda maritima* in salt marshes. J Veg Sci 13:115–122

Urabe J, Suzuki T, Nishita T, Makino W (2013) Immediate ecological impacts of the 2011 Tohoku earthquake tsunami on intertidal flat communities. PLoS ONE 8:e62779

Warren RS, Fell PE, Rozsa R, Brawley AH, Orsted AC, Olson ET, Swamy V, Niering WA (2002) Salt marsh restoration in Connecticut: 20 years of science and management. Restor Ecol 10:497–513

Watkinson AR, Davy AJ (1985) Population biology of salt marsh and sand dune annuals. Vegetatio 62:487–497

Yamamuro M (1996) Kanchoiki no teiseidoubutsu (translated title: Macrozoobenthos in estuaries). In: Saijo Y, Okuda S (eds) Kasen kanchoiki–Sono shizen to henbou (translated title: Estuaries–the nature and changes). Nagoya University Press, Nagoya, pp 151–172 (in Japanese)

Yamochi S (2013) Effects of desiccation and salinity on the outbreak of a green tide of *Ulva pertusa* in a created salt marsh along the coast of Osaka Bay, Japan. Estuar Coast Shelf Sci 116:21–28

Yokoya NS, Kakita H, Obika H, Kitamura T (1999) Effects of environmental factors and plant growth regulators on growth of the red alga *Gracilaria vermiculophylla* from Shikoku Island, Japan. Hydrobiologia 398(399):339–347

Chapter 8
Impacts of the Tsunami Disaster Caused by the Great East Japan Earthquake on the Benthic Communities and Their Recovery in Tidal Flats of Matsukawaura Lagoon, Soma, Japan

Takao Suzuki

Abstract Benthic communities and the physical properties of the bottom sediments were examined for 3 years after the Great East Japan Earthquake and subsequent tsunamis on 11 March 2011 at several sites of tidal flats in Matsukawaura Lagoon, Soma, Japan. The bottom sediments were changed mostly to sandy after the tsunamis except at some sites in the innermost parts of the lagoon, where the sediments became rather muddy due to the insufficient exchange of seawater. In 2011 just after the tsunamis, the species richness and abundance of benthic animals were almost halved compared with those before the tsunamis. However, a number of benthic species were recovered within 1 or 2 years after the tsunamis, although the species compositions of the benthic communities were temporally changed during the study periods. In particular, polychaetes showed early recovery while crustaceans, gastropods, and bivalves remained numerically far from recovery. According to census data collected for Matsukawaura Lagoon, many species inhabiting the tidal flats, including not only sessile and creeping species but also crawling and embedded species, could not successfully colonize in regions where embankments were rapidly constructed after the tsunamis in order to protect the farms and residential areas from future tsunamis. This study suggests that most benthic species have the potential to recover their populations rapidly after a tsunami if the seawater is well exchanged and the conditions of the sediment remain healthy. Thus, the conservation of potential habitats is essential to promote the recovery of benthic communities in this lagoon.

Keywords Benthic community • Tidal flat • Species richness • Recovery after tsunamis • Matsukawaura • Conservation of environment

---

T. Suzuki (✉)
Graduate School of Life Sciences, Tohoku University,
Aoba 6-3, Aramaki, Aoba-ku, Sendai 980-8578, Japan
e-mail: takaos@m.tohoku.ac.jp

J. Urabe, T. Nakashizuka (eds.), *Ecological Impacts of Tsunamis on Coastal Ecosystems*, Ecological Research Monographs,
DOI 10.1007/978-4-431-56448-5_8

8.1 Introduction

Tidal flats located along the Pacific coast of the Tohoku district of Japan, especially from Minami-sanriku to Sendai Bay, suffered heavy damages after the massive tsunamis associated with the Great East Japan Earthquake on 11 March 2011 (Urabe et al. 2013). However, the physical force of the tsunamis as gauged by, for example, their inundation heights, varied among the sites according to the topographies and coastal features (Suzuki 2011). Therefore, the ecological impact of the tsunamis on the benthic communities was likely site specific.

Because tidal flats are generally productive ecosystems, they are the preferred habitats for a number of benthic organisms and thus contribute to the purification of flowing water from surrounding areas. In addition, tidal flats are often used for marine resources cultivation and various recreational activities, including fishing. The ecosystem of the tidal flats thus provides various services for human society. However, to fully utilize these ecosystem services, it is essential to examine the ecological impacts of the tsunami disturbance on the various intertidal flats and to clarify whether their communities have recovered.

Matsukawaura is a large lagoon located at the south end of the Sendai Bay (Soma, Fukushima Prefecture); it has several tidal flat areas and provides various ecosystem services to the local communities, as mentioned above. One month after the Great East Japan Earthquake, I visited Matsukawaura Lagoon and found that the degree of the tsunami disturbance appeared to differ among sites within the lagoon: some tidal flats seemed to have disappeared completely or partially, while in others the sediment conditions seemed to have changed. Since the physical and chemical conditions including the water retention and salinity were relatively homogeneous within the lagoon, the variations in the degree of the tsunami disturbance provided a chance to examine how the response and recovery of the benthic communities differ depending on the ecological impacts of the tsunami disturbance.

To clarify this potential association between the response and recovery of the benthic communities and the degree of tsunami disturbance, I conducted a series of census surveys for 3 years after the Great East Japan Earthquake at various tidal flats in Matsukawaura Lagoon, where I had made similar surveys in 2004 (Suzuki 2005). Since the lagoon is used for fishing, includes popular tourism sites, and is managed by the Fukushima Prefectural Nature Park, this study will provide essential information to promote not only the resilience of the lagoon ecosystem but also the revival of the local economy.

8.2 Materials and Methods

In Matsukawaura Lagoon, I established 13 stations (Fig. 8.1) with different degrees of disturbance caused by the tsunamis and monitored the recovery process of the benthic communities. In each of these 13 stations, I investigated the quality of the

Fig. 8.1 Map of Matsukawaura Lagoon, Soma, Fukushima Prefecture, Japan, showing the stations where quantitative samplings were made. Note that seawalls at the eastern position of St. K were destroyed by the tsunamis in 11 March 2011. The disturbance induced by the tsunamis was huge at Sts. A, K, and L, intermediate at Sts. C, D, E, F, G, I, and J, and small at Sts. B, HS (shore site of St. H), and HO (offshore site of St. H) (The photo was taken just after the Great East Japan Earthquake occurred on 11 March, 2011)

bottom sediments and species composition of the benthic communities using a surf-fishing boat in September 2012, March 2013, September 2013, and April 2014. In addition, at four of the stations (Sts. A, B, D, and HS), I investigated the species composition of the benthic animals in September 2011, just 6 months after the tsunamis. Because 11 of these stations (or all but Sts. K and L) were the same locations of my earlier investigation in June 2004 (Suzuki 2005) and in March 2012 (Suzuki 2013), I was able to compare the sediments and species before and after the tsunamis and then going forward for 3 years after the tsunamis.

At the samplings, I used a core sampler (15 cm in diameter, 30 cm in depth) or Ekman-Birge grab (bottom sampler) for collecting benthic animals. At each station, I collected four samples of the bottom sediments and sieved each sample with a 1 mm mesh. Then, the animals collected on the mesh were transferred into a plastic bag for fixation with neutralized formalin (10 %) and brought to a laboratory for sorting, identification, and enumeration.

In this study, I categorized the bottom sediments into four classes: sand, muddy sand, mud, and soft mud. Then, I recorded the sediment class at each site when benthic animals were collected. In addition, I measured the thickness of the bottom sediments by inserting a bamboo stick with a scale. The data were used to examine the accumulation of reductive mud (mud and soft mud). The redox potential was also measured using an ORP meter (TRX-999, TOKO). In addition, I collected sediment samples for measuring silt-clay content and organic matter content in the laboratory.

The core and grab samplers used in this study could not efficiently collect highly mobile animals such as brachyurans and low-density animals such as large mollusks. Therefore, apart from these samplings, I established 20 sites around the shore of Matsukawaura in August 2014 and examined the presence of such species by means of qualitative census surveys. Note that, among these sites, ten sites were artificially disturbed during the period from fall 2012 to spring 2013 due to the construction of embankments. Since similar census surveys were carried out in 2012 (Fukushima Prefecture 2013) when the embankment constructions had not yet started, it was possible to elucidate the influences of the embankment construction on living benthic animals.

8.3 Results and Discussion

8.3.1 Impacts of the Tsunamis

Visual inspection of the sites suggested that Sts. A, K, and L were subjected to the greatest disturbance, since the tsunamis had destroyed the seawall at or near to these stations, and the bottom sediments were greatly disturbed, mostly carried away and replaced by sand that was probably deposited by the tsunamis. The disturbance to Sts. C, D, E, F, and G appeared to be more moderate, probably because the water depths increased due to clearance of the bottom sediments and/or subsidence, which resulted in a decrease in the area of tidal flats. Stations I and J were also moderately disturbed. At these stations, however, the bottom conditions were actually improved, because the muddy sediments were replaced by sand. Accordingly, the water depth became more shallow, and the tidal flat area was expanded. Compared with these sites, the disturbance seemed to be even more limited at Sts. B, HS (located on the shore of St. H), and HO (located offshore of St. H), since erosion of the bottom sediments was limited.

From summer 2011 to spring 2013, Fukushima Prefecture conducted digging at the tidal flats to remove debris from the bottom sediments (uprooted pine trees, fishing nets and gear, and shipwrecks that had been broken up and washed inland by the tsunamis) and restore the waterway and then replaced the bottom sediments. In fall 2012, restoration of the embankments that had been destroyed by the tsunamis began along the shore. Due to the embankment restoration, there was also some construction-related disturbance at the investigated tidal flats, although the degree of such disturbance differed among the stations.

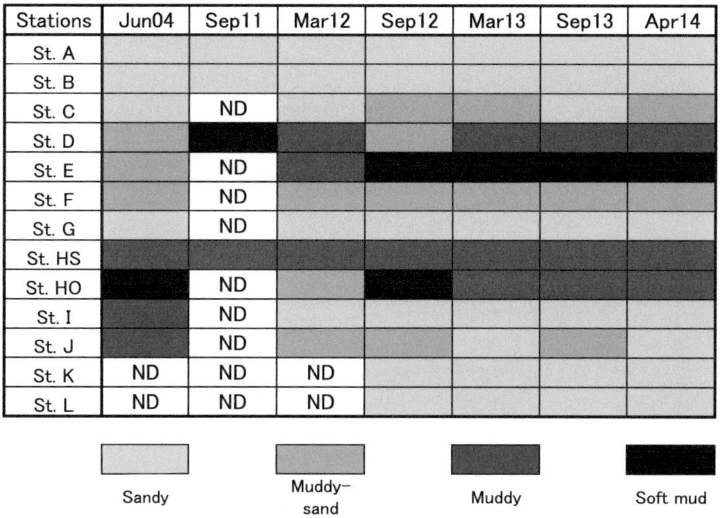

Fig. 8.2 Property of the bottom sediments at 13 stations of Matsukawaura Lagoon observed in each year after the tsunamis. *ND* not determined

8.3.2 Changes in the Bottom Sediments

Changes in the properties of the bottom sediments and thickness of the reductive mud (mud and soft mud) are summarized in Figs. 8.2 and 8.3, respectively. In more than half of the stations, large changes were found in the bottom properties after the tsunamis. When we visited in 2004, I categorized the sediment as mud or soft mud at Sts. I and J. However, on the one hand, most of the soft mud at the bottom was washed away and replaced with sand that was probably deposited by the tsunamis. On the other hand, at Sts. D and E, which were located in the southern innermost area, I found thickly accumulated mud after the tsunamis, which were likely deposited due to insufficient exchange of seawater. These changes were also detected in the thickness of bottom mud. Stations A, K, and L were greatly disturbed by the tsunamis, and most of the bottom sediments were changed to sand. The muddy areas around Sts. HS and HO, which were located in the western innermost zone, were relatively less disturbed by the tsunamis, as evidenced by the fact that only a small portion of the mud sediments had washed away. At other stations (Sts. B, C, F, and G), there were no major changes in the properties of the bottom sediments.

In general, the bottom sediments at Matsukawaura Lagoon were mostly changed to sand just after the disturbance caused by the tsunamis. Thereafter, however, the bottom sediments became muddy at some stations in the innermost zone, because mud particles were deposited due to the inefficient exchange of seawater. The changes in the redox potential, silt-clay contents, and organic matters of the sediments showed similar patterns throughout the stations examined.

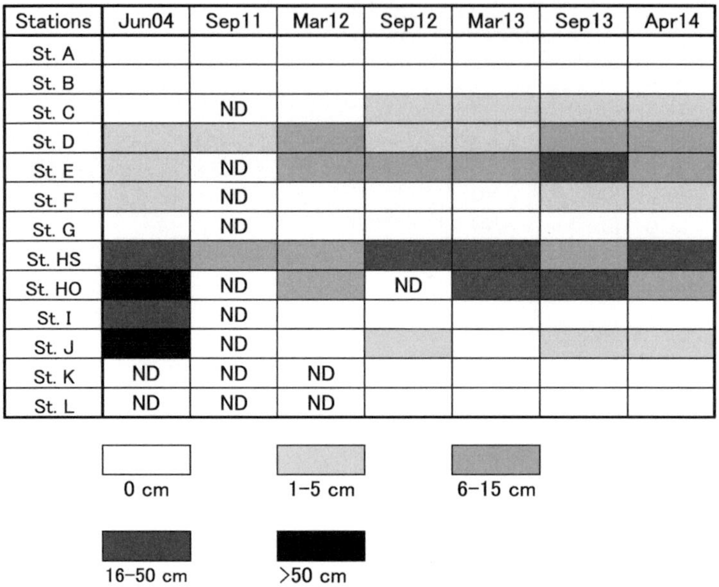

Fig. 8.3 Thickness of the reductive mud (mud and soft mud) accumulated on the bottom sediment at 13 stations in each year after the tsunamis. *ND* not determined

8.3.3 Benthic Communities

Changes in the species richness and abundances of benthic invertebrate animals as the average population densities based on four samplings at each of the 13 stations are shown in Figs. 8.4 and 8.5, respectively. In September 2011, when 6 months had passed since the tsunamis were induced by the Great East Japan Earthquake, both the species richness and abundances were almost halved compared with those before the tsunamis at Sts. A, B, D, and HS. At these stations, most of the previously observed gastropods and bivalves were absent. However, polychaetes were found abundantly, probably because some species bred and increased the size of their populations over the 6 months after the tsunamis. At 1 year after the tsunamis, gastropods and bivalves began to be observed, although their numbers were still limited. Polychaete species were still abundantly found in spring 2012 at all 13 stations, while crustaceans were less abundant.

In Matsukawaura Lagoon, I found a total of 63 species in March 2012, 62 in September 2012, 56 in March 2013, 59 in September 2013, and 58 in April 2014 just 3 years after the tsunamis. These numbers indicated that the level of species richness was almost the same as before the tsunamis, i.e., there were 67 species in June 2004. However, the community compositions of the benthic animals changed over time. As for the ten most representative species based on the combined yearly average densities of the 13 stations, two crustaceans (Corophiidae sp., *Grandidierella*

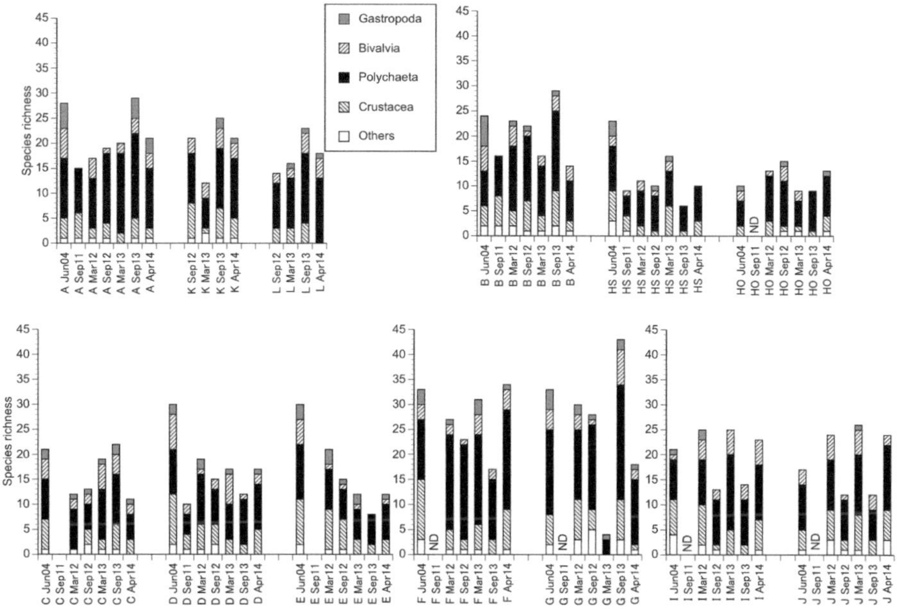

Fig. 8.4 Species richness of benthic invertebrate animals at 13 stations of Matsukawaura Lagoon. The disturbance induced by the tsunamis was huge at Sts. A, K, and L, small at Sts. B, HS, and HO, and intermediate at the other stations. *ND* not determined

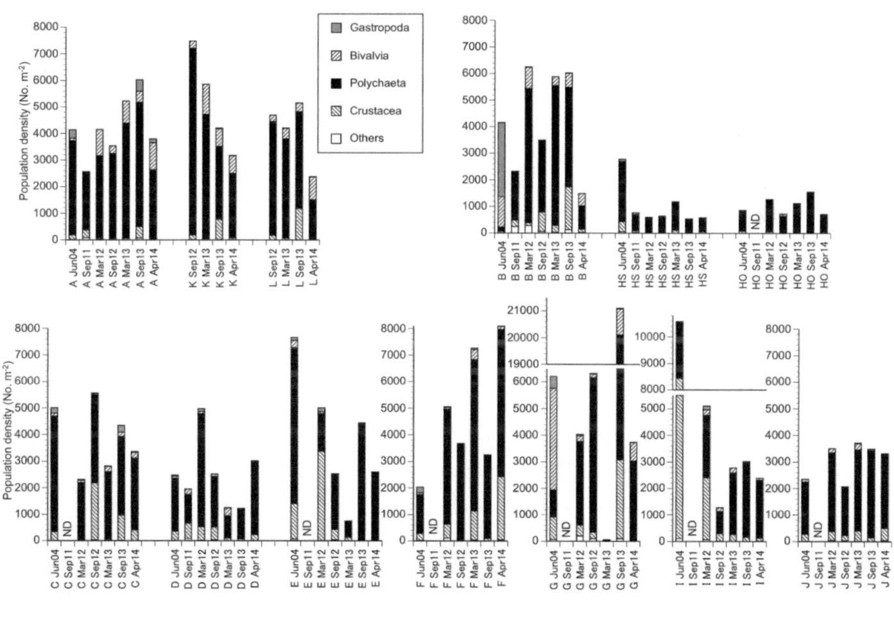

Fig. 8.5 Temporal changes in the abundances of the benthic invertebrate animals at 13 stations in Matsukawaura Lagoon. *ND* not determined

japonica), two gastropods (*Batillaria attramentaria*, *Assiminea* sp.), and a bivalve (*Ruditapes philippinarum*) were selected in June 2004 before the tsunamis. The other five species were all polychaetes.

In March 2012, however, only a single crustacean (Corophiidae sp.) and a bivalve (*Nuttallia japonica*) made the list, with the remaining species being polychaetes. Thereafter, from September 2012 to April 2014, only one to two crustacean, gastropod, and bivalve species were among the ten most representative species. Indeed, after the tsunamis the species that accounted for more than 10 % of total individuals in the benthic communities throughout the lagoon were all polychaetes, such as *Rhynchospio* sp. and *Pseudopolydora* cf. *kempi* in March 2012, *Hediste* spp. in September 2012, *Armandia* cf. *amakusaensis*, *Nephtys polybranchia*, and *Cirriformia* cf. *comosa* in March 2013, *Cirriformia* cf. *comosa* and *Pseudopolydora* cf. *kempi* in September 2013, and *Cirriformia* cf. *comosa*, *Scoletoma nipponica*, and *Heteromastus* sp. in April 2014. Thus, although polychaetes were numerically abundant, the dominant species changed year by year.

In contrast, the populations of crustaceans, gastropods, and bivalves were numerically limited compared with those before the tsunamis, indicating that the benthic animal community in Matsukawaura Lagoon was still far from recovery even in 2014 (Fig. 8.4). The temporal variation of the dominant polychaete species and limited abundance of crustaceans and mollusks suggest that the bottom environments were unstable even when several years had passed after the tsunamis. This instability may have been caused by anthropogenic activities such as the debris removal from the bottom sediments or digging of the bottom sediments to create waterways for ships in the lagoon.

For example, in March 2013, just before our survey, the tidal flat at St. G was greatly disturbed by the piling up of sediment that had been dug up to create the waterway. When benthic animals at St. G were examined in March 2013, I found that the abundance of most species was greatly decreased compared with that in the previous year. At 6 months after this event, in September 2013, *Pseudopolydora* cf. *kempi* and *Pseudopolydora paucibranchiata* (polychaeta) were abundant. However, with the exception of this station, the abundance of polychaetes was almost equivalent to that before the tsunamis (Fig. 8.5).

Spatial and temporal variations in the species evenness of the benthic communities were assessed using Pielou's index of equitability (J′). The J′ value was relatively high in 2012 and successive years at Sts. A, B, F, G, I, and J, which were located nearly in the north central area of Matsukawaura Lagoon (Fig. 8.6 left). In particular, at Sts. A and G, which were close to the waterway to the Pacific Ocean, the J′ value of evenness indices tended to increase after the tsunamis, suggesting that the conditions at these stations became more favorable to many benthic animal species. On the other hand, at Sts. C, D, E, HS, and HO, which were located in the innermost area and thus characterized by less efficient exchange of seawater, the evenness indices fluctuated greatly across time after the tsunamis (Fig. 8.6 right). At Sts. HS and HO, although the evenness indices were high in April 2014, the species richness was still low (Fig. 8.4). More remarkably, the values of the evenness indices were low in April 2014 at St. C and in September 2013 and April 2014 at St.

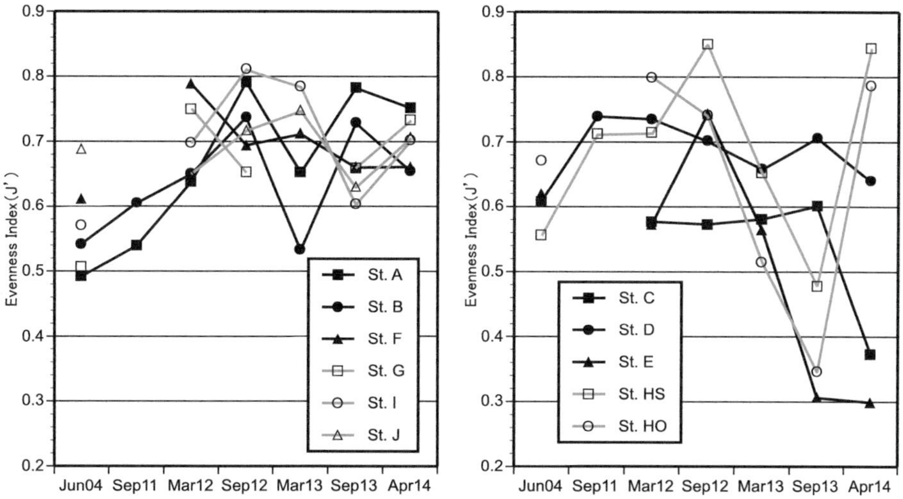

Fig. 8.6 Temporal changes in the evenness index (J′) of benthic communities at 11 stations. Stations A, B, F, G, I, and J were located nearly in the north central area of Matsukawaura Lagoon, and Sts. C, D, E, HS, and HO were located in the innermost southern or western areas

E. In these stations, muddy sediments were accumulated at the bottom (Fig. 8.2), probably due to the low water exchange as mentioned above. Thus, such changes in the sediment conditions may have deteriorated the habitats of many benthic animals.

8.3.4 Habitat Conditions for the Large-Scale Benthos

In August 2014, I made qualitative census surveys at 20 sites in Matsukawaura Lagoon to cover benthic animals that may not have been detected in the quantitative samplings reported above. A total of 72 species were recorded in these census surveys, which included 37 species not collected by the quantitative samplings. Most of these species were crustaceans, gastropods, and bivalves. In marked contrast to the results of the quantitative sampling, only three species of polychaetes were present in the surveys.

When I performed the census surveys in 2014, the construction of an embankment to protect the inland areas from future tsunamis was either underway or already completed at ten sites. Fortunately, Fukushima Prefecture (2013) conducted an identical census survey at these sites in 2012 when the embankment was not yet constructed. Therefore, I compared the number of benthic species between sites where the construction of an embankment was done or underway and sites where no embankment was constructed using data obtained in 2012 and 2014 (Fig. 8.7). In 2012, no significant difference was detected in the number of species between sites with and without the construction of the embankment. However, in 2014, the num-

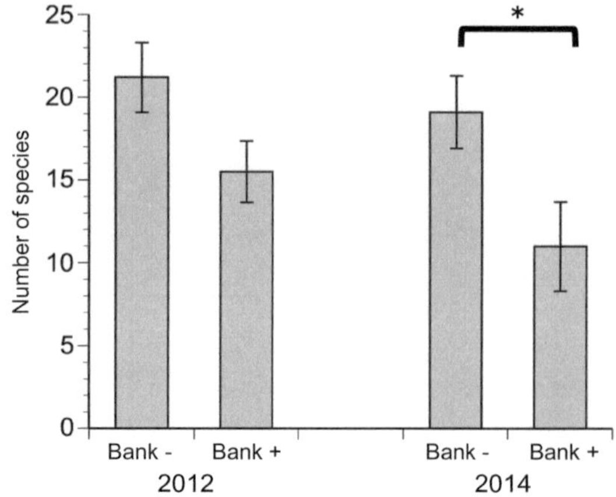

Fig. 8.7 Number of benthic animal species found in the qualitative census surveys at the sites with (Bank +) and without an embankment (Bank –) in 2012 and 2014. Significant differences between these sites at $p<0.05$ (t-test) are denoted with an *asterisk*

ber of species was significantly lower in the sites with an embankment than in those without an embankment.

Moreover, when the census surveys in 2012 and 2014 were compared, it was clear that in 2014, the number of the sites had decreased not only for sessile and creeping species, which inhabit the cracks in rock or concrete bricks, but also for species that crawl and embed in tidal flat sediments (Fig. 8.8). These results suggest that some benthic species could not live in the areas where the embankments were constructed. It is inevitable that the constructions of infrastructures such as embankments will be accompanied by the loss of some benthic species due to habitat destruction or deterioration.

8.3.5 Future Prospects for the Conservation of Tidal Flats

In this study, the benthic communities were shown to recover to a large extent at 1 year after the disturbance caused by the tsunamis. However, the recovery of crustaceans, gastropods, and bivalves was insufficient in terms of the number of individuals. In addition, the dominant polychaete species changed from year to year, although these species were numerically abundant. These results suggest that the recovery process of the benthic communities of Matsukawaura Lagoon following the tsunami disturbance is still ongoing.

This study also showed that within Matsukawaura Lagoon, the community composition differed largely between the sites. In general, regardless of the strength of disturbance by the tsunamis, benthic communities were poorly developed at the sites where seawater exchange was limited and muddy sediments were accumulated, even if these stations were not hugely disturbed by the tsunamis directly. Moreover, species richness was dramatically decreased at sites where embankments were constructed.

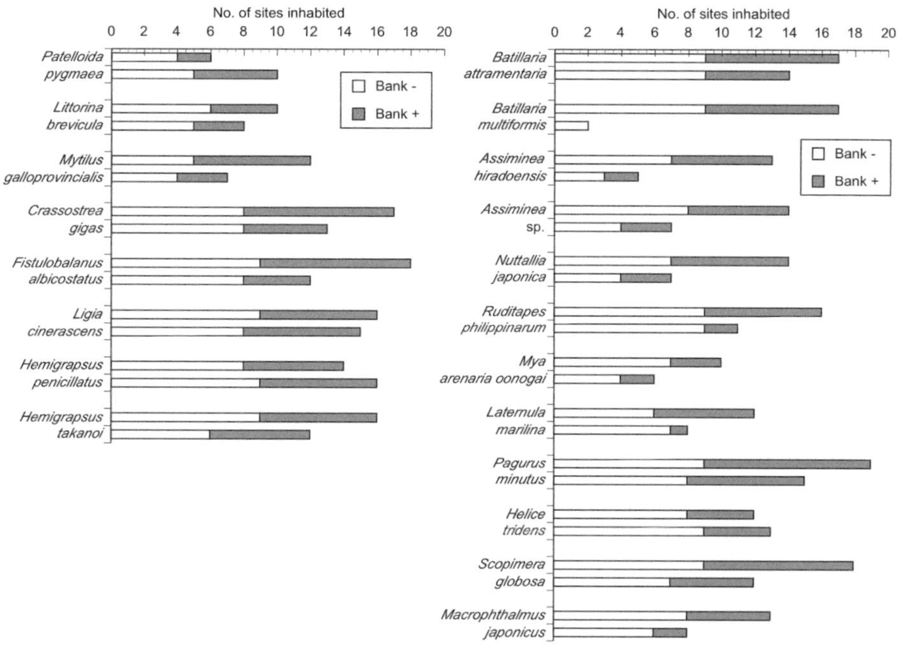

Fig. 8.8 Number of sites found in a given species. The species occurring at ten or more sites either in the 2012 or 2014 surveys are shown. For each species, the upper bar denotes the number of sites in 2012 and the lower one the number of sites in 2014. The sites with (Bank +) and without an embankment (Bank −) are shown by filled and white bars, respectively. *Left panel*: sessile and creeping species, *right panel*: crawling and embedded species

To recover the benthic communities or to maintain their high species richness even after a tsunami disturbance, restoration and conservation of the benthic habitats proved to be essential. Based on these results, the following recommendations can be made:

(a) Avoid artificial disturbances to areas where communities with high species richness are established or are likely to be established.
(b) Secure the continuity between the tidal flats and the inshore seawater.
(c) Secure the seawater exchange.
(d) Maintain the refuge sites of the benthic animals and secure the waterways on which the benthos might migrate during the construction of infrastructures.
(e) Monitor the benthic communities.

Acknowledgments I thank T. Kazama, T. Nishita, Kei, Yamaki, Ken, Yamaki, and T. Hirama of the Community Ecology Laboratory of Tohoku University and K. Watanabe and S. Nemoto for their support in the fieldworks. I also thank A. Sato for steering the boat during the field research. This study was partly supported by a Grant-in-Aid for Scientific Research from the Ministry of Education, Culture, Sports, Science and Technology, Japan (No. 24570018).

References

Fukushima Prefecture, Nature conservation Section (2013) Report on the basic studies of the local bio-diversity, Fukushima Prefecture, 2012. 149p. (in Japanese)

Suzuki T (2005) Benthic animals (Benthic communities and bottom sediments in Matsukawaura). In: Reports for the integrated research studies on Matsukawaura important wetland. Ed by Nature Conservation Section, Fukushima Prefecture, pp 55–83. (in Japanese)

Suzuki T (2011) Environmental changes in tidal flats and influences on benthic invertebrates caused by the earthquake and tsunami at Eastern Japan. J Japan Soc Water Environ 34:395–399 (in Japanese)

Suzuki T (2013) Natural environments of Matsukawaura and influences by the disaster from the viewpoint of the benthic animals. Report on the Nature and Livelihood Recovery Project, WWF Japan, pp 20–33. (in Japanese)

Urabe J, Suzuki T, Nishita T, Makino W (2013) Immediate ecological impacts of the 2011 Tohoku earthquake tsunami on intertidal flat communities. PLoS One 8(5):e62779. doi:10.1371/journal.pone.0062779

Chapter 9
Colonization of the Restored and Newly Created Tidal Flats by Benthic Animals in the Sanriku Region of Northern Japan

Masatoshi Matsumasa and Kyoko Kinoshita

Abstract Tsunamis and subsidence caused by the Great East Japan Earthquake in 2011 created and restored some intertidal areas along the Sanriku coastline of northern Japan. Otomo-ura is one such area in Iwate Prefecture. The tidal flat of Otomo-ura had been reclaimed for farmland 50 years prior but has "returned to the sea" after the tsunamis. Our 3-year-long qualitative and quantitative investigation of benthic fauna after the tsunamis reveals that a variety of animals have quickly colonized the newly created habitat. In August 2012, 2013, and 2014, the number of species observed was 27, 51, and 59, respectively. Rapid colonization was also observed for another newly created habitat in the Unosumai River estuary, which is situated about 40 km north of Otomo-ura. We find that that habitat heterogeneity between the peripheral and central areas and microhabitats provided by primary (e.g., sand and cobbles) and secondary substrates (e.g., oyster's and algal bodies) both contributed to increased species richness and population density. Although direct-developing species, which lack planktonic larval stages, had hardly been expected to colonize the restored and newly created vacant tidal flats, a direct-developing gastropod *Batillaria attramentaria* has been observed in Otomo-ura since August 2014. Colonization by the direct-developing gastropod and DNA analyses for 15 populations, including one from Otomo-ura, indicate that regional management of source populations will be necessary to prevent colonization by *Euspira fortunei*, another direct-developing alien predator, and to diminish the effects of parasitic *Perkinsus* on the host clam population.

M. Matsumasa (✉)
Department of Biology, Center for Liberal Arts and Sciences, Iwate Medical University, 2-1-1 Nishitokuta, Yahaba, Iwate 028-3694, Japan
e-mail: mmasa@iwate-med.ac.jp

K. Kinoshita
Organization of Revitalization for Sanriku-region, Iwate University, 3-18-8 Ueda, Morioka, Iwate 020-8550, Japan
e-mail: kkyoko@iwate-u.ac.jp

J. Urabe, T. Nakashizuka (eds.), *Ecological Impacts of Tsunamis on Coastal Ecosystems*, Ecological Research Monographs,
DOI 10.1007/978-4-431-56448-5_9

Keywords Colonization • Direct development • Habitat heterogeneity • Host-parasite relationships • Larval supply • Newly created habitat • Reclaimed land • Species composition • Species richness • Tsunami

9.1 Introduction

The southern part of the Sanriku coastline in the northeastern Japan consists of many groups of rias, which are low, partially submerged river valley structures that branch out, treelike, into open sea (Fig. 9.1). This characteristic topography makes Sanriku prone to wave amplification during tsunamis. The tsunamis and the subsidence due to the Great East Japan Earthquake that occurred on 11 March 2011 severely damaged the coastline of Sanriku and created some new and restored

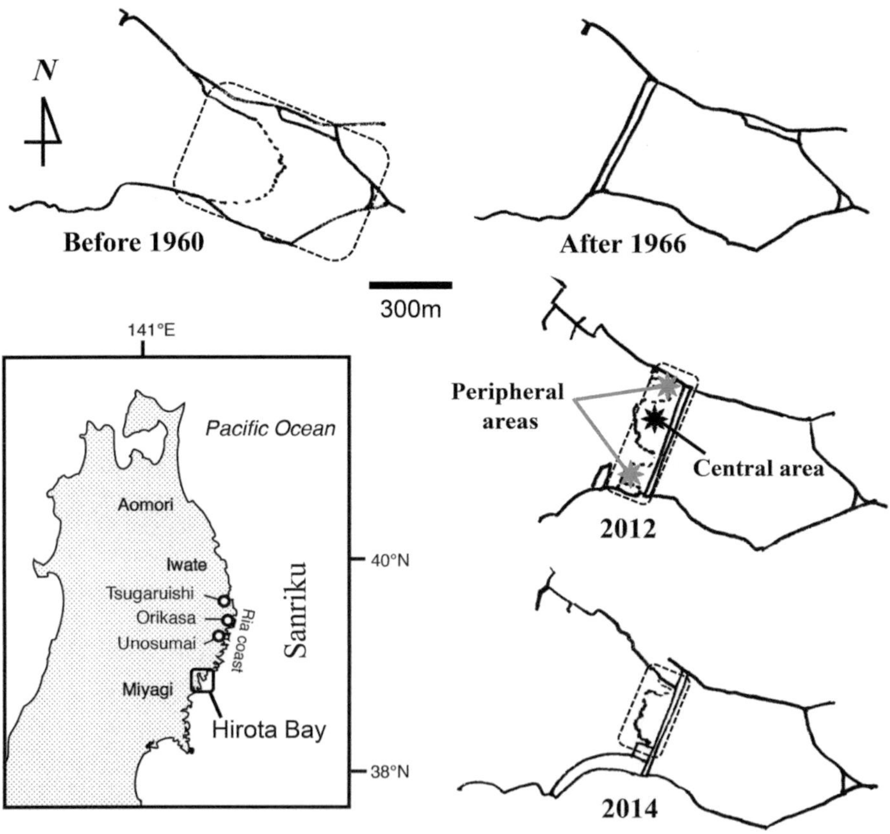

Fig 9.1 Alterations of the shape of Otomo-ura in Hirota Bay. *Boxes* with *dotted lines* roughly indicate intertidal areas. *Inlet* shows the location of Hirota Bay in the Sanriku region, northern Japan

intertidal areas there (Matsumasa et al. 2013, 2015; Kanaya et al. accepted). Such intertidal areas have the potential to be rich ecosystems associated with a variety of life, as long-standing intertidal rocky and boulder shores or tidal flats and marshes. The development of ecosystems in intertidal areas where land once existed likely progresses with colonization by coastal and brackish animals and plants in the newly created and restored vacant habitats. Although these processes have repeatedly occurred along the coastline of Japan in geological history, they have seldom been observed on the ecological timescales of human (i.e., several decades) because major earthquakes and tsunamis happen so rarely (i.e., several hundred to thousand years). Learning from the colonization of restored and newly created intertidal areas helps us to understand how intertidal animals and plants respond to major low-frequency disturbances, how their communities are organized, and how to construct guidelines for the management of coastal ecosystems after natural disasters. In this chapter, we summarize and discuss the ecological implications of the colonization patterns of a variety of benthic animals, such as clams, snails, and crabs, in the restored and newly created tidal flats resulting from the 2011 Great East Japan Earthquake and associated large tsunamis.

9.2 Life Forms and Dispersal Abilities of Tidal Benthic Animals

The colonization patterns of animals in restored and newly created habitats might be closely related to their mobility and dispersal ability (Matsumasa et al. 2013). In coastal and estuarine habitats, for example, adults of the blue swimming crab, *Portunus pelagicus*, migrate between estuaries and offshore areas. It has been reported that the distance of migration ranges from ca. 3 to 13 km in the Sarawak coastal waters of the South China Sea (Ikhwanuddin et al. 2012). In the Gamo Lagoon of Miyagi Prefecture in northeastern Japan, the adults of the mud crab, *Helice tridens*, migrate during summer from reed marshes in the upper-shore burrowing area to the lower-shore area and wander the water-logged tidal flat in the lagoon, which is 860 m long and no more than 250 m wide, (Kurihara et al. 1988). Adults of the hard clam, *Meretrix lusoria*, occasionally drift between tidal flats by using mucous cords to float (Nakamura 2013; Henmi et al. 2014). Mobile epibenthic and infaunal animals, like these decapod crustaceans and bivalves, could quickly colonize the restored and newly created tidal flats if some of the nearby adults survived the tsunamis.

In marked contrast to mobile animals, adults of sessile species, such as the Pacific oyster, *Crassostrea gigas*, and the blue mussel, *Mytilus galloprovincialis*, could not move into the restored and newly created tidal flats because they are unable to leave their hard substrates, such as the rocks and shells of conspecifics. Sedentary infaunal bivalves, such as *Mya* (*Arenomya*) *arenaria oonogai* and *Laternula* (*Exolaternula*) *marilina*, move little horizontally and have difficulty submerging themselves into sand and mud once they have been removed from the sediment. Therefore, colonization by adults of sessile and sedentary animals of the restored and newly created tidal flats is hardly expected, but almost all of these animals, including most mobile animals, have larval dispersal stages in their life cycles. It is known that most marine invertebrates have larval stages and travel relatively long distances in a variety of aquatic habitats. Larvae that have 1–2 weeks and 0.5–3 months planktonic larval stages might be transported approximately 10 km and 100 km (see Barnes 1991), respectively. Success/failure and the timing of colonization in such sessile and sedentary species would depend on the availability of larval supplies from source populations.

However, some members of benthic animals on tidal flats do not have any larval stage in their life cycles, and the juveniles hatch with the same morphology as the adults. The marine mud snail *Batillaria attramentaria* (= *cumingi*) and the predatory sea snail *Euspira fortunei* are such representatives (Adachi and Wada 1999 for *B. attramentaria*; Tomiyama et al. 2011 for *E. fortunei*, see also Ohtsuki et al. 2016 in this volume) in northeastern Japan. This reproductive strategy, referred to as "direct development," would be a disadvantage in colonizing the restored and newly created habitats. Peracarid crustaceans such as amphipods and tanaidacea are also direct-developing animals, but their small body sizes allow the adults as well as juveniles to be carried for a distance by water currents when they swim into the water column from their bottom substrates. In this study, we focus on the colonization patterns of a variety of benthic animals in the restored tidal area of Otomo-ura in Hirota Bay, Iwate Prefecture, and compare them to those in another newly created tidal flat in the estuary of the Unosumai River, Otsuchi Bay, in the same prefecture (see Chap. 10) in order to clarify the common characteristics of colonization by benthic animals in both restored and newly created vacant habitats.

9.3 Otomo-ura: A Restored Vacant Tidal Flat

Otomo-ura is one of the restored intertidal areas in Hirota Bay of Rikuzentakata City, Iwate Prefecture, in northeastern Japan. The tidal flat of Otomo-ura had been reclaimed as a farmland 50 years prior. However, large tsunamis with inundation heights of about 13–15 m struck the area during the 2011 earthquake, resulting in the land being "returned to the sea" (Fig. 9.1; Matsumasa et al. 2013). Shortly after the earthquake, a breakwater was built to the north and to the south by piling up sandbags about 150 m landward from the former concrete breakwater that had been destroyed by the tsunamis. As a result, the intertidal area of about 45,000 m^2 remained, leading to the declaration by the government offices of Rikuzentakata City and Iwate Prefecture that the seaward area of Otomo-ura would be restored to an intertidal flat. The bottom substrate near both the northern and southern ends of the intertidal area ("peripheral areas") had been clayey sand, and that of the central part ("central area") between the peripheral areas has been cobbles with sandy patches. Unfortunately, almost all of the peripheral areas were reclaimed for road building during the summer of 2014, and about one third of the intertidal area has been lost.

9.4 Benthic Animals in Otomo-ura

9.4.1 Sampling Methods

To find benthic animals of a variety of mobility types and life forms (epifauna and infauna), we carried out both quantitative and qualitative samplings as a part of research programs by Biodiversity Center of Japan (2014, 2015). In 2013 and 2014, we quantitatively sampled the bottom sediment (to 20 cm) using a core sampler (diameter: 15.4 cm) and collected small epifaunal and infaunal animals using a 1 mm mesh sieve. Two points were placed in the central area, and one point was placed in each peripheral area of the northern and southern ends of the tidal flat. Then, three cores were taken for each point (Fig. 9.1). In the qualitative method, two persons observed for 15 min the bottom surface around the point where core samples had been taken and recorded sessile species on and mobile animals under cobbles and boulders. Observers also sometimes dug up the bottom using trowels to find deeper burrowers and infauna such as mud shrimp, *Upogebia* species, and a bivalve, *M. oonogai*. The observations and samplings were conducted during a low tide of the spring tide in August of 2013 and 2014. Similarly, we made observations and samplings during a low tide of the spring tide in August of 2012. However, in 2012 we observed benthic animals in 12 quadrats (25×25 cm^2) that were randomly placed for each point. For species that were difficult to identify in the field, we took photos of the quadrat surface and then collected the animals for species identification in the laboratory. In addition, we were unable to obtain any quantitative samples for the central area in 2012 because the bottom surface was covered with cobbles. However, many juveniles (shell length: ca. 2–5 cm) of the pacific oyster *C. gigas* were observed on those cobbles (Matsumasa et al. 2013, 2014), so we counted the oysters in eight quadrats (20×20 cm^2) during daytime low tide on 15 September 2012.

9.4.2 Changes in Species Richness

Eighty-five species of benthic animals were recorded using both quantitative and qualitative methods in the Otomo-ura tidal flat. The number of species observed in 2012, 2013, and 2014 was 27, 51, and 59, respectively. Our quantitative investigation in 2012 was performed in only the peripheral area because the bottom surface of the central area was obstructed by cobbles, as mentioned previously. In any event, the use of such a method in 2012 would have overlooked animals living under cobbles and boulders, in addition to deeper burrowers, that might not have been

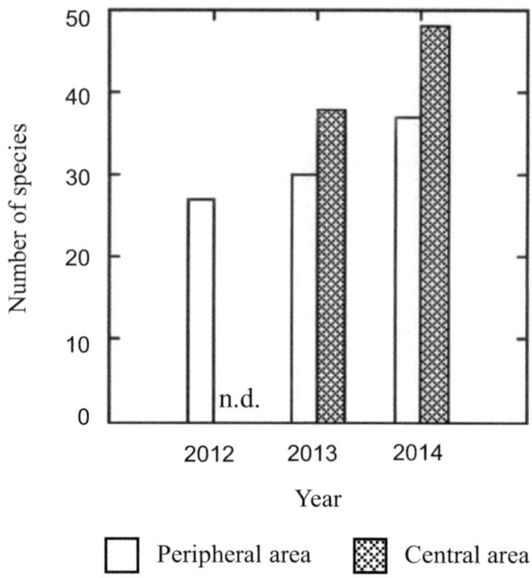

Fig 9.2 Yearly changes in the number of benthic animal species on the tidal flats of the peripheral and central areas of Otomo-ura

overlooked in 2013 and 2014 (see above). However, species richness increased in both the peripheral and central areas from 2013 to 2014 (Fig. 9.2), showing that the colonization by benthic animals was in progress in the restored tidal area. Since the number of species collected by the quantitative method for the peripheral area changed little (21, 20, 17 in 2012, 2013, and 2014, respectively), the increase in the total number of species there was mainly due to the increase of benthic animals collected by only the qualitative method, such as a chiton *Acanthochitona* sp. and gastropod species *Omphalius rusticus* and *Batillaria attramentaria*. In contrast, the number of species in the central area increased from 38 in 2013 to 48 in 2014, and both results included ten species collected by only the qualitative methods. The increase in the total number of species in the central area was mainly due to the increase of infaunal animals collected by the quantitative method, such as polychaetes *Glycera nicobarica* and *Pseudopolydora* spp. and small peracarid crustaceans *Ampithoe* sp. and *Melita* sp.

Figure 9.3 shows species discovery curves depicting the cumulative number of species recorded in ascending order for six cores. In the central area, the increase in species richness estimated by the quantitative method was obvious. The three curves for the peripheral area and the one curve for the central area in 2013 increased almost linearly with the increase in relative area, but only the curve for the central area in 2014 had an inflection point and leached 33 species, which was the highest value among the curves. This change in shape of curves for the central area is probably because the habitat heterogeneity of the area increased from 2013 to 2014. The bottom substrate in the peripheral area was clayey sand, while that of the central area consists mainly of cobbles associated with patchy deposition of fine sand. Sessile organisms, such as the oyster *C. gigas*, a barnacle *Amphibalanus amphitrite*, and algae, were observed on the cobbles. In particular, the oyster was conspicuous

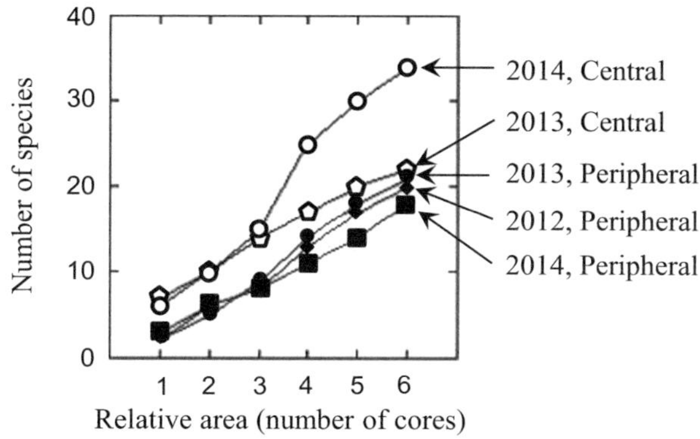

Fig 9.3 The relationship between species richness and relative area in quantitative samples from the peripheral and central areas of Otomo-ura

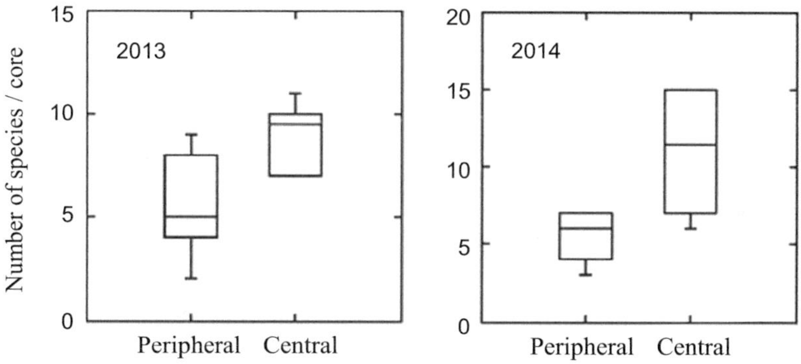

Fig 9.4 Number of benthic animal species collected using the quantitative method in the peripheral and central areas of Otomo-ura

in the lower part of the central area, and the estimated density (mean ± SE) was 540.75 ± 37.475 individuals per m^2 ($n=8$) on 15 September 2012. The growth of sessile organisms, including *C. gigas*, could provide heterogeneous habitat by causing the deposition of sand for small animals, which responded specifically to the microhabitats provided by the primary substrates (i.e., cobbles and deposited sand) and secondary substrates (i.e., sessile organisms) (Matsumasa 1994). This might also affect the species richness on a finer scale since the number of species per core was significantly higher in the central area than in the peripheral area in both 2013 and 2014 (*U*-test, $P<0.05$; Fig. 9.4).

9.4.3 *Species Composition*

Although species richness in the peripheral area was lower than that in the central area, species composition indicates that the peripheral area supplied a unique habitat to benthic animals that had not been found in the central area. The reed-marsh mud crab *Helice tridens* inhabited the upper part of the peripheral area. The estuarine ragworm *Hediste* sp. and the acorn barnacle *Fistulobalanus albicostatus* were also typical inhabitants of the peripheral area. These species are brackish water representatives, so the freshwater inflows in the peripheral area might have been an important factor shaping the species composition there. On the other hand, major species that were typical to the central area, such as the mud snail *Nassarius* (*Hima*) *hypolius*, the tellin clam *Macoma incongrua*, the caridean shrimp *Crangon* sp., and the pebble crab *Gaetice depressus*, are more polyhaline. Of the 85 species that were found in the Otomo-ura tidal flat in this study, 17 were observed only in the peripheral area, and 23 were observed only in the central area. The peripheral and central areas are both therefore considered to be important in supporting the biodiversity of tidal flats in this region. However, as mentioned above, almost all of the peripheral area was reclaimed for road building during the summer of 2014.

9.4.4 Abundance of Major Infaunal Animals

The reclaimed peripheral area had included the burrowing area of the reed-marsh mud crab *H. tridens*. The reclamation might have some negative effects on this crab population. We adopted the qualitative method to find animals under cobbles and boulders, highly mobile animals, and deeper burrowing animals, including the mud crab, which is the important component linking the upper intertidal burrowing area with the lower intertidal flat (Takeda and Kurihara 1987; Takeda et al. 1988; Kurihara et al. 1988), but our method could not evaluate the population densities of those species. Our methods also could not clarify the change in abundance of one of the most dominant species, the Pacific oyster *C. gigas*, which is an important fishery resource (Matsumasa et al. 2013, 2014, 2015). Although our sampling methods had these limitations, the abundances of some benthic animals that had been collected by the quantitative method using a core sampler could be compared among years and between areas. The Manila clam *Ruditapes philippinarum* is one of five such dominant species. Although the effect of "year" on its abundance was marginal [one-way *ANOVA* for log (N+1) transformed data; $P=0.0575$], it still increased from 2012 to 2013 and remained at a similar level in 2014 in the peripheral area (Fig. 9.5). Its abundance in the central area (Fig. 9.6) also did not differ significantly (alpha level: 0.05) between 2013 and 2014, and comparisons of its abundance between the peripheral and the central areas for each year did not show any significant difference. However, the density of the polychaete *Heteromastus* species clearly increased from 2012 to 2014. The effect of year was significant [one-way *ANOVA* for log (N+1); $P=0.0103$], and the Bonferroni test indicated that the difference between its abundance in 2012 and 2014 was significant ($P=0.0089$) in the peripheral area. Also, in the central area (Fig. 9.6), the abundance of the polychaete species showed a significant difference between years (t-test, $P=0.0253$) and was not significantly different from those in the peripheral area. Another polychaete *Perinereis* species did not show any significant differences in its abundance among

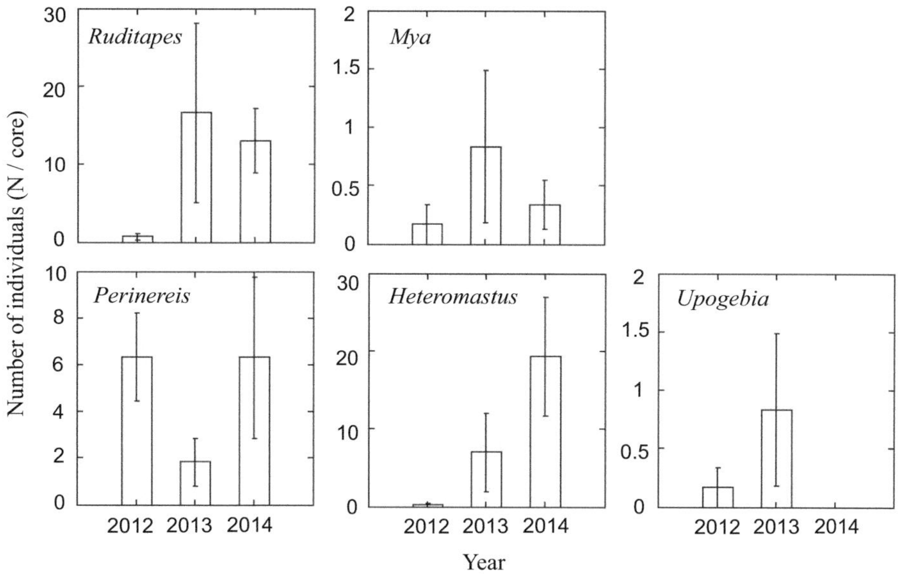

Fig 9.5 Changes in density for five dominant species collected using the quantitative method in the peripheral area of Otomo-ura

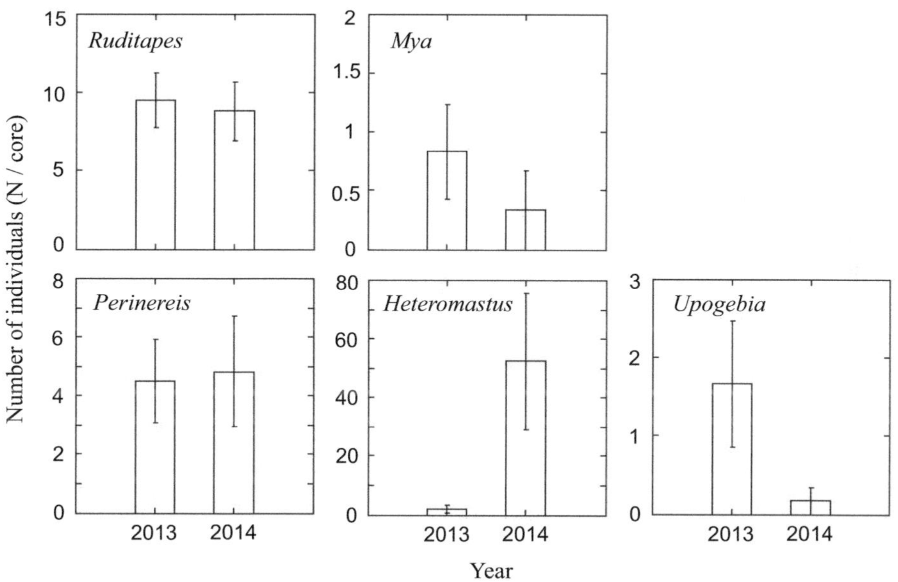

Fig 9.6 Changes in density for five dominant species collected using the quantitative method in the central area of Otomo-ura

the 3 years or between the two areas and appeared relatively constant in core samples for every year and area. We could not find any significant differences in the abundances of the bivalve *Mya* (*Arenomya*) *arenaria oonogai* and the decapod crustacean *Upogebia* species among the years or between the areas, but their abundances tended to decrease from 2013 to 2014 (Figs. 9.5 and 9.6). Since the adults of the relatively large infaunal bivalve and the crustacean deep burrower are difficult to catch using our core sampler (see Cross et al. 2012 for *Mya*; Kinoshita 2002, Kinoshita and Itani 2005 for *Upogebia*), more specific investigations are needed for these large benthic animals in the future.

On the whole, the bottom sediment of the restored tidal flat, which initially would have been nearly lifeless substrate, has been inhabited by infaunal animals, and some dominants increased in number after 2012. No significant spatial and temporal differences in abundance were detected for the five dominant species, indicating that both peripheral and central areas would have been equally suitable for colonization by infaunal animals had adequate substrates been present.

9.5 Colonization by the Direct-Developing Gastropod *Batillaria attramentaria*

As mentioned earlier, colonization by direct-developing species such as the marine mud snail *B. attramentaria* and the predatory sea snail *Euspira fortunei* would require more time than colonization by species that have larval stages in their life cycles. Actually, neither the mud snail nor the predatory snail were observed until August 2013 in the tidal flat of Otomo-ura or in another newly created tidal flat in the estuary of the Unosumai River, Otsuchi Bay (39°20′19.79″ N, 141°53′45.53″), which is situated about 40 km north of Otomo-ura (see Chap. 10). However, the mud snail *B. attramentaria* was first found in August 2014 at both the peripheral and central areas in Otomo-ura (Matsumasa et al. 2015). To inquire into the source of the snail, we tried genotyping ten individuals of this species collected at Otomo-ura with 14 microsatellite loci (Itoh et al. 2013; Miura et al. 2014) from nuclear DNA and the nucleotide sequences of a 1020 bp region of the mitochondrial gene for cytochrome *c* oxidase subunit I (COI) (Matsumasa et al. 2015). The result for the nucleotide sequence of the COI gene showed that all of the individuals from Otomo-ura exhibited the haplotypes of the "Tsushima group," which are commonly found in northern Japan (Kojima et al. 2004). A phylogenetic tree by the neighbor-joining method using Nei's D_A with 15 populations from various places in northern Japan indicated that the population of Otomo-ura was differentiated from the other populations but was closest to that from Ofunato Bay, north of Hirota Bay. The results of assignment tests based on microsatellite data also showed that the genetic distinctness of the Otomo-ura population from the other populations was very clear except for the two populations of Ofunato Bay and Kesen-numa Bay, north and south of Hirota Bay, respectively. Therefore, the Otomo-ura mud snails might be immigrants from the source populations in Hirota Bay or the neighboring bays. Since no data was available for the populations in these areas before the tsunami, more detailed investigations into the populations in these areas are needed to help clarify where the direct-developing mud snail came from and how they reached the restored tidal flat.

The alien predatory snail *E. fortunei*, another direct-developing species, has been introduced with its prey Manila clam *R. philippinarum*, which is an important fishery resource, to northern Japan (Okoshi 2004) and has damaged the clam populations. Fortunately, this predator, which was unintentionally introduced to the Sanriku region, has been found neither in the restored tidal flat of Otomo-ura nor the newly created flat in the Unosumai River estuary. On these tidal flats, the populations of the prey Manila clam are free from predation by the snail *E. fortunei*. It is important not to transport the clam to these tidal flats to keep these habitats free from the alien predator.

9.6 Colonization by the Protozoa *Perkinsus* Parasitizing the Manila Clam

Just like in the above predator-prey relationships between the predatory snail *E. fortunei* and the prey clam *R. philippinarum*, parasite-free populations are also likely to develop at tidal flats that have been restored and newly created by the tsunami if only larvae of the clam reach those vacant habitats. The host-parasite relationship between the clam *R. philippinarum* and the protozoan parasite *Perkinsus* is one of the suitable subjects of researches to examine this possibility. Since the prevalence of the protozoan parasite is thought to be caused mainly by transporting infected clams (Hamaguchi et al. 2002), *Perkinsus*-free populations are likely to develop at restored and newly created tidal flats when only larvae of the clam reach such habitats. *Perkinsus* parasites infect the clam after its settlement, not larva in the planktonic stage. Therefore, from April to June 2014, we investigated the prevalence and infection intensity of the protozoan parasite in the Manila clam on nine tidal flats including the flat in Otomo-ura and the other two tidal flats newly created by the tsunami and subsidence using Ray's fluid thioglycollate medium assay (Choi et al. 1989; Umeda and Yoshinaga 2012, Kinoshita and Matsumasa submitted). The prevalence was less than 30 % in both the restored Otomo-ura flat and the two newly created tidal flats, and the values were much lower than those in the previously existing tidal flats (>95 %) where clam seeds had been released before the disaster. However, the *Perkinsus* infection in clams was confirmed for all of the restored and newly created tidal flats. The results indicate that the zoospores of *Perkinsus* could reach the restored and newly created tidal flats from the neighboring preexisting tidal flats because *Perkinsus* is thought to be transmitted mainly via zoospores, which are released from the dead host to seawater (Auzoux-Bordenave et al. 1995).

Park and Choi (2001) reported that the infection intensity was much lower in small clams than in large clams in Korean waters. In contrast, Kinoshita and

Matsumasa (submitted) found a significant negative correlation between the shell length of the clam and infection intensity on the tidal flat in Otomo-ura. These organisms should be monitored to deepen our understanding on effects of the parasite on calm populations.

9.7 Lessons from the Restored and Newly Created Tidal Flats

We could do little in the face of the disaster, but even our limited investigations into the colonization patterns of benthic animals on the restored and newly created tidal flats give us some insight into the responses of benthic animals to the disaster and how we should try to conserve them. First, fortunately, colonization of vacant habitats by benthic animals occurred much more quickly than was expected. When the tidal area in Otomo-ura was observed on 13 October 2011, 7 months after the tsunamis caused by the Great East Japan Earthquake, many juveniles of the pacific oyster *Crassostrea gigas*, a very important fishery species, had already covered the surfaces of the remaining embankment and cobbles on the flat (Matsumasa et al. 2013, 2014, 2015). The abundance of juvenile pacific oyster on the tidal flat was estimated to be 540.75 ± 37.475 individuals per m^2 ($n = 8$) on 15 September 2012. Larval supplies from source populations around Otomo-ura during summer might have brought about the quick and conspicuous settlement of the restored tidal flat. The results of our quantitative and qualitative samplings showed a constant colonization by a variety of benthic animal species. Although the transportation of animals by the tsunamis is an interesting possible way to break into the vacant, restored habitat (i.e., adults had been transported by the tsunamis into new habitats where they reproduced), the settlement of larva via their passive migrations would be a more important route, at least for sedentary infaunal bivalves, such as *Mya* (*Arenomya*) *arenaria oonogai* and *Laternula* (*Exolaternula*) *marilina*. This is because the adults of those species could not survive the huge physical disturbance of tsunamis: they move little horizontally and have difficulty submerging themselves in sand and mud once they have been removed from the sediment. Also, the shell of *L. marilina* is easily broken by even slight external pressure. Therefore, it is likely that source populations of such species were able to escape the disturbance to their habitat and supply their larvae to Otomo-ura. The complex topography of the rias might also be advantageous for harboring such populations, though its topography makes Sanriku prone to wave amplification during tsunamis.

The quick colonization of filter-feeding species, including the oyster *C. gigas*, on restored and newly created tidal flats points may have prompted pollution control by removing suspended matter (e.g., organic particles, planktons, etc.) and nutritional cycling by removing such organic matters from the water column and by excreting feces, pseudo-feces, and ammonia into the environments. The Pacific oyster, together with the Manila clam *Ruditapes philippinarum* in Otomo-ura, must have played an important role in such water pollution control and nutritional cycling in

Hirota Bay, particularly when almost all of the cultured oysters and sandy habitats of the clam in Hirota Bay were lost by the tsunamis and subsidence. In Matsukawa Lagoon, Soma, Fukushima Prefecture (Suzuki 2016 in this volume), these two bivalve species of filter feeders absorbed as much nitrogen as would be found in the whole water of the lagoon a few days before the tsunamis (Kohata et al. 2003). As mentioned above, the abundance of juvenile oyster (shell length: ca 2–5 cm) in the lower part of the Otomo-ura central area was estimated to be about 500 individuals per square meter in September 2012. Based on an extrapolation from the equation by Kobayashi et al. (1997), if we assume that the filtration rate of an individual of the juvenile oyster is about 0.5 l per hour at 20 °C and that the individuals were uniformly placed in the lower part of the central area of 2625 m^2, we can estimate roughly that juvenile oysters were filtering 7875 m^3 of water per day (high tide: 12 h). This estimation may be an underestimation because there were a lot of oysters on the surface of the remaining embankment and the boulders in the peripheral area, and the vertical distribution of these oysters is biased in favor of the lower intertidal to upper subtidal zones (i.e., the feeding duration was >12 h). It would therefore be beneficial to clarify the contribution of filter-feeding species, including the oyster *C. gigas*, to water pollution control and nutritional cycling in Hirota Bay to gain a better understanding of their ecological roles in the restored tidal flat in Otomo-ura.

Species composition of Otomo-ura differed between the peripheral and central areas as shown above. Therefore, both areas contributed to the species diversity of benthic animals in this flat ecosystem and should have been preserved (see 9.4). Unfortunately, almost all of the peripheral area has been reclaimed from the sea for building roads, and the upper intertidal habitat, including the burrowing area of the mud crab *Helice tridens*, has been lost. Our results suggested that sedimentation in the cobble field and the growth of sessile organisms on cobbles and boulders were important for increasing species richness because they enhanced habitat heterogeneity. Because the reclamation of the peripheral areas have likely altered water currents, wave power, sediment stability, and the growth of sessile organisms, it is critically important that we monitor the effects of human activities on these benthic animal communities and put these results to wise use for the future management of intertidal ecosystems.

In the restored tidal flat of Otomo-ura, none of the direct-developing gastropod *Batillaria attramentaria* had been observed before August 2013, and another direct-developing predatory snail *Euspira fortunei* has not been found yet (see 9.5). Genetic analyses indicate that the mud snail *B. attramentaria* of Otomo-ura has probably come from populations in nearby Hirota Bay or in the neighboring bays. This fact suggests a possibility that the unwelcome alien predator *E. fortunei* might immigrate to this area in the near future. To prevent colonization by this alien predator on the Otomo-ura flat, where the Manila clam *R. philippinarum* is considered to be a potential fishery resource, it is essential not to bring the Manila clam to Otomo-ura or to the other habitats in Hirota Bay and the neighboring bays, because the predatory snail *E. fortunei* is usually introduced unintentionally with its major prey, the Manila clam. Another unfavorable cryptic inhabitant, *Perkinsus* species, the protozoan parasite of the Manila clam, was found in 2014, though its prevalence was lower than it was in the previously existing tidal flats (see 9.6). In order to diminish the harmful effects of *Perkinsus* on the clam population in Otomo-ura, simultaneous management of all of the populations in Hirota Bay, including Otomo-ura, is critical because the probability of colonization/invasion of the protozoan parasite into new habitats is a function of the prevalence and infection intensity of the parasite in source populations.

In conclusion, it is our responsibility to transmit the lessons from low-frequency, major disturbances to those who might be susceptible to such natural disturbances in the future.

Acknowledgments We are very grateful to the staff of the Iwate Fisheries Technology Center, Iwate Prefecture, and the Miyako Fisheries Research and Development Center, Iwate Prefecture, for their kindness and cooperation. The authors also thank Dr. Jotaro Urabe for his valuable comments on our manuscript. The study was partly conducted as a project of "The Ecosystems Monitoring Survey of the Pacific Coastal Areas of the Tohoku Region" by the Biodiversity Center of Japan, Nature Conservation Bureau, Ministry of the Environment. Finally, the authors pray for the victims' souls, that they may rest in peace, and hope for people's safety from natural disturbances in the future.

References

Adachi N, Wada K (1999) Distribution in relation to life history in the direct-developing gastropod *Batillaria cumingi* (Batillariidae) on two shores of contrasting substrata. J Molluscan Stud 65:275–287

Auzoux-Bordenave S, Vigario AM, Ruano F, Domart-Coulon I, Doumenc D (1995) In vitro sporulation of the clam pathogen *Perkinsus atlanticus* (Apicomplexa, Perkinsea) under various environmental condition. J Shellfish Res 14:469–475

Barnes RSK (1991) Reproduction, life history and dispersal. In: Barnes RSK, Mann KH (eds) Fundamentals of aquatic ecology, 2nd edn. Blackwell Scientific Publication, Oxford, pp 145–171

Biodiversity Center of Japan (2014) The ecosystem monitoring survey of the Pacific coastal areas of the Tohoku region in FY2013. Biodiversity center of Japan, Nature conservation bureau, Ministry of the environment (in Japanese). http://www.shiokaze.biodic.go.jp/25sokuhou.html. Accessed on 14 June 2015

Biodiversity Center of Japan (2015) The ecosystem monitoring survey of the Pacific coastal areas of the Tohoku region in FY2014. Biodiversity center of Japan, Nature conservation bureau, Ministry of the environment (in Japanese). http://www.shiokaze.biodic.go.jp/26sokuhou.html. Accessed on 14 June 2015

Choi K-S, Lewis DH, Powell EN, Ray SM (1989) The energetic coat of *Perkinsus marinus* parasitism in oysters: quantification of the thioglycollate method. J Shellfish Res 8:125–131

Cross ME, Lynch S, Whitaker A, O'Riordan RM, Culloty SC (2012) The reproductive biology of the softshell clam, *Mya arenaria*, in Ireland, and the possible impacts of climate variability. J Mar Biol 2012, Article ID 908163:9. http://dx.doi.org/10.1155/2012/908163

Hamaguchi M, Sasaki M, Ushuki H (2002) Prevalence of a *Perkinsus* protozoan in the clam *Ruditapes philippinarum* in Japan. Jpn J Benth 57:168–176 (in Japanese with English abstract)

Henmi Y, Kobayashi S, Yamaguchi J, Hashiguchi M (2014) Recruitment and movement of the hard clam *Meretrix lusoria* in a tidal river of northern Kyushu, Japan. Fish Sci 80:705–714

Ikhwanuddin M, Nurfaseha AH, Abol-Munafi AB, Shabdin ML (2012) Movement patterns of blue swimming crab *Portunus pelagicus* in the Sarawak coastal water, South China Sea. JSSM 7:1–8

Itoh H, Kamimura S, Takeshima H, Kojima S (2013) Isolation and characterization of microsatellite loci in the Asian horn snail *Batillaria attramentaria* (Mollusca, Gastropoda). Mol Ecol Res 2013 (Molecular Ecology Resources Primer Development Consortium et al. Permanent Genetic Resources added to Molecular Ecology Resources Database 1 February 2013; 31 March 2013. Mol Ecol Res 13:760–762)

Kanaya G, Suzuki T, Kinoshita K, Matsumasa M, Yamada K, Seike K, Okkoshi K, Miura O, Nakai S, Sato-Okoshi W, Kikuchi E (accepted) Disaster-induced changes in coastal wetlands and soft-bottom habitats – an overview on 2011 tsunami and Great East Japan Earthquake. Biol Int

Kinoshita K (2002) Burrow structure of the mud shrimp *Upogebia major* (Decapoda: Thalassinidea: Upogebiidae). J Crustac Biol 22:474–480

Kinoshita K, Itani G (2005) Interspecific differences in the burrow morphology between the sympatric mud shrimps, *Austinogebia narutensis* and *Upogebia issaeffi* (Crustacea: Thalassinidea: Upogebiidae). J Mar Biol Assoc UK 85:943–994

Kobayashi M, Hofmann EE, Powell EN, Klinck JM, Kusaka K (1997) A population dynamics model for the Japanese oyster, *Crassostrea gigas*. Aquaculture 149:285–321

Kohata K, Hiwatari T, Hagiwara T (2003) Natural water-purification system observed in a shallow coastal lagoon: Matsukawa-ura, Japan. Mar Pollut Bull 47:148–154

Kojima S, Hayashi I, Kim D, Iijima A, Furota T (2004) Phylogeography of an intertidal direct-developing gastropod, *Batillaria cumingi*, around the Japanese Islands. Mar Ecol Prog Ser 276:161–172

Kurihara Y, Sekimoto K, Miyata M (1988) Wandering behaviour of the mud-crab *Helice tridens* related to evasion of cannibalism. Mar Ecol Prog Ser 49:41–50

Matsumasa M (1994) Effect of secondary substrate on associated small crustaceans in a brackish lagoon. J Exp Mar Biol Ecol 176:245–256

Matsumasa M, Obata F, Tohyama K (2013) Life on the Sanriku coast: how estuaries are recovering from the 2011 tsunami. PuLSe: News Linn Soc Lond Bot 19:6–7

Matsumasa M, Kishi T, Endo T (2014) Collecting oyster spats in the northern part of Sanriku region for establishing a robust culture system against Tsunami and infectious diseases. Asian Marine Biology Symposium 2014, Oct 1–4, Jeju Korea: 78

Matsumasa M, Kinoshita K, Itoh H, Kojima S (2015) Between tides on Sanriku coast: the vulnerability and robustness to catastrophic, natural disturbances. DNA Polymorphism 23:9–16 (in Japanese)

Miura O, Makino W, Urabe J (2014) Characterization and cross-species utility of polymorphic microsatellite markers in the Asian mud snails *Batillaria attramentaria* and *B. multiformis*. Conserv Genet Resour 6:461–463

Nakamura Y (2013) Secretion of a mucous cord for drifting by the clam *Meretrix lusoria* (Veneridae). Plankton Benth Res 8:31–45

Ohtsuki H, Suzuki T, Kinoshita K, Kanaya G, Hirama T, Sato S, Shibata K, Okoshi K and Urabe J (2016) Genetic structures of Euspira fortunei metapopulation in the north east coast of Japan after tsunamis of the 3.11 Great East Japan Earthquake. In: Urabe, Nakashizuka (eds) Ecological impacts of tsunamis on coastal ecosystems: lessons from the Great East Japan Earthquake, Springer, pp 209–222

Okoshi K (2004) Alien species introduced with imported clams: the clam-eating moon snail *Euspira fortunei* and other unintentionally introduced species. Jpn J Benth 59:74–82 (in Japanese with English abstract)

Park K-I, Choi K-S (2001) Spatial distribution of the protozoan parasite *Perkinsus* sp. found in the Manila clams, *Ruditapes philippinarum*, in Korea. Aquaculture 203:9–22

Suzuki T. (2016) Impacts of the tsunami disaster caused by the 3.11 Great East Japan Earthquake on the benthic communities and their recovery in tidal flats of Matsukawaura Lagoon, Soma,

Japan. In Urabe, Nakashizuka (eds) Ecological impacts of tsunamis on coastal ecosystems: lessons from the Great East Japan Earthquake, Springer, pp 105–116

Takeda S, Kurihara Y (1987) The effects of burrowing of *Helice tridens* (De Haan) on the soil of a salt-marsh habitat. J Exp Mar Biol Ecol 113:79–89

Takeda S, Matsumasa M, Kurihara K (1988) Seasonal changes in the stomach contents of the burrowing mud-crab, *Helice tridens* (De Haan). Bull Mar Biol Station Asamushi Tohoku Univ 18:77–86

Tomiyama T, Suzuki T, Sato T, Kato Y, Kameiwa S, Sugibayashi Y, Okoshi K (2011) Unintentional introduction and the distribution of the nonindigenous moonsnail *Euspira fortunei* in Matsukawaura Lagoon, Japan. Bull Jpn Soc Sci Fish (= Nippon Suisan Gakkaishi) 77:1020–1026, in Japanese with English abstract

Umeda K, Yoshinaga T (2012) Development of real-time PCR assays for discrimination and quantification of two *Perkinsus* spp. in the Manila clam *Ruditapes philippinarum*. Dis Aquat Organ 99:215–225

Chapter 10
Effects of the Great East Japan Earthquake on Intertidal Macrobenthos in Iwate Prefecture

Kyoko Kinoshita and Masatoshi Matsumasa

Abstract To determine the effects of the Great East Japan Earthquake and resulting tsunamis on the distribution of marine benthic macrofauna throughout Iwate Prefecture, the fauna of three tidal flats was compared before and after this catastrophic event. Because this prefecture has a history of tsunami events, floodgates and seawalls were established along its coastline following the tsunami in 1933 as a disaster prevention strategy. Consequently, the intertidal environment throughout the region had been already modified before the 2011 tsunamis. Despite this former loss of habitat, a number of benthic invertebrates occurred in the tidal areas before the intertidal disturbance due to the 2011 tsunamis. More importantly, 52–76 % of taxa of benthic invertebrates previously recorded in this region were found in the tidal area after the tsunamis' disturbance. Additionally, several taxa that are considered threatened (according to the Japanese Association of Benthology) appeared in these and newly created tsunami-flooded habitats after 2011. Our data suggest that the marine macrobenthos in this region, including these threatened taxa, were tolerant of the tsunami disturbance or comprised opportunistic taxa capable of rapidly colonizing new habitats. However, the recovery patterns of benthic invertebrates varied across sites: remarkable reduction of tidal flat decreased its biological diversity, but creation of new tidal habitats strikingly increased taxa richness. The reconstruction of coastal structures should be planned considering preservation of the remaining and newly created tidal flats. Future disaster risk management should take into consideration both the maintenance costs of these structures and the effects they have on biodiversity and ecosystem functioning in intertidal flats.

K. Kinoshita (✉)
Organization of Revitalization for Sanriku-region, Iwate University,
3-18-8 Ueda, Morioka, Iwate 020-8550, Japan
e-mail: kinoshita.kyoko@mbn.nifty.com

M. Matsumasa
Department of Biology, Center for Liberal Arts and Sciences, Iwate Medical University,
2-1-1 Nishitokuta, Yahaba, Iwate 028-3694, Japan

J. Urabe, T. Nakashizuka (eds.), *Ecological Impacts of Tsunamis on Coastal Ecosystems*, Ecological Research Monographs,
DOI 10.1007/978-4-431-56448-5_10

Keywords Benthic invertebrate • Disaster management conservation • Iwate • Marine biodiversity • Seawall • Tidal flat

10.1 Introduction

Tsunamis, most notably those of 1896 (Meiji Sanriku), 1933 (Showa Sanriku), and 1960 (Chile), have repeatedly struck the northwestern Pacific Ocean coastal areas of Japan, disturbing ecosystems, particularly those in the coastal terraces and ria coasts of Iwate Prefecture (Fig. 10.1). The high waves and flooding associated with catastrophic tsunami events have greatly affected local towns and villages along the Iwate coast, resulting in countermeasures such as construction of floodgates and seawalls to limit future damage by tsunamis. Although these structures may have acted to save lives, they also encroached upon the seashore area and deprived tidal habitats of marine organisms.

Tsunamis and its countermeasures such as construction of seawalls are not the only causes of disturbance to marine organisms in this region. For example, the Tsugaruishi and Orikasa Rivers were both routinely dredged to facilitate upstream migration of commercially important chum salmon (Sakai et al. 1991). The nature and frequency of these and other disturbances, natural or otherwise, have likely influenced not only the coastal geography but also the marine organisms in these regions.

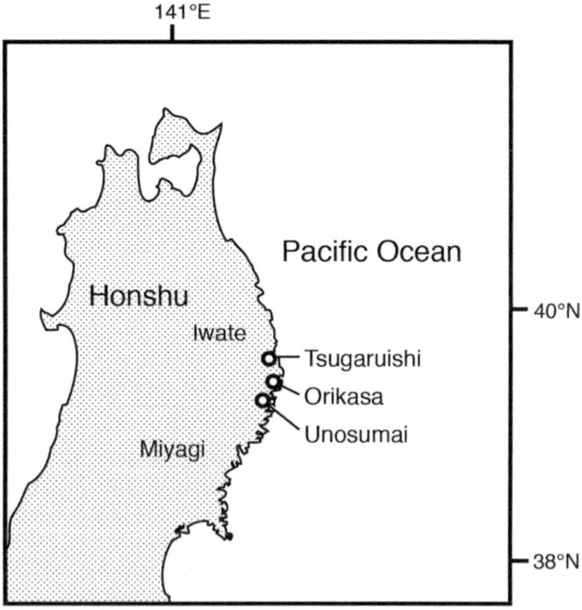

Fig. 10.1 Map showing study area

In this study, we described the tidal flat communities in Iwate Prefecture before and after the Great East Japan Earthquake in 2011. We first demonstrated physical impacts of tsunamis caused by the Great East Japan Earthquake to the coast of the Iwate Prefecture and then evaluated the effects of the tsunamis' disturbance on macrobenthos by comparing taxa composition of benthic communities before and after the disturbance at three intertidal flats. Finally, we discussed potential effects of future seawall constructions on the coastal communities and possible *alternative* strategies to conserve marine biodiversity in this region.

10.2 Survey Sites and Method

Surveys were conducted in tidal flats of Tsugaruishi (39°35′N, 141°56′E), Orikasa (39°26′N, 141°57′E), and Unosumai (39°19–20′N, 141°53′E) estuaries. The largest tidal flat in Iwate Prefecture is located in the Tsugaruishi area. The Orikasa area was a renowned clam-gathering field, and the Unosumai area was surrounded by sandy beaches and 6.4-m-high seawalls, with seawater mixing slightly with river water, before the tsunamis.

The pre-tsunami surveys were conducted in tidal flats of the Tsugaruishi and Orikasa areas in August 2002 and in tidal flats of the Unosumai area in September 2003, as a part of the 7th National Survey on the Natural Environment (Biodiversity Center of Japan 2007). The post-tsunami surveys were performed in July or August from 2011 to 2014 at the Tsugaruishi and Orikasa areas and in August from 2012 to 2014 at the Unosumai area. During each survey, one to three transect lines were established from the top to the bottom of the tidal flat, along which one to three sampling stations were set. The number of transect lines and sampling stations was adjusted according to the surface area of the tidal flat. The Great East Japan Earthquake in 2011 caused large tsunamis and land subsidence, with waves reaching heights of 11.2 m in Tsugaruishi, 8.0 m in Orikasa, and 14.7 m in Unosumai (Haraguchi and Iwamatsu 2013). Accordingly, in some cases, we could not use the transect lines established in the pre-tsunami surveys. In such cases, we established new transect lines for the post-tsunami surveys.

The sampling in both the pre- and post-tsunami surveys was made by at least two researchers during daytime low-tide periods. In each survey, a quadrat of 25 m^2 was placed on the sediment at each station. For each quadrat, macro-epibenthos (>5 mm) found at the sediment surface by naked eye during a 5-min observation conducted by the two researchers was identified and recorded. Then, one researcher dug up the sediment in the quadrat to a 20-cm depth using a shovel (head length and width: 345 and 139 mm, respectively) and the other researcher(s) collected the macroinfauna observed during a 10-min period and then identified and recorded the specimens.

Prior to 2011, sediments were visually characterized as gravel, sand, or mud. After 2012, sediment cores were taken at each sampling station using a plastic pipe (inside diameter, 5 cm) to a 5-cm depth. In the laboratory, the samples were wet-sieved through a 2-mm and a 0.063-mm mesh, dried, and weighed. Sedimentary properties were described by weight proportion of three grain sizes, i.e., gravel (>2 mm), sand (0.063–2 mm), and mud (<0.063 mm).

Salinity was measured at the sampling areas during sampling in the post-tsunami surveys after 2012 using a salinity refractometer (S/Mill-E, Atago, Tokyo, Japan).

10.3 Results

10.3.1 Tsugaruishi Area

Prior to the tsunamis in 2011, eight stations were monitored at the Tsugaruishi area (Table 10.1; Fig. 10.2, Stations 1–8). Because three of those stations, which are located along the northernmost coastline, were constantly submerged after the tsunamis (Fig. 10.2, Stations 1–3), three new stations were established (Fig. 10.2, Stations 9–11) and monitored through 2012 and 2013, except for temporally submerged station (Fig. 10.2, Station 9) when we carried out our surveys. In 2014, since five of these stations were lost due to seawall reconstruction (Fig. 10.2, Stations 7–11), six new stations were established (Fig. 10.2, Stations 12–17). Consequently, only two stations could be monitored continuously throughout the surveys conducted in this study (Fig. 10.2, Stations 4 and 5). Sediment in the Tsugaruishi area consisted of gravel and sand (Fig. 10.3), and new muddy stations were lost due to seawall reconstruction in 2014. Salinity in this area ranged from 14 % to 27 % (Table 10.1).

Table 10.1 Number of transect lines and stations and salinity at each survey area

		Number of				
Site	Year	Transect lines (lines)		Stations (Stns)		Salinity (‰)
Tsugaruishi	2002	3	(a–c)	8	(1–8)	nd
	2011	3	(b–d)	8	(4–11)	nd
	2012	3	(b–d)	6	(4–5, 7–8, 10–11)	27
	2013	3	(b–d)	7	(4–5, 7–11)	14
	2014	3	(b, e–f)	8	(4–5,12–17)	24
Orikasa	2002	3	(a–c)	8	(1–8)	nd
	2011	1	(c)	2	(6, 8)	nd
	2012	1	(c)	2	(6, 8)	27
	2013	1	(c)	1	(8)	24
	2014	1	(c)	1	(8)	16
Unosumai	2003	3	(a–c)	7	(1–7)	nd
	2012	2	(d–e)	3	(8–10)	15–25
	2013	2	(d–e)	3	(8–10)	14–17
	2014	2	(e–f)	3	(10–12)	8–18

nd no data

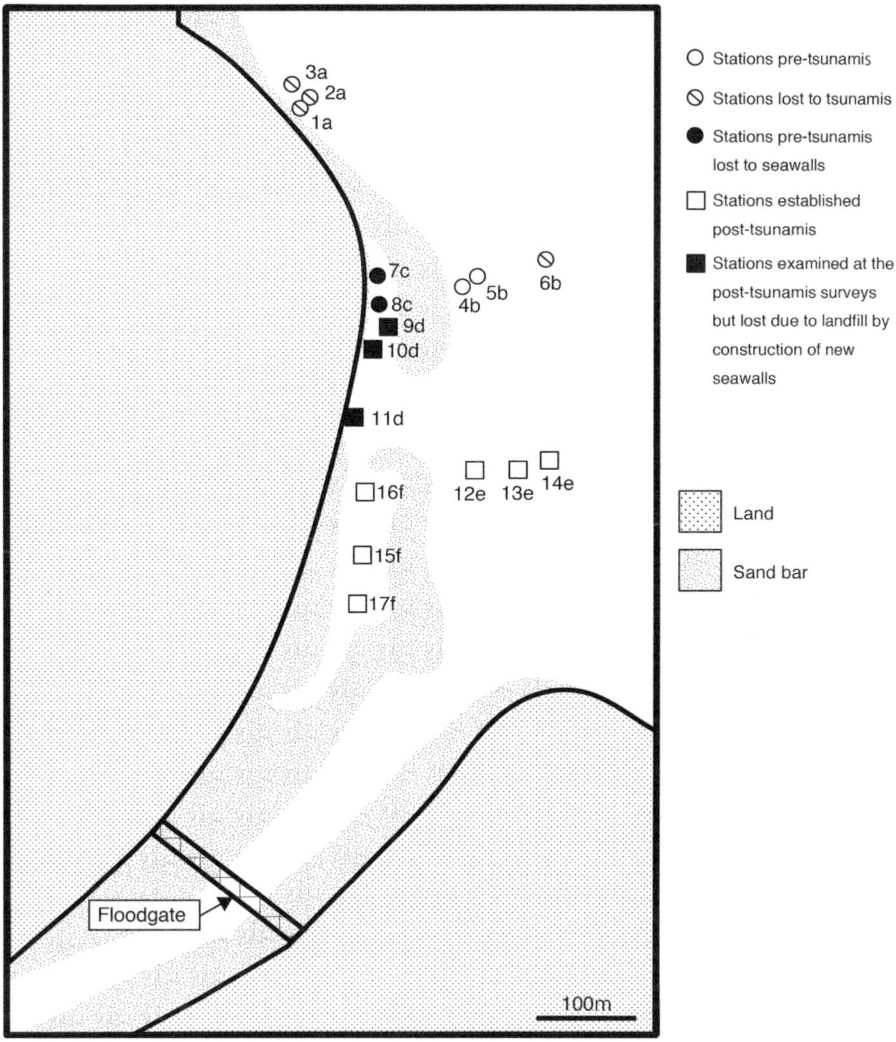

Fig. 10.2 Map showing census transect lines and survey stations in Tsugaruishi

Before the Great East Japan Earthquake, 39 benthic-invertebrate taxa were recorded from the Tsugaruishi area (Table 10.2). Following the tsunamis, 51 taxa were identified during the 4-year survey (23–32 taxa per year), including *Hemigrapsus takanoi*, a recently described cryptic crab species (Asakura 2006). Of these, 30 taxa had been recorded in the pre-tsunami survey; thus, 76.9 % of taxa recorded prior to the tsunamis were confirmed in surveys following the tsunamis.

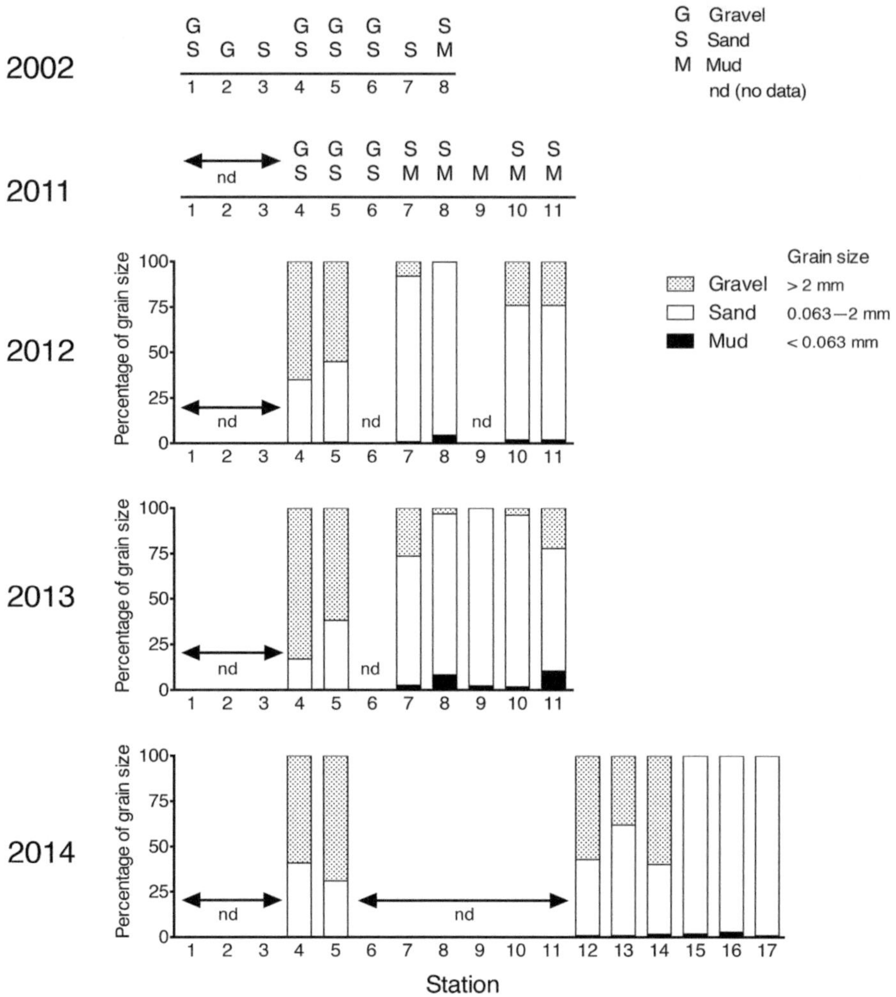

Fig. 10.3 Change in sedimentary characteristics before and after the 2011 tsunami disaster at Tsugaruishi

10.3.2 Orikasa Area

The tsunamis and land subsidence considerably reduced the tidal flat at the Orikasa area. Prior to the tsunamis, surveys were conducted at eight stations, six of which were permanently submerged following the tsunamis (Table 10.1; Fig. 10.4, Stations 1–5, 7). From 2013, surveys were conducted only at one station (Fig. 10.4, Station 8). Sediments at Orikasa consisted mainly of gravel (Fig. 10.5), with more sandy stations being submerged or gone. Salinity in this area ranged from 16‰ to 27‰ (Table 10.1).

Table 10.2 Macrobenthos at Tsugaruishi before and after the 2011 tsunamis: + present, – absent

Phylum	Class	Taxa	Year				
			2002	2011	2012	2013	2014
Cnidaria	Anthozoa	*Synandwakia hozawai*	+	–	–	–	–
Nemertea	Anopla	*Nemertea* spp.	–	–	+	–	–
Mollusca	Gastropoda	*Nipponacmea habei*	+	+	–	–	–
		Nipponacmea nigrans	+	+	+	+	+
		Batillaria cumingi	+	+	+	+	+
		Littorina brevicula	+	+	+	+	+
		Littorina sitkana	+	–	–	–	–
		Assiminea sp.	+	–	+	–	+
		Laguncula pulchella	+	–	–	+	+
	Bivalvia	*Musculista senhousia*	+	–	+	+	–
		Mytilus galloprovincialis	+	+	+	–	+
		Crassostrea gigas	+	+	+	+	–
		Pseudocardium sachalinense	–	+	–	–	–
		Mactra chinensis	–	–	+	–	+
		Nuttallia japonica	+	+	+	+	+
		Phacosoma japonicum	+	+	–	–	–
		Ruditapes philippinarum	+	+	+	+	+
		Solen strictus	–	–	+	–	–
		Mya arenaria oonogai	+	–	+	–	–
		Panopea japonica	+	–	–	–	–
		Laternula marilina	+	–	+	+	+
Annelida	Polychaeta	*Glycera nicobarica*	+	+	+	+	+
		Glycera spp.	+	–	+	+	+
		Ceratonereis erythraeensis	+	+	+	+	+
		Hediste spp.	–	+	+	+	+
		Neanthes virens	–	–	–	–	+
		Nereis vexillosa	+	–	–	+	+
		Perinereis nuntia brevicirris	+	–	+	+	+
		Diopatra sugokai	+	–	–	–	+
		Cirriformia cf. *comosa*	–	–	–	+	–
		Heteromastus sp.	+	+	+	+	+
	Oligochaeta	*Pontodrilus litoralis*	–	–	+	–	–
		Pachydrilus nipponicus	–	+	–	–	–

(continued)

Table 10.2 (continued)

Phylum	Class	Taxa	Year 2002	2011	2012	2013	2014
Arthropoda	Maxillopoda	*Amphibalanus amphitrite*	+	–	–	+	+
		Amphibalanus improvisus	+	+	–	–	–
		Balanus albicostatus	–	–	+	+	+
		Balanus glandula	–	+	+	+	+
	Malacostraca	*Ampithoe* spp.	+	–	+	+	+
		Corophium spp.	+	+	–	–	–
		Jassa sp.	+	–	–	–	–
		Pontogeneia rostrata	+	–	–	–	–
		Eogammarus possjeticus	–	–	–	–	+
		Melita sp.	+	+	+	–	+
		Ptilohyale barbicornis	–	–	–	+	–
		Hyale sp.	+	–	+	+	+
		Idotea ochotensis	+	–	–	–	–
		Synidotea laevidorsalis	+	–	–	–	–
		Ligia exotica	–	+	–	–	–
		Gnorimosphaeroma spp.	+	+	+	–	+
		Nihonotrypaea harmandi	+	–	–	–	–
		Nihonotrypaea japonica	+	–	–	–	–
		Upogebia major	+	–	+	–	–
		Palaemon serrifer	–	+	–	–	–
		Pagurus minutus	–	–	–	–	+
		Eriocheir japonicus	–	+	–	–	–
		Hemigrapsus penicillatus	+	–	+	+	+
		Hemigrapsus takanoi[a]	–	–	–	+	+
		Ocypode stimpsoni	–	–	–	–	+
		Deiratonotus cristatus	–	–	+	+	–
Echinodermata	Holothuroidea	Synaptidae sp.	–	–	–	–	+
Total			39	23	30	25	32

[a]This species was identified as *H. penicillatus* in 2002 but later recorded as *H. takanoi* by Asakura (2006)

In the pre-tsunami survey, we recognized 32 taxa in the Orikasa area (Table 10.3), but in the post-tsunami survey, we found 26 taxa, of which 15 appeared for the first time after the tsunamis (9–15 taxa per year). Thus, only 34.4 % of taxa recorded prior to the tsunamis were confirmed in surveys conducted after the tsunamis. Nine of the 17 taxa that had been recorded in pre-tsunami surveys (i.e., 52.9 %) were confirmed when we compared the number of taxa recorded in pre- and post-tsunami surveys for only two stations for which we could obtain data from both surveys (Table 10.3; Fig. 10.4, Stations 6 and 8).

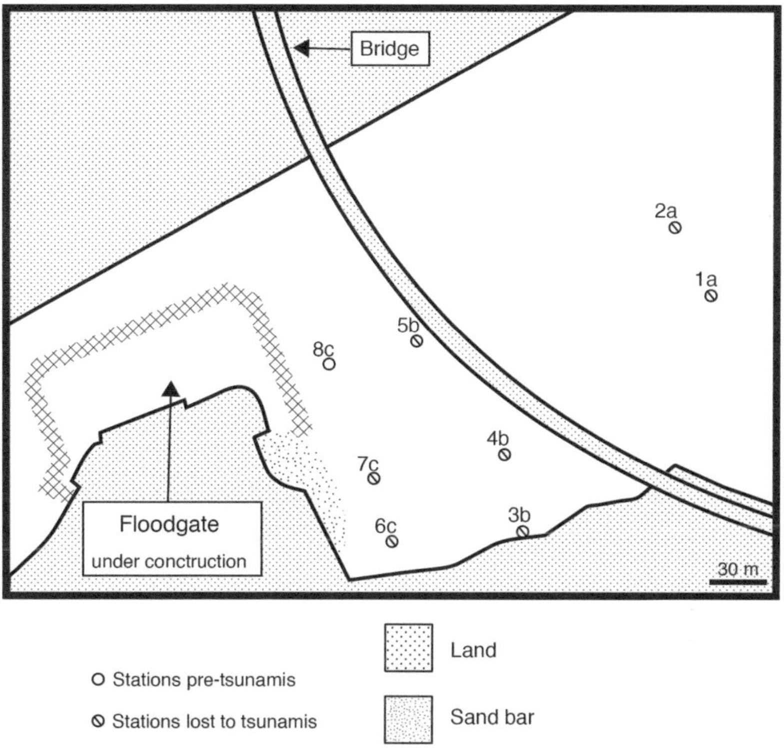

Fig. 10.4 Map showing census transect lines and survey stations in Orikasa

10.3.3 Unosumai Area

The tsunamis caused by the Great East Japan Earthquake considerably modified the tidal flats in Unosumai, causing significant damages that extended along the river and into the town. All pre-tsunami survey sites were lost during the tsunamis in 2011 (Fig. 10.6a, Stations 1–7). The post-tsunami surveys were conducted at three new sites, including a new brackish habitat in 2012 and 2013 (Fig. 10.6b, Stations 8–10). Because the new brackish habitat was restored to land in late 2013, which was then used for seawall construction (Fig 10.6b, Stations 8 and 9), we surveyed additional stations in 2014 (Fig. 10.6b, Stations 11 and 12). Before the tsunamis, sediment at Unosumai consisted mainly of sand (Fig. 10.7), but after the tsunamis, the sediment at the study area comprised gravel and mud. Salinity in this area ranged from 8‰ to 25‰ (Table 10.1).

During the pre-tsunami surveys, 13 taxa were recorded at Unosumai (Table 10.4), but during the post-tsunami surveys, we found 34 taxa and detected a notable increase in the number of bivalves, barnacles, and amphipods (16–23 taxa per year). Ten of the 34 taxa had been recorded in the pre-tsunami survey. Hence, 76.9 % of taxa recorded prior to the tsunamis were found in surveys following the tsunamis.

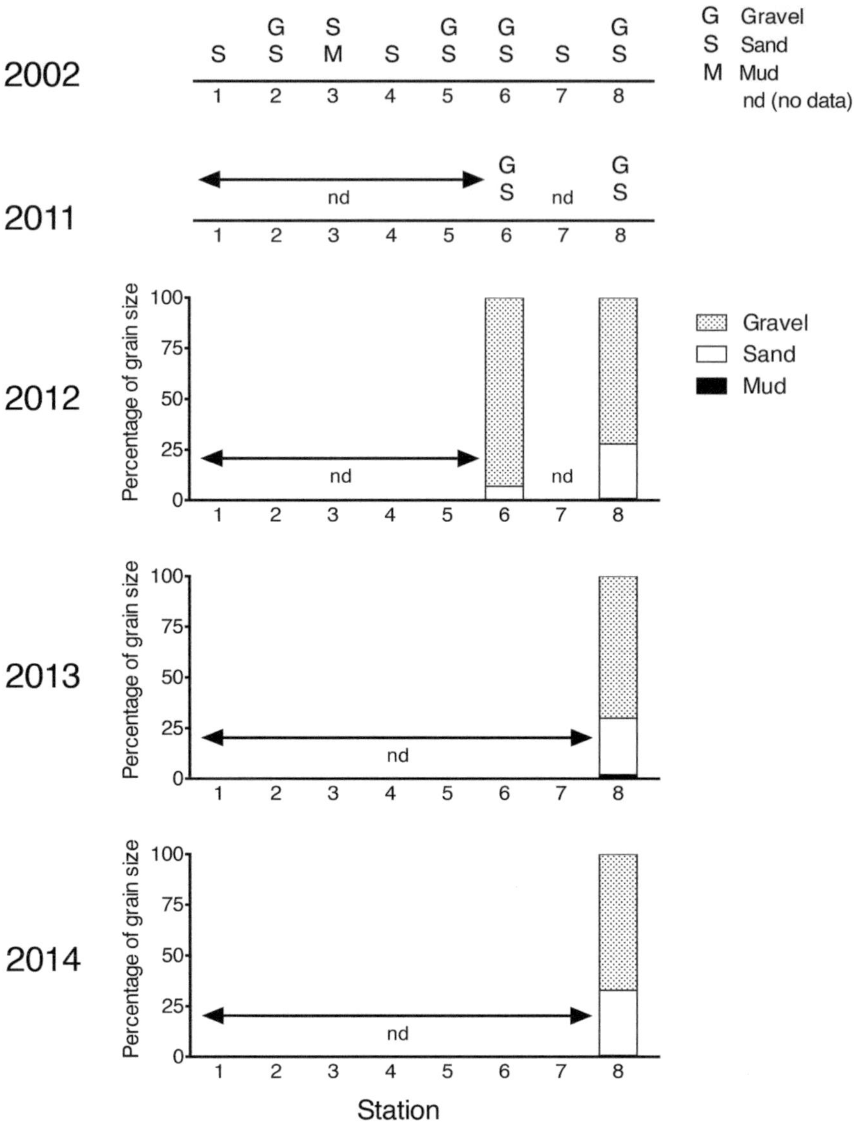

Fig. 10.5 Change in sedimentary characteristics before and after the 2011 tsunami disaster at Orikasa

10.4 Discussion

The tsunamis and land subsidence caused by the Great East Japan Earthquake in 2011 reduced the extent of tidal flat habitats throughout Iwate Prefecture. However, the degree of reduction in tidal flats and recovery pattern of macrobenthos varied in each area. The Tsugaruishi area had the largest tidal flat in Iwate Prefecture

(Biodiversity Center of Japan 2007), where remained as the largest one even after the disaster. Although some of the sampling stations were lost after the disaster, we were able to confirm the presence of some of previously observed taxa at new stations. Accordingly, 76.9 % of taxa recorded prior to the tsunamis were confirmed in surveys following the tsunamis. In contrast, in Orikasa area, only 34.4 % of the taxa recorded prior to the tsunamis were appeared in surveys conducted after the tsunamis. This incongruence in the results may have been caused by different number of stations in the pre- and post-tsunami surveys. Since tidal flats shrank remarkably as the subtidal zone expanded in this area, we compared the number of taxa recorded in pre- and post-tsunami surveys for only two stations for which we could obtain data from both surveys (Fig. 10.4, Stations 6 and 8, Table 10.3) and confirmed the presence of 9 of the 17 taxa that had been recorded in pre-tsunami surveys

Table 10.3 Macrobenthos at Orikasa before and after the 2011 tsunamis: + present, – absent

			Year				
Phylum	Class	Taxa	2002	2011	2012	2013	2014
Cnidaria	Anthozoa	*Anthopleura* sp.	–	–	+	–	–
Mollusca	Gastropoda	*Nipponacmea habei**	+	+	–	–	–
		Nipponacmea nigrans	–	–	+	+	+
		*Patelloida heroldi**	+	–	–	–	+
		Euchelus sp.	–	–	+	–	–
		Omphalius rusticus	–	–	–	–	+
		Monodonta labio	–	–	–	+	+
		*Batillaria cumingi**	+	+	+	+	+
		Nassarius hypolius	+	–	+	–	+
		Haloa japonica	–	–	+	–	–
		Laguncula pulchella	–	–	–	+	+
	Bivalvia	*Mytilus galloprovincialis*	–	+	+	+	+
		Crassostrea gigas	–	–	–	+	–
		Macoma incongrua	+	–	–	–	–
		*Nuttallia japonica**	+	–	–	–	–
		*Ruditapes philippinarum**	+	–	+	+	+
Annelida	Polychaeta	*Glycera alba*	+	–	–	–	+
		Neanthes virens	+	–	–	–	–
		Nereis multignatha	–	–	+	–	–
		Nereis vexillosa	–	–	–	–	+
		Polydora sp.*	+	–	+	–	–
		Ctenodrilidae sp.*	+	–	–	–	–
		Heteromastus spp.*	+	–	–	–	–
		Arenicola brasiliensis	+	–	–	–	–
		Ampharetidae sp.	+	–	–	–	–
	Hirudinea	*Trachelobdella okae*	+	–	–	–	–

(continued)

Table 10.3 (continued)

Phylum	Class	Taxa	Year 2002	2011	2012	2013	2014
Arthropoda	Maxillopoda	*Amphibalanus improvisus*	+	+	+	+	–
		Balanus glandula*	–	+	+	+	+
	Malacostraca	*Grandidierella japonica**	+	–	–	–	–
		Corophium sp.	+	–	–	–	–
		Jesogammarus sp.*	+	–	–	–	–
		Melita sp.*	+	–	–	–	–
		Cyathura sp.*	+	–	–	–	–
		*Gnorimosphaeroma chinense**	+	–	–	–	–
		Ione cornuta	+	–	–	–	–
		Ligia cinerascens	–	–	–	+	–
		Ligia exotica	–	+	–	–	–
		Alpheidae sp.	+	–	–	–	–
		Crangon sp.	+	–	–	–	–
		Nihonotrypaea harmandi	+	–	–	–	–
		Nihonotrypaea japonica	+	–	–	–	–
		Upogebia major	+	–	–	–	–
		*Upogebia yokoyai**	+	+	+	–	+
		*Pagurus minutus**	+	+	–	–	+
		Portunus trituberculatus	+	–	–	–	–
		*Hemigrapsus penicillatus**	+	+	+	–	+
Echinodermata	Holothuroidea	Synaptidae sp.	–	–	+	–	–
Total			32	9	15	10	15

The *asterisk* indicates that taxa have been collected at stations 6 and 8 of pre-survey

(i.e., 52.9 %). Therefore, if the tidal flat habitats are restored in this area, those habitats might harbor more diverse invertebrates than the present situation with a limited tidal area. In contrast, we found increased number of taxa after the tsunamis compared to that in the pre-tsunami period in the Unosumai area (Table 10.4), where the new brackish habitat emerged inside the river mouth (Fig. 10.6). Various taxa were recorded in this new habitat created by the tsunamis in 2011 (i.e., Stations 8 and 9, Fig. 10.6), including the bivalve *Mya arenaria oonogai*, a near-threatened taxon according to the Japanese Association of Benthology (2012). The post-tsunami salinity around these stations ranged from 8 ‰ to 25 ‰ (Table 10.1). Appearance of new habitats where composed of heterogeneous sediments with gravels suitable for attachment of sessile organisms and the increased salinity could induce change in the structure of macrobenthic assemblages. These results indicated that natural-driven physical disturbance contributed to the increase in richness of taxa. Several studies have described the possible effects of tsunamis on macrobenthic diversity. For example, Lomovasky et al. (2011) showed that a 2007 tsunami in Peru

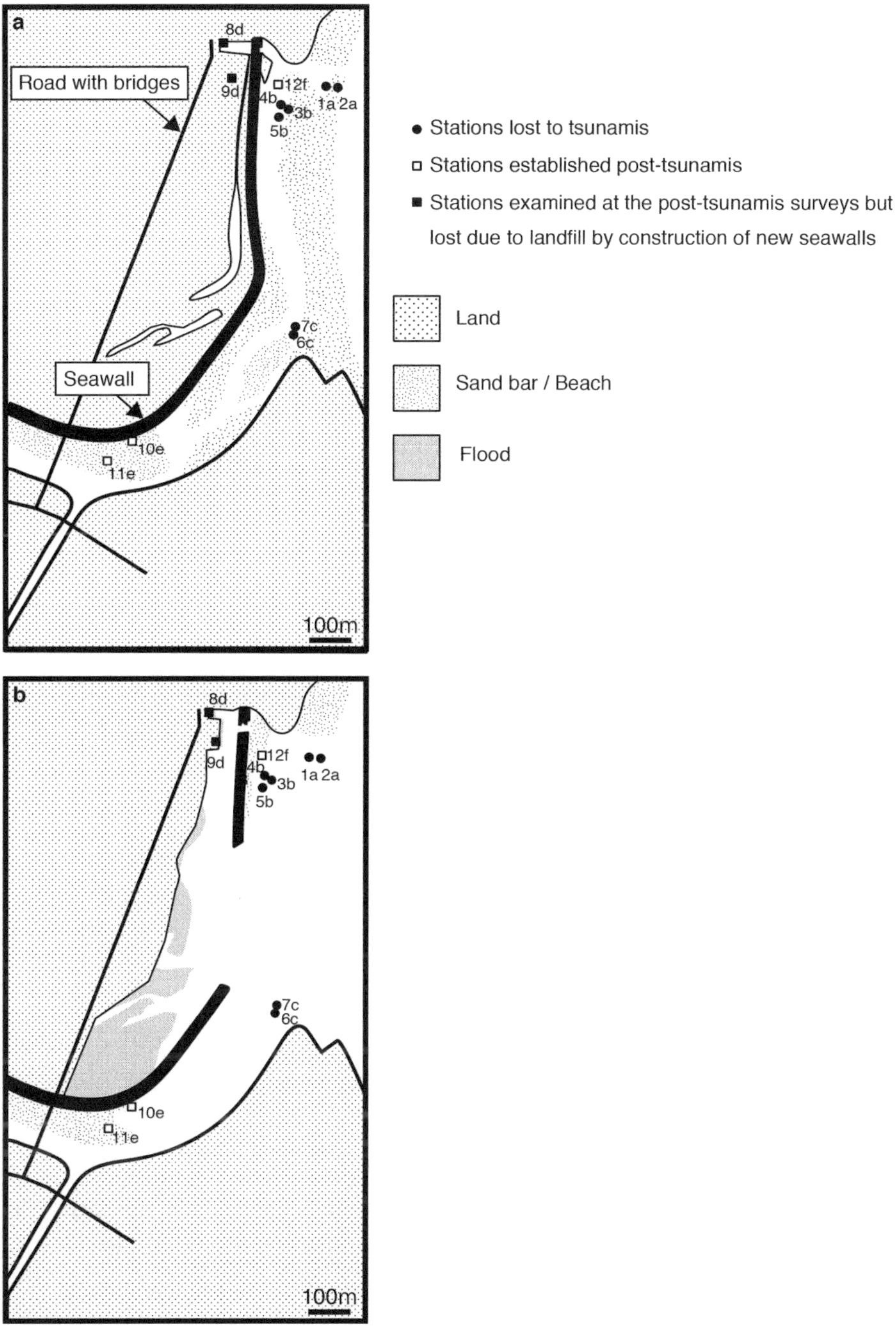

Fig. 10.6 Map showing census transect lines and survey stations in Unosumai. (**a**) Pre-tsunami; (**b**) post-tsunami

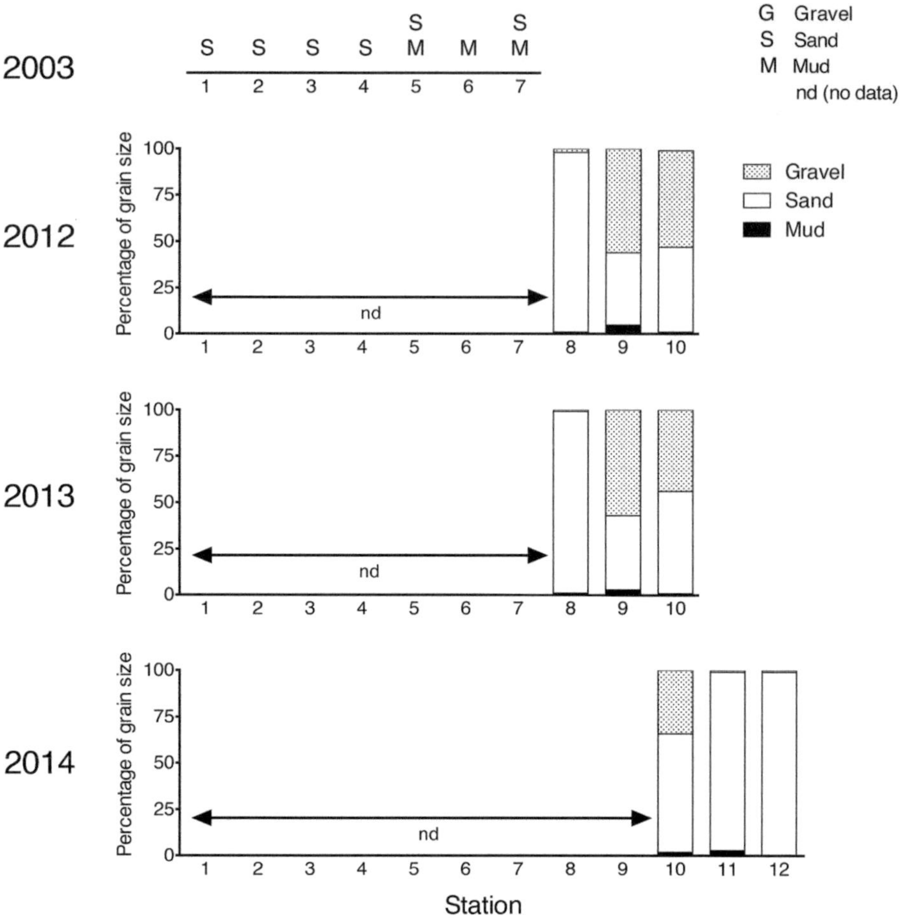

Fig. 10.7 Change in sedimentary characteristics before and after the 2011 tsunami disaster at Unosumai

promoted colonization of an infaunal gastropod *Nassarius gayi*. Although many infaunal animals disappeared after the tsunami, the gastropod appeared and maintained its abundance throughout the study period (about 4 months) (Lomovasky et al. 2011). Urabe et al. (2013) reported that the tsunamis triggered by the 2011 Tohoku earthquake likely transported some subtidal taxa into the intertidal zone. Unfortunately, newly created tidal flats in the Unosumai area were lost in 2014 during seawall reconstruction, and therefore, we were unable to monitor the persistence of taxa in this habitat.

On the whole, our results indicate that 52–76 % of benthic animal taxa reported from sediments prior to the tsunamis were found in the post-tsunami survey when tidal flat habitats had remained in the estuaries. The Iwate coastline had been affected by large and infrequent catastrophic events such as the tsunamis caused by

Table 10.4 Macrobenthos at Unosumai before and after the 2011 tsunamis: + present, – absent

Phylum	Class	Taxa	Year			
			2003	2012	2013	2014
Mollusca	Gastropoda	*Nipponacmea nigrans*	–	+	+	+
		Assiminea sp.	+	+	+	+
		Littorina brevicula	–	+	+	–
	Bivalvia	*Mytilus galloprovincialis*	+	+	–	+
		Crassostrea gigas	–	+	+	–
		Nuttallia japonica	+	+	+	+
		Ruditapes philippinarum	–	+	+	–
		Gomphina melanegis	–	–	–	+
		Mya arenaria oonogai	–	+	–	–
		Laternula marilina	–	+	+	–
Annelida	Polychaeta	*Glycera* sp.	–	+	–	–
		Hediste spp.	+	+	+	+
		Nereis vexillosa	–	–	–	+
		Perinereis nuntia brevicirris	–	–	+	–
		Tylorrhynchus heterochaetus	+	+	–	+
		Pseudopolydora sp.	–	+	–	–
		Heteromastus sp.	+	–	–	+
Arthropoda	Maxillopoda	*Amphibalanus improvisus*	–	+	–	+
		Balanus albicostatus	–	+	+	+
		Balanus glandula	–	+	+	+
		Megabalanus rosa	–	–	–	+
	Malacostraca	*Grandidierella* sp.	–	–	+	+
		Corophiidae gen. sp.	–	–	–	+
		Eogammarus possjeticus	–	–	–	+
		Melita setiflagella	–	+	–	+
		Ptilohyale sp.	–	–	–	+
		Haustorioides sp.	+	–	–	–
		Platorchestia sp.	+	–	–	+
		Excirolana chiltoni	+	–	–	–
		Gnorimosphaeroma sp.	+	+	–	+
		Ligia cinerascens	–	+	+	+
		Ligia exotica	+	–	–	–
		Crangon sp.	–	+	–	–
		Upogebia yokoyai	+	+	+	–
		Hemigrapsus penicillatus	–	+	+	+
		Hemigrapsus takanoi	–	–	–	+
		Deiratonotus cristatus	+	–	+	–
Total			13	22	16	23

the Great East Japan Earthquake in 2011 and smaller-scale annual disturbances such as estuarine dredging to facilitate salmon migration (Sakai et al. 1991). Therefore, most of the fauna of this region may already have been tolerant to disturbance or otherwise characterized by opportunistic species.

Before the tsunamis in 2011, the thalassinid shrimps (*Nihonotrypaea japonica*, *N. harmandi*, *Upogebia major*, and *U. yokoyai*) occurred in Orikasa tidal flats (Biodiversity Center of Japan 2007). However, in the post-tsunami surveys, we found only *U. yokoyai* in that area. As the pre-tsunami surveys were performed in 2002 and the area had experienced other disturbances (e.g., construction of a floodgate) prior to the tsunamis, the loss of any species cannot be conclusively attributed to impacts of the tsunamis in 2011. Kanaya et al. (2012) showed that the thalassinid shrimps were among the most tsunami-affected taxa in Gamo Lagoon, Miyagi Prefecture. Both *Nihonotrypaea* species also disappeared from our post-tsunami survey sites at Tsugaruishi (Fig. 10.2).

Reconstruction of the seawalls is planned as an effective means of managing and minimizing future disaster risks. In the Iwate region, construction of coastal structures such as seawalls and floodgates began in 2012 (Sakaguchi 2014). The height of many of the planned seawalls is greater than it was before the disaster. In our study area, the height of seawalls will be an imposing 10.4 m in Tsugaruishi, 9.7 m in Orikasa, and 14.5 m in Unosumai (Iwate Prefecture 2013). Furthermore, floodgates will be established in each of the three areas. The construction of high seawalls and floodgates can have a negative impact on benthic animals by reducing the habitats and changing the salinity. We have identified several taxa that are considered threatened throughout Japan in our study area: a brackish-water crab *Deiratonotus cristatus* (vulnerable taxon), an anemone *Synandwakia hozawai*, a bivalve *M. arenaria oonogai*, and a palolo worm *Tylorrhynchus osawai* (Japanese Association of Benthology 2012). *D. cristatus* and *T. osawai* prefer habitats with low salinity, which are often appeared at the construction site of floodgates. Therefore, these threatened taxa may be lost during the reconstruction of the seawall and floodgates throughout the region.

As a global trend, ecosystem-based flood defense has been attracting attention instead of the conventional coastal engineering such as building of seawalls, dykes, and embankments. Recent research suggests that the development and restoration of coastal ecosystems (wetlands, saltmarsh, and tidal flats) can provide protection against flooding (e.g., Gedan et al. 2011; Temmerman et al. 2013; Ferrario et al. 2014). Generally, these ecosystem-based flood defense methods apparently do not impair coastal habitats suitable to many benthic organisms and may be cost-effective systems compared to the conventional coastal engineering. In Japan, restoration of tidal flats has been conducted in some regions to recover the ecosystem function and biodiversity (Kokubo and Yamada 2011). Our data indicate that conventional coastal engineering can likely make it impossible to return coastal communities disturbed by tsunamis. Ecosystem-based flood defense should be considered to reduce risk of tsunami disaster without decreasing biodiversity of coastal ecosystems in Japan.

Acknowledgments The authors thank the following people for their cooperation, without whom this project and manuscript would not have been materialized. Special thanks go to the late Y Maki for his support. S Kajiwara, A Takehara, T Uchino, and the students of Iwate University assisted with the field observations.

References

Asakura A (2006) The story behind the discovery of *Hemigrapsus takanoi* Asakura and Watanabe, 2005, a cryptic species of crab, previously thought to be the common Japanese intertidal crab *H. penicillatus* (Decapoda: Brachyura: Grapsoidea). Taxa 21:33–39

Biodiversity Center of Japan (2007) The 7th National survey on the national environment: Shallow sea survey (tidal flats). Biodiversity Center of Japan, Nature Conservation Bureau, Ministry of the Environment. http://www.biodic.go.jp/kiso/34_higat_f.html. Accessed 1 Apr 2013

Ferrario F, Beck MW, Storlazzi CD et al (2014) The effectiveness of coral reefs for coastal hazard risk reduction and adaptation. Nat Commun 5:3794. doi:10.1038/ncomms4794

Gedan KB, Kirwan ML, Wolanski E et al (2011) The present and future role of coastal wetland vegetation in protecting shorelines: answering recent challenges to the paradigm. Clim Change 106:7–29. doi:10.1007/s10584-010-0003-7

Haraguchi T, Iwamatsu A (2013) Detailed maps of the impacts of the 2011 Japan tsunami. Kokon Shoin Publishers Ltd, Tokyo

Iwate Prefecture (2013) Restoration of tsunami measures facilities, maintenance areas (Tsunamitaisaku sisetu no fukkyu to seibi kasyo) (In Japanese). http://www.pref.iwate.jp/kasensabou/kasen/fukkyuu/tsunami/index.html. Accessed 20 Feb 2016

Japanese Association of Benthology (ed) (2012) Threatened animals of Japanese tidal flats: Red data book of seashore benthos. Tokai University Press, Kanagawa

Kanaya G, Suzuki T, Maki H, Nakamura Y, Miyajima Y, Kikuchi E (2012) Effects of the 2011 tsunami on the topography, vegetation, and macrobenthic fauna in Gamo Lagoon, Japan. Jpn J Benthol 67:20–32

Kokubo H, Yamada H (2011) Evaluation of tidal flat restoration effect in the coastal unused reclaimed area by promoting tidal exchange in Ago Bay. J Japan Soc Civil Eng Ser B2 (Coast Eng) 67:I 956–I 960, http://doi.org/10.2208/kaigan.67.I_956

Lomovasky BJ, Firstater FN, Salazar AG, Mendo J, Iribarne OO (2011) Macro benthic community assemblage before and after the 2007 tsunami and earthquake at Paracas Bay, Peru. J Sea Res 65:205–212. doi:10.1016/j.seares.2010.10.002

Sakaguchi N (2014) Reviewing arguments on the construction of seawalls in Iwate Nippo Newspaper articles from March 2011 to March 2014. J Policy Stud 16:1–17

Sakai S, Ishikawa K, Sasamoto M, Hirayama K (1991) River mouth fluctuation of small river in Iwate. Proc Hydraul Eng 35:217–224, doi.org/10.2208/prohe.35.217

Temmerman S, Meire P, Bouma TJ, Herman PMJ, Ysebaert T, De Vriend HJ (2013) Ecosystem-based coastal defence in the face of global change. Nature 504:79–83. doi:10.1038/nature12859

Urabe J, Suzuki T, Nishita T, Makino W (2013) Immediate ecological impacts of the 2011 Tohoku earthquake tsunami on intertidal flat communities. PLoS One 8:e62779. doi:10.1371/journal.pone.0062779

Chapter 11
Ecological Responses of Macrobenthic Communities in Tidal Flats to Disturbances by the Great East Japan Earthquake

Tatsuki Nishita, Wataru Makino, Takao Suzuki, and Jotaro Urabe

Abstract We investigated changes in taxon richness and composition of macrobenthic communities for four successive years after the tsunamis caused by the Great East Japan Earthquake at six intertidal flats in Sendai Bay, where monitoring had been ongoing before the tsunami. The field surveys showed that although taxon richness decreased in most of intertidal flats immediately after the tsunami struck, it subsequently increased within 1–2 years due to the colonization of many taxa, including those that had not existed in these tidal flats before the tsunamis. However, by 2014 (i.e., 4 years after the tsunami), taxon richness had decreased again to pre-tsunami levels. In addition, taxon composition was close to that found before the tsunami struck, especially in intertidal flats that were subjected to large disturbances by the tsunamis. Thus, a number of opportunistic taxa that had not been observed previously appeared after the Great East Japan Earthquake, but gradually decreased as the community structure recovered. This study also suggests that taxon composition of the macrobenthic communities changed year to year to some extent regardless of the tsunami disturbance. Our findings imply that since macrobenthic communities in intertidal flats are somewhat dynamic, they are not likely to attain the exact same structure as before the tsunamis, even though taxon composition is largely shaped by site-specific environmental conditions.

Keywords Benthic animals • Community structure • Natural disturbance • Recovery process • Sendai Bay • Succession • Stability • Tidal flat

---

T. Nishita • W. Makino • T. Suzuki • J. Urabe (✉)
Graduate School of Life Sciences, Tohoku University,
Aoba 6-3, Aramaki, Aoba-ku, Sendai 980-8578, Japan
e-mail: urabe@m.tohoku.ac.jp

J. Urabe, T. Nakashizuka (eds.), *Ecological Impacts of Tsunamis on Coastal Ecosystems*, Ecological Research Monographs,
DOI 10.1007/978-4-431-56448-5_11

11.1 Introduction

In the ecological context, disturbance is defined as an intermittent event that upsets the structures of populations, communities, and ecosystems and alters the amount of resources and substrates or the state of the environmental conditions (White and Pickett 1985). Since disturbance reduces population abundance of existing species and impedes interspecific competition, it may promote colonization of species that were not present before the perturbation (Pickett 1980; Sousa 1984; Bongers and Bongres 1998; Pierce et al. 2007; Göthlich and Oschlies 2015). As a result, the species diversity of a community may increase with disturbances that are small or moderate in degree and frequency, although diversity may eventually decrease under a long-term stable environment (Connell 1978; White and Pickett 1985; Collins et al. 1995; Roxburgh et al. 2004).

Where a disturbance occurs repeatedly and at shorter time spans than organisms' longevity, species that are competitively inferior but have the ability to withstand the physical forces of the disturbance may predominate (Turner and Dale 1998). However, where large-scale disturbances occur at time intervals much longer than organisms' life spans, most species may disappear instantaneously (see review in Mori 2010). In such a situation, it can take a long time for the community to recover after the disturbance (Turner and Dale 1998). In addition, species composition and structure of the community may change substantially compared with that before the disturbance. Thus, long-term monitoring of communities after a large-scale disturbance may provide a good opportunity for clarifying ecological processes and key factors determining community structures and thus number of coexistent species (Levin et al. 1996; Perkol-Finkel and Airoldi 2010; Botter-Carvalho et al. 2011; Joydas et al. 2012; Kanaya et al. 2012).

To date, several studies have examined the ecological impacts of large-scale disturbances on aquatic communities after the occurrence of rare events such as extreme flooding (Cardoso et al. 2008), drought (Pillay and Perissinotto 2008; Stubbington et al. 2015), hurricanes (Balthis et al. 2006; Engle et al. 2009), and oil spills (De la Huz et al. 2005; Schlacher et al. 2011). These studies showed that such disturbances dramatically altered community structures by reducing species richness and population abundance. The striking of huge tsunamis is another rare but large-scale disturbance to coastal ecosystems. Historically, tsunamis have occurred repeatedly and likely affected coastal communities for long periods of time (Krishnankutty 2006; Bourrouilh-Le Jan et al. 2007). The ecological effects of tsunami disturbance on coastal macrobenthic communities have been studied in mangrove forests (Fujioka et al. 2008), seagrass beds (Nakaoka 2006; Whanpetch et al. 2010), subtidal soft bottoms (Jaramillo et al. 2012), and intertidal flats (Lomovasky et al. 2011; Kanaya et al. 2012; Miura et al. 2012; Urabe et al. 2013). These studies showed that the ecological impacts of a tsunami on coastal communities vary depending on type of ecosystem, taxa, and species-specific traits. However, these previous studies examined changes in community structures immediately after the tsunamis or for only 1 year afterwards. In some habitats, community composition may recover soon by recolonization of previously distributed species, but more time may be needed to recover the previous state in other habitats, depending on environmental conditions and scale of the disturbance experienced. To assess ecological impacts of rare and large-scale disturbances on communities, therefore, it is essential to monitor changes in community structures in various habitats for several years at least.

The Great East Japan Earthquake that occurred on 11 March 2011 was one of the largest in recent history, registering 9.0 on the moment magnitude scale (Mw). The resulting large tsunamis (the 2011 Tohoku earthquake tsunamis) struck the Pacific coastline of eastern Japan, including Sendai Bay and the Sanriku Ria coast, with a tsunami inundation height that was often larger than 10 m and the highest run-up wave of 40.1 m (Mori et al. 2012). In a previous study, we assessed the ecological impacts of the tsunamis caused by the Great East Japan Earthquake on the macrobenthic communities in several intertidal flats of Sendai Bay by comparing the taxon richness and community composition before and immediately after the tsunami struck and showed that the magnitude of changes in community structures was greater for intertidal flats that experienced larger inundation heights (Urabe et al. 2013). However, it remains unclear if the communities strongly disturbed by the tsunamis have changed since then. In this study, therefore, we examined temporal changes in these disturbed communities for 4 years after the tsunamis. Specifically, we examined if the community response to the disturbances by the tsunamis differed among the intertidal flat communities and if the taxon compositions of these communities returned to those before the tsunamis within 4 years.

11.2 Materials and Methods

11.2.1 Census Sites

Field census surveys were conducted at six intertidal flats—Kastura Is, Hitsugaura, Sokanzan, Matsukawaura (Unoo area), Gamo, and Torinoumi—located in the coastal area of Sendai Bay (Fig. 11.1). One census site was set in each of the first four tidal flats, while the last two flats had two sites (A and B) each. Accordingly, we conducted census surveys at eight sites in total (Fig. 11.1). The Kastura Is is the foreshore intertidal flat facing Matsushima Bay on the inner most part of Sendai Bay. The Hitsugaura and Sokanzan flats are also foreshore intertidal flats and are located in the innermost area of Matsushima Bay, while the Gamo, Torinoumi, and Matsukawaura-Unoo flats are located inside of respective lagoons. Sites A and B of the Gamo flat were located at the east and west side of Gamo lagoon, respectively. Site A of the Torinoumi flat was the intertidal flat located near the mouth of the Torinoumi lagoon opening to Sendai Bay, while site B was set at a distance from the lagoon mouth. The Matsukawaura-Unoo flat was located at the northeast part of the lagoon. The exposed area at each flat varied from 0.5 to 7 ha. The whole exposed

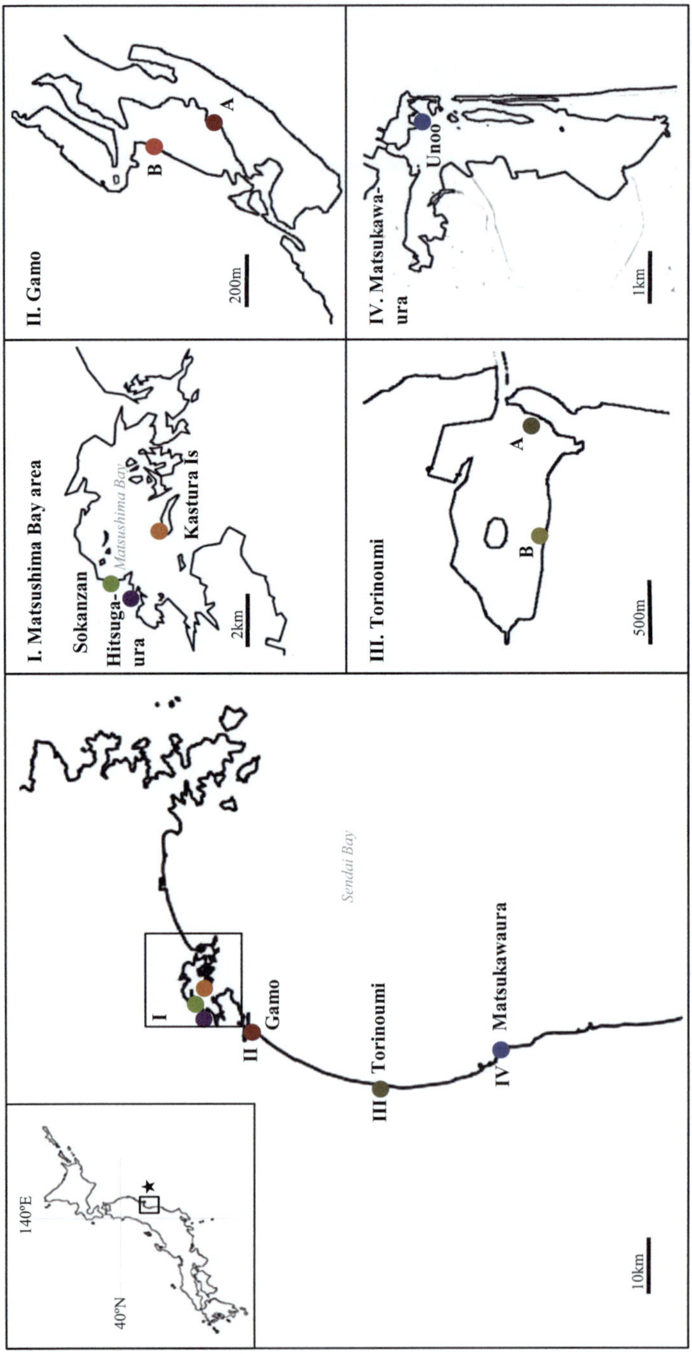

Fig. 11.1 Maps of Tohoku area with Sendai Bay showing the locations of intertidal flats examined

Table 11.1 Longitude and latitude (expressed in the decimal system) of the census sites examined, approximate surface area censused, and census date after the tsunamis

	Latitude	Longitude	Area	Census date		
Observational flat	°N	°E	ha	2012 survey	2013 survey	2014 survey
Kastura Is	38.330	141.126	2	8/4	7/27	7/26
Hitsugaura	38.351	141.052	4	5/26	5/25	5/31
Sokanzan	38.353	141.059	1	5/26	5/25	5/31
Gamo A	38.255	141.013	2	6/9	6/8	6/14
Gamo B	38.257	141.013	2	6/9	6/8	6/14
Torinoumi A	38.031	140.916	2.5	6/10	6/9	6/15
Torinoumi B	38.031	140.906	2	6/10	6/9	6/15
Matsukawaura-Unoo	37.819	140.982	7	6/23	6/22	5/17

area of the Kastura Is, Hitsugaura, and Sokanzan flats was estimated to be 2, 4, and 1 ha, respectively (Table 11.1). The Gamo flat had a total exposed area of ca. 10 ha (Kanaya this volume), and we conducted the census on about ca. 2 ha for sites A and B. At the Torinoumi flat, which had a total exposed area of ca. 60 ha, ca. 2–2.5 ha were surveyed at each site. Matsukawaura is one of the largest lagoons on Sendai Bay, with the total surface area of ca. 738 ha. In this lagoon, we surveyed ca. 7 ha at the Unoo tidal flat among the total exposed area of ca. 350 ha.

11.2.2 Census Method

At each census site, field surveys were conducted once in May–August every year from 2011 to 2014. Because not all animals were identified to the species level (see below), community characteristics were expressed in terms of taxon richness and composition. Taxon richness and community composition of macrobenthos on intertidal flats were examined during the low tide using the survey method described in Suzuki and Sasaki (2010). Briefly, 12 researchers walked randomly for 15 min over the entire area of a given observational flat, collecting the epibenthic animals (epifauna) they encountered. Each researcher also dug 15 holes, approximately 15 cm in diameter and 20 cm in depth, with shovels and visually searched and collected endobenthic animals (infauna). The animals that were identified easily at the species level on-site were returned to the observational flat after recording, but those that could not be identified were fixed with 10 % formalin for later microscopic examination in the laboratory. For animals that could not be identified at species level, identification was made to genus or the finest taxonomic level. Throughout the censuses, a fixed researcher always confirmed taxon identifications to minimize or equalize artifacts caused by misidentification. Data from the 12 researchers were pooled to determine the taxon richness and composition of each observational flat.

Data on the taxon richness and composition in each flat before the Great East Japan Earthquake in 2011 (pre-tsunami period), as well as those obtained immediately after the resulting tsunamis (already shown in a previous study; Urabe et al. 2013), were also used in the present study.

11.2.3 Statistical Analysis

11.2.3.1 Annual Changes in Taxon Richness

A generalized linear mixed model (GLMM) was used to evaluate the year-to-year variation in taxon richness from the pre-tsunami period to 4 years after. In this analysis, the observational years were treated as a fixed effect and the observational flats as a random effect to exclude potential influences of site-specific factors. The model assumed a Poisson distribution and was constructed with a log-link function. Explanatory variables were treated as categorical data. The Wald test was used to assess the significance of the explanatory variable.

11.2.3.2 Time-Course Trajectory of Taxon Composition

Similarity (S) in taxon composition in different years at a given flat was measured by the Jaccard index (Legendre and Legendre 1998) as follows:

$$S = c / (a + b - c),$$

where a and b are taxon richness before and after the tsunamis in 2011, respectively, and c is number of taxa commonly found in both years. Then the dissimilarity index, defined as 1-S, was estimated. The results were visualized on a two-dimensional plane with the aid of nonmetric multidimensional scaling (nMDS), where observations with smaller dissimilarity indices were plotted closer to each other and vice versa (Legendre and Legandre 1998).

Significant difference in the taxon composition before (~2010) and after the tsunamis (2011, 2012, 2013 and 2014) was tested with the aid of bootstrap simulations (Legendre and Legandre 1998). In these simulations, the probability of occurrence at a site in all five surveys (one pre-tsunami and four post-tsunami surveys) was estimated for each taxon. For example, if a taxon appeared only once in the five census surveys, its probability of occurrence was 0.2, while the probability was 1 if it occurred in all five surveys. We then created two hypothetical bootstrapping communities where the presence/absence of each taxa in each community was randomly assigned using the probability of occurrence estimated for a given taxa, and calculated the dissimilarity index mentioned above between the two hypothetical communities. This procedure was iterated 9,999 times to quantify the potential degree of variation under a null hypothetical condition. The confidence interval of the dissimilarity index was also estimated, and if the observed dissimilarity index between the pre-tsunami year and one of the post-tsunami years was larger than the 95 % of this confidence interval, it was judged that the difference in taxon composition between these 2 years was significant and had not emerged by chance. On the other hand, if the dissimilarity index between these 2 years was included in the 95 % of this confidence interval, we considered that taxon composition in the given post-tsunami years had returned to the indigenous level before the tsunamis. In this analysis, the alpha level of statistical significance was corrected according to the sequential Bonferroni adjustment, because the dissimilarity index before and after the tsunami struck was estimated for four combinations per observational flat. All of the statistical tests were performed with the aid of the statistical package R (ver. 2.15.1).

11.3 Results

11.3.1 Taxon Richness

When data from all the intertidal flats were pooled, a total 186 taxa were found. Before the tsunamis caused by the Great East Japan Earthquake in 2011, 108 taxa were confirmed in the same set of intertidal flats. After the tsunamis, the number of taxa found in the flats was 97, 99, 127, and 108 in 2011, 2012, 2013, and 2014, respectively. Among these taxa, 52 were confirmed consecutively from the pre-tsunami years until 2014 in at least one of the census sites. In particular, crustaceans *Hemigrapsus penicillatus* and *Pagurus minutus*, and a gastropod *Batillaria attramentaria*, were confirmed in more than six census sites every year. In this study, the number of taxa commonly found before and after the tsunamis was 92, while the number found only after the tsunamis was 78. Accordingly, 16 taxa present before the 2011 tsunami struck were not found after. These included a crustacean *Chiromantes haematocheir*; gastropods *Cecina manchurica*, *Assiminea yoshidayukioi*, and *Crepidula onyx*; and a bivalve *Xenostrobus atrata*. In contrast, a polychaeta *Armandia amakusaensis* and a crustacean *Portunus* (*Portunus*) *trituberculatus* that were not present before the tsunamis were commonly found in 2012 and 2013 at many census sites, although by 2014 they had disappeared at most sites. In addition, a crustacean *Charybdis japonica*, polychaeta *Tylorrhynchus osawai* and a species of Nephtyidae, a bivalve *Mya* (*Arenomya*) *arenaria oonogai*, piscine *Luciogobius guttatus*, and species of Gobiidae were found in at least one census site in each of the post-tsunami years, although these taxa had never been observed before the tsunamis in 2011.

Before the tsunamis, an average of 35.5 taxa (range 18–53) were observed per census site (Fig. 11.2). In the year 2011, i.e., immediately after the tsunami struck, the taxon richness was reduced in seven out of eight census sites, and the average taxon richness per census site fell to 27.1 (range 16–57) (Fig. 11.2). According to the GLMM, this reduction of taxon richness immediately after the tsunamis was

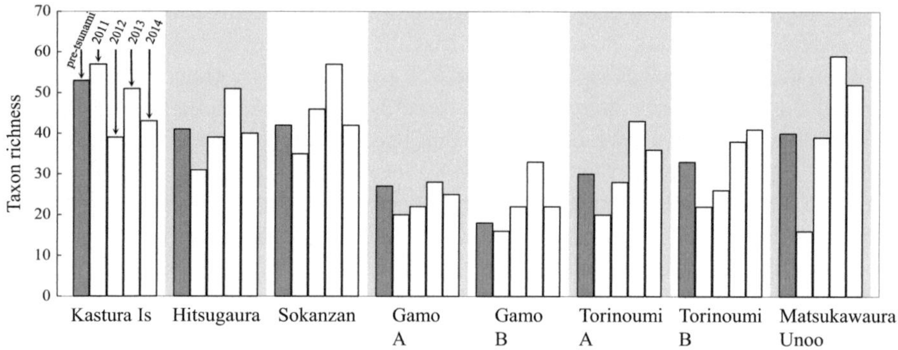

Fig. 11.2 Annual changes in taxon richness of macrobenthos at intertidal flats in Sendai Bay. Taxon richness values before the tsunami struck in 2011 are denoted by *gray bars* and post-tsunami values by successive *white bars* in order of research years

Table 11.2 Results of a generalized linear mixed model to examine potential factors for year-to-year variation in macrobenthos taxon richness, from pre-tsunami period through 4 years after

Fixed effect	Coefficient	Standard error	z	Wald test
Intercept	3.54	0.11	32.96	<0.001
Census year				
2011	–0.27	0.09	–2.98	0.003
2012	–0.08	0.09	–0.98	0.325
2013	0.24	0.08	2.99	0.003
2014	0.06	0.08	0.70	0.482

In the model, the taxon richness was the response variable, while the census years were an explanatory variable as a fixed factor. The observational flats were also added to an explanatory variable as a random factor in order to exclude potential influence of observational flat-specific factors such as substrate environments on the analysis

The Wald test was used to assess the significance of each explanatory variable

significant (Table 11.2). In 2012, however, taxon richness increased in all census sites except the Kastura Is. Accordingly, average taxon richness per site was 32.6 (range 22–46), not significantly different from the level before the tsunami (Table 11.2). In 2013, taxon richness was larger than the previous year in all census sites (Fig. 11.2), and average richness per site reached 45.0 (range 28–59), significantly larger than those recorded before the tsunami struck (Table 11.2). In 2014, the average taxon richness per site was reduced to 37.6 (range 22–52), which again was not significantly different from the level before the 2011 tsunami (Table 11.2).

11.3.2 Taxon Composition

Before the tsunami struck in 2011, taxon composition of macrobenthic communities was largely different among the intertidal flats examined (Fig. 11.3a, b). For example, at the Hitsugaura, Sokanzan, and Kastura Is, macrobenthic animals that prefer relatively high salinity, such as *Cirriformia* cf. *comosa*, *Reticunassa festiva*, and *Cyclina sinensis*, were commonly found, while at the Torinoumi and Gamo, animals that prefer relatively low salinity, such as *Hediste* spp., *Deiratonotus cristatus*, and *Assiminea hiradoensis*, were found (Fig. 11.3a, b). The impact of the tsunami disturbance on, and subsequent time-course trajectory of, taxon composition of the communities also varied among census sites as described below (Fig. 11.3a).

In the Sokanzan where the taxon composition immediately after the tsunami struck (2011) was not significantly different from the pre-tsunami composition, taxon composition in 2012, 2013, and 2014 also did not different from pre-tsunami composition (Fig. 11.4: $p>0.05$). In the Hitsugaura, on the other hand, pre-tsunami taxon composition was statistically significantly different from that in 2012 and 2014, but not in 2013. It should be noted that in these intertidal flats, the dissimilarity indices for the taxonomic composition among years were relatively smaller than those observed in other intertidal flats (Fig. 11.3).

Indeed, in the rest of the intertidal flats, the taxon composition in 2011 was significantly different from those observed before the tsunamis (Figs. 11.3a and 11.4). However, even in these intertidal flats, difference in taxon composition against the pre-tsunami year became smaller with the elapsed time (Fig. 11.3a). Indeed, taxon compositions in 2012 and thereafter were not significantly different from that in the pre-tsunami year at site A of Gamo, sites A and B of Torinoumi, and Matsukawaura-Unoo (Fig. 11.4: $p>0.05$). On the other hand, taxon composition of macrobenthic communities in the Kastura Is and site B of Gamo in 2012–2014 were continuously different from those before the tsunami struck (Fig. 11.4: $p<0.05$). In Kastura Is, the dissimilarity indices against the pre-tsunami composition reached their largest value in 2013 (Fig. 11.4). In site B of Gamo, the dissimilarity indices against the pre-tsunami composition were relatively large compared with those observed in other census sites (Fig. 11.4), and the taxon composition there tended to be similar to site A of the Gamo flat year by year (Fig. 11.3a).

11.4 Discussion

In this study, we investigated changes in taxon richness and composition of macrobenthic communities for four successive years after the tsunamis caused by the Great East Japan Earthquake at a total of eight census sites in six intertidal flats along Sendai Bay where there were monitoring data on these communities before the tsunamis (Urabe et al. 2013). In seven out of eight census sites, taxon richness decreased immediately after the tsunami struck. Moreover, taxon composition immediately after the tsunami struck was significantly different at three intertidal flats, Gamo, Torinoumi, and Matsukawaura-Unoo, as demonstrated by the nMDS diagram shown in Fig. 11.3. Since the magnitude of dissimilarity in taxon composition before and immediately after the tsunamis could not be explained by chance,

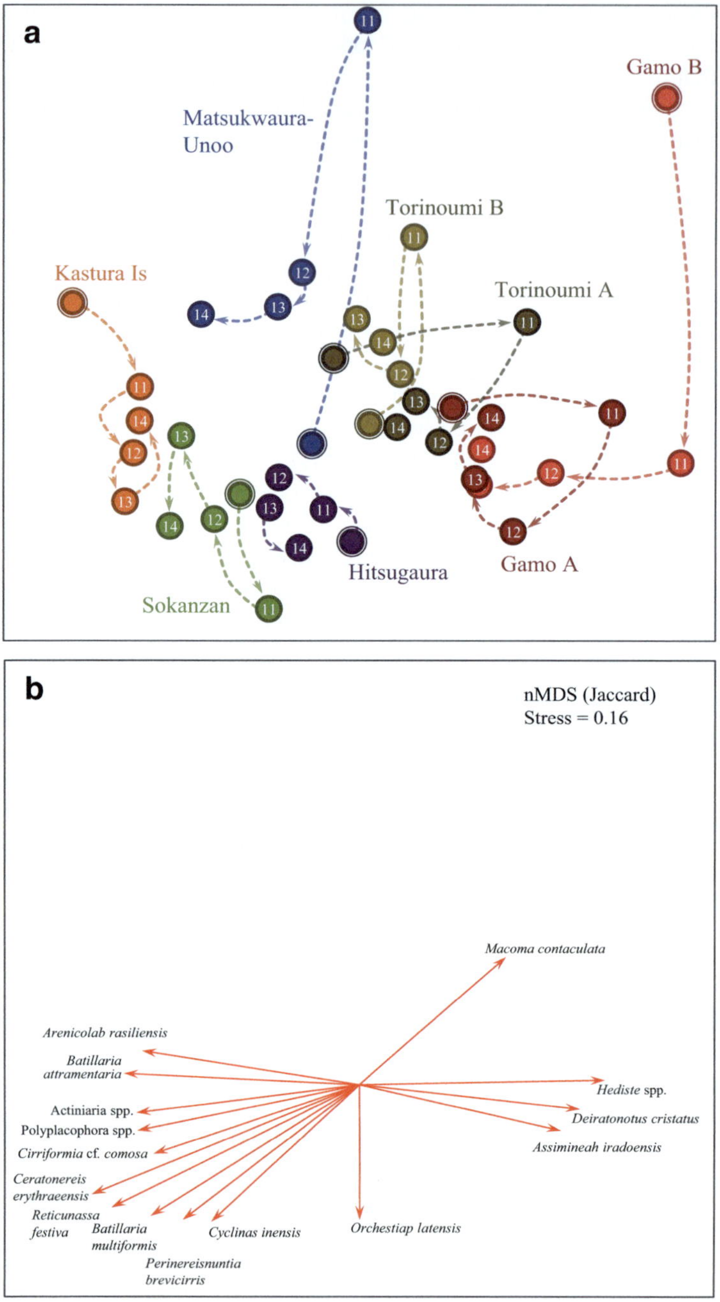

Fig. 11.3 (**a**) The nMDS diagram showing spatial difference of taxon richness among the intertidal flats and temporal changes in the taxon composition of each flat. The numbers inside the *colored circles* denote the year of observation (i.e., 11, 12, 13, and 14 denote years 2011, 2012, 2013, and 2014, respectively). Plots without numbers represent the original (i.e., pre-tsunami) taxon composition for each observational flat. (**b**) The nMDS diagram showing vectors representing relative loading factors for the first 15 taxa

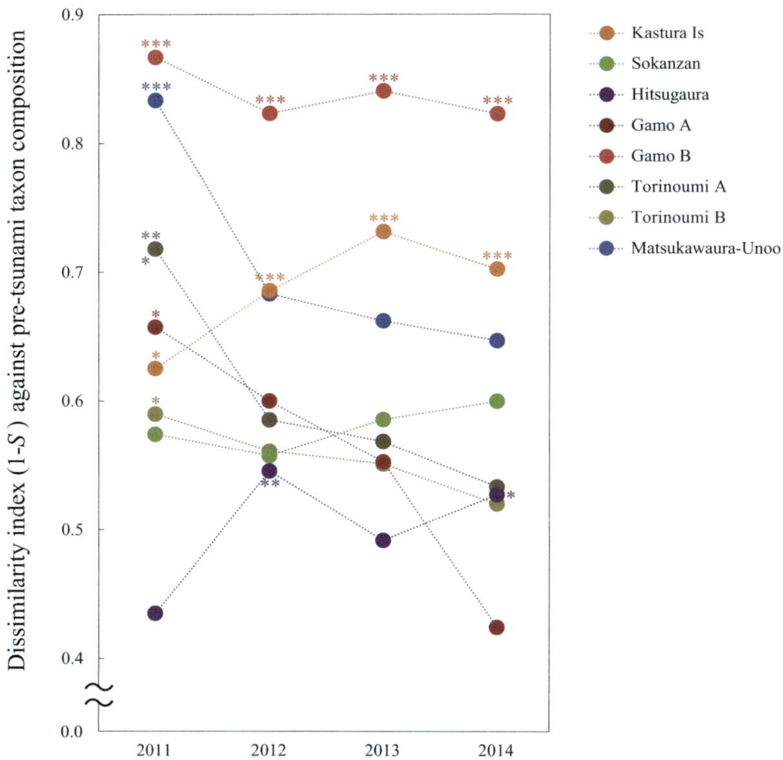

Fig. 11.4 Time-course changes in the dissimilarity indices for taxon compositions before and after the tsunamis in each observational flat. *Asterisks* denote that the taxon compositions were significantly different compared to before the tsunami.*$p<0.05$, **$p<0.01$, ***$p<0.001$ (after the sequential Bonferroni corrections)

the large changes in 2011 at these intertidal flats were apparently due to the disturbance of the tsunami.

Significant difference in taxonomic composition assessed by dissimilarity indices was also found in the Kastura Is. However, at this site, taxon richness increased immediately after the tsunamis. In Hitsugaura and Sokanzan, although the taxon richness decreased, the value of the dissimilarity index in taxon composition before and immediately after the tsunami was explained by chance, suggesting that taxon composition immediately after the tsunami was not largely changed. The intertidal flat examined in Kastura Is is located behind the island's hill against the southward ocean, and inundation heights of the tsunamis were less than 5 m at the Hitsugaura and Sokanzan flats. Therefore, we could expect the macrobenthic communities to be less damaged by the tsunamis in these flats (Urabe et al. 2013).

In 2012, however, taxon richness at the seven sites where it decreased immediately after the tsunami struck returned to the pre-tsunami levels. Taxon richness continued to increase even in 2013 at these flats, resulting in higher taxon richness than before the tsunamis. This finding suggests that the tsunamis caused by the

Great East Japan Earthquake temporally created a number of vacant niches by reducing the abundance and taxon richness of indigenous animals. Accordingly, these niches may have been occupied by taxa not frequently observed before the tsunami. This inference is supported by the colonization of, for example, a polychaeta *Armandia amakusaensis*. This species had never been observed in the study sites before the tsunamis, but was found commonly in the six census sites afterward, although it disappeared in 2014. According to Posey et al. (1996) and Okey (1997), *A. amakusaensis* is likely an opportunistic species in coastal ecosystems including intertidal flats since it increases the population size after disturbance events.

Inference that the increase in taxon richness in 2012 and 2013 was due to colonization of opportunistic taxa was also supported by the fact that in the most intertidal flats taxon richness decreased to pre-tsunami levels in 2014. Theoretically, the number of opportunistic species increases immediately after a disturbance, but declines when indigenous species recover (Levin et al. 1996, Joydas et al. 2012). The decrease in taxon richness in 2014 may have reflected the recovery of the intertidal flat communities toward their pre-tsunami levels.

The recovery of these communities was also supported by the temporal trajectory of taxon composition. As mentioned above, taxon composition was greatly altered immediately after the tsunami struck in the census sites at Gamo, Torinoumi, and Matsukawaura-Unoo. However, the taxon composition of these flats returned to pre-2011 levels in 2012, except for site B of Gamo. This finding suggests that the macrobenthic communities were in the process of recovery toward their original state. At the Torinoumi and Matsukawaura-Unoo flats, however, taxon compositions in 2012 and 2013 tended to move away from those before the tsunami struck. These results suggest that the communities may not have completely recovered to their pre-tsunami state. Indeed, taxon compositions changed somewhat haphazardly for 4 years after the tsunami at Kastura Is, Sokanzan, and Hitsugaura, where notable effects of the tsunamis were not detected in species richness and taxon composition. The results imply that taxon composition changed year to year to some extent regardless of the tsunami disturbance. The macrobenthic communities are dynamic and therefore may never attain the exact same structures as before the tsunamis in any of intertidal flats.

In site B of Gamo, taxon composition differed substantially from that in site A before the tsunamis, but were close to that in site A year to year afterward. One possible reason for this is that the tsunamis caused a drastic change in the environmental conditions of site B. Although site A of Gamo was exposed to sandy beach and located on a mixture of sandy and muddy sediments, site B was surrounded by dense reed marshes before the 2011 Tohoku earthquake tsunami. When the tsunami struck, almost all the reed marshes were lost in Gamo (Kanaya et al. 2012, 2016). The mashes had not yet recovered at all even by the time of the 2014 survey, and thus, the environmental conditions were similar between sites B and A of Gamo. Accordingly, the taxon composition of macrobenthic community gradually became more similar betweenn these sites after the tsunamis. This argument implies that although taxon compositions of the macrobenthic communities in intertidal flats are somewhat dynamic, and thus partly stochastically determined, they are also largely shaped by the site-specific environmental conditions of the flats.

Acknowledgments We thank T. Uchino, G. Kanaya, and K. Kinoshita, Wetlands International Japan, undergraduate students in the biology course at Tohoku University, and all the participants of the citizen research program by Earthwatch Japan for their field support. This study was financially supported by Mitsui & Co., Ltd. Environment Fund (F11-F1-020), Keidanren Committee on Nature Conservation, and the Tohoku Ecosystem-Associated Marine Sciences (TEAMS) project.

References

Balthis WL, Hyland JL, Bearden DA (2006) Ecosystem responses to extreme natural event: impact of three sequential hurricanes in fall 1999 on sediment quality and condition of benthic fauna in the Neuse River estuary, North Carolina. Environ Monit Assess 119:367–389

Bongers T, Bongers M (1998) Functional diversity of nematodes. Appl Soil Ecol 10:239–251

Botter-Carvalho ML, Carvalho PVVC, Santos PJP (2011) Recovery of macrobenthos in defaunated tropical estuarine sediments. Mar Pollut Bull 62:1867–1876

Bourrouilh-Le Jan FG, Beck C, Gorsline DS (2007) Catastrophic events (hurricanes, tsunami and others) and their sedimentary records: introductory notes and new concepts for shallow water deposits. Sed Geol 199:1–11

Cardoso PG, Raffaelli D, Lillebø AI, Verdelhos T, Pardal MA (2008) The impact of extreme flooding events and anthropogenic stressors on the macrobenthic communities' dynamics. Estuar Coast Shelf Sci 76:553–565

Collins SL, Glenn SM, Gibson DJ (1995) Experimental analysis of intermediate disturbance and initial floristic composition: decoupling cause and effect. Ecology 76:486–492

Connell JH (1978) Diversity in tropical rain forests and reefs. Science 199:1302–1310

De la Huz R, Lastra M, Junoy J, Castellanos C, Viéitez JM (2005) Biological impacts of oil pollution and cleaning in the intertidal zone of exposed sandy beaches: preliminary study of the "Prestige" oil spill. Estuar Coast Shelf Sci 65:19–29

Engle VD, Hyland JL, Cooksey C (2009) Effects of hurricane Katrina on benthic macroinvertebrate communities along the northern Gulf of Mexico coast. Environ Monit Assess 150:193–209

Fujioka Y, Tabuchi R, Hirata Y, Yoneda R, Patanaponpaiboon P, Poungparn S, Shibuno T, Ohba H (2008) Disturbance and recovery of mangrove forests and macrobenthic communities in Andaman Sea, Thailand following the Indian Ocean Tsunami. In: Proceedings of the 11th international coral reef symposium, pp 1225–1229

Göthlich L, Oschlies A (2015) Disturbance characteristics determine the timescale of competitive exclusion in a phytoplankton model. Ecol Model 296:126–135

Jaramillo E, Dugan JE, Hubbard DM, Melnick D, Manzano M, Duarte C, Campos C, Sanchez R (2012) Ecological implications of extreme events: footprints of the 2010 earthquake along the Chilean Coast. PLoS One 7:1–8

Joydas TV, Qurban MA, Al-Suwailem A, Krishnakumar PK, Nazeer Z (2012) Macrobenthic community structure in the northern Saudi waters of the Gulf, 14 years after the 1991 oil spill. Mar Pollut Bull 64:325–335

Kanaya G, Suzuki T, Maki H, Nakamura Y, Miyajima Y, Kikuchi E (2012) Effects of the 2011 tsunami on the topography, vegetation, and macrobenthic fauna in Gamo Lagoon, Japan. Jpn J Benthol 67:20–32

Kanaya G, Suzuki T, Kanou K, Kondoh T, Sato-Okoshi W, Kikuchi E (2016) Ecological consequences of the tsunamis caused by the Great East Japan Earthquake and subsequent disturbance events in a shallow brackish lagoon in Sendai Bay, Japan. In Urabe, Nakashizuka (eds)

Ecological impacts of tsunamis on coastal ecosystems: Lessons from the Great East Japan Earthquake, Springer, pp 85–104
Krishnankutty N (2006) Effects of 2004 tsunami on marine ecosystems – a perspective from the concept of disturbance. Curr Sci 90:772–773
Legendre P, Legendre L (1998) Numerical ecology. Elsevier, Amsterdam, p 853
Levin LA, Talley D, Thayer G (1996) Succession of macrobenthos in a created salt marsh. Mar Ecol Prog Ser 141:67–82
Lomovasky B, Firstater FN, Salazar AG, Mendo J, Iribarne OO (2011) Macro benthic community assemblage before and after the 2007 tsunami and earthquake at Paracas Bay, Peru. J Sea Res 65:205–212
Miura O, Sasaki Y, Chiba S (2012) Destruction of population of *Batillaria attramentaria* (caenogastropoda: batillariidae) by tsunami waves of the 2011 Tohoku earthquake. J Molluscan Stud 78:377–380
Mori A (2010) Disturbance ecology clarifies the non-equilibrium nature of forest ecosystems. Japanese J Ecol 60:19–39 (in Japanese with English abstract)
Mori M, Takahashi T, The 2011 Tohoku Earthquake Tsunami Joint Survey Group (2012) Nationwide post event survey and analysis of the 2011 Tohoku earthquake tsunami. Coast Eng J 54:1–27
Nakaoka M, Tanaka Y, Mukai H, Suzuki T, Aryuthaka C (2006) Tsunami impacts on biodiversity of seagrass communities in the Andaman Sea, Thailand: (1) seagrass abundance and diversity. The Nagisa World Congress, pp 49–56
Okey TA (1997) Sediment flushing observations, earthquake slumping, and benthic community changes in Monterey Canyon head. Cont Shelf Res 17:877–897
Perkol-Finkel D, Airoldi L (2010) Loss and recovery potential of marine habitats: an experimental study of factors maintaining resilience in subtidal algal forests at the Adriatic Sea. PLoS One 5:1–11
Pickett STA (1980) Non-equilibrium coexistence of plants. Bull Torrey Bot Club 107:238–248
Pierce S, Luzzaro A, Caccianiga M, Ceriani RM, Cerabolini B (2007) Disturbance is the principal α-scale filter determining niche differentiation, coexistence and biodiversity in an alpine community. J Ecol 95:698–706
Pillay D, Perissinotto R (2008) The benthic macrofauna of the St. Lucia Estuary during the 2005 drought year. Estuar Coast Shelf Sci 77:35–46
Posey M, Lindberg W, Alphin T, Vose F (1996) Influence of storm disturbance on an offshore benthic community. Bull Mar Sci 59:523–529
Roxburgh SG, Shea K, Wilson JB (2004) The intermediate disturbance hypothesis: patch dynamics and mechanisms of species coexistence. Ecology 85:359–371
Schlacher TA, Holzheimer A, Stevens T, Rissik D (2011) Impacts of the 'Pacific Adventurer' oil spill on the macrobenthos of subtropical sandy beaches. Estuar Coasts 34:937–949
Sousa WP (1984) The role of disturbance in natural communities. Ann Rev Ecol Syst 15:353–391
Stubbington R, Boulton AJ, Little S, Wood PJ (2015) Changes in invertebrate assemblage composition in benthic and hyporheic zones during a severe supraseasonal drought. Freshw Sci 34:344–354
Suzuki T, Sasaki M (2010) Civil procedure for researching benthic invertebrate animals inhabiting tidal flats in eastern Japan. Plankton Benthos Res 5:221–230
Turner MG, Dale VH (1998) Comparing large, infrequent disturbances: what have we learned? Ecosystems 1:493–496
Urabe J, Suzuki T, Nishita T, Makino W (2013) Immediate ecological impacts of the 2011 tohoku earthquake tsunami on intertidal flat communities. PLoS One 8(5):1–6
Whanpetch N, Nakaoka M, Mukai H, Suzuki T, Nojima S, Kawai T, Aryuthaka C (2010) Temporal changes in benthic communities of seagrass beds impacted by a tsunami in the Andaman Sea, Thailand. Estuar Coast Shelf Sci 87:246–252
White PS, Pickett STA (1985) Natural disturbance and patch dynamics: an introduction. In: Pichett STA, White PS (eds) The ecology of natural disturbance and patch dynamics. Academic, New York, pp 3–13

Chapter 12
The Effects of Liquefaction, Tsunami, and Land Subsidence on Intertidal Mollusks Following the Great East Japan Earthquake

Kenji Okoshi

Abstract The Great East Japan Earthquake that occurred on 11 March 2011 caused a great deal of damage to the organisms that inhabit the coast of the Tohoku district of Japan. In this paper, I mainly focus on the earthquake's impact on bivalves that inhabit the sandy intertidal zone, with specific emphasis on the short-term effects from liquefaction and the tsunamis and the long-term effects of land subsidence. As previous studies have not discussed the details of the separate physical events of this type of natural disaster, this study divided it into three main events: the initial earthquake including liquefaction, subsequent tsunamis, and land subsidence. During the liquefaction period, bivalves were brought to the surface with jets of water and were then swept away by the tsunamis. As such, both the magnitude of the tsunamis and length of the liquefaction period were factors critical to the survival of clams. The tsunamis have also contributed to dispersion of benthic organisms. For example, the redistribution of oysters across the mud bottom increased their range. In addition to these events, land subsidence resulted in marked changes to the vertical position of the intertidal zone. These large changes in the physical environment cannot recover in the short term, and the readjustment of tidal communities is expected to take several decades. However, land subsidence has also formed new intertidal zones, which will ultimately promote the colonization of various organisms. At some sites where the silt content of the bottom sediments increased due to land subsidence, the dominant species have changed from those that prefer a sandy bottom to those that prefer a muddy bottom. These disturbances to the ecosystem have destabilized these bivalve species resulting in fluctuations in abundance and community structure for at least 3 years following the earthquake. Finally, continual monitoring studies are needed to understand the ecological impacts resulting from the earthquake, and the artificial seawall and tidal flat that have been constructed since the quake, on benthic organisms inhabiting the coastline of the Tohoku district.

K. Okoshi (✉)
Laboratory of Marine Ecology, Graduate School of Science, Toho University, Miyama 2-2-1, Funabashi, Chiba 274-8510, Japan
e-mail: kenji.okoshi@env.sci.toho-u.ac.jp

J. Urabe, T. Nakashizuka (eds.), *Ecological Impacts of Tsunamis on Coastal Ecosystems*, Ecological Research Monographs,
DOI 10.1007/978-4-431-56448-5_12

Keywords Mollusk • Earthquake • Tsunami • Liquefaction • Subsidence • Disturbance • Tidal flat

12.1 Introduction

The Great East Japan Earthquake that occurred on 11 March 2011 caused significant damage to the organisms that inhabit the coast of the Tohoku district of Japan (Suzuki 2011; Miura et al. 2012; Okoshi 2012a, b, 2013; Seike et al. 2013; Takami et al. 2013; Urabe et al. 2013; Kanaya et al. 2014; Okoshi et al. 2014; Okoshi 2015). The impacts of the earthquake and subsequent tsunamis on commercially important mollusk species, such as abalone, oysters, and scallops, have been of particular concern. However, far less is known about the post-quake impact on commercially important species, such as the Manila clam *Ruditapes philippinarum* and other species that appear in the Red Data list as endangered species. Before the earthquake, I conducted a series of field surveys along the Pacific coast of the Tohoku district over several years to examine ecological impacts of the invasive moon snail *Euspira fortunei* (*Laguncula pulchella*) on commercially important clams (Okoshi 2004; Okoshi and Sato-Okoshi 2011; see also Ohtsuki et al. 2016). The survey covered coastal areas from the inner bay of Soma City, Fukushima Prefecture, to Ishinomaki City, Miyagi Prefecture. During these surveys, I also examined other benthic organisms, especially mollusks. When the 2011 Great East Japan Earthquake occurred, I was undertaking fieldwork on a tidal flat as part of these field surveys. My experience with these past field surveys provided me a unique opportunity to examine the effects of the 2011 Great East Japan Earthquake on a variety of molluscan species. I, therefore, continued to study molluscan species in the Pacific coast of the Tohoku district after the earthquake. In this paper, I focus primarily on bivalves that inhabit the sandy intertidal zone, describing the short-term effects of liquefaction and the tsunamis and the long-term effects from land subsidence on these bivalve populations. I also briefly discuss the possible impacts resulting from the post-quake construction of seawalls and artificial tidal flats.

12.2 Twenty-Minute Post-Quake Event

The earthquake and subsequent tsunamis did not arrive at the coast simultaneously. There was an interval of approximately 10–20 min between the earthquake and the arrival of the first waves depending on the distance of the shore from the earthquake's epicenter off the coast of Sanriku. The areas located further from the epicenter experienced a longer delay and were more severely subjected to the post-quake events prior to the arrival of the tsunami wave. Many questions remain in regard to what happened to the coastline during this period.

At 14:46 p.m. when the earthquake occurred, the Pacific coastline from Tohoku to the Tokyo Bay was almost at the maximum neap low tide. As such, at the time of the earthquake, the sandy flats were emerged. When the earthquake occurred, I was performing field work at the Yatsu tidal flats, which is located at the most inner part of Tokyo Bay. Three minutes prior to the earthquake, the water was clear and I could observe the bottom through the surface of the water (Okoshi 2012b, 2015). During the earthquake, the sediment surface of the tidal flats moved in a wavelike fashion, causing a rippling/sparkling effect as the sunlight reflected off the wet tidal flat surface. The movement of sediments was not uniform, meaning that different parts of the tidal flat were influenced differently by the shaking. The ducks and gulls on the tidal flat took flight and then landed once the earthquake had subsided.

Liquefaction occurred just after the earthquake before the arrival of the tsunami. Liquefaction is a phenomenon where the strength and stiffness of a soil are weakened by earthquake shaking. Water spurted from cracks in the soil, which looked like fountains. We observed the emergence of many rifts and jets of water around the Yatsu tidal flats (Okoshi 2012a, b, 2013; Okoshi et al. 2014; Okoshi 2015). Many cracks caused by the earthquake were also observed around the Yatsu and the Sanbanze tidal flats, which connect to the Yatsu tidal flats via two channels. Twenty minutes after the earthquake, the water had changed from clear to muddy around the Yatsu tidal flats, and this water flowed to the Sanbanze tidal flats through these two channels. There were many sand boils that erupted on the Sanbanze tidal flats after the earthquake (Okoshi 2013, 2015, Fig. 12.1), which are thought to be caused by ground liquefaction.

According to the Kahoku Shinpo newspaper article of 25 August 2011, bivalves like the sand gaper clam *Mya arenaria oonogai*, which are normally distributed in deep mud, were found at the sediment surface in Gamo tidal flats in Miyagi Prefecture during liquefaction. I found the same phenomena at Matsukawaura Lagoon (Fig. 12.2) in Fukushima Prefecture in 8 April 2011 (Okoshi 2012a, b, 2013; Okoshi et al. 2014; Okoshi 2015). It is likely that these bivalves were pushed out from the deep sediments by these jets of water, leaving many of the bivalves that live submerged in the sediments along the Pacific coast of Tohoku exposed on the tidal flat surface prior to the arrival of the tsunamis. Given the weight of the large sand gaper clam foot is less than 1 % of their total weight, individual clams of large size would not be able to bury themselves immediately once brought to the surface (Okoshi unpublished data). Hence, a number of infaunal organisms that were left exposed on the tidal flat surface are likely to be swept away by the tsunamis.

Many broken shells of the trough shell *Mactra veneriformis* were observed along the coastline of the Sanbanze tidal flats in April 2011 (Okoshi 2015). However, only a few broken shells of the Manila clam *Ruditapes philippinarum* were found at the same time and area. While the shell sizes of these two species are similar, Manila clams have thicker and heavier shells than *Mactra*. It is likely that the heavy thickened shell was less prone to collapse and more resist to physical forces induced by the earthquake shaking. In addition, individuals with heavy shells are more likely to resist the upward movements of seawater in the sediment, which enabled them to escape from being washed away by the subsequent tsunamis at the Sanbanze tidal flats.

Fig. 12.1 Sand boils that erupted in Sanbanze tidal flats in Tokyo Bay after the earthquake (From Okoshi 2013, 2015)

Fig. 12.2 Damaged bivalves found in Matsukawaura Lagoon after the 2011 Great East Japan Earthquake. Soil liquefaction due to the earthquake pushed up many underground organisms such as *Mya arenaria oonogai* to the sediment surface. More than half of the clams survived (8th of April 2011)

Thus, the difference in shell thickness and weight might have biased the survival rates toward the Manila clam compared to the trough shell. Unlike the sand gaper clam, the Manila clam can bury itself in the sand immediately after being unearthed. Given the lag time between earthquake and tsunamis, depending on the distance from the quake epicenter to the coast, highly mobile species such as the Manila clam may have been more able to avoid damage and survive compared to slower moving species. This implies that the vulnerability of bivalve species to earthquake events, including liquefaction and tsunamis, is dependent on a combination of animal's microhabitat, mobility, and shell size, thickness, and weight.

It may also be considered that the impact on clams with different morphologies and existing in different niches of the tidal flats will be affected differently depending on the various degrees of severity and combination of earthquake, post-quake events, and tsunami. Additionally, the impact on the Manila clam would be less in the case where only an earthquake occurs and not a tsunami, while the sand gaper clam may only be affected by liquefaction. The large tsunamis caused by the massive earthquake that occurred in Chile in 1960 resulted in extensive damage to coastal villages in Japan when these reached the Japanese islands about 1 day after the earthquake (Kato et al. 1961). Unfortunately, no reports were made on the impact on marine organisms. Owing to the distance of this earthquake's epicenter from Japan, there was no liquefaction or land subsidence along the coast of Japan. The separate impacts of earthquakes and tsunamis have often been confused with respect to their influence on marine life, but it is necessary to consider each separately.

12.3 Impacts and Processes of Tsunami Events

Tsunamis caused by the Great East Japan Earthquake displaced many subtidal organisms to the land. On the coast of the Ishinomaki Bay, the Purple Washington Clam *Saxidomus purpurata*, which inhabits the sandy subtidal zone, was washed inland with subtidal flounders by the large waves of the tsunamis. The horse mussel *Modiolus kurilensis* was also swept to the supertidal zone still attached to rocks by their byssus threads (Okoshi 2012b, Fig. 12.3). As such, the tsunami waves displaced marine organisms originally living in the subtidal zone to inter- and super tidal zones, as reported in various studies (Suzuki 2011; Okoshi 2012b; Urabe et al. 2013). This redistribution of species complicates evaluations of species richness and diversity as displaced individuals may take time to die. Researchers should therefore be aware of the normal vertical distribution range for a given species until the new zonation stabilizes. Conversely, the strong undertow generated by the outgoing tsunamis resulted in intertidal and subtidal organisms being swept away to the deep sea. For example, after the tsunamis, a large number of sea urchins that normally inhabited shallow coastal waters were found alive in the deeper water at Oshika Peninsula (Takami et al. 2013). Marine organisms that were displaced inland were more likely to die, while those swept offshore or into deeper waters by the undertow appear to have a greater chance of survival.

Fig. 12.3 The horse mussel *Modiolus kurilensis* (*arrow*) exposed and still stuck to the rocks by byssus threads (7 April 2011)

The tsunamis directly affected benthic organisms by destroying or redistributed benthic substrate. Oysters have lost mobility and live cemented mostly to hard substrates. Because mud and sand particles are very small, there are few firm bases for the attachment of oyster larvae in intertidal mud flats (Komatsu et al. 2002). The Japanese oyster *Crassostrea gigas* is a reef-building oyster. Huge oyster reefs that were developed at the mouth of the Matsukawaura Lagoon (Fig. 12.4) were destroyed by the tsunamis, and pieces of the reefs, with living oysters, were spread around the muddy and sandy lagoon bottoms. As the results, new oyster reefs had formed on these redistributed pieces after the tsunami (Fig. 12.4). For the past 3,500 years, massive earthquakes, like the Great East Japan Earthquake in 2011, have occurred at least seven times (source: The Japan Times News, Jan. 27, 2012). Thus, earthquakes with tsunamis have occurred in ~500 year cycles along the Pacific coast of northern Japan (Okoshi 2015) and may have contributed to the creation of new habitats for sessile organisms such as the oyster. Conversely, the coast of China and the Korean Peninsula are rarely subjected to tsunamis ("List of historical tsunamis" in Wikipedia), so few opportunities would exist for this form of dispersal in these areas. Even within the same species, whether living in the place where the tsunami is coming or not has a possibility to affect the process of expansion of distribution areas.

The earthquake and tsunamis resulted in the destruction of a large number of houses and buildings, which collapsed and were swept out to sea as debris. The movements of debris by the tsunami waves were likely to have crushed submerged rocks and stripped them of oysters (Okohsi 2012b, Fig. 12.5). It is therefore possible that many marine organisms died from being crushed by this debris.

Fig. 12.4 (**a**) Huge oyster reefs in the mouth of the bay part of Matsukawaura Lagoon in Fukushima before the tsunami (2009); (**b**) A small oyster reef constructed in the inner part of Matsukawaura Lagoon (2015)

Some species of infaunal bivalves seemed to be affected by the disturbance caused by the tsunamis. In Matsukawaura Lagoon, Soma City, tsunamis ca.9 m in height struck (Ozaki 2011). The tsunamis swept away the sandy bottom with many infaunal organisms (Fig. 12.2). The infaunal bivalves exposed by this force were probably not killed instantly. Some adult individuals of deep burrowing bivalves *Macoma contaculata* and *M. arenaria oonogai* survived on the tidal surface in

Fig. 12.5 Dead oyster shells that were likely broken by repeated collisions with rubble. Many left-hand valves without the right-hand valve attached to rock were observed after the tsunami

Matsukawaura Lagoon at least until the end of June 2011 when we made a field census survey (Okoshi 2012b, 2013, 2015). Unlike the Manila clam and sand gaper clam, new recruitment of juvenile *M. contaculata* was not confirmed in the 3 years after the earthquake. These bivalves may have been buried by the sand and then died (Okoshi 2013). This possibility is inferred based on my own observations and also suggests that the tsunamis affected benthic organisms through various processes: some were killed because the tsunamis exposed them by removing sediment and washing them away, while others were killed by being buried by the displaced sediments.

12.4 Land Subsidence: Long-Term Events

Land subsidence results in a downward displacement of the intertidal zone. It is not clear how the sudden occurrence of land subsidence damages marine organisms, but organisms adapted to living within specific vertical zones are disrupted by such changes. At Oshika Peninsula, the land subsided as much as 1 m due to the Great East Japan Earthquake. At Mangoku-ura Lagoon near Oshika Peninsula, land subsidence of ca. 80 cm was observed after the earthquake and the intertidal zone was shifted to the subtidal zone. I examined species that disappeared and those that successfully colonized this transition area of Mangoku-ura Lagoon for 3 years after the earthquake. The observation showed that ecological effects of the land subsidence varied among species and locations. As expected, the impact of land subsidence was more severe for sessile organisms compared with mobile organisms. Although many bivalves were damaged by the liquefaction and tsunamis as described above, recruitments of the juveniles was observed after July 2011 within the tidal zone of the Mangoku-ura Lagoon. Typically, larval oysters are known to attach to the parent adhesion layer at intertidal rocky shores. Due to the land subsidence, the roots and trunks of cedar trees originally grown along shoreline sunk below the seawater and were used as settlement substrata by larval oysters and barnacles (Fig. 12.6). Accordingly, although adult oysters found within the subtidal zone had originally occupied the intertidal zone, juvenile oysters have been able to colonize substrates that had originally been supratidal. Thus, adult and juvenile oysters exhibited an unusually overlapped distribution in the growing season of 2011 (Fig. 12.7). The oysters born before 2011 continue to grow on the subtidal zone, but monitoring is need to determine the fate of individuals growing on the new substrata, such as the trunks of trees.

The snails *Littorina brevicula* and *Batillaria attramentaria*, which are mobile benthic organisms, were found on the shores that were previously land before the earthquake (Okoshi 2012b, 2013). During the low tide, tide pools appeared at a part of newly formed tidal zone at Ohama in Mangoku-ura Lagoon. The tide pools are functionally important places for some aquatic organisms because it is an interface between terrestrial and marine environments. However, the tide pools and their peripheries are highly vulnerable to soil erosion from the surrounding land by meteorological disturbances such as heavy storms and typhoons. Therefore, it was interesting to examine what species were colonized in these new habitats. In September 2011, a large typhoon struck Ohama, resulting in soil entering the tide pools from the surrounding land and changing the environmental conditions drastically. *B. attramentaria* and *Assiminea* spp. were found in the tide pools at Ohama extensively after the typhoon. Many juveniles of the Venus clam *Cyclina sinensis* and some Manila clams were also found in and around the tide pools (Okoshi 2013, 2015). This demonstrates that these species could promptly colonize newly formed intertidal zones after the earthquake. It is expected that these newly formed intertidal zones may be environmentally unstable and change continually and gradually as the physical and chemical conditions such as particle size composition, salinity, and nutrient supplies change due to meteorological events like typhoon and anthropogenic activities.

According to an nMDS analysis on the molluscan communities in Mangoku-ura Lagoon (Okoshi et. al. submitted), the species composition gradually changed where the intertidal zone was turned into the subtidal due to the land subsidence. The dominant species has been changed from species that prefer sandy bottom environments (*R. philippinarum*) to those prefer muddy bottom systems (*M. incongrua* and *C. angustus*) owing to the increase in the silt content of the sediment. Populations of the mussel species, *Musculista senhousia*, numerically dominated on

Fig. 12.6 (**a**) A newly formed intertidal zone due to subsidence by ca. 80 cm in Mangoku-ura Lagoon. (**b**) Close up of the area denoted by the white circle in panel a, showing colorization by oysters and barnacles on roots of cedar trees after the summer in 2011. At the time of high tide they were exposed to the sea

the sediments in some parts of Mangoku-ura Lagoon after the earthquake but showed a large fluctuation in their abundance. Mass colonization and mortality of this mussel species occurred every year after the earthquake. Seaweeds also smothered some species of bivalves that live in the sediment and killed them (Fig. 12.8). There is a possibility that the algal community structure in Mangoku-ura Lagoon has changed due to the subsidence (Norio Kikuchi, personal communication). According to our field surveys in Mangoku-ura Lagoon for the 3 years after the

Fig. 12.7 (**a**) A number of small oyster spats (*arrow*) attached to a big rock, which were newly settled in summer 2011. Area of the oyster larval attachment became elevated after the summer of 2011. The area of 1-year or older oyster attachment was below the sea surface after subsidence. (**b**) Oysters grown on the same rock in summer 2015

earthquake, species composition and abundance of bivalves changed every year, indicating that the bivalve community was compositionally unstable. It is essential to monitor these organisms continually to uncover the long-term effects of the land subsidence caused by the earthquake on aquatic organisms (Okoshi 2015).

In Mangoku-ura Lagoon, sediments from the mountains have been transported artificially to the bottom from October 2013, which kept the depth of the littoral zone shallow and created tidal flats where the Manila clam population are colonized

Fig. 12.8 (**a**) Mussel *Musculista senhousia* found abundantly in June 2015 in Mangoku-ura Lagoon after the subsidence. (**b**) A number of mussels and other infaunal clams were killed in July 2015 due to smothering of seaweed

and developed, which are an important fishery resource for professional fishermen (Okoshi 2015). The construction of tsunami seawalls was planned and has begun along the Pacific coast of the Tohoku regions. New breakwaters have been created by reclaiming the shoreline of Matsukawaura Lagoon in Fukushima (Fig. 12.9). Given these significant natural and artificial alterations to the coastline of East Japan, long-term monitoring is needed to evaluate the continued changes that are expected following these events. Such information is important in order to monitor the health of marine communities along the coast of northeast Japan.

Fig. 12.9 A new seawall constructed around Matsukawa-ura Lagoon. The *white line* indicates the position of the old seawall before the earthquake

Acknowledgments I am grateful to members of the Laboratory of Marine Ecology, Graduate School of Science, Toho University, for their cooperation and assistance during sample collection and field observations. I am also grateful to Dr. J. Shaw of the University of Western Australia for his critical reading and comments on the earlier and final versions of the manuscript. This study was partially supported by a research grant from the Tohoku Ecosystem-Associated Marine Sciences (TEAMS) research program of the Ministry of Education, Culture, Sports, Science and Technology (MEXT).

References

Japan times News. http://www.japantimes.co.jp/news/2012/01/27/national/past-3500-years-saw-seven-m9s/#.VqIzULlf2Uk

Kato Y, Suzuki Z, Nakamura K, Takagi A, Emura K, Ito M, Ishida H (1961) The Chile tsunami of May 24, 1960 observed along the Sanriku coast, Japan, the Chilean tsunami of May 24, 1960. The committee for the field investigation of the Chilean tsunami of 1960, pp 107–125

Kanaya G, Maki H, Suzuki T, Sato-Okoshi W, Kikuchi E (2014) Tsunami-induced changes in a shallow brackish lagoon ecosystem (Gamo Lagoon) in Sendai Bay, Japan. Glob Environ Res 18:35–46

Komatsu T, Chinzei K, Zakhera MS, Matsuoka H (2002) Jurassic soft-bottom oyster *Crassostrea* from Japan. Palaeontology 45:1037–1048

Miura O, Sasaki Y, Chiba S (2012) Destruction of populations of *Batillaria attramentaria* (Caenogastropoda: Batillariidae) by tsunami waves of the 2011 Tohoku earthquake. J Mollusc Stud 78:377–380

Ohtsuki H, Suzuki T, Kinoshita K, Kanaya G, Hirama T, Sato S, Shibata K, Okoshi K and Urabe J (2016) Genetic structures of *Laguncula pulchella* metapopulation in the north east coast of Japan after the tsunamis caused by the Great East Japan Earthquake. In: Urabe, Nakashizuka (eds) Ecological impacts of tsunamis on coastal ecosystems: lessons from the Great East Japan Earthquake, Springer, pp 209–222

Okoshi K (2004) Alien species introduced with imported clams: the clam-eating moon snail *Euspira fortunei* and other unintentionally introduced species. Jpn J Benthol 59:74–82 (in Japanese with English abstract)

Okoshi K (2012a) How do the 3.11 tsunami affect the intertidal animals? Jpn J Benthol 66:117–119 (in Japanese)

Okoshi K (2012b) Effect of earthquake and tsunami to the coastal marine life. In: Iwatsuki K, Domoto A (eds) A catastrophic disaster and biodiversity. Biodiversity Network Japan, Japan, pp 20–25 (in Japanese)

Okoshi K (2013) Changes in the distribution and abundance of intertidal bivalves in sandy shore one and a half years after the Pacific coast of Tohoku earthquake. J Jpn Soc Water Environ 36:44–48 (in Japanese)

Okoshi K, Sato-Okshi W (2011) *Euspira fortunei*: biology and fisheries science of invasive species. Kouseishakouseikaku Press, Tokyo, p 225 (in Japanese)

Okoshi K, Suzuki M, Maruyama Y, Shinohara W (2014) Traces of the earthquake and tsunami observed in shells. Chikyu Mon 36:42–46 (in Japanese)

Okoshi K (2015) Impact of repeating massive earthquakes on intertidal mollusk community in Japan. In: Ceccaldi HJ, Hénocque Y, Koike Y, Komatsu T, Stora G, Tusseau-Vuillemin MH (eds) Marine productivity: perturbations and resilience of socio-ecosystems. Springer, Switzerland, pp 55–62

Ozaki T (2011) Outline of the 2011 off the Pacific coast of Tohoku earthquake (Mw 9.0) –tsunami warnings/advisories and observations. Earth Planets Space 63:827–830

Seike K, Shirai K, Kogure Y (2013) Disturbance of shallow marine soft-bottom environments and megabenthos assemblages by a huge tsunami induced by the 2011 M9.0 Tohoku-Oki earthquake. PLoS One 8:e65417

Suzuki T (2011) Environmental changes in tidal flats and influences on benthic invertebrates caused by the earthquake and tsunami at eastern Japan. J Jpn Soc Water Environ 34:395–399 (in Japanese)

Takami H, Won NI, Kawamura T (2013) Impacts of the 2011 mega-earthquake and tsunami on abalone *Haliotis discus hannai* and sea urchin *Strongylocentrotus nudus* populations at Oshika Peninsula, Miyagi, Japan. Fish Oceanogr 22:113–120

Urabe J, Suzuki T, Nishita T, Makino W (2013) Immediate ecological impacts of the 2011 Tohoku earthquake tsunami on intertidal flat communities. PLoS One 8, e62779

Chapter 13
Ecological Impacts and Recovery of Molluskan Populations After the Tsunami Struck Around Matsushima Bay and Sendai Bay, Miyagi Prefecture, Northeastern Japan

Shin'ichi Sato and Tomoki Chiba

Abstract We investigated impacts of the Great East Japan Earthquake and Tsunami on population dynamics of molluskan species and the recovery processes on the Tona coast in Matsushima Bay and shallow sea in Sendai Bay, Miyagi Prefecture, northeastern Japan. On the Tona coast, *Ruditapes philippinarum*, *Macoma incongrua*, *Pillucina pisidium*, and *Batillaria cumingii* were always common for 10 years before the tsunami (2001–2010), although *R. philippinarum* and *M. icongrua* decreased in number as a result of selective predation after arrival of an alien predator *Laguncula pulchella* in 2002. After the tsunami, all of the above five species drastically decreased in July 2011, but *R. philippinarum* and *M. incongrua* rapidly recovered from 2012 to 2013. Furthermore, *Musculista senhousia* and *Mya (Arenomya) arenaria oonogai*, those were low density before the tsunami, temporarily appeared on the Tona coast in July 2011 and then decreased till June 2013. These results suggest that the molluskan populations on the Tona coast are damaged by the tsunami and the other opportunistic species increased instead of them, but they quickly recovered for 2 or 3 years. By contrast, in the shallow sea of Sendai Bay, only *Raetellops pulchellus* increased at 1 year after the tsunami (July to November 2012), but mean individual density of bivalves decreased 2 years after the tsunami (March 2013). In the shallow sea, the recovery processes of molluskan populations may be slower than those in the coastal area.

Keywords Earthquake tsunami • Matsushima Bay • Molluskan population • *Macoma incongrua* • *Musculista senhousia* • *Ruditapes philippinarum* • Sendai Bay • Tona coast

---

S. Sato (✉)
The Tohoku University Museum, Aoba 6-3, Aramaki, Aoba-ku, Sendai 980-8578, Japan

Present address: Institute of Geosciences, Shizuoka University, Oya 836, Shizuoka 422-8529, Japan
e-mail: sato.shinichi.c@shizuoka.ac.jp

T. Chiba
Institute of Geology and Paleontology, Graduate School of Science, Tohoku University, Aoba 6-3, Aramaki, Aoba-ku, Sendai 980-8578, Japan

J. Urabe, T. Nakashizuka (eds.), *Ecological Impacts of Tsunamis on Coastal Ecosystems*, Ecological Research Monographs,
DOI 10.1007/978-4-431-56448-5_13

13.1 Introduction

On March 11, 2011, the Great East Japan Earthquake (M 9.0) occurred off Miyagi Prefecture, northeastern Japan (Fig. 13.1a), and the massive tsunami attacked the wide area of the Pacific coasts from the Tohoku to the Kanto regions (Mori et al. 2011). This tsunami strongly damaged human communities and also heavily impacted coastal ecosystems (e.g., Hayasaka et al. 2012; Kanaya et al. 2012, 2014; Miura et al. 2012; Seike et al. 2013; Takami et al. 2013; Urabe et al. 2013). Ecological impacts of tsunamis on the benthic communities were also reported at many regions in the world (e.g., Whanpetch et al. 2010; Lomovasky et al. 2011; Jaramillo et al. 2012). However, it is very rare that the ecological data have been collected for long period spanning before and after the catastrophic events.

The Tona coast (Fig. 13.1b), Higashi-Matsushima City, Miyagi Prefecture, where we surveyed the benthic communities every year during 2001 and 2010 (Sato et al. 2012; Chiba and Sato 2013), was also physically disturbed by the huge tsunami in March 2011. The height of tsunami around the sampling area was 3.4 m (Mori et al. 2011). Around the Tona coast, the tsunami from the seaside (Matsushima Bay in Fig. 13.1b) was not so high (1.36 m, Mori et al. 2011), because many islands around the mouth of Matsushima Bay weakened the tsunami (Fig. 13.1b). However, the huge tsunami arrived at the Tona coast from the landside through the Nobiru coast (arrow in Fig. 13.1b). The height of tsunami around the Nobiru coast was 6.93 m (Mori et al. 2011). Therefore, many cars and broken dikes were flowed into the Tona coast (Fig. 13.2). Furthermore, the depth of land subsidence that occurred around the sampling area was –0.38 to –0.43 m (Geographical Survey Institute of Japan 2011), and most sampling stations of the Tona coast became not exposed even at low tide after the tsunami (Fig. 13.1d).

We investigated the density fluctuation of molluskan populations between 2001 and 2010 on the Tona coast tidal flat. Since 2001, *Ruditapes philippinarum*, *Macoma incongrua*, *Pillucina pisidium*, and *Batillaria cumingii* have been always common, and then *Laguncula pulchella* (formerly known as *Euspira fortunei*, Torigoe and Inaba 2011) as an alien predator for them appeared in 2002 (Sato et al. 2012). These species were always dominant till 2010, before the tsunami. Thus, the main objective of the present study is to document the density fluctuation in these abundant molluskan populations after the tsunami developed by the Great East Japan Earthquake. Based on these data, we discuss ecological impacts of the earthquake and tsunami on population dynamics of dominant molluskan populations.

Fig. 13.1 (continued) attacked and passed over the land to the Tona coast (see *text*). The *black circles* in (**c**) and (**d**) represent the locations of the 16 sampling stations, with their station numbers, on the Tona coast. *Broken lines* in (**c**) and (**d**) mean low water line of spring tides in June 2001 and July 2011, respectively. *Gray zone inside of the broken line* in (**d**) indicates deeper area (>1 m in depth) because of the broken dikes. (**e**) Map showing sampling site in the shallow sea of Sendai Bay. The *black circles* in (**e**) represent the locations of the 18 sampling stations, with their station numbers. *Broken line* in (**e**) means bathymetric line of 20 m in depth

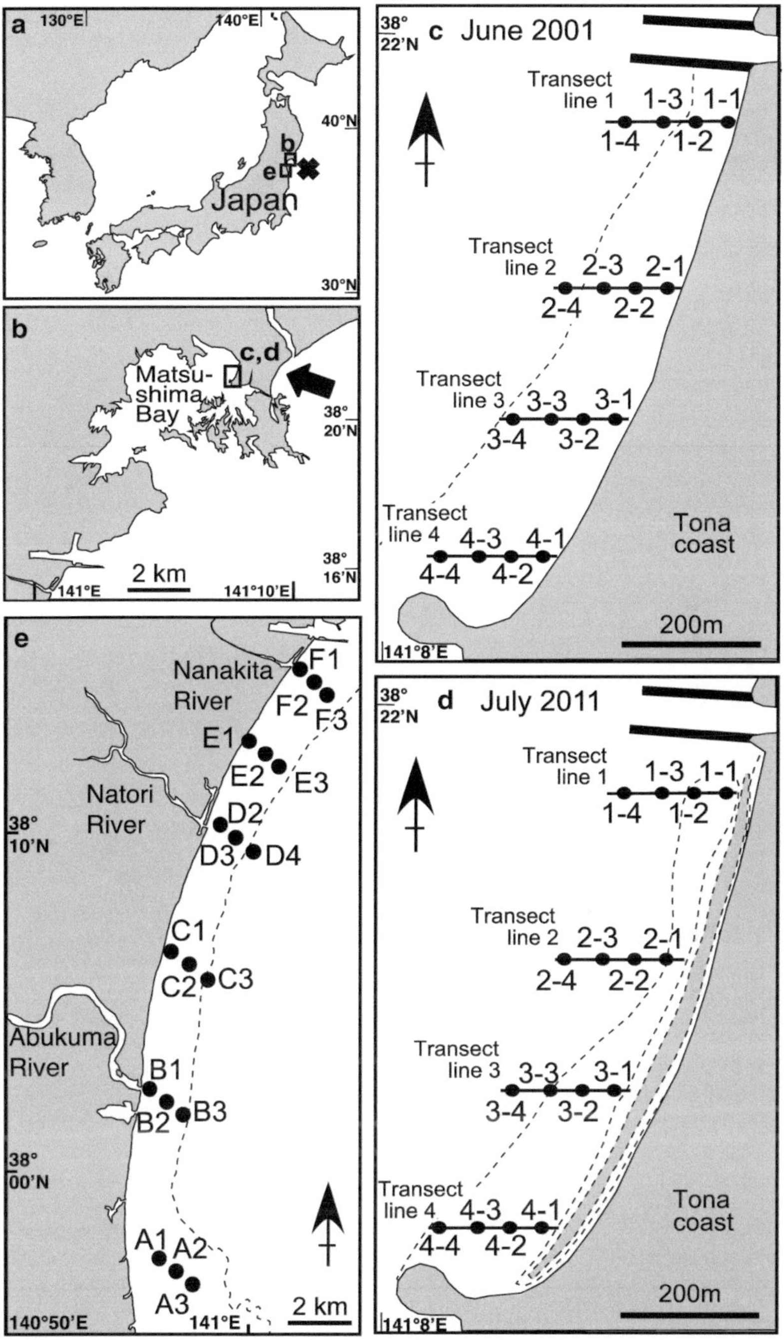

Fig. 13.1 (**a**) Locality map of study areas and the epicenter of the Great East Japan Earthquake (×). (**b–d**) Map showing sampling site around the Tona coast tidal flat in Matsushima Bay, Miyagi Prefecture, northeastern Japan (Sato et al. 2012). The *black box* in (**b**) indicates the sampling site on the Tona coast. The *arrow* in (**b**) shows the Nobiru coast, where the huge tsunami

Fig. 13.2 Photographs around the Tona coast after the tsunami (taken by S Sato in June 2011). (**a**) and (**b**) show the southern and northern part of the sampling site, respectively

In addition, we investigated temporal change of bivalves in shallow sea of Sendai Bay from March 2012 to November 2013 (Fig. 13.1e). Using these data, we compared the recovery processes of molluskan populations after the tsunami between the coastal tidal flat and shallow sea.

13.2 Materials and Methods

On the Tona coast, quadrat sediment samples were collected annually from 16 fixed stations during June 2001 and June 2014 (Fig. 13.1c, d). Four transect lines named as the transect lines 1–4 were established at 250-m intervals along the coast, and four sampling stations were fixed on each transect line at 50-m intervals from the land (Fig. 13.1c). Each sampling station was located using a global positioning system (GPS) receiver (Pockenavi Mini, Empex Ltd.), and samples from one quadrat (25 × 25 cm) were obtained by digging to a depth of 20 cm using a shovel. The samples were sieved with a 2-mm mesh, and living benthic animals were removed from the debris in the laboratory.

We also periodically monitored the number of species and mean densities of bivalves in 18 fixed stations around the shallow sea (about 5–30 m in depth) of Sendai Bay between March 2012 and November 2013 (Fig. 13.1e). In each sampling station, sediment samples were taken using an Ekman-Birge grab (sampling area: 0.02 m^2) or a Smith-McIntyre bottom sampler (sampling area: 0.05 m^2) from a fishing vessel (see Table 13.3). The samples collected were sieved with a 1-mm mesh, and living benthic animals were removed from the debris in the laboratory. All living mollusks were counted by species and preserved in 70 % ethanol. Water temperature, salinity, and saturation ratio of dissolved oxygen (DO) of the bottom water layers were measured at each sampling station by a multiple water-quality sonde with standardized water (Pro2030, YSI Inc.).

13.3 Results

13.3.1 Tona Coast in Matsushima Bay

Before the tsunami, five species (i.e., *R. philippinarum*, *M. incongrua*, *P. pisidium*, *B. cumingii*, and *L. pulchella*) mostly dominated for 10 years (Table 13.1). In June 2001, living individuals of *R. philippinarum*, *M. incongrua*, *P. pisidium*, and *B. cumingii* were collected in large numbers from the 16 sampling stations of the Tona coast (Table 13.1, Fig. 13.3). One individual of *L. pulchella* first appeared in June 2002, and the population of this species rapidly increased from 2002 to 2004 (Table 13.1, Fig. 13.3b). Then, the populations of *R. philippinarum* and *M. incongrua* decreased significantly from 2001 to 2004 (Fig. 13.3c, d). However, *P. pisidium* population gradually increased during this period (Fig. 13.3e). After *L. pulchella* decreased from 2004 to 2005, densities of *R. philippinarum* and *M. incongrua* have temporarily increased (Table 13.1, Fig. 13.3). The density of *B. cumingii* gradually decreased from 2001 to 2010, and the density of *P. pisidium* decreased from 2005 to 2008 (Table 13.1, Fig. 13.3a and e).

Table 13.1 Individual density (1/16 m^{-2}) of the seven dominant molluskan species (*Batillaria cumingii*, *Laguncula pulchella*, *Ruditapes philippinarum*, *Macoma incongrua*, *Pillucina pisidium*, *Musculista senhousia*, and *Mya* (*Arenomya*) *arenaria oonogai*) and others collected from each sampling station around the Tona coast tidal flat in Matsushima Bay from June 2001 to June 2014

Transect line	Transect line 1				Transect line 2				Transect line 3				Transect line 4				
Station no.	1–1	1–2	1–3	1–4	2–1	2–2	2–3	2–4	3–1	3–2	3–3	3–4	4–1	4–2	4–3	4–4	Total
a. *Batillaria cumingii*																	
June 11, 2001	8	9	0	7	12	5	5	4	49	6	9	4	0	6	19	11	154
June 29, 2002	20	3	1	–	1	4	3	5	27	0	1	0	1	8	6	3	83
July 5, 2003	26	1	4	–	5	2	1	–	15	2	1	0	1	2	9	12	81
July 5, 2004	13	4	2	1	13	1	1	–	0	6	6	–	6	5	3	7	68
June 25, 2005	11	0	0	0	1	0	1	5	24	5	4	0	8	3	7	1	70
July 15, 2006	21	1	0	0	1	2	1	2	23	3	3	0	1	4	12	5	79
July 2, 2007	10	0	0	2	3	0	1	2	2	1	0	8	1	1	0	4	35
June 23, 2008	18	3	1	1	0	2	0	0	1	1	0	0	1	0	1	0	29
July 11, 2009	10	0	1	2	1	0	0	0	0	0	0	15	0	2	2	1	34
June 28, 2010	1	0	0	1	5	3	10	1	0	0	0	14	0	0	1	3	39
July 2–15, 2011	0	0	0	0	0	0	0	0	0	0	0	0	1	0	0	0	1
June 4, 2012	0	0	0	0	0	0	0	0	0	0	0	0	0	0	0	0	0
June 24, 2013	0	0	0	0	0	0	0	0	0	0	0	0	0	0	0	0	0
June 13, 2014	0	0	0	0	0	0	0	0	0	0	0	0	0	0	0	0	0
b. *Laguncula pulchella*																	
June 11, 2001	0	0	0	0	0	0	0	0	0	0	0	0	0	0	0	0	0
June 29, 2002	0	0	0	–	0	0	0	0	0	0	0	0	1	0	0	0	1
July 5, 2003	0	0	0	–	0	0	0	–	0	0	0	0	1	4	3	1	9
July 5, 2004	0	0	0	0	0	2	0	–	0	4	0	–	0	2	3	1	12
June 25, 2005	0	1	0	1	0	0	0	0	0	0	0	0	0	1	2	0	5
July 15, 2006	0	0	0	0	0	0	0	1	0	1	0	0	1	1	0	0	4
July 2, 2007	0	0	0	0	0	0	0	1	2	0	0	0	0	0	0	0	3

June 23, 2008	0	0	0	0	1	0	0	0	0	0	1	0	0	1	0	0	3
July 11, 2009	0	1	0	0	0	0	0	0	0	0	0	0	0	2	0	0	3
June 28, 2010	1	0	0	0	1	0	0	0	1	0	0	6	0	3	2	1	15
July 2–15, 2011	0	0	0	0	0	0	0	0	0	0	0	0	1	0	0	0	1
June 4, 2012	0	0	0	1	0	0	0	0	0	0	0	0	1	0	0	0	2
June 24, 2013	0	2	0	0	0	0	0	0	0	1	0	0	0	1	1	3	8
June 13, 2014	0	0	0	0	0	0	0	0	0	0	1	0	1	0	0	1	3
c. *Ruditapes philippinarum*																	
June 11, 2001	1	1	0	3	3	3	10	3	0	5	5	7	0	10	42	71	164
June 29, 2002	2	2	1	–	2	0	0	0	0	0	0	0	1	4	2	0	14
July 5, 2003	1	0	1	–	0	0	0	–	0	0	2	0	0	0	0	1	5
July 5, 2004	3	0	0	0	1	0	0	–	2	0	0	–	2	0	0	0	8
June 25, 2005	0	0	0	0	0	0	0	0	43	0	0	0	1	0	0	0	44
July 15, 2006	0	0	0	0	0	0	0	0	10	0	0	0	0	0	0	0	10
July 2, 2007	0	0	0	0	0	0	0	0	0	0	0	0	1	0	0	0	1
June 23, 2008	0	0	0	0	0	0	1	0	1	0	1	0	1	0	7	2	13
July 11, 2009	0	0	0	0	0	0	0	0	0	0	0	0	0	0	3	1	4
June 28, 2010	1	0	0	0	1	0	0	0	0	0	1	14	0	4	0	2	23
July 2–15, 2011	0	0	0	0	1	0	0	0	3	10	1	0	0	0	0	0	15
June 4, 2012	0	4	9	8	3	7	22	12	5	12	6	0	7	4	2	5	106
June 24, 2013	0	1	0	0	0	0	0	1	2	2	1	2	0	0	7	0	16
June 13, 2014	0	1	0	1	1	11	1	1	13	13	1	4	15	2	3	7	74
d. *Macoma incongrua*																	
June 11, 2001	2	10	2	2	0	1	4	2	0	1	1	2	0	0	3	1	31
June 29, 2002	1	3	1	–	0	0	0	0	0	0	2	3	0	0	2	0	12
July 5, 2003	0	3	3	–	2	0	0	–	0	0	0	0	0	1	0	1	10
July 5, 2004	0	0	0	0	0	0	0	–	0	0	0	–	0	1	0	0	1

(continued)

Table 13.1 (continued)

Transect line	Transect line 1				Transect line 2				Transect line 3				Transect line 4				
Station no.	1–1	1–2	1–3	1–4	2–1	2–2	2–3	2–4	3–1	3–2	3–3	3–4	4–1	4–2	4–3	4–4	Total
June 25, 2005	0	0	1	1	2	1	0	1	1	0	0	0	2	0	3	1	13
July 15, 2006	0	0	0	1	0	0	0	1	0	0	0	0	0	0	0	0	2
July 2, 2007	0	0	0	1	0	1	0	0	0	0	0	0	1	0	0	0	3
June 23, 2008	0	0	0	0	0	3	3	2	0	1	0	0	2	1	0	1	13
July 11, 2009	1	0	0	2	1	1	0	3	0	0	1	0	0	0	1	0	10
June 28, 2010	0	1	0	2	0	0	5	0	0	0	2	3	0	1	1	1	16
July 2–15, 2011	0	0	0	1	0	0	0	0	0	0	0	0	0	0	0	0	1
June 4, 2012	9	5	8	6	10	8	5	1	0	27	4	5	3	0	2	10	103
June 24, 2013	2	13	12	6	2	5	3	1	2	8	4	35	1	1	22	6	123
June 13, 2014	3	7	6	2	4	4	1	1	3	4	1	5	2	3	6	5	57
e. *Pillucina pisidium*																	
June 11, 2001	0	0	0	0	1	0	1	1	0	1	1	3	0	3	1	5	17
June 29, 2002	1	3	0	–	0	1	0	0	0	2	3	4	1	2	2	2	21
July 5, 2003	0	0	1	–	2	1	3	–	0	1	3	6	1	0	2	2	22
July 5, 2004	2	0	1	0	6	2	11	–	0	4	0	–	2	3	0	5	36
June 25, 2005	1	0	0	0	8	8	3	7	0	5	11	3	2	2	2	3	55
July 15, 2006	0	0	0	0	2	2	0	0	0	3	0	0	0	3	9	0	19
July 2, 2007	0	0	0	1	2	2	0	0	1	1	4	0	0	2	1	3	17
June 23, 2008	0	0	0	0	6	2	0	0	0	0	0	0	2	0	1	4	15
July 11, 2009	0	0	0	0	2	6	2	0	8	2	2	7	0	0	0	1	30
June 28, 2010	0	0	0	0	1	3	2	0	3	2	6	5	0	2	2	2	28
July 2–15, 2011	0	0	0	0	0	0	0	0	0	0	0	0	0	0	0	0	0
June 4, 2012	0	0	0	1	0	0	1	0	0	0	0	0	0	1	0	0	3
June 24, 2013	0	1	0	0	0	0	0	0	0	0	3	0	1	0	0	0	5
June 13, 2014	0	0	0	0	1	1	1	0	1	1	0	0	0	1	0	0	6

f. *Musculista senhousia*																	
June 11, 2001	0	0	0	0	0	0	0	1	2	2	0	0	0	1	2	0	8
June 29, 2002	0	0	0	–	0	0	0	0	0	0	0	0	0	0	0	0	0
July 5, 2003	0	0	0	–	0	0	0	–	0	0	0	0	0	0	0	0	0
July 5, 2004	0	0	0	0	0	0	0	–	0	0	0	–	0	0	0	0	0
June 25, 2005	0	0	0	0	2	0	0	0	1	0	0	0	0	0	0	0	3
July 15, 2006	0	0	0	0	0	0	0	0	0	0	0	0	0	0	0	0	0
July 2, 2007	0	0	0	0	0	0	0	0	0	0	0	0	0	0	0	0	0
June 23, 2008	0	0	0	0	0	3	1	1	0	0	1	1	1	0	0	0	8
July 11, 2009	0	0	0	0	0	0	0	0	0	1	1	0	0	0	0	1	3
June 28, 2010	1	0	0	0	0	0	0	1	0	0	0	57	0	0	1	0	60
July 2–15, 2011	19	0	0	0	1	3	1	2	471	141	7	1	0	0	0	0	646
June 4, 2012	1	9	0	0	1	137	40	8	13	23	38	0	20	1	6	56	353
June 24, 2013	0	0	0	0	0	0	0	0	0	0	0	0	0	0	1	0	1
June 13, 2014	0	1	2	1	0	1	0	2	31	2	0	9	0	0	0	0	49
g. *Mya (Arenomya) arenaria oonogai*																	
June 11, 2001	1	0	0	0	0	0	0	1	0	0	0	0	0	0	0	0	2
June 29, 2002	0	0	0	–	0	0	0	0	0	0	0	0	0	0	0	0	0
July 5, 2003	0	0	0	–	0	0	0	–	0	0	0	0	0	0	0	0	0
July 5, 2004	0	0	0	0	0	0	0	–	0	0	0	–	0	0	0	0	0
June 25, 2005	0	0	0	0	0	0	0	0	0	0	1	0	0	0	0	0	1
July 15, 2006	0	0	0	0	0	0	0	1	0	0	0	0	0	0	0	0	1
July 2, 2007	0	0	0	0	0	0	0	0	0	0	0	0	0	0	0	0	0
June 23, 2008	0	0	0	0	0	0	0	0	0	0	0	0	0	0	0	0	0
July 11, 2009	0	0	0	0	0	0	0	0	0	0	0	0	1	0	0	0	1
June 28, 2010	0	0	0	2	0	0	0	0	0	0	0	1	0	0	2	0	5

(continued)

Table 13.1 (continued)

Transect line	Transect line 1				Transect line 2				Transect line 3				Transect line 4				
Station no.	1–1	1–2	1–3	1–4	2–1	2–2	2–3	2–4	3–1	3–2	3–3	3–4	4–1	4–2	4–3	4–4	Total
July 2–15, 2011	1	2	0	1	1	0	1	5	6	7	5	3	0	0	0	0	32
June 4, 2012	0	2	0	3	0	0	1	0	0	0	0	0	0	0	0	0	6
June 24, 2013	0	2	0	0	0	1	1	1	0	1	1	2	0	1	1	0	11
June 13, 2014	1	1	0	1	1	1	1	0	0	0	0	1	0	0	1	0	8
h. Other mollusks																	
June 11, 2001	0	1	2	0	0	0	1	0	3	1	1	0	1	1	2	1	14
June 29, 2002	2	9	0	–	0	1	0	0	0	2	1	0	0	0	0	0	15
July 5, 2003	4	2	0	–	2	1	0	–	0	1	1	0	0	1	2	2	16
July 5, 2004	0	2	0	3	0	3	3	–	0	1	1	–	1	0	0	0	14
June 22, 2005	0	3	1	3	1	2	2	1	1	1	1	0	1	0	1	1	19
July 15, 2006	3	1	0	0	9	0	0	1	4	1	0	0	1	2	4	2	28
July 2, 2007	1	0	0	0	2	0	1	0	1	0	4	1	1	1	3	1	16
June 23, 2008	2	0	0	0	1	0	1	1	0	0	0	0	0	0	1	1	7
July 11, 2009	1	0	1	0	0	1	2	1	0	2	1	2	2	2	2	3	20
June 28, 2010	0	0	0	0	0	1	0	0	27	4	0	0	2	2	5	1	42
July 2–15, 2011	1	0	0	0	1	0	0	0	0	8	4	0	0	0	0	0	14
June 4, 2012	2	4	0	1	0	0	0	3	0	1	4	0	1	9	1	7	33
June 24, 2013	0	0	0	0	1	0	2	3	1	4	2	0	0	8	9	9	39
June 13, 2014	0	0	0	1	3	4	6	10	7	9	8	3	4	4	7	6	72

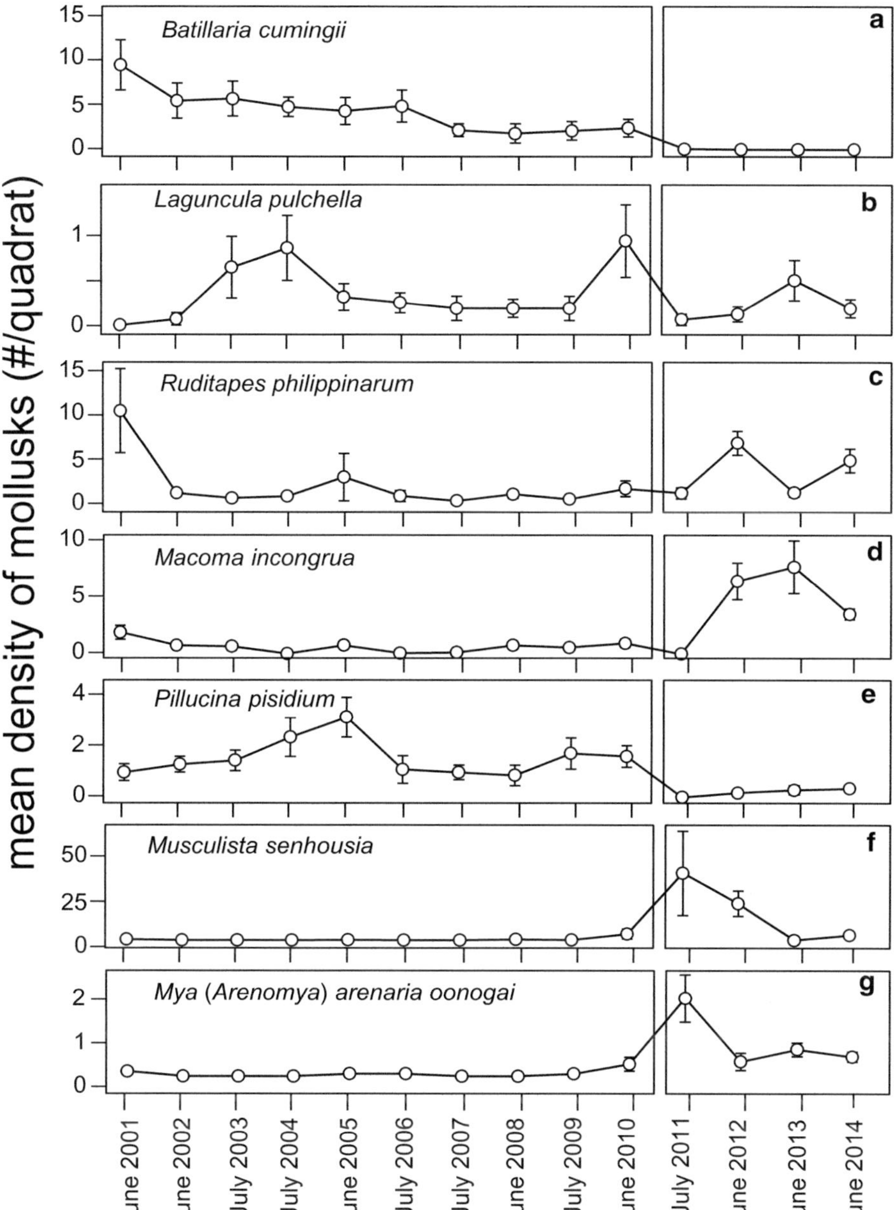

Fig. 13.3 Temporal fluctuations in mean individual densities (1/16 m^2) of the seven dominant mollusk species (i.e., (**a**) *Batillaria cumingii*, (**b**) *Laguncula pulchella*, (**c**) *Ruditapes philippinarum*, (**d**) *Macoma incongrua*, (**e**) *Pillucina pisidium*, (**f**) *Musculista senhousia*, and (**g**) *Mya* (*Arenomya*) *arenaria oonogai*) collected from quadrat samples (partly from Sato et al. 2012). Each bar represents SE on the mean. The *gap* between June 2010 and July 2011 means that the Great East Japan Earthquake occurred during this interval

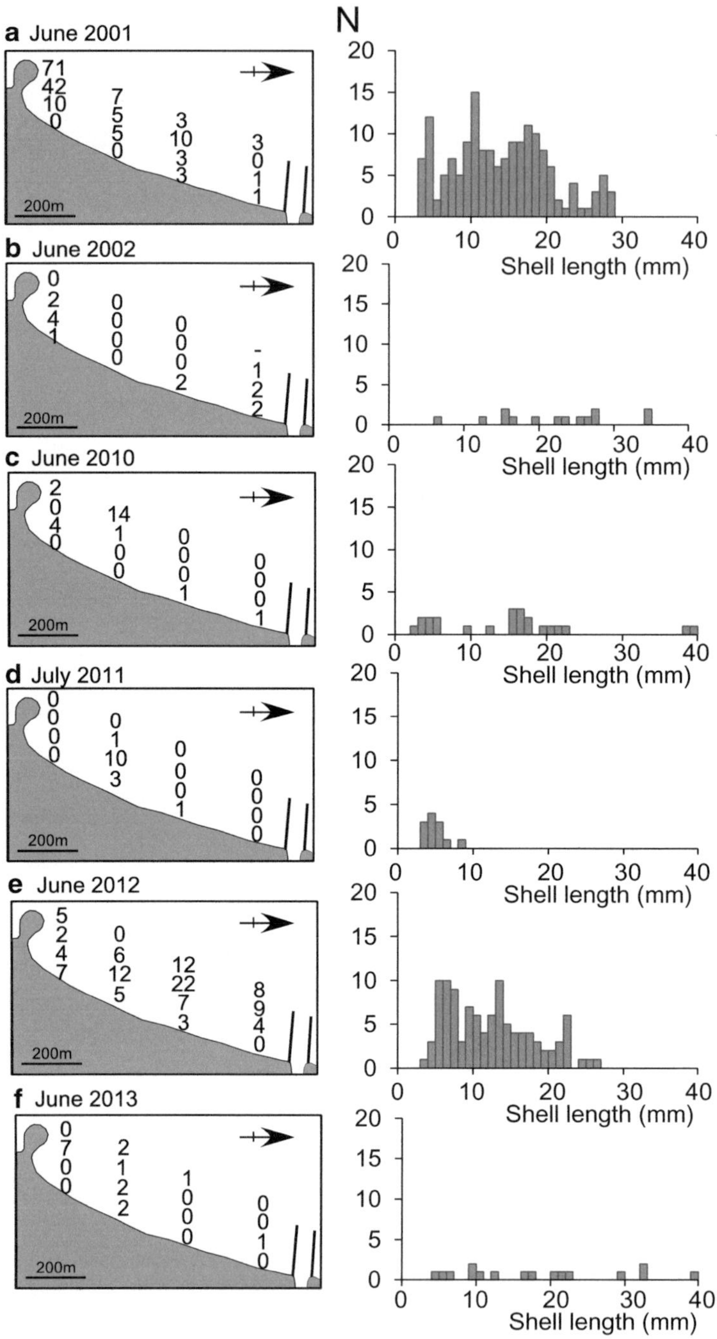

Fig. 13.4 Temporal fluctuations in the distribution of *Ruditapes philippinarum* on the Tona coast from June 2001 to June 2013. *Allow* means true north. Individual density per 1/16 m^2 at each sampling station (*left*) and frequency distribution of shell length among the individuals collected from all sampling stations (*right*) are shown in each sampling date (i.e., (**a**) June 2001, (**b**) June 2002, (**c**) June 2010, (**d**) July 2011, (**e**) June 2012, and (**f**) June 2013)

After the tsunami, all of the above five species decreased in individual number or disappeared in 2011, but some of them rapidly recovered the populations from 2012 to 2013 (Table 13.1, Fig. 13.3). In July 2011, *B. cumingii*, *L. pulchella*, and *M. incongrua* were collected with only one individual from the 16 sampling stations, and *P. pisidium* were not collected at all (Table 13.1). *R. philippinarum* was still collected in July 2011, but these individuals were less than 10 mm in shell length (Fig. 13.4d). In June 2012, many juveniles of *R. philippinarum* and *M. incongrua* with shell length <10 mm were collected from the most sampling stations (Figs. 13.4e and 13.5e). In particular, the density of *M. incongrua* was much greater than those before the tsunami (Fig. 13.3d), and most of them were less than 15 mm in shell length (Fig. 13.5e). In June 2013, the population of *L. pulchella* increased again (Fig. 13.3b), and then *R. philippinarum* temporarily decreased (Fig. 13.3c). However, only *M. incongrua* still increased in June 2013 (Fig. 13.3d), with many recruited juveniles at the most sampling stations (Fig. 13.5f). *B. cumingii* was not collected at all between June 2012 and June 2014, and few individuals of *P. pisidium* were collected during this period (Table 13.1, Fig. 13.3).

In contrast, *Musculista senhousia* and *Mya (Arenomya) arenaria oonogai*, those were low density before the tsunami, temporarily increased in July 2011 (Table 13.1, Fig. 13.3f and g). During 2001 and 2009, the mean densities of these two species were less than one individual per quadrat (Fig. 13.3f and g). In June 2011, many juveniles (shell length <10 mm) were collected mainly from the stations near the land (Figs. 13.6b and 13.7b). However, they rapidly decreased or disappeared after June 2012 (Fig. 13.3f and g). In *M. senhousia*, juveniles were still in large density in June 2012, but they almost disappeared in June 2013 (Fig. 13.6). In *Mya (A.) a. oonogai*, only few large individuals (shell length >35 mm) were collected in June 2012, but small individuals (shell length <10 mm) appeared again in June 2013 (Fig. 13.7c and d).

13.3.2 Shallow Sea in Sendai Bay

In March 2012, 1 year after the tsunami by the Great East Japan Earthquake, nine bivalve species such as *Raetellops pulchellus*, *Donax kiusiuensis*, *Moerella jedoensis*, and *Axinopsida subquadrata* were collected from the 13 sampling stations in Sendai Bay (Table 13.2). Among them, only *R. pulchellus* rapidly increased the density from July to November 2012 and then decreased from March to November 2013 (Fig. 13.8). Mean individual density of bivalves also suddenly increased in 2012 and decreased in 2013; however, a number of bivalve species did not change during this interval (Fig. 13.8). Instead of *R. pulchellus*, *D. kiusiuensis* and *A. subquadrata* increased the densities from March 2013 (Table 13.2).

In the shallow sea of Sendai Bay, *R. pulchellus* and *M. jedoensis* were collected mainly from the muddy sediments around the mouth of Abukuma River, and *D.*

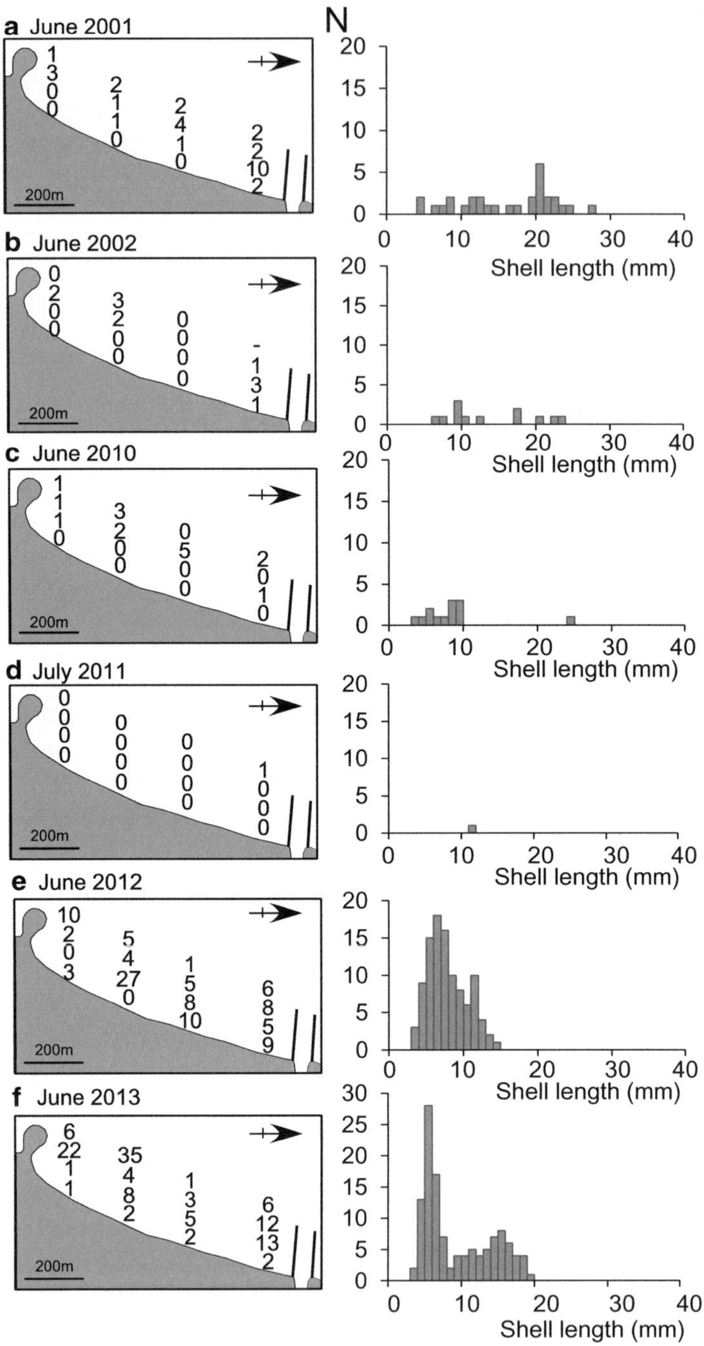

Fig. 13.5 Temporal fluctuations in distribution of *Macoma incongrua* on the Tona coast from June 2001 to June 2013. *Allow* means true north. Individual density per 1/16 m^2 at each sampling station (*left*) and frequency distribution of shell length among the individuals collected from all sampling stations (*right*) are shown in each sampling date (i.e., (**a**) June 2001, (**b**) June 2002, (**c**) June 2010, (**d**) July 2011, (**e**) June 2012, and (**f**) June 2013)

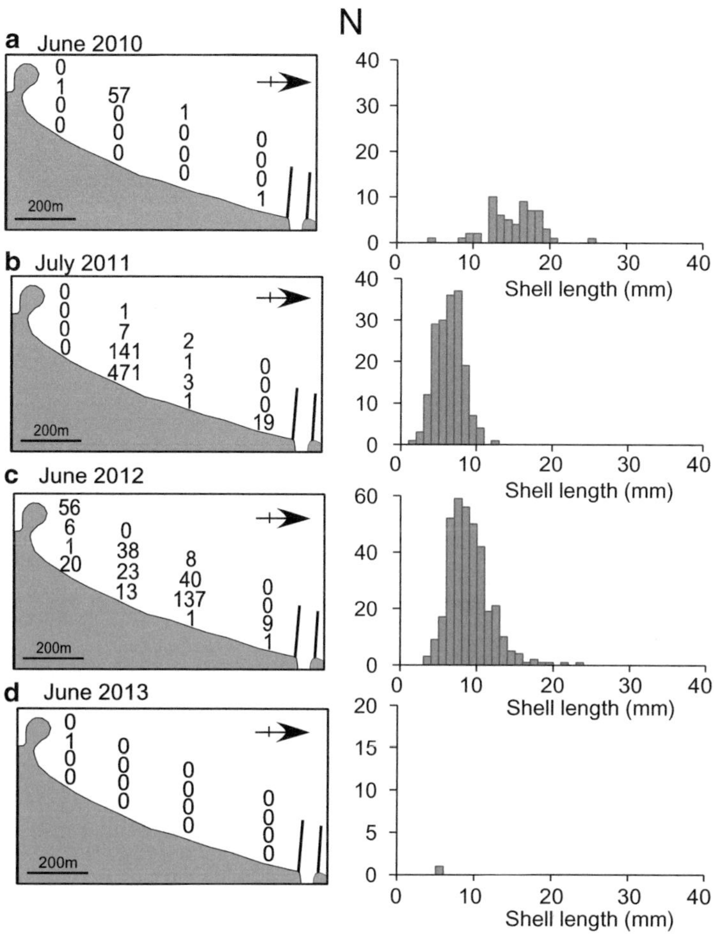

Fig. 13.6 Temporal fluctuations in distribution of *Musculista senhousia* on the Tona coast from June 2010 to June 2013. *Allow* means true north. Individual density per 1/16 m^2 at each sampling station (*left*) and frequency distribution of shell length among the individuals collected from all sampling stations (except for st. 3–1 in June 2011) (*right*) are shown in each sampling date (i.e., (**a**) June 2010, (**b**) July 2011, (**c**) June 2012, and (**d**) June 2013)

kiusiuensis and *Callithaca adamsi* usually distributed around the sandy beach between the Nanakita River and Natori River (Table 13.2; Fig. 13.9). Around the mouth of Abukuma River, densities of bivalves decreased from 2012 to 2013, but around the mouth of Natori and Nanakita rivers, *D. kiusiuensis* increased the density during this interval (Fig. 13.9).

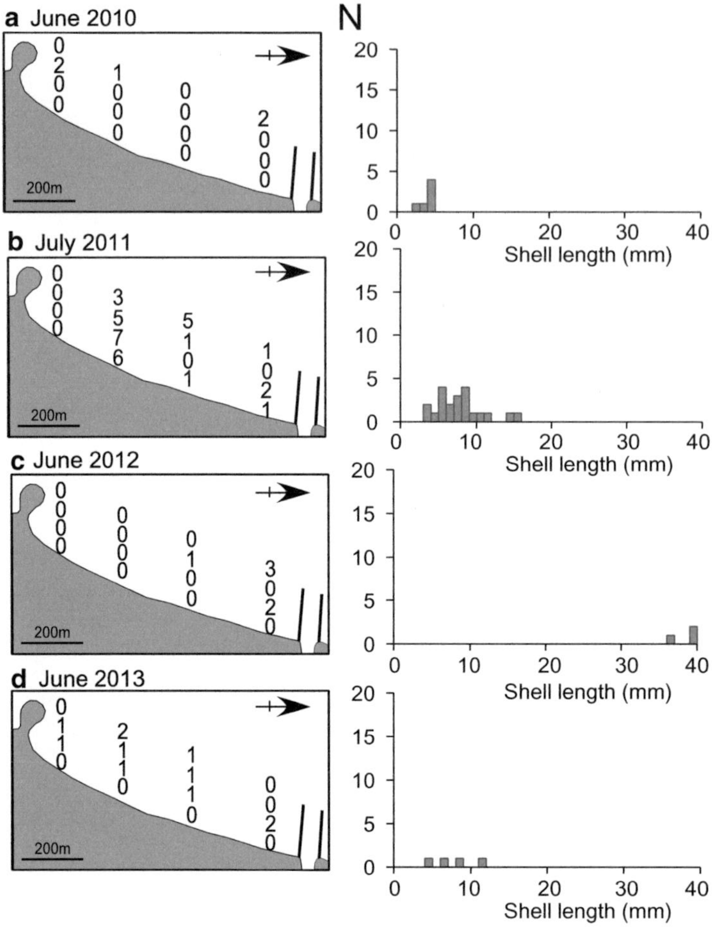

Fig. 13.7 Temporal fluctuations in distribution of *Mya* (*Arenomya*) *arenaria oonogai* on the Tona coast from June 2010 to June 2013. *Allow* means true north. Individual density per 1/16 m² at each sampling station (*left*) and frequency distribution of shell length among the individuals collected from all sampling stations (*right*) are shown in each sampling date (i.e., (**a**) June 2010, (**b**) July 2011, (**c**) June 2012, and (**d**) June 2013)

13.4 Discussion

On the Tona coast, *R. philippinarum*, *M. incongrua*, *P. pisidium*, *B. cumingii*, and *L. pulchella* were always the dominant species before the Great East Japan Earthquake (Fig. 13.3). Among them, *L. pulchella* is known as an alien predator for the bivalve species around the Tohoku region (Sakai 2000; Okoshi 2004; Okoshi and Sato-Okoshi 2011; Ohtsuki et al. 2016, this volume). Our past experiments revealed that

Table 13.2 Mean individual density (m^{-2}) of the seven dominant bivalve species (*Yoldia notabilis*, *Axinopsida subquadrata*, *Raetellops pulchellus*, *Theora fragilis*, *Moerella jedoensis*, *Donax kiusiuensis*, and *Callithaca adamsi*) and others collected from each sampling station around the shallow sea in Sendai Bay from March 2012 to November 2013

Station no.	A1	A2	A3	B1	B2	B3	C1	C2	C3	D2	D3	D4	E1	E2	E3	F1	F2	F3	Total
a. *Yoldia notabilis*																			
March 21–22, 2012	0	17	–	0	0	0	17	20	–	0	0	0	0	0	–	0	–	–	53
July 18–20, 2012	0	0	33	0	0	0	20	133	20	0	0	0	0	0	0	0	0	0	207
Nov. 13–14, 2012	0	0	0	0	0	0	67	17	0	0	0	17	0	–	–	0	0	–	100
March 18–19, 2013	0	0	0	0	0	0	33	17	50	0	17	0	0	–	–	0	0	–	117
July 16–17, 2013	0	0	0	0	0	17	17	33	33	0	0	0	0	0	0	0	0	0	100
Nov. 5–6, 2013	0	0	0	0	0	0	17	17	0	0	0	0	0	0	0	0	0	0	33
b. *Axinopsida subquadrata*																			
March 21–22, 2012	13	0	–	0	0	50	33	40	–	17	17	0	0	0	–	0	–	–	169
July 18–20, 2012	20	0	50	0	0	50	60	0	40	0	17	33	0	0	0	0	0	0	270
Nov. 13–14, 2012	0	0	0	0	0	17	0	50	33	0	17	0	0	–	–	0	0	–	117
March 18–19, 2013	0	0	0	0	0	50	67	100	133	0	267	0	0	–	–	0	0	–	617
July 16–17, 2013	0	0	0	0	0	217	50	0	167	0	0	0	0	0	17	0	0	0	450
Nov. 5–6, 2013	0	0	0	0	0	0	33	17	17	0	0	0	0	0	0	0	0	0	67
c. *Raetellops pulchellus*																			
March 21–22, 2012	13	0	–	0	33	50	2483	40	–	0	0	0	0	0	–	0	–	–	2619
July 18–20, 2012	1060	0	17	0	26,167	250	0	183	160	33	500	50	0	17	50	0	0	0	28,487
Nov. 13–14, 2012	200	0	0	2800	20,767	83	450	0	0	0	0	0	0	–	–	0	17	–	2,4317
March 18–19, 2013	400	0	0	67	3300	0	183	83	17	0	0	0	0	–	–	0	0	–	4050
July 16–17, 2013	17	0	0	0	233	0	17	0	0	33	0	0	17	0	0	0	0	0	317
Nov. 5–6, 2013	0	0	0	0	683	0	0	0	33	0	0	0	0	0	0	0	0	0	717

(continued)

Table 13.2 (continued)

Station no.	A1	A2	A3	B1	B2	B3	C1	C2	C3	D2	D3	D4	E1	E2	E3	F1	F2	F3	Total
d. *Theora fragilis*																			
March 21–22, 2012	0	0	–	0	0	17	17	0	–	0	0	0	0	0	–	0	–	–	33
July 18–20, 2012	0	0	0	0	0	0	0	50	0	17	117	0	0	0	33	0	0	33	250
Nov. 13–14, 2012	0	0	0	0	0	0	0	0	0	0	0	0	0	–	–	0	0	–	0
March 18–19, 2013	0	0	0	0	0	0	0	0	0	0	0	0	0	–	–	0	0	–	0
July 16–17, 2013	0	0	0	0	0	17	0	0	0	0	0	0	0	0	17	0	0	0	33
Nov. 5–6, 2013	0	0	0	0	0	0	0	0	0	0	0	0	0	0	0	0	0	0	0
e. *Moerella jedoensis*																			
March 21–22, 2012	25	0	–	0	33	100	67	20	–	0	0	0	0	0	–	0	–	–	245
July 18–20, 2012	0	0	17	0	33	50	160	33	0	0	17	0	0	0	0	0	0	0	310
Nov. 13–14, 2012	17	0	0	17	0	50	67	33	17	0	33	0	0	–	–	0	0	–	233
March 18–19, 2013	0	0	0	0	17	17	67	0	0	0	0	0	0	–	–	0	0	–	100
July 16–17, 2013	17	0	0	0	0	50	17	0	0	0	0	0	0	0	0	0	0	0	83
Nov. 5–6, 2013	0	0	0	0	33	0	0	0	0	0	0	0	0	0	0	0	0	0	33
f. *Donax kiusiuensis*																			
March 21–22, 2012	0	0	–	0	0	0	0	0	–	17	0	0	700	0	–	414	–	–	1131
July 18–20, 2012	0	0	0	0	0	0	0	0	0	0	0	0	50	0	0	17	0	0	67
Nov. 13–14, 2012	0	0	0	0	0	0	0	0	0	0	0	0	0	–	–	683	17	–	700
March 18–19, 2013	0	0	0	0	0	0	0	0	0	83	0	0	1983	–	–	33	0	–	2100
July 16–17, 2013	0	0	0	50	0	0	0	0	0	0	0	0	0	0	0	0	0	0	50
Nov. 5–6, 2013	0	0	0	200	0	0	0	0	0	100	0	0	1617	0	0	83	0	0	2000

g. *Callithaca adamsi*																			
March 21–22, 2012	0	0	0	0	0	0	0	0	–	0	0	0	0	0	–	0	–	–	0
July 18–20, 2012	0	0	33	0	0	0	0	17	0	0	0	0	0	0	0	0	0	0	50
Nov. 13–14, 2012	0	33	0	0	0	50	50	0	50	0	33	33	0	–	–	0	0	–	250
March 18–19, 2013	0	0	17	0	0	17	0	67	50	17	0	17	0	–	–	0	0	–	183
July 16–17, 2013	0	0	20	0	0	17	17	0	0	0	0	0	0	0	17	0	17	0	87
Nov.5–6, 2013	0	0	0	0	0	0	0	0	0	0	0	0	0	0	17	0	0	0	17
h. Other bivalves																			
March 21–22, 2012	0	0	–	0	0	33	0	0	–	0	0	0	67	0	–	86	–	–	186
July 18–20, 2012	0	0	0	0	0	0	0	0	20	0	0	0	0	0	0	0	0	0	20
Nov. 13–14, 2012	0	33	0	0	0	0	0	0	0	0	0	17	0	–	–	0	0	–	50
March 18–19, 2013	0	0	17	0	0	0	0	67	17	17	0	0	0	–	–	0	0	–	117
July 16–17, 2013	0	0	20	0	0	0	0	33	0	0	0	17	0	0	0	0	0	0	70
Nov.5–6, 2013	0	0	0	0	0	0	0	0	0	0	0	0	0	0	0	83	0	0	83

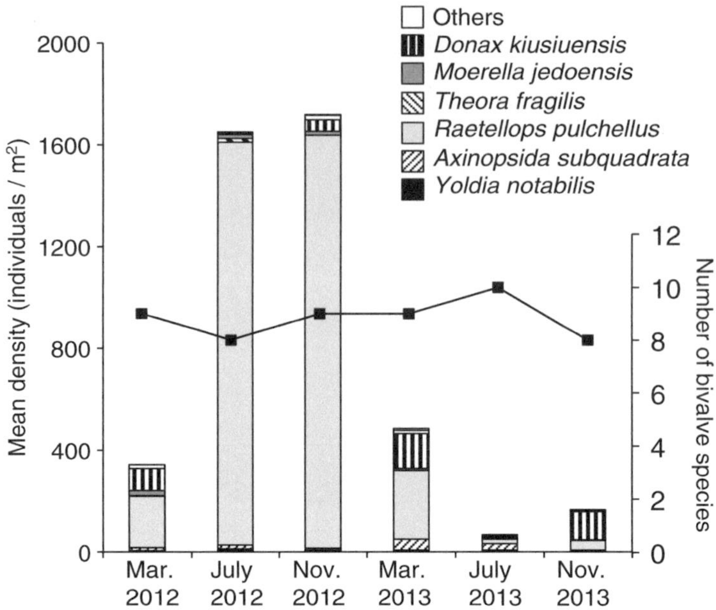

Fig. 13.8 Temporal change of number of bivalve species and mean individual density (m^{-2}) of each species collected from the 13–18 sampling stations around the shallow sea in Sendai Bay from March 2012 to November 2013

L. pulchella preferred to consume *R. philippinarum* and *M. incongrua* rather than *P. pisidium* (Sato et al. 2012). Therefore, *R. philippinarum* and *M. icongrua* decreased in number as a result of selective predation after the first appearance of *L. pulchella* in 2002. The naticid predator *L. pulchella* produces characteristic countersunk drill holes on prey shells (Chiba and Sato 2012). Using the predatory drill hole on dead shells, Chiba and Sato (2013) estimated that more than 35 % of *R. philippinarum* and 20 % of *M. incongrua* died because of naticid predation after 2004.

P. pisidium did not show a rapid decrease in abundance from 2001 to 2004 (Fig. 13.3e), because this species is less vulnerable to predation by *L. pulchella* (Sato et al. 2012; Chiba and Sato 2013). However, the density of *P. pisidium* decreased from 2005 to 2008 (Fig. 13.3e). Proportion of drilled dead shells for *P. pisidium* significantly increased after *R. philippinarum* and *M. incongrua* had been almost completely consumed (Chiba and Sato 2013). These results suggest that *L. pulchella* focused attacks on *P. pisidium* in 2004, when *R. philippinarum* and *M. incongrua* had become scarce due to naticid predation (Sato et al. 2012; Chiba and Sato 2013).

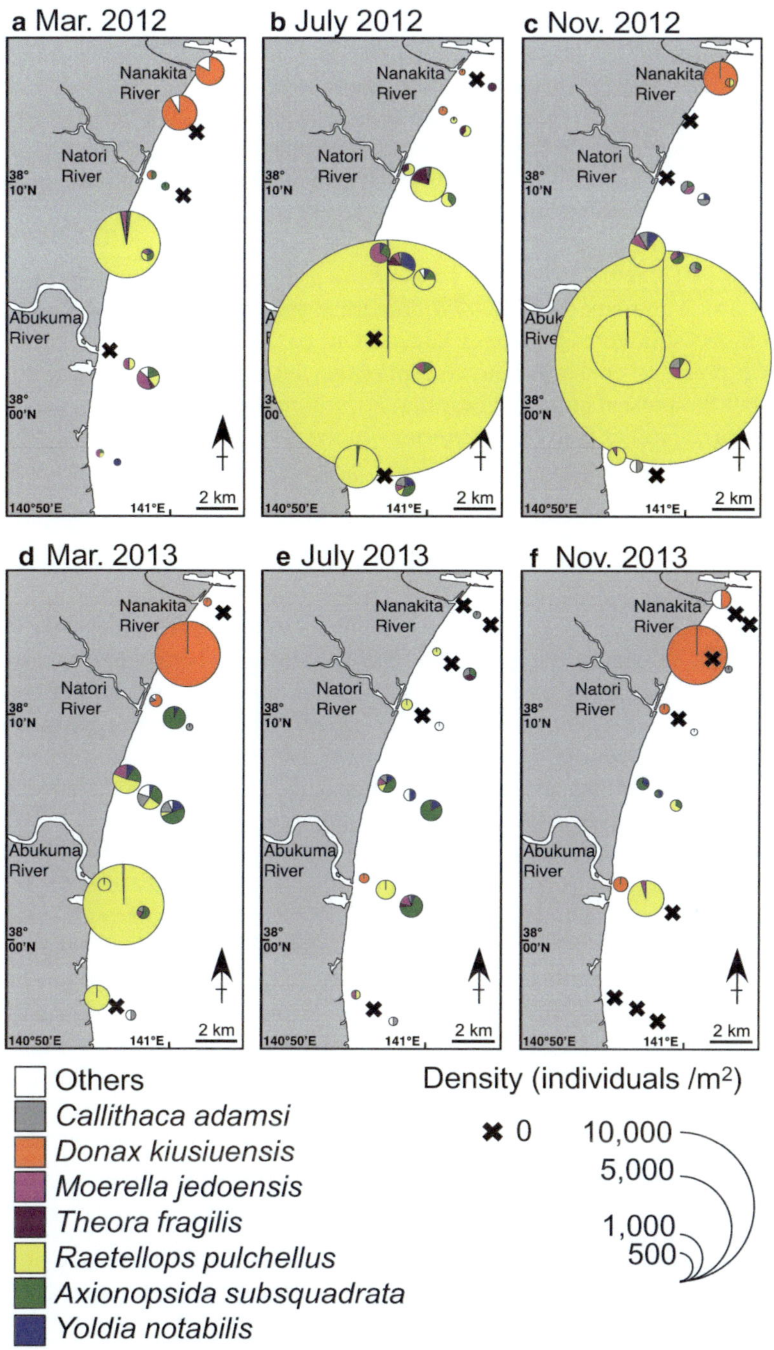

Fig. 13.9 Temporal change of distribution pattern of bivalve species collected from the 13–18 sampling stations around the shallow sea in Sendai Bay from March 2012 to November 2013. *Size of each circle* indicates mean individual density of bivalves (m^{-2}) in each sampling station, and the *cross* means no bivalves collected from the station. *Allow* means true north

After the tsunami, juveniles of *R. philippinarum* were found in July 2011, and many juveniles of *M. incongrua* also appeared from June 2012 (Figs. 13.4e and 13.5e). Around Miyagi Prefecture, spawning season of *R. philippinarum* is known from June to October (Toba et al. 1992; Kanazawa and Sato 2008). Therefore, some planktonic larva of *R. philippinarum* settled on the Tona coast from June to July 2011. Although there are no exact data about spawning season of *M. incongrua*, this species seemed to spawn and settle on the Tona coast during the summer in 2011 and spring in 2012.

In June 2013, naticid predator *L. pulchella* increased again on the Tona coast (Fig. 13.3b). Simultaneously, *R. philippinarum* population rapidly decreased, but *M. incongrua* still increased (Fig. 13.3c, d). Our past experiments confirmed that *L. pulchella* prefer to consume both *R. philippinarum* and *M. incongrua* (Sato et al. 2012), but the present study suggest that *R. philippinarum* may be more preferred by *L. pulchella* rather than *M. incongrua*. On the contrary, in June 2014, *M. incongrua* and *L. pulchella* decreased and *R. philippinarum* increased again (Table 13.1; Fig. 13.3).

Among the five dominant species before the tsunami, *B. cumingii* and *P. pisidium* have not recovered till June 2014 (Table 13.1, Fig. 13.3). According to Furota et al. (2002), *B. cumingii* spawns isolated eggs from which juveniles hatch out directly. After the tsunami, we observed many individuals of this species that were distributed in patch outside of the quadrat on the Tona coast. Therefore, it takes longer time to increase in abundance compared with the other species having planktonic larvae stage, but *B. cumingii* will recover steadily on the Tona coast in the near future. On *P. pisidium*, there are no data about life history, but the present study clarified that this species also slowly recovers against the catastrophic events.

By contrast, *M. senhousia* and *Mya* (*A.*) *a. oonogai* populations have low density before the Great East Japan Earthquake, but rapidly increased in July 2011 (Fig. 13.3f and g). The juveniles of *M. senhousia* had still large density in June 2012, but they almost disappeared in June 2013 (Fig. 13.6). Before the tsunami, *M. senhousia* distributed mainly on the subtidal zone of this coast (Fig. 13.6a). After the tsunami, most sampling stations of the Tona coast were not exposed even at low tide because of land subsidence (Fig. 13.1d), and *M. senhousia* became to settle on the stations near the land. However, it gradually changed to expose at low tide due to sand deposition within 3 years after the tsunami, and *M. senhousia* may settle again around the subtidal zone of this coast.

In *Mya* (*A.*) *a. oonogai*, many juveniles appeared in July 2011, but only few large individuals were collected in June 2012 (Fig. 13.7). According to Goshima (1982), *Mya* (*A.*) *a. oonogai* in Hakata Bay grows to 37.9 mm in shell length at 1 year, and large individuals (>40 mm in shell length) can dwell in sediments deeper than 100 mm. Therefore, on the Tona coast, juveniles of this species settled in 2011 have grown to >35 mm in shell length till June 2012. Further, these large individuals dwelled deeper than 100 mm in sediments. Thus, we might not have collected all individuals from quadrat samples using a shovel. The possibility suggests that we may underestimate the density of this species before and after the tsunami.

In the shallow sea of Sendai Bay, only *R. pulchellus* increased temporarily from July to November 2012 and then decreased from March to November 2013 (Fig. 13.8). After March 2013, the mean individual density of bivalves also decreased, although a number of bivalve species did not change during this interval (Fig. 13.8). Similar phenomena were also found in the shallow sea of Isahaya Bay, Kyushu, southwestern Japan, after the reclamation dike shut off (Yamamoto et al. 2015). Around the outer area of the reclamation dike in Isahaya Bay, *Theora fragilis* increased after the water gates temporarily opened during April and May 2002, but rapidly decreased after the water gates closed again (Yamamoto et al. 2015). *R. pulchellus* is regarded as an opportunistic species on the muddy sediments in the shallow sea as in the case of *T. fragilis*. Therefore, these species temporarily increased after catastrophic events such as tsunami and reclamation works of the bay.

On the Tona coast, three of the five species which were dominant before the tsunami disaster have already recovered 2 years after the tsunami (Fig. 13.3). However, in the shallow sea of Sendai Bay, only one species temporarily increased, then decreased during the 2 years after the tsunami, and the mean individual density of bivalves has been still low till November 2013 (Fig. 13.8). These results suggest that recovery processes of molluskan populations after the tsunami are different between the coastal area and shallow sea. In Sendai Bay, *D. kiusiuensis* gradually increased 2 years after the tsunami around the sandy beach between the Nanakita River and Natori River (Fig. 13.9). The dead shells of this species were abundantly found around the Gamo coast in Sendai Bay after the tsunami (Fujiwara et al. 2014). Therefore, this species was common in the shallow sea of Sendai Bay before the tsunami, and it began to recover 2 years after the tsunami. The other bivalve species have not yet recovered till November 2013 around the shallow sea in Sendai Bay (Figs. 13.8 and 13.9). Although there are few published data about densities of benthic animals before the tsunami in Sendai Bay, recovery speed of molluskan populations in the shallow sea may be slower than those in the coastal area such as the Tona coast.

Acknowledgments This study was supported in part by grants from Tohoku Ecosystem-Associated Marine Sciences. The authors are grateful to J. Urabe (Tohoku University) for the invitation to contribute on this chapter. We thank H. Hasegawa (Hosho High School) for providing the valuable data. We also would like to thank the staff of the Naruse Branch (Higashi-Matsushima City) and Sendai Branch of Japan Fisheries Cooperative for their help in collecting samples. The 2011 Tohoku earthquake and tsunami struck Higashi-Matsushima City and Sendai City and caused serious damage there. We sincerely pray for a quick recovery.

Appendix

Table 13.3 Sampling conditions of each station and data of temperature, salinity, and DO of the bottom water layers around the shallow sea of Sendai Bay

St.	Sampling date and time		Latitude	Longitude	Depth of water (m)	No. of samples taken	Bottom water layers		
							Temp. (°C)	Salinity	DO (mg/L)
A1	3/22/2012	9:10	N37°57.991′	E140°56.008′	15.8	4	6.1	29.3	5.31
A2	3/22/2012	9:55	N37°57.771′	E140°56.499′	17.6	3	6.2	29.2	6.50
B1	3/22/2012	11:30	N38°02.537′	E140°56.067′	7.3	3	6.1	27.5	6.49
B2	3/22/2012	11:55	N38°02.291′	E140°56.530′	15.7	3	6.1	29.3	6.45
B3	3/22/2012	12:25	N38°02.018′	E140°56.977′	20.1	3	6.1	28.5	6.68
C1	3/22/2012	13:15	N38°06.525′	E140°57.054′	20.2	3	6.2	29.0	6.66
C2	3/22/2012	13:45	N38°06.257′	E140°57.512′	20.9	2.5[a]	6.1	28.8	5.90
D2	3/21/2012	8:08	N38°10.272′	E140°58.415′	10.1	3	5.2	29.3	6.19
D3	3/21/2012	8:40	N38°10.048′	E140°59.054′	19.9	3	6.0	29.4	6.12
D4	3/21/2012	9:25	N38°09.756′	E140°59.497′	21.3	3	6.1	27.0	4.63
E1	3/21/2012	10:40	N38°12.758′	E140°59.470′	4.5	3	5.9	29.4	5.73
E2	3/21/2012	11:05	N38°12.506′	E141°00.008′	18.4	3	6.0	29.4	6.39
F1	3/21/2012	11:50	N38°15.113′	E141°00.923′	4.0	3.5	5.9	29.3	5.80
A1	7/18/2012	10:05	N37°57.965′	E140°56.025′	15.2	2.5[a]	14.1	29.9	7.06
A2	7/18/2012	10:40	N37°57.720′	E140°56.532′	17.3	3	14.1	29.8	7.31
A3	7/18/2012	11:15	N37°59.984′	E140°56.084′	17.6	3	13.8	29.9	7.34
B1	7/18/2012	16:15	N38°02.519′	E140°56.013′	7.5	2.5[a]	16.8	29.6	6.64
B2	7/18/2012	16:00	N38°02.287′	E140°56.512′	15.1	3	14.5	30.1	7.58
B3	7/18/2012	15:40	N38°02.001′	E140°57.027′	19.9	3	13.5	29.9	6.48
C1	7/19/2012	7:55	N38°06.480′	E140°57.001′	18.1	2.5[a]	14.1	29.9	5.14
C2	7/19/2012	8:15	N38°06.234′	E140°57.521′	19.5	3	13.8	30.0	7.47

C3	7/19/2012	8:30	N38°05.997′	E140°58.014′	20.8	2.5a	13.1	29.5	8.15
D2	7/19/2012	13:30	N38°10.073′	E140°58.494′	15.0	3	14.9	30.7	7.84
D3	7/19/2012	13:15	N38°10.017′	E140°59.051′	19.1	3	13.5	28.9	4.97
D4	7/19/2012	13:00	N38°09.768′	E140°59.533′	20.2	3	13.4	27.9	3.40
E1	7/20/2012	9:35	N38°12.732′	E140°59.534′	6.4	3	22.4	26.2	7.16
E2	7/20/2012	9:22	N38°12.549′	E141°00.029′	17.6	3	13.8	30.0	5.03
E3	7/20/2012	9:10	N38°11.991′	E141°00.501′	18.1	3	13.7	29.6	3.32
F1	7/20/2012	7:05	N38°15.010′	E141°01.246′	3.7	3	22.0	26.0	7.49
F2	7/20/2012	7:20	N38°14.519′	E141°01.548′	17.2	3	13.7	30.0	2.49
F3	7/20/2012	7:32	N38°14.428′	E141°02.031′	18.5	3	13.6	28.7	3.39
A1	11/13/2012	8:49	N37°58.041′	E140°55.995′	15.4	3	17.5	25.6	7.08
A2	11/13/2012	9:08	N37°57.735′	E140°56.514′	18.1	3	17.4	25.2	7.01
A3	11/13/2012	9:36	N37°57.520′	E140°57.003′	18.1	3	17.5	29.2	6.32
B1	11/13/2012	10:50	N38°02.393′	E140°56.206′	10.9	3	17.5	30.0	6.72
B2	11/13/2012	10:33	N38°02.242′	E140°56.507′	15.5	3	17.5	29.8	5.92
B3	11/13/2012	10:07	N38°01.991′	E140°56.976′	19.1	3	17.4	29.8	7.44
C1	11/13/2012	11:21	N38°06.441′	E140°57.077′	18.8	3	17.3	30.1	5.27
C2	11/13/2012	11:46	N38°06.240′	E140°57.510′	20.7	3	17.4	30.1	5.94
C3	11/13/2012	12:07	N38°06.018′	E140°57.985′	21.7	3	17.4	30.3	6.30
D2	11/13/2012	13:20	N38°10.110′	E140°58.559′	16.1	3	17.1	30.2	7.01
D3	11/13/2012	13:01	N38°09.986′	E140°58.995′	21.1	3	17.3	30.3	5.85
D4	11/13/2012	12:42	N38°09.760′	E140°59.510′	21.1	3	17.3	30.2	5.91
E1	11/14/2012	7:26	N38°12.712′	E140°59.703′	14.6	3	16.7	30.2	6.87
F1	11/14/2012	8:34	N38°14.930′	E141°01.364′	8.6	3	16.2	30.0	7.45
F2	11/14/2012	8:55	N38°14.764′	E141°01.508′	15.3	3	16.2	30.1	7.41
A1	3/19/2013	10:24	N37°58.001′	E140°56.007′	16.2	3	8.7	31.5	9.05
A2	3/19/2013	10:11	N37°57.729′	E140°56.512′	18.4	3	8.8	31.5	7.45

(continued)

Table 13.3 (continued)

St.	Sampling date and time		Latitude	Longitude	Depth of water (m)	No. of samples taken	Bottom water layers		
							Temp. (°C)	Salinity	DO (mg/L)
A3	3/19/2013	9:54	N37°57.473′	E140°57.020′	17.8	3	8.9	31.5	8.55
B1	3/19/2013	9:15	N38°02.491′	E140°56.004′	8.2	3	8.6	31.5	9.19
B2	3/19/2013	8:37	N38°02.251′	E140°56.511′	16.4	3	8.5	31.4	9.27
B3	3/19/2013	8:11	N38°01.992′	E140°56.996′	20.5	3	8.5	30.5	8.72
C1	3/19/2013	11:57	N38°06.541′	E140°57.007′	18.6	3	8.5	31.4	9.24
C2	3/19/2013	11:34	N38°06.210′	E140°57.757′	19.8	3	8.4	31.3	9.56
C3	3/19/2013	11:13	N38°05.991′	E140°58.006′	21.3	4	8.5	31.2	9.69
D2	3/19/2013	12:26	N38°10.259′	E140°58.511′	11.5	4	8.6	31.2	9.04
D3	3/19/2013	12:40	N38°10.006′	E140°59.010′	19.1	3	8.7	31.2	9.10
D4	3/18/2013	9:15	N38°09.729′	E140°59.507′	21.7	3	8.5	31.8	8.51
E1	3/18/2013	8:48	N38°12.702′	E140°59.510′	7.7	3	8.2	31.2	9.32
F1	3/18/2013	7:44	N38°14.895′	E141°01.257′	7.7	4	8.2	31.4	9.35
F2	3/18/2013	7:17	N38°14.749′	E141°01.501′	16.3	3	7.9	31.3	8.59
A1	7/16/2013	8:50	N37°58.032′	E140°55.997′	16.3	3	16.8	30.1	8.64
A2	7/16/2013	8:35	N37°57.763′	E140°56.643′	18.3	3	12.6	30.3	6.64
A3	7/16/2013	8:10	N37°57.447′	E140°57.193′	17.3	2.5	15.8	30.0	9.37
B1	7/16/2013	9:40	N38°02.515′	E140°56.075′	8.2	3	22.5	28.1	7.63
B2	7/16/2013	9:25	N38°02.248′	E140°56.496′	16.8	3	16.1	30.2	8.57
B3	7/16/2013	9:15	N38°02.015′	E140°57.146′	21.0	3	15.2	30.4	9.01
C1	7/16/2013	10:25	N38°06.386′	E140°57.208′	19.4	3	16.5	30.3	8.46
C2	7/16/2013	10:15	N38°06.175′	E140°57.638′	20.7	3	13.6	30.0	9.11
C3	7/16/2013	10:05	N38°05.898′	E140°57.911′	21.5	3	13.4	30.6	9.01
D2	7/16/2013	11:12	N38°10.249′	E140°58.496′	15.4	3	18.7	29.8	7.50
D3	7/16/2013	11:02	N38°09.975′	E140°59.203′	20.0	3	12.1	30.8	4.22

D4	7/16/2013	10:55	N38°09.775′	E140°59.514′	21.1	3	12.1	30.6	7.82
E1	7/17/2013	8:25	N38°12.733′	E140°59.717′	14.6	3	12.0	14.8	3.50
E2	7/17/2013	8:35	N38°12.496′	E140°59.914′	18.1	3	11.9	30.6	4.73
E3	7/17/2013	8:50	N38°12.508′	E141°00.458′	19.3	3	11.7	30.7	2.96
F1	7/17/2013	7:53	N38°14.946′	E141°01.262′	6.0	3	21.5	27.1	6.84
F2	7/17/2013	7:42	N38°14.732′	E141°01.513′	15.5	3	12.1	30.6	3.98
F3	7/17/2013	7:25	N38°14.486′	E141°02.024′	18.1	3	12.3	30.2	0.10
A1	11/5/2013	8:25	N37°58.009′	E140°56.042′	15.8	3	18.5	30.3	5.45
A2	11/5/2013	8:45	N37°57.743′	E140°56.497′	17.2	3	18.6	30.3	5.41
A3	11/5/2013	8:58	N37°57.751′	E140°57.030′	18.7	3	18.5	30.3	6.56
B1	11/5/2013	9:30	N38°02.534′	E140°56.024′	7.1	3	17.5	N.D.	N.D.
B2	11/5/2013	9:45	N38°02.220′	E140°56.637′	16.5	3	18.3	N.D.	N.D.
B3	11/5/2013	10:10	N38°01.991′	E140°56.991′	18.7	3	18.2	N.D.	N.D.
C1	11/5/2013	10:45	N38°06.510′	E140°56.981′	17.9	3	18.2	N.D.	N.D.
C2	11/5/2013	11:00	N38°06.231′	E140°57.502′	19.6	3	18.0	N.D.	N.D.
C3	11/5/2013	11:20	N38°05.991′	E140°58.006′	21.1	3	17.9	N.D.	N.D.
D2	11/5/2013	11:45	N38°10.249′	E140°58.467′	11.4	3	18.7	N.D.	N.D.
D3	11/5/2013	12:15	N38°09.927′	E140°59.025′	19.5	3	N.D.	N.D.	N.D.
D4	11/5/2013	12:00	N38°09.747′	E140°59.519′	20.4	3	18.3	N.D.	N.D.
E1	11/6/2013	8:15	N38°12.755′	E140°59.514′	8.0	3	17.8	29.7	4.23
E2	11/6/2013	8:30	N38°12.506′	E141°00.011′	17.8	3	18.5	30.2	4.22
E3	11/6/2013	8:40	N38°12.235′	E141°00.540′	19.1	3	18.4	30.3	3.37
F1	11/6/2013	9:05	N38°14.984′	E141°01.186′	4.0	3	18.7	30.0	2.91
F2	11/6/2013	9:20	N38°14.751′	E141°01.521′	15.6	3	18.7	30.3	3.02
F3	11/6/2013	9:30	N38°14.516′	E141°02.016′	18.1	3	18.6	30.1	2.74

N.D. No data because the multiple water-quality meter was broken
[a]Sediment sample was taken by Smith-McIntyre bottom sampler

References

Chiba T, Sato S (2012) Size-selective predation and drillhole site selectivity in *Euspira fortunei* (Gastropoda: Naticidae): implications for ecological and paleoecological studies. J Mollus Stud 78:205–212. doi:10.1093/mollus/eys002

Chiba T, Sato S (2013) Invasion of *Laguncula pulchella* (Gastropoda: Naticidae) and predator–prey interactions with bivalves on the Tona coast, Miyagi prefecture, northern Japan. Biol Invasions 15:587–598. doi:10.1007/s10530-012-0310-1

Fujiwara O, Tanigawa K, Sato S (2014) Molluscan shell concentration by great tsunami: with special reference to the 2011 Tohoku-oki event. Earth Mon 36:36–41 (in Japanese)

Furota T, Sunobe T, Arita S (2002) Contrasting population status between the planktonic and direct-developing batillariid snails *Batillaria multiformis* (Lischke) and *B. cumingi* (Crosse) on an isolated tidal flat in Tokyo Bay. Venus 61:15–23

Geographical Survey Institute of Japan (2011) Land subsidence caused by 2011 Tohoku earthquake (in Japanese). http://www.gsi.go.jp/sokuchikijun/sokuchikijun40003.html. Accessed 27 Feb 2013

Goshima S (1982) Population dynamics of the soft clam, *Mya arenaria* L., with special reference to its life history pattern. Publ Amakusa Mar Biol Lab 6:119–165

Hayasaka D, Shimada N, Konno H, Sudayama H, Kawanishi M, Uchida T, Goka K (2012) Floristic variation of beach vegetation caused by the 2011 Tohoku-oki tsunami in northern Tohoku, Japan. Ecol Eng 44:227–232. doi:10.1016/j.ecoleng.2012.03.014

Jaramillo E, Dugan JE, Hubbard DM, Melnick D, Manzano M, Duarte C, Campos C, Sanchez R (2012) Ecological implications of extreme events: footprints of the 2010 earthquake along the Chilean coast. PLoS One 7(5), e35348. doi:10.1371/journal.pone.0035348

Kanaya G, Suzuki T, Maki H, Nakamura Y, Miyajima Y, Kikuchi E (2012) Effects of the 2011 tsunami on topography, vegetation, and macrobenthic fauna in Gamo Lagoon, Japan. Jpn J Benthol 67:20–32 (in Japanese with English Abstract)

Kanaya G, Maki H, Suzuki T, Sato-Okoshi W, Kikuchi E (2014) Tsunami-induced changes in a shallow brackish lagoon ecosystem (Gamo Lagoon) in Sendai Bay, Japan. Glob Environ Res 18(2014):35–46

Kanazawa T, Sato S (2008) Environmental and physiological controls on shell microgrowth pattern of *Ruditapes philippinarum* (bivalvia: veneridae) from Japan. J Mollus Stud 74:89–95. doi:10.1093/mollus/eym049

Lomovasky BJ, Firstater FN, Salazar AG, Mendo J, Iribarne OO (2011) Macro benthic community assemblage before and after the 2007 tsunami and earthquake at Paracas bay, Peru. J Sea Res 65:205–212. doi:10.1016/j.seares.2010.10.002

Miura O, Sasaki Y, Chiba S (2012) Destruction of populations of *Batillaria attramentaria* (Caenogastropoda: Batillariidae) by tsunami waves of the 2011 Tohoku earthquake. J Mollus Stud 78:377–380. doi:10.1093/mollus/eys025

Mori N, Takahashi T, Yasuda T, Yanagisawa H (2011) Survey of 2011 Tohoku earthquake tsunami inundation and run-up. Geophys Res Lett 38:L00G14. doi:10.1029/2011GL049210

Ohtsuki H, Suzuki T, Kinoshita K, Kanaya G, Hirama T, Sato S, Shibata K, Okoshi K and Urabe J (2016) Genetic structures of Euspira fortunei metapopulation in the north east coast of Japan after tsunamis of the 3.11 Great East Japan Earthquake. In: Urabe, Nakashizuka (eds) Ecological impacts of tsunamis on coastal ecosystems: lessons from the Great East Japan Earthquake, Springer, pp 209–222

Okoshi K (2004) Alien species introduced with imported clams: the clam-eating moon snail *Euspira fortunei* and other unintentionally introduced species. Jpn J Benthol 59:74–82 (in Japanese with English abstract)

Okoshi K, Sato-Okoshi W (2011) *Euspira fortunei*: biology and fisheries science of invasive species. Kouseishakouseikaku Press, Tokyo, pp 1–225 (in Japanese)

Sakai K (2000) Predation of the moon snail *Neverita didyma*, on the Manila clam *Ruditapes philippinarum*, at the culture ground in Mangoku-ura Inlet. Bull Miyagi Prefect Fish Res Dev Center 16:109–111 (in Japanese)

Sato S, Chiba T, Hasegawa H (2012) Long-term fluctuation in mollusk populations before and after the appearance of an alien predator *Euspira fortunei* (Gastropoda: Naticidae) on the Tona coast, Miyagi Prefecture, northern Japan. Fish Sci 78:589–595. doi:10.1007/s12562-012-0491-7

Seike K, Shirai K, Kogure Y (2013) Disturbance of shallow marine soft-bottom environments and megabenthos assemblages by a huge tsunami induced by the 2011 M9.0 Tohoku-Oki earthquake. PLoS One 8(6):e65417. doi:10.1371/journal.pone.0065417

Takami H, Won N, Kawamura T (2013) Impacts of the 2011 mega-earthquake and tsunami on abalone *Haliotis discus hannai* and sea urchin *Strongylocentrotus nudus* populations at Oshika Peninsula, Miyagi, Japan. Fish Oceanogr 22:113–120. doi:10.1111/fog.12008

Toba M, Natsume Y, Yamakawa H (1992) Sexual maturation and spawning of Japanese littleneck *Tapes philippinarum* collected from Tokyo bay. Fish Eng 29:47–53 (in Japanese)

Torigoe K, Inaba A (2011) Revision on the classification of recent Naticidae. Bull Nishinomiya Shell Mus 7:1–133

Urabe J, Suzuki T, Nishita T, Makino W (2013) Immediate ecological impacts of the 2011 Tohoku earthquake tsunami on intertidal flat communities. PLoS One 8(5), e62779. doi:10.1371/journal.pone.0062779

Whanpetch N, Nakaoka M, Mukai H, Suzuki T, Nojima S, Kawai T, Aryuthaka C (2010) Temporal changes in benthic communities of seagrass beds impacted by a tsunami in the Andaman sea, Thailand. Estuar Coast Shelf Sci 87:246–252. doi:10.1016/j.ecss.2010.01.001

Yamamoto A, Sato S, Azuma M (2015) Faunal changes of benthic animals around the outer area of the Isahaya reclamation dike before and after the temporally opening of the water gates: especially on *Devonia semperi* (Ohshima, 1930) distributed near the northern water gate. Molluscan Divers 4:29–37

Chapter 14
Genetic Structures of *Laguncula pulchella* Metapopulations Along the Northeast Coast of Japan After the Tsunamis Caused by the Great East Japan Earthquake

Hajime Ohtsuki, Takao Suzuki, Kyoko Kinoshita, Gen Kanaya, Takuto Hirama, Shin'ichi Sato, Kiyotaka Shibata, Kenji Okoshi, and Jotaro Urabe

Abstract The carnivorous snail (*Laguncula pulchella*) was once considered an endangered species in Japan because its habitat was limited to western Japan. However, alien *L. pulchella* populations were transported from China and the Korean Peninsula to the Pacific coast of northeast Japan, presumably via anthropogenic activities. This study examined the genetic structures of the invasive *L. pulchella* populations in various coastal estuaries of Sendai Bay and Sanriku Ria areas to determine how the metapopulations were affected by the tsunamis caused by the Great East Japan Earthquake. The genetic compositions of these carnivorous snails significantly varied between the Sendai Bay and Sanriku areas. The result suggests that local *L. pulchella* populations were relatively isolated, and their genetic structure

H. Ohtsuki (✉) • T. Suzuki • T. Hirama • J. Urabe
Graduate School of Life Science, Tohoku University, Aoba 6-3, Aramaki, Aoba-ku, Sendai 980-8578, Japan
e-mail: ohtsuki-hajime@m.tohoku.ac.jp

K. Kinoshita
Organization of Revitalization for Sanriku-region, Iwate University, 3-18-8 Ueda, Morioka, Iwate 020-8550, Japan

G. Kanaya
Center for Regional Environmental Research, National Institute for Environmental Studies, 16-2 Onogawa, Tsukuba 305-8506, Japan

S. Sato
Institute of Geosciences, Shizuoka University, Oya 836, Shizuoka 422-8529, Japan

K. Shibata
School of Science and Engineering, Ishinomaki Senshu University, Minamizakai, Ishinomaki 986-8580, Japan

K. Okoshi
Laboratory of Marine Ecology, Graduate School of Science, Toho University, Miyama 2-2-1, Funabashi, Chiba 274-8510, Japan

J. Urabe, T. Nakashizuka (eds.), *Ecological Impacts of Tsunamis on Coastal Ecosystems*, Ecological Research Monographs,
DOI 10.1007/978-4-431-56448-5_14

was minimally affected by the tsunamis. In addition, their genetic compositions were locally fixed. These results suggest that difference in the genetic composition among *L. pulchella* populations reflects that of artificial transport source.

Keywords Biological invasion • Artificial transportation • Fishery • Mitochondrial DNA • Gene flow

14.1 Introduction

Biological invasion is a major global problem because it directly or indirectly occurs due to anthropogenic activities, such as international transportation (Vitousek et al. 1997). A number of studies have documented that the establishment of successive nonnative species populations can cause serious ecological impacts on associated ecosystems. For example, invasive species have decreased native species populations and caused extinctions via competitive interactions and predation (Mooney and Cleland 2001; Pimentel et al. 2005; Branco et al. 2015). These impacts decrease biological diversity in affected regions. In addition, the impacts often include economic damages associated with commercially important species, such as agricultural or fisheries species (Pejchar and Mooney 2009).

The carnivorous snail *Laguncula pulchella* (*Euspira fortunei*) (Fig. 14.1a) is now recognized as an invasive species that has impacted tidal flat ecosystems in northeast Japan (Okoshi 2004). Native *L. pulchella* populations exist in the shallow coastal ecosystems of western Japan. *Laguncula pulchella* was once considered an endangered species in Japan (World Wide Fund for Nature Japan 1996; Fukuda 2000; Japanese Association of Benthology 2012) because their original habitat was highly limited in western areas of Japan, and few individuals were collected in population studies (Okutani 2000). However, the species was frequently found in several areas of northeast Japan during the 1990s, including areas where the snails had not been previously collected. Because a number of the carnivorous snail populations have been recorded at various locations along the Pacific coast of northeast Japan (Okoshi and Sato-Okoshi 2011), *L. pulchella* have significantly expanded their range over the past 20 years. According to Okoshi and Sato-Okoshi (2011), the populations are not native and were likely transported from other Asian coasts, such as the east coast of China or Korean Peninsula, via international shellfish trade.

Japan imports marine products from various areas of the Asian continent. The Asari or Manila clam *Ruditapes philippinarum* is commonly used in Japanese dishes and has been caught by commercial fishing vessels at various coastal locations in Japan. However, the clams are often imported from China and Korea and transported alive to Japan when domestic catch rates are low. The clams imported from these countries were previously kept in "aquafarms" along the Japanese coast. Fisheries companies and/or fisheries associations managed the aquafarms until the clams were sold to Japanese markets for production and recreational clamming

Fig. 14.1 (**a**) *Laguncula pulchella*. (**b**) An empty *Ruditapes philippinarum* shell with feeding area (indicated by *arrow*) drilled by *L. pulchella*

("Shiohigari" in Japanese). The aquafarms encompassed extensive tidal flats, and the imported live clams were often accompanied by other organisms that were unintentionally shipped. Invasive *L. pulchella* were likely one of these unintentional species (Okoshi and Sato-Okoshi 2011).

Sato et al. (2012) and Chiba and Sato (2013) recently reported that invasive *L. pulchella* populations dramatically affect the bivalve communities in northeast Japan tidal flats (Tomiyama et al. 2011). *Laguncula pulchella* drill holes around clam shells and eat the soft tissue (Fig. 14.1b), typically decreasing *R. philippinarum* populations. Accordingly, many commercial and recreational clamming estuaries were forced to close because the *R. philippinarum* populations suffered catastrophic damages due to predation by the invasive *L. pulchella* populations. Local fisheries associations have attempted to exterminate *L. pulchella* by removing the individuals and egg masses deposited in the estuary, but these techniques produced poor results.

It is unclear how and when *L. pulchella* have expanded their range. The snails were likely introduced via *R. philippinarum* transportation (Okoshi 2004; Okoshi and Sato-Okoshi 2011). According to trade statistics from the Ministry of Finance, Japan mainly imports *R. philippinarum* from China. However, the *R. philippinarum* commercial transportation routes are complicated and not officially disclosed (Hamaguchi 2009). Therefore, little is known about the spatial and temporal characteristics of *L. pulchella* transportation. In addition to commercial transportation, *L. pulchella* may have expanded their distribution to northeast Japan after they were transported elsewhere in Japan. The recent distribution rate must be accurately assessed because the predatory snail is harmful to the commercial *R. philippinarum* fishery.

The Great East Japan Earthquake occurred on March 11, 2011. Following the earthquake, the resulting large tsunamis (the 2011 Tohoku earthquake tsunami) struck the Pacific coastline of eastern Japan, including Sendai Bay and the Sanriku Ria coast. This natural event may have disturbed local *L. pulchella* populations at various sites, affecting population dynamics and distributions. This study attempts to establish the origin of *L. pulchella* in northeast Japan and determine if natural events, such as the tsunamis caused by the Great East Japan Earthquake, contributed to the expansion of their distribution range. For these objects, the genetic structures of several local *L. pulchella* populations were examined in Sendai Bay and Sanriku Ria coast estuaries.

14.2 Materials and Methods

14.2.1 Sampling Sites

Laguncula pulchella adult individuals were collected at several Sendai Bay and Sanriku Ria coast sites (Table 14.1 and Fig. 14.2). Sendai Bay samples were collected at Matsukawaura Lagoon (Fukushima Pref.), Torinoumi Lagoon (Miyagi Pref.), and small Matsushima Bay estuaries (Miyagi Pref.). Sanriku Ria coast (Iwate Pref.) samples were collected at Miyako Bay and Yamada Bay. These individuals were mainly collected after the Great East Japan Earthquake. Egg mass and adult samples from Jeju Island (Korea) were used for comparison.

Table 14.1 Sampling site (geographical name and abbreviation used in Fig. 14.2), year, and number of individuals (N)

Prefecture or country	Site		Abbr.	Year	N
Fukushima	Matsukawaura Lagoon	Isobe	Is	2013	29
		Udagawa	Ud	2013	23
		Kawaguchimae	Ka	2009, 2010	32
Miyagi	Matsushima Bay	Hatsutsuura	Ha	2013	22
	Mainland coast	Hitsugaura	Hi	2013	6
		Sokanzan	So	2012, 2013	6
		Tona Coast	Tn	2013	3
				2004	8
	Urato Islands	Mahanashi Is.	Mh	2013	4
		Nono Is.	Nn	2013	10
		Katsura Is.	Kt	2013	15
	Torinoumi Lagoon		To	2012, 2013, 2014	10
				2010	16
Iwate	Miyako Bay		Mi	2013,2014	12
	Yamada Bay		Ya	2014, 2015	25
Korea	Jeju Island		Je	2014	10

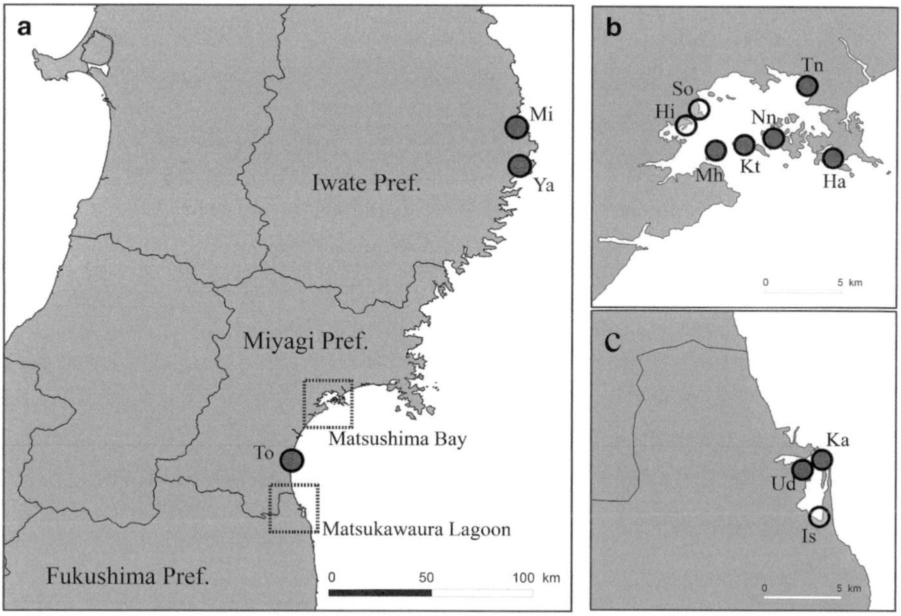

Fig. 14.2 (**a**) Map of Tohoku Region sampling sites. *Dotted squares* represent the enlarged areas in (**b**) Matsushima Bay and (**c**) Matsukawaura Lagoon. *Filled circles* indicate *Ruditapes philippinarum* fisheries. *Open circles* indicate sites that have never been stocked with *R. philippinarum*. The site abbreviations are shown in Table 14.1. These maps were drawn using the information supplied by GADM, the Database of Global Administrative Areas (http://www.gadm.org/)

14.2.2 Genetic Analysis

Laguncula pulchella soft tissues or eggs were used for DNA extraction. Less than 10 mg of soft foot tissue was removed from the adult samples for DNA extraction. DNA were extracted using a DNeasy(R) Blood & Tissue Kit (Qiagen) per the manufacturer's instructions. A single egg from each egg mass was used per individual. DNA extraction was performed using a QuickExtract™ solution (Epicentre). Samples were incubated with 50 µl of Q. E. solution at 65 °C for 2 h and 95 °C for 20 min. These extracted DNA were stored at −20 °C until use.

PCR was used to amplify the cytochrome c oxidase subunit I (COI) region in the mitochondrial DNA. Approximately 700 bp from the COI region was amplified using universal primers LCO1490 and HCO2198 (Folmer et al. 1994). Each 10 µL reaction consisted of 1.0 µL of extracted DNA (1/1–1/20 dilution), 0.25 units of TaKaRa ExTaq™ (TaKaRa), 0.8 µL of 2.5 mM of each dNTP, 1.0 µL of 10× ExTaq™ buffer, and 0.2 µM of each primer. The thermal cycle condition was as follows: a 2 min initial cycle at 95 °C, 35 cycles of 98 °C for 10 s, 52 °C for 30 s and 72 °C for 1 min, and a 10 min final cycle at 72 °C. All amplified samples were purified using ExoSAP-IT(R) (Affymetrix) and sequenced using the BigDye™ Terminator v3.1

Cycle Sequencing Ready Reaction Kit (Applied Biosystems). Each sequencing reaction was purified using the BigDye(R) XTerminator™ Purification Kit (Applied Biosystems). DNA sequencing was performed using an ABI PRISM(R) 3100 Avant Genetic Analyzer (Applied Biosystems).

The sequences obtained from each sample were aligned using ClustalW (Thompson et al. 1994). Ambiguous regions at both ends of sequences were visually eliminated. Unique haplotypes were identified and used for further analyses. A haplotype network was constructed to visualize the genetic distance among the haplotypes using Network 4.6 (Fluxus Technology Ltd. 2012; Forster et al. 1996). Two sequences deposited in GenBank, HQ834111 (Zou et al. 2001) and JX503048 (Kim et al. unpublished), were also included in the haplotype network.

14.3 Results

14.3.1 Haplotypes by COI

616 bp sequences of the COI gene were examined from the collected samples. Samples collected in Japan after the earthquake (165 individuals) contained 15 haplotypes (Table 14.2). Each unique haplotype was denoted as F (found mainly in Fukushima Pref.), M (Miyagi Pref.), or I (Iwate Pref.). One haplotype was frequently observed in the samples and denoted as C. The Korean samples (ten individuals) contained two haplotypes (X1 and X2) that were not found in Japan.

Table 14.2 Haplotype compositions after the Great Earthquake

Site	F1	F2	F3	F4	F5	F6	F7	M1	M2	M3	M4	I1	I2	I3	X1	X2	C	Total
Is	10	3	2	0	1	0	0	1	0	0	0	0	0	0	0	0	12	29
Ud	5	2	1	1	3	1	1	0	0	0	0	0	0	0	0	0	9	23
(Sum)	15	5	3	1	4	1	1	1	0	0	0	0	0	0	0	0	21	52
Ha	0	1	0	0	0	0	0	21	0	0	0	0	0	0	0	0	0	22
Hi	0	0	0	0	0	0	0	2	0	0	0	0	0	0	0	0	4	6
So	0	0	0	0	0	0	0	2	0	0	0	0	0	0	0	0	4	6
Tn	0	0	0	0	0	0	0	1	0	0	0	0	0	0	0	0	2	3
Mh	0	0	0	0	0	0	0	0	1	3	0	0	0	0	0	0	0	4
Nn	0	0	0	0	0	0	0	3	0	0	0	0	0	0	0	0	7	10
Kt	1	0	0	0	0	1	0	6	0	0	1	0	0	0	0	0	6	15
(Sum)	1	1	0	0	0	1	0	35	1	3	1	0	0	0	0	0	23	66
To	0	0	0	0	0	0	0	3	0	0	0	0	0	0	0	0	7	10
Mi	0	5	0	0	0	0	0	0	0	0	0	2	0	3	0	0	2	12
Ya	0	0	0	0	0	0	0	0	0	0	0	24	1	0	0	0	0	25
Je	0	0	0	0	1	1	0	0	0	0	0	0	0	0	3	2	3	10

Refer to Table 14.1 for site abbreviations

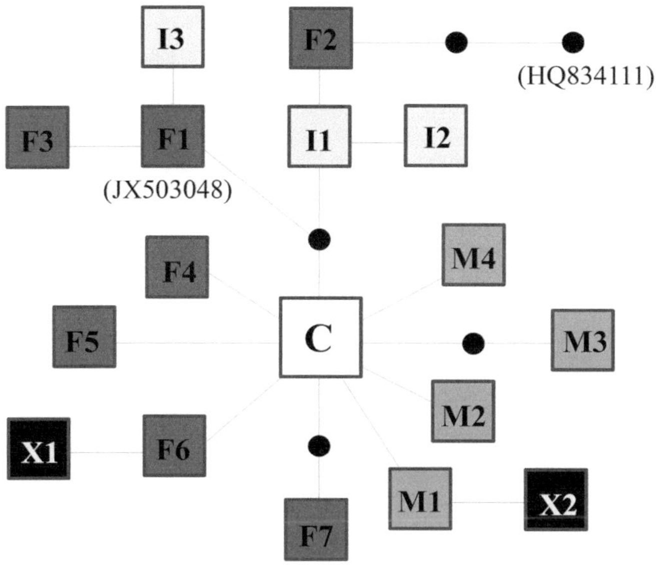

Fig. 14.3 Haplotype network by COI sequence. *Squares* and *circles* represent haplotypes detected and not detected in this study, respectively. *Lines* between haplotypes indicate single nucleotide substitutions

A haplotype network was constructed using the observed haplotypes, including two from the GenBank database (Fig. 14.3). The network illustrates that multiple haplotypes were genetically directly or indirectly derived from haplotype C. All other haplotypes possessed less than four nucleotide substitutions compared with haplotype C. Haplotype F1 matched JX503048 in the GenBank database.

14.3.2 Genetic Structures of Local Populations

Haplotype compositions of *Laguncula pulchella* populations after the earthquake are shown in Table 14.2 and Fig. 14.4. Haplotype C appeared at all sampling sites, with the exception of Yamada Bay, and dominated the Matsukawaura Lagoon, Matsushima Bay, and Torinoumi Lagoon samples. However, other haplotypes were less common and found at limited sites. Haplotype F1 was found most frequently in Matsukawaura Lagoon with haplotype F2, which also appeared in Miyako Bay. Haplotype M1 was frequently found in Matsushima Bay and Torinoumi Lagoon. Haplotypes M2, M3, and M4 were found only in Matsushima Bay. Miyako Bay and Yamada Bay (Iwate Pref.) populations exhibited significantly different haplotype compositions compared with these sites. The *L. pulchella* population in Miyako Bay displayed unique haplotypes I1 and I3. The Yamada Bay population was mainly composed of haplotype I1, as well as one individual with haplotype I2. Haplotypes C, F5, and F6 were detected at Jeju Island (Korea).

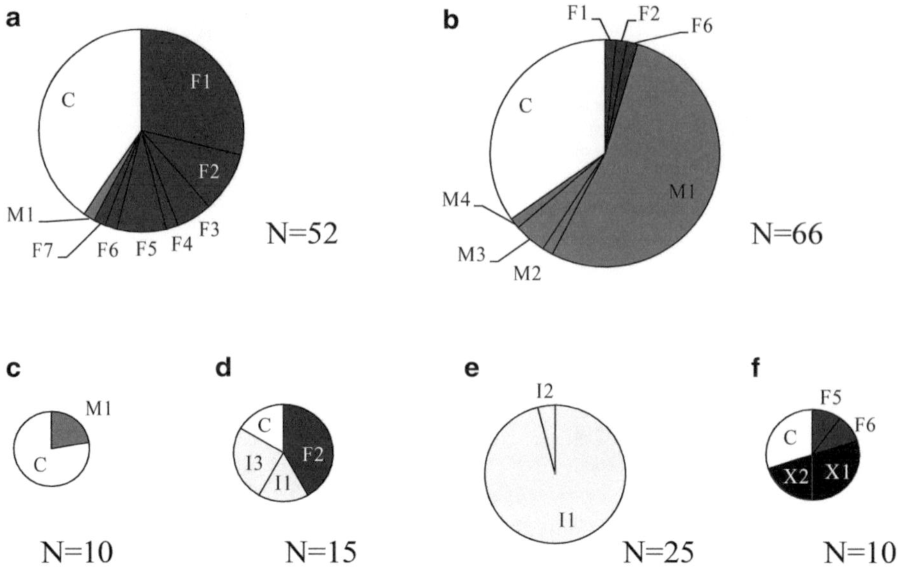

Fig. 14.4 Haplotype compositions from (**a**) Matsukawaura Lagoon, (**b**) Matsushima Bay, (**c**) Torinoumi Lagoon, (**d**) Miyako Bay, (**e**) Yamada Bay, and (**f**) Jeju Island

The haplotype compositions within Matsukawa Lagoon were similar between different sites regardless characteristics (Table 14.2 and Fig. 14.5). Local fishermen intensively use Udagawa as a *Ruditapes philippinarum* fishery and began stocking imported individuals in the 1990s. Isobe is located approximately 5 km from Udagawa, but is not used a fishery, implying that *L. pulchella* were not artificially released via imported clams. However, *L. pulchella* are abundant at Isobe. In addition, both areas exhibited similar haplotype compositions. Similarly, the Soukanzan and Hitsugaura areas in Matsushima Bay have not been used as *R. philippinarum* fisheries. However, *L. pulchella* was identified at these sites, with post-tsunami samples consisting of haplotypes C and M1. These haplotypes were also recorded at other Matsushima Bay sites where imported *R. philippinarum* individuals were stocked.

14.3.3 Comparison of Haplotype Compositions Before and After the 2011 Tsunamis

Several *L. pulchella* individuals were collected from Matsukawaura Lagoon, Torinoumi Lagoon, and the Tona Coast of Matsushima Bay before the 2011 tsunamis. The samples were then stored in refrigerators. These samples were used to compare the haplotype compositions before and after the tsunamis (Table 14.3 and Fig. 14.6). Overall, the haplotype compositions did not differ before and after the

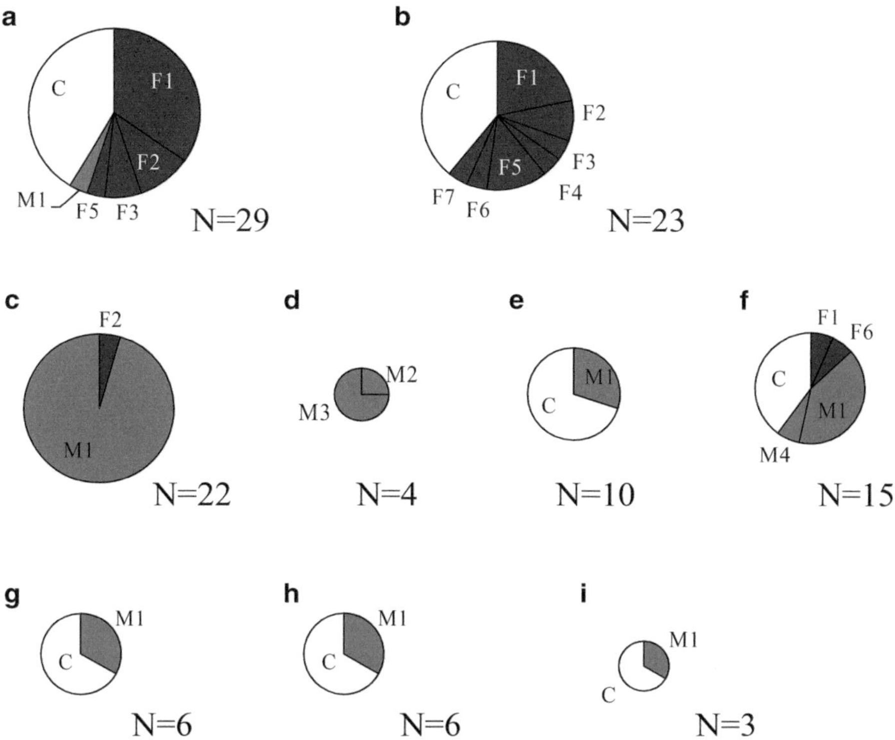

Fig. 14.5 Haplotype compositions from Matsukawaura Lagoon (**a**, **b**) and Matsushima Bay (**c**–**i**) internal sites. (**a**) Isobe, (**b**) Udagawa, (**c**) Hatsutsuura, (**d**) Mahanashi Is., (**e**) Nono Is., (**f**) Katsura Is., (**g**) Sokanzan, (**h**) Hitsugaura and (**i**) Tona Coast

Table 14.3 Haplotype compositions before the Great Earthquake

Site	F1	F2	F4	F5	F8	F9	M1	C	Total
Ka	10	2	1	1	1	1	4	12	32
Tn	0	2	0	0	0	0	2	4	8
To	0	0	0	0	0	0	6	10	16

Refer to Table 14.1 for site abbreviations

tsunami at any of the study sites. Haplotype C was found in all populations both before and after the tsunami. Six pre-tsunami haplotypes from Matsukawaura Lagoon were also identified after the tsunamis. Other unique haplotypes, such as F8 and F9, were also recorded before the tsunamis. According to the haplotype network, these two individuals possess two nucleotide substitutions compared with C (Fig. 14.3). The Torinoumi Lagoon population was composed of haplotypes C and M1 before and after the tsunamis. Haplotypes C and M1 were also common in the Tona Coast population before and after the tsunami, while F2 was only recorded before the earthquake.

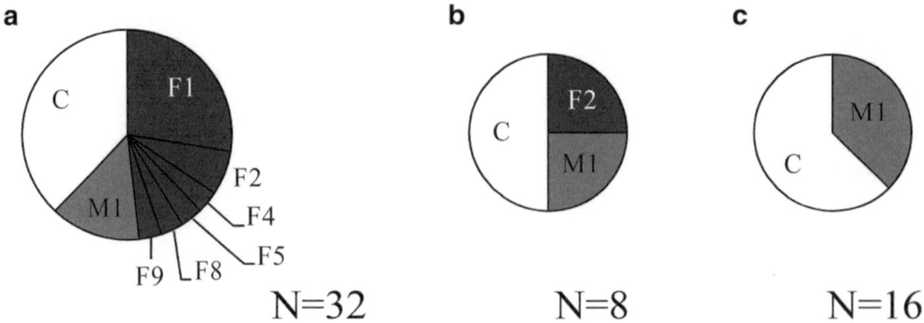

Fig. 14.6 Haplotype compositions before the Great Earthquake. (**a**) Kawaguchimae (Matsukawaura Lagoon), (**b**) Tona Coast (Matsushima Bay), and (**c**) Torinoumi Lagoon

14.4 Discussion

14.4.1 Haplotype Composition and Gene Flow

This study showed that the *Laguncula pulchella* haplotype compositions differed among the study sites. Although most populations contained haplotype C, other haplotypes were generally found at different sites. For example, haplotype M1 was common at Matsushima Bay and Torinoumi Lagoon sites, but almost never found at other sites. In addition, *L. pulchella* populations from Miyako Bay and Yamada Bay did not share the same haplotypes despite being separated by a relatively small distance. This result indicates that differences in the haplotype composition among local populations were not determined by distance. In addition, these results suggest that little or no gene flow has occurred among the local populations, and the current *L. pulchella* range expansion in Japan cannot be explained by natural migration.

One of the objectives of this study was to assess *L. pulchella* distribution variations due to the tsunamis generated by the Great East Japan Earthquake, which occurred in March 2011. A number of studies have demonstrated the ecological impacts of the tsunamis on benthic communities along the Pacific coast of northeast Japan (Seike et al. 2013; Urabe et al. 2013). Fortunately, this study was able to analyze several *L. pulchella* specimens collected before the tsunamis at Matsukawaura Lagoon, Torinoumi Lagoon, and the Tona Coast of Matsushima Bay. The haplotype compositions of these did not significantly differ before and after the tsunamis. This result suggests that the tsunami effects on the local populations were not large enough to alter the dominant haplotype at each site. However, some haplotypes at Matsukawaura Lagoon appeared before the tsunamis, but were not collected after the tsunamis; other haplotypes were only found in populations after the tsunamis. In addition, haplotype F2 from the Tona Coast was observed only before the tsunamis. These results suggest that the 2011 tsunamis altered or decreased the genetic diversities of the local populations based on the temporal snail density declines. The genetic *L. pulchella* compositions at these sites should be continually monitored to verify this theory.

14.4.2 Origin and Invasion Process

The degree of genetic differentiation between native and invasive populations remains unclear because samples could not be obtained from native *L. pulchella* habitats in Japan. However, a study showed that the genetic structure of *Batillaria cumingi*, a gastropod species inhabiting tidal flats, was genetically separated into two lineages, including a Tsushima group distributed along the Japan Sea and Korean Peninsula and a Kuroshio group distributed along the Pacific coast of Japan (Kojima et al. 2004). These two groups were likely differentiated by geographic and oceanic events that occurred before the last glacial period. Thus, it is likely that native *L. pulchella* populations in Japan have differentiated from other populations throughout history. However, several haplotypes that were recorded in Japan (C, F5 and F6) were also found at Jeju Island (Korea). In addition, haplotype F1 in this study was genetically identical to a specimen recorded in Korea (GenBank accession number JX503048). Moreover, all of the haplotypes recorded in this study possessed less than four nucleotide substitutions compared with haplotype C. These findings suggest that *L. pulchella* were genetically enveloped by the same group and originated from other Asian populations.

Although some haplotypes collected in Japan were similar to those from Korea, the source of these invasive populations remains unclear because *L. pulchella* is widely distributed along the east coast of Asia. According to Okoshi (2004) and Okoshi and Sato-Okoshi (2011), a number of live *L. pulchella* individuals were found in bulk *Ruditapes philippinarum* shipments that were imported by domestic fishermen. These clams were stocked in Sendai Bay and Sanriku Ria coast tidal flat aquafarms during the 1990s. However, no record exists pertaining to where these clams were collected before being shipped to Japan. Genetic population structures from across the continent must be examined to reveal the source of the Sendai Bay and Sanriku Ria coast *L. pulchella* populations.

Nevertheless, haplotype composition variations among the study sites reflect different clam trade routes. *Laguncula pulchella* haplotype C was abundant in Matsukawaura Lagoon, Torinoumi Lagoon, and Matsushima Bay and also observed in Miyako Bay. However, haplotype compositions generally differed among these sites. *Laguncula pulchella* populations in Yamada Bay and Miyako Bay shared only one haplotype, despite being separated by a relatively small distance. These results suggest that fishermen at each site use unique trade channels and obtain clams at different times or from different distributers.

Laguncula pulchella directly develops embryos and experiences no planktonic stage. Therefore, their distribution is largely determined by anthropogenic activities, as confirmed by the site-specific haplotype compositions in this study. However, this study also shows that the snails possess a limited ability to expand their distribution. *Laguncula pulchella* were found at Matsukawaura Lagoon and Matsushima Bay sites that have never been used for clam fishing. The haplotype compositions at these sites were similar to sites used as stocked *R. philippinarum* aquafarms and located within 10 km. Okoshi and Sato-Okoshi (2011) reported that *L. pulchella* can move via floating under the water (Okoshi and Sato-Okoshi 2011). This migration technique is likely limited within clam embayments and lagoons. However, haplotype composition similarities between the Matsukawaura Lagoon and Matsushima Bay sites suggest that natural migrations of at least 10 km are possible.

14.5 Conclusion

This study demonstrated that the *Laguncula pulchella* haplotype compositions differed among the bays and lagoons of northeast Japan, suggesting artificial transportation associated with clam fisheries as the cause for the snail's increased distribution. The tsunamis induced by the Great East Japan Earthquake in 2011 minimally affected the genetic structures of the local populations. Large *L. pulchella* populations still exist at some sites (Sato and Chiba 2016). Active migration and natural events contributed to the range expansion. However, significant, rapid expansions are generally due to anthropogenic activities. Removing invasive *L. pulchella* individuals and eggs from tidal flats is important to decrease and eventually alleviate these snail populations, protecting the domestic *R. philippinarum* fishery.

Acknowledgments This study was financially supported by the Tohoku Ecosystem-Associated Marine Sciences, MEXT, Japan.

References

Branco S, Videira N, Branco M, Paiva MR (2015) A review of invasive alien species impacts on eucalypt stands and citrus orchards ecosystem services: towards an integrated management approach. J Environ Manag 149:17–26

Chiba T, Sato S (2013) Invasion of *Laguncula pulchella* (Gastropoda: Naticidae) and predator-prey interactions with bivalves on the Tona coast, Miyagi prefecture, northern Japan. Biol Invasions 15:587–598

Fluxus Technology Ltd. (2012) fluxus-engineering.com, Free Phylogenetic Network Software. http://www.fluxus-engineering.com/sharenet.htm. Accessed 20 Aug 2015

Folmer O, Black M, Hoeh W, Lutz R, Vrijenhoek R (1994) DNA primers for amplification of mitochondrial cytochrome c oxidase subunit I from diverse metazoan invertebrates. Mol Mar Biol Biotechnol 3:294–299

Forster P, Harding R, Torroni A, Bandelt HJ (1996) Origin and evolution of native American mtDNA variation: a reappraisal. Am J Hum Genet 59:935–945

Fukuda H (2000) Gastropods I: general view. In: Sato M (ed) Life in Ariake Sea: biodiversity in tidal flats and estuaries. Kaiyusha, Tokyo, pp 100–137 (in Japanese)

Hamaguchi M (2009) Where does *Ruditapes philippinarum* come from? In: Plankton Society of Japan, Japanese Association of Benthology (ed) Marine aliens introduced by human activities and their impacts on ecosystems and industries. Tokai University Press, Kanagawa, pp 228–230 (in Japanese)

Japanese Association of Benthology (Ed) (2012) Threatened animals of Japanese tidal flats: Red Data Book of seashore benthos. Tokai University Press, Kanagawa (in Japanese)

Kojima S, Hayashi I, Kim D, Iijima A, Furota T (2004) Phylogeography of an intertidal direct-developing gastropod *Batillaria cumingi* around the Japanese Islands. Mar Ecol Prog Ser 276:161–172

Mooney HA, Cleland EE (2001) The evolutionary impact of invasive species. Proc Natl Acad Sci U S A 98:5446–5451

Okoshi K (2004) Alien species introduced with imported clams: the clam-eating moon snail *Euspira fortunei* and other unintentionally introduced species. Jpn J Benthol 59:74–82 (in Japanese with English abstract)

Okoshi K, Sato-Okoshi W (2011) *Euspira fortunei*: biology and fisheries science of invasive species. Kouseishakouseikaku Press, Tokyo (in Japanese)

Okutani T (ed) (2000) Marine mollusks in Japan. Tokai University Press, Kanagawa (in Japanese)

Pejchar L, Mooney HA (2009) Invasive species, ecosystem services and human well-being. Trends Ecol Evol 24:497–504

Pimentel D, Zuniga R, Morrison D (2005) Update on the environmental and economic costs associated with alien-invasive species in the United States. Ecol Econ 52:273–288

Sato S, Chiba T (2016) Ecological impacts and recovery of molluscan populations after the tsunami struck around Matsushima Bay and Sendai Bay, Miyagi Prefecture, northeastern Japan. In: Urabe J, Nakashizuka T (eds) Ecological impacts of tsunamis on coastal ecosystems: Lessons from the Great East Japan Earthquake. Springer, pp 179–208

Sato S, Chiba T, Hasegawa H (2012) Long-term fluctuations in mollusk populations before and after the appearance of the alien predator *Euspira fortunei* on the Tona coast, Miyagi Prefecture, northern Japan. Fish Sci 78:589–595

Seike K, Shirai K, Kogure Y (2013) Disturbance of shallow marine soft-bottom environments and megabenthos assemblages by a huge tsunami induced by the 2011 M9.0 Tohoku-Oki earthquake. PLoS One 8, e65417

Thompson JD, Higgins DG, Gibson TJ (1994) CLUSTAL W: improving the sensitivity of progressive multiple sequence alignment through sequence weighting, position-specific gap penalties and weight matrix choice. Nucleic Acids Res 22:4673–4680

Tomiyama T, Suzuki T, Sato T, Kato Y, Kameiwa S, Sugibayashi Y, Okoshi K (2011) Unintentional introduction and the distribution of the nonindigenous moonsnail *Euspira fortunei* in Matsukawaura Lagoon, Japan. Bull Jpn Soc Sci Fish 77:1020–1026 (in Japanese with English abstract)

Urabe J, Suzuki T, Nishita T, Makino W (2013) Immediate ecological impacts of the 2011 Tohoku Earthquake tsunami on intertidal flat communities. PLoS One 8, e62779

Vitousek PM, D'Antonio CMD, Loope LL, Rejmanek M, Westbrooks R (1997) Introduced species: a significant component of human-caused global change. N Z J Ecol 21:1–16

World Wide Fund for Nature Japan (1996) WWF Japan science report, vol 3. World Wide Fund for Nature Japan, Tokyo (in Japanese)

Zou S, Li Q, Kong L (2001) Additional gene data and increased sampling give new insights into the phylogenetic relationships of Neogastropoda, within the caenogastropod phylogenetic framework. Mol Phylogenet Evol 61:425–435

Chapter 15
Impacts of Fuel Spills Caused by the Great East Japan Earthquake and Tsunami on the Subtidal Soft-Bottom Communities of a Semi-enclosed Bay Located on the Sanriku Coast

Gen Kanaya, Hideaki Maki, Fumihiro Chiba, Kazuki Miura, Shin-ichi Fukuchi, Hisao Sasaki, and Osamu Nishimura

Abstract We assessed the ecological consequences of the fuel spills and the subsequent conflagration that resulted from the tsunamis caused by the Great East Japan Earthquake. Analyses were performed at a range of subtidal stations (water depth: 8.2–37.2 m) located in Kesennuma Bay, Japan, during the period 2011–2014. The sediments contained high concentrations of polycyclic aromatic hydrocarbons (PAHs) (mean total PAH concentrations: 18,023–89,197 μg kg^{-1} dry weight). Most PAHs in the sediment of the inner bay (57–82 %) were derived largely from petrogenic sources, but at other stations most (55–86 %) had pyrogenic origins. Levels of petrogenic PAHs decreased more rapidly than those of pyrogenic PAHs over 4 years, possibly because of microbial degradation. PAH contents in the sediments peaked at 8–15 cm depth and were much reduced near the sediment/water interface. The macrozoobenthic community was largely dominated by opportunistic polychaetes; other taxa, including amphipods and bivalves (but not *Theora fragilis*), were less abundant. The PAHs often exceeded the effects range median (ERM) in US Sediment Quality Guidelines, indicating elevated ecotoxicological risk levels for the benthic invertebrates. Nevertheless, negative impacts of PAHs were not detected in the dominant macrozoobenthos. Thus, the dominant taxa were potentially

G. Kanaya (✉) • H. Maki
National Institute for Environmental Studies, 16-2 Onogawa, Tsukuba 305-8506, Japan
e-mail: gen@nies.go.jp

F. Chiba • K. Miura • S.-i. Fukuchi
Miyagi Prefectural Institute of Public Health and Environment,
4-7-2 Saiwai-cho, Miyagino-ku, Sendai 983-0836, Japan

H. Sasaki • O. Nishimura
School of Engineering, Tohoku University,
Aoba 6-6-04, Aramaki, Aoba-ku, Sendai 980-8579, Japan

J. Urabe, T. Nakashizuka (eds.), *Ecological Impacts of Tsunamis on Coastal Ecosystems*, Ecological Research Monographs,
DOI 10.1007/978-4-431-56448-5_15

tolerant to PAH pollution. Sediment in the bay currently contains high levels of PAHs, especially in thesubsurface layer. Therefore, long-term monitoring is required to fully understand the ecological consequences (for the coastal soft-bottom communities) of the 2011 fuel spill and subsequent conflagration.

Keywords Great East Japan Earthquake • Tsunami • Polycyclic aromatic hydrocarbons (PAHs) • Fuel spill • Macroinvertebrate • Subtidal soft-bottom community • Sediment • Sanriku coast

15.1 Introduction

Polycyclic aromatic hydrocarbons (PAHs) are pollutants that occur in the atmosphere, terrestrial soils, and marine sediments worldwide. They are produced by a variety of sources, including motor exhausts, steel and coke industries, oil spills, and biomass burning (Colombo et al. 1989; Oug et al. 1998; Samanta et al. 2002). PAHs persist over protracted periods in natural environments because their hydrophobic properties confer chemical stability and refractoriness (Samanta et al. 2002; Johnsen et al. 2005). Since PAHs are highly toxic to aquatic invertebrates (Gómez Gesteira and Dauvin 2000; Page et al. 2002; Engraff et al. 2011), environmental contamination with these compounds often induces large changes in the structure of marine macroinvertebrate assemblages (Oug et al. 1998; Belan 2003; Nikitik and Robinson 2003; Joydas et al. 2012).

Oil spills are among the major sources of PAHs in marine environments. They often have negative impacts on local faunas (Nikitik and Robinson 2003; Joydas et al. 2011, 2012). In general, spilled oil is removed or degraded through physical dispersion, weathering, evaporation, and microbial decomposition (Boopathy et al. 2012; Bacosa and Inoue 2015). However, a certain proportion may persist for decades in the sediments. Thus, complete recovery of environmental conditions and benthic community structure after an oil spill is generally a lengthy process (e.g., 5–15 years, Nikitik and Robinson 2003; Joydas et al. 2011; but see Page et al. 2002).

Responses of macroinvertebrates toward PAH toxicity vary among taxonomic groups. Previous studies have shown that amphipods are sensitive to PAH pollution and disappear shortly after an oil spill, but polychaetes are generally more tolerant (Gómez Gesteira and Dauvin 2000; Nikitik and Robinson 2003). Thus, PAH-tolerant polychaetes proliferate in benthic communities after an oil spill, while PAH-sensitive taxa, such as amphipods, decline or completely disappear until PAH levels fall below ecophysiological threshold levels (Grassle and Grassle 1974; Oug et al. 1998; Nikitik and Robinson 2003; Joydas et al. 2011). Accordingly, existing macrozoobenthic assemblage structures may be strongly modified by oil spill events (Joydas et al. 2011, 2012).

The Pacific coast of northeastern Japan was struck by huge tsunamis (wave heights reached 20 m) caused by the Great East Japan Earthquake on 11 March 2011. The waves caused major physical disturbances in shallow coastal habitats

(Haraguchi et al. 2013; Seike et al. 2013; Kanaya et al. 2014) that led to structural changes in the biotic communities, including those on sandy shores (Hayasaka et al. 2012), intertidal flats (Kanaya et al. 2014), and subtidal soft-bottoms (Seike et al. 2013). The earthquake, the tsunamis, and the ensuing conflagration caused releases of chemical pollutants, including heavy metals, oils, and pyrogenic PAHs, into coastal waters (Sera et al. 2012; Kanaya et al. 2014; Bacosa and Inoue 2015). These chemicals were potential ecotoxicological risk factors for the biota in impacted habitats. However, the ecological impacts of the chemical pollutants released by the tsunami disaster, particularly those affecting marine benthic invertebrates, have yet to be fully addressed.

In this study, we assessed the ecological impacts of PAH contamination during the period 2011–2014 in six subtidal stations (water depths: 8.2–37.2 m) located in Kesennuma Bay on the southern Sanriku ria coast. We examined spatiotemporal variations in sediment PAH levels and macrozoobenthic community structure at the stations. The main objectives of the study were to (i) measure spatiotemporal variations in the quantity and quality of PAHs in the sediments, (ii) identify the potential ecotoxicological risks of PAHs for macroinvertebrates based on published toxicity thresholds, and (iii) detect the impacts of PAH pollution on benthic macroinvertebrate community structure in Kesennuma Bay.

15.2 Methods

15.2.1 Study Site

Kesennuma Bay (34 km^2) is located on the southern Sanriku coast of northeastern Japan (Fig. 15.1). A narrow waterway, the Oshima Channel (water depth: 30–40 m), connects the west and east bays (water depths generally <15 m and 20–40 m in the two bays, respectively) of this water body. Kesennuma Bay was an important aquaculture ground for oysters, tunicates, and scallops. Before the tsunami, the inner portion of the bay was eutrophicated (Igarashi et al. 1978), and the sediments were largely covered with organically enriched mud. Red tide and hypoxia events often occurred in warmer months of the year (Ito 2008; Yamaoka et al. 2011).

On 11 March 2011, huge tsunamis with maximum inundation depths of 16.6 m struck the bay. The fuel storage tanks in Kesennuma City (Fig. 15.1) were destroyed and a total of 12,810 kL of fuel (mostly light diesel fuel for fishing boats) were discharged into the bay (Masuda 2011), where they caught fire and caused massive sediment contamination with both petrogenic and pyrogenic PAHs. We collected sediment and macrozoobenthos at one to five stations (St.) in 2011–2014. Sediment and macrozoobenthos were sampled at St. 2 in May, September, and December 2011. Samples were collected at St. 2–6 in May 2012, September 2012, and January 2013. In August 2014, samples were collected at St. 1, 2, 4, and 6. In May 2012, sampling was performed at St. 6', which was located about 1 km south of St. 6. Water depth was shallower at St. 1 and 3 (8.2–12.3 m) than at the other stations (30.8–35.0 m, Table 15.1). The sediments were muddy at all stations (silt–clay: 63–98 %).

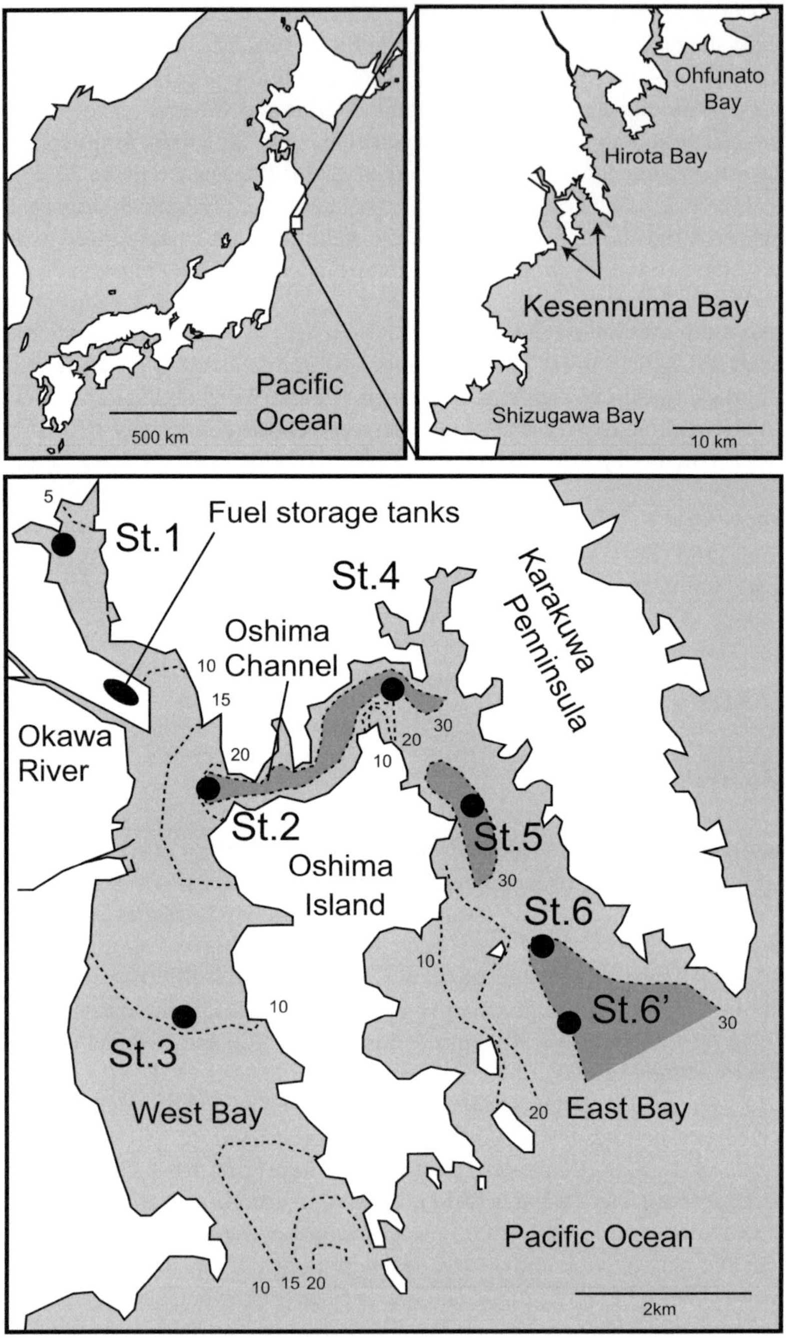

Fig. 15.1 Locations of sampling stations in Kesennuma Bay, Japan. Water depths (m) before the 2011 tsunami are shown

Table 15.1 Water depth and characteristics of sediments collected from 0 to 5 cm deep at each station. Values for the sampling occasions are means (SD)

	St. 1	St. 2	St. 3	St. 4	St. 5	St. 6[a]
	$n = 1$	4	3	4	3	4
Depth (m)	8.2	37.2 (1.7)	12.3 (0.4)	32.3 (1.1)	30.8 (0.8)	32.1 (2.3)
SC (%)	98	91 (12)	89 (9)	92 (8)	94 (5)	87 (16)
Eh (mV)	−121	−158 (35)	−45 (72)	−23 (76)	37 (76)	16 (84)
TOC (%)	2.5	3.1 (0.2)	2.5 (0.6)	2.6 (0.5)	2.4 (0.4)	3.1 (0.6)
C/N_{atom}	14.8	13.7 (0.7)	12.5 (3.8)	13.2 (0.3)	13.9 (1.5)	13.3 (2.6)
$\delta^{13}C$ (‰)	−24.3	−24.2 (0.1)	−18.9 (4.4)	−23.9 (0.4)	−21.7 (3.1)	−21.1 (3.7)
$\delta^{15}N$ (‰)	3.5	5.0 (1.0)	6.6 (0.5)	5.2 (0.9)	5.5 (0.3)	5.2 (0.6)

Eh was measured at 5 cm depth in the sediment
[a]Samples taken from Stn 6′ in May 2012
SC silt–clay, *TOC* total organic carbon

15.2.2 Water and Sediment Quality Measures

At each station, we measured water temperature (WT), salinity, and dissolved oxygen (DO) at intervals of ca. 0.2 m upward from the seafloor using a water quality profiler (DataSonde 4a or 5; Hydrolab, HACH). Secchi depth was measured with a Secchi disk on each sampling occasion in 2012 and 2013.

Sediment was collected with an Ekman-Birge grab sampler (400 cm^2 in area, 0–10 cm deep, Rigo-sha). At St. 6 in May 2012, we used a different grab sampler (15 cm diameter, 15 cm deep; KS sampler, Rigo-sha). For PAH analyses, we collected one grab sample from each station in the period 2011–2013 and three grab samples from each station in 2014. Sediment was transferred into a bucket; a subsample for PAH analyses was taken using a spatula and transferred into a plastic bottle, sealed with a cap, and placed in a cool box. The sediment was frozen at −20 °C until further analyses were performed. Subsamples for analyses of silt–clay, total organic carbon (TOC), total nitrogen (TN), C/N_{atom}, and carbon and nitrogen stable isotope ratios ($\delta^{13}C$ and $\delta^{15}N$) were also taken as the same manner. Redox potential (Eh) was measured using a handheld ORP meter (RM-30P, TOA DKK) inserted to 5 cm depth in the sediment. Additional cores (10 cm diameter, 50 cm length) were also taken from St. 2 (May and September 2012) and St. 4 (May 2012) using a pendant core sampler (HR type, Rigo-sha) to determine the depth profiles of PAHs. Each sediment core was sliced at 1 cm intervals to a depth of 16 or 21 cm and examined as detailed above.

PAH contents were determined by gas chromatography-mass spectrometry (GCMS-QP5050A, Shimadzu) following the methods of Ministry of the Environment (2003), with some modifications. Sediment samples were extracted with acetone and dichloromethane, and the resulting extracts were dehydrated with anhydrous sodium sulfate and cleaned and fractionated by silica gel column chromatography. Fractions for aromatic hydrocarbon measurements and the internal standards (triphenylene-d10 and hexamethylbenzene) were separately injected into the GC-MS through a fused silica capillary column in the splitless mode. Each peak of PAHs was identified and quantified in the selected ion monitoring mode. We quantified 29 PAHs in total (see Table 15.2), which summed to give a total PAH (TPAH) value. We also identified (Σalkylated PAHs + dibenzothiophene [DBT]) as petrogenic PAHs and (TPAHs – petrogenic PAHs) as pyrogenic PAHs, since PAHs in diesel fuel oil mostly comprise alkylated PAHs and DBT (95 % of TPAHs, see Table 15.2).

Sediments for silt–clay and organic matter analyses were dried at 70 °C for 48 h. Silt–clay content was determined by wet sieving through a 0.063 mm mesh. For organic matter analyses, we treated sediment with 1M HCl. TOC, TN, $\delta^{13}C$, and $\delta^{15}N$ were determined using a mass spectrometer-elemental analyzer (EA-300/Delta XP, Finnigan Mat). Isotope ratios were expressed in delta notation: $d^{13}C \text{ or } d^{15}N(‰) = (R_{sample} / R_{standard} - 1) \times 1000$ where R is $^{13}C/^{12}C$ or $^{15}N/^{14}N$ for $\delta^{13}C$ or $\delta^{15}N$. Pee Dee belemnite (PDB) and atmospheric N_2 were used as reference compounds for $\delta^{13}C$ and $\delta^{15}N$, respectively. D-Alanine was used as a working standard. C/N_{atom} was calculated from TOC and TN contents.

15.2.3 Macrozoobenthos

Macrozoobenthos was collected together with sediment samples. Although we attempted to collect samples using the pendant core sampler, the procedure was not successful in most cases. Thus we used the Ekman-Birge grab sampler instead. The sediment was sieved through a 1 mm mesh in situ and the retained fraction was fixed in 5 % neutralized formalin containing Rose Bengal stain. Animals were sorted, transferred into 70 % ethanol, and identified under a dissecting stereomicroscope. Wet weights were measured after ethanol was removed by dabbing on a paper towel.

Differences in sampling apparatuses can induce biases in sampling efficiency; grab samplers underestimate the abundance of deep-dwelling taxa relative to core samplers. See Appendix Table 15.4 for the sampling apparatus used on each sampling occasion. In the 2011 sampling at St. 2, no macrofauna species were encountered.

15.2.4 Data Analyses

Relationships among the macrozoobenthos and environmental variables were examined by canonical correspondence analyses (CCA) executed by CANOCO software (ter Braak and Smilauer 1998). CCA is a nonlinear eigenvector ordination that constrains the axes to linear combinations of environmental variables. It can be used to determine the environmental characteristics of each sample and the habitat preferences of each taxon. A forward selection procedure tested the significance of each environmental variable at $P<0.1$. Parameters included in the analysis were the density of 22 common taxa that occurred in >4 samples (occasions and/or locations) and values of 14 normalized environmental variables (TPAHs, petrogenic PAHs,

Table 15.2 Contents of polycyclic aromatic hydrocarbons (PAHs) in the upper sediment layer (0–10 cm deep) in the period 2012–2013. Values are means of three sampling occasions (May, September, and January) except for Stn 1 of which value is at August 2014. The PAH composition of diesel fuel oil used for ship propulsion is provided

PAHs	Contents in sediment (μg kg^{-1})						US SQG[c]		Diesel fuel oil
	St. 1	St.2	St.3	St.4	St.5	St.6	ERL	ERM	(μg g^{-1})
Acenaphthylene	47	168	79	364	228	462	44	640	174
Acenaphthene	32	280	27	258	144	263	16	500	943
Naphthalene	37	122	10	211	146	253	160	2100	2083
$\Sigma C_{1\text{-}4}$ Naphthalenes	197	18,625	386	9137	1572	2250			102,458
Dibenzothiophene (DBT)	54	122	28	222	122	246			1180
$\Sigma C_{1\text{-}3}$ DBTs	4114	13,243	320	5444	1422	1824			11,647
Anthracene	151	309	177	575	450	728	85	1100	228
Phenanthrene	238	324	534	849	846	1334	240	1500	2291
$\Sigma C_{1\text{-}3}$ Phenanthrenes	4377	29,600	1528	12,207	3826	6577			10,955
Fluorene	82	112	55	281	213	387	19	540	832
$\Sigma C_{1\text{-}2}$ Fluorenes	3724	11,154	219	4275	980	1842			7839
Fluoranthene	1120	1489	1788	3669	3215	**5116**	600	5100	73
Pyrene	1703	2106	2115	**4188**	**3908**	**5513**	665	2600	451
Benz [a] anthracene	684	1242	1329	**3084**	**2178**	**5430**	261	1600	9.5
Chrysene	602	1249	1225	2733	1984	**4840**	384	2800	18
Benzo[b,k]fluoranthene	1395	1408	1369	3515	2290	5578			0.41
Benzo [a] pyrene	860	**1939**	**2070**	**5364**	**3989**	**8975**	430	1600	1.5
Benzo [e] pyrene	978	1643	1570	3728	2820	6276			7.2
Indeno[1,2,3-cd]pyrene	611	1160	991	2892	1657	4295			1.1
Dibenz[a,h]anthracene	**320**	**378**	250	**856**	**491**	**1290**	63	260	2.2
Benzo[ghi]perylene	641	1125	1063	2810	1860	4327			n.d.
TPAHs	21,968	**89,197**	18,023	**69,363**	36,201	**71,605**	4022	44,792	141,192

(continued)

Table 15.2 (continued)

PAHs	Contents in sediment ($\mu g\ kg^{-1}$)						US SQG[c]		Diesel fuel oil
	St. 1	St.2	St.3	St.4	St.5	St.6	ERL	ERM	($\mu g\ g^{-1}$)
Petro. PAHs[a]	12,412	72,743	2479	31,286	7920	12,740			134,078
% to TPAHs	57	82	14	45	22	18			95
Pyro. PAHs[b]	9555	16,454	15,544	38,077	28,281	58,865			7114
% to TPAHs	43	18	86	55	78	82			5
ΣUSEPA 16PAHs	8523	13,408	13,082	31,649	23,599	48,790			

Contents above the ERM are in bold

[a]Petrogenic PAHs (Σalkylated PAHs + DBT)

[b]Pyrogenic PAHs (TPAHs – petrogenic PAHs)

[c]NOAA (1999) Sediment Quality Guidelines

ERL effects range low, *ERM* effects range median

pyrogenic PAHs, water depth, salinity, WT, DO, silt–clay, Eh, TOC, TN, C/N_{atom}, $\delta^{13}C$, and $\delta^{15}N$) collected on four sampling occasions during the period 2012–2014. Density of macrozoobenthos was square-rooted to reduce the heterogeneity of variance.

15.3 Results

15.3.1 Water Quality

Bottom temperature was lowest in January 2013 (7.5–8.7 °C), highest in September 2012 and August 2014 (17.7–21.6 °C), and intermediate in May 2012 (9.8–11.0 °C) (Fig. 15.2). Bottom salinity varied only slightly through space and time (in the range 33.2–34.2). Secchi depth was lower in the west bay (St. 2–3) than in the east bay (St. 4–6). It was highest in January 2013 at all stations (8.0–13.5 m) and lowest in September 2012 especially in the inner bay (St. 2–3; 3.5–4.5 m). DO varied over time, with lowest values in September 2012 and August 2014 (4.0–7.5 mg L^{-1}), intermediate values in May 2012 (7.2–8.4 mg L^{-1}), and highest values in January 2013 (9.6–9.9 mg L^{-1}). The vertical profiles of WT, salinity, and DO in August 2014 differed among the stations (Fig. 15.3). The thermocline formed at ca. 3–5 m depth at all stations. Salinity was lowest near the surface (<4 m depth) and uniformly high in deeper layers. DO was highest near the surface (<3 m) at St. 1–2, but not at other stations. DO fell markedly toward the bottom at St. 1 (<4.0 mg L^{-1}).

15.3.2 Sediment Characteristics

Sediment was muddy at all stations, with a mean silt–clay content of 87–98 % (Table 15.1). Eh was lower in the inner bay (St. 1–2; <–100 mV) than at other stations (> –45 mV). TOC ranged from 2.4 % to 3.1 %; C/N_{atom} was higher in the inner bay (St. 1) than at other stations. $\delta^{13}C$ was lowest at St. 1–2 (< −24 ‰), intermediate at St. 4 (−23.9 ‰), and higher at other stations (> −21.7 ‰). $\delta^{15}N$ was lower in the inner bay (St. 1, 2: 3.5–5.0 ‰) than at other stations (5.2–6.6 ‰).

15.3.3 PAHs

The contents and compositions of PAHs in the sediments differed among the stations (Table 15.3). On average, TPAH levels were highest at St. 2 (89,197 μg kg^{-1}), intermediate at St. 4 and 6 (69,363–71,605 μg kg^{-1}) and 5 (36,201 μg kg^{-1}), and lowest at St. 1 and 3 (18,023–21,968 μg kg^{-1}). PAHs in sediments were mostly petrogenic at St. 1 and 2 (57–82 %), but derived more from pyrogenic sources at other

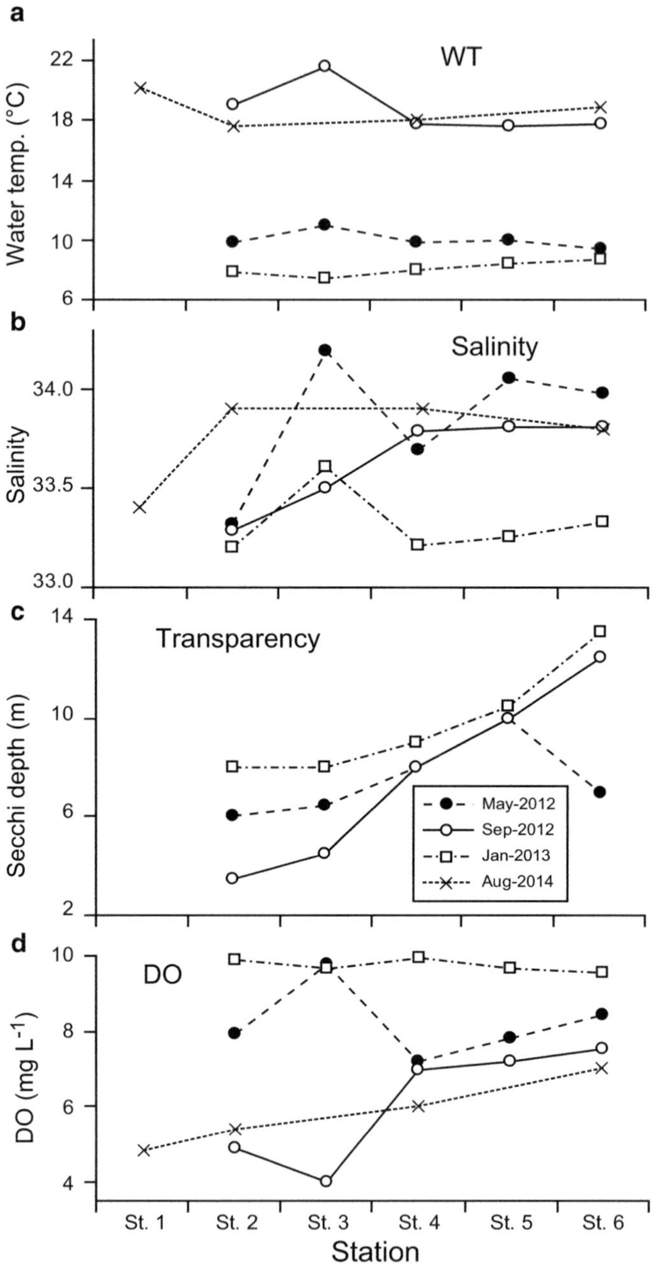

Fig. 15.2 Water temperature (WT) (**a**), salinity (**b**), transparency (**c**), and dissolved oxygen (DO) (**d**) at the stations in May 2012, September 2012, January 2013, and August 2014. WT, salinity, and DO were measured 1 m above the seafloor

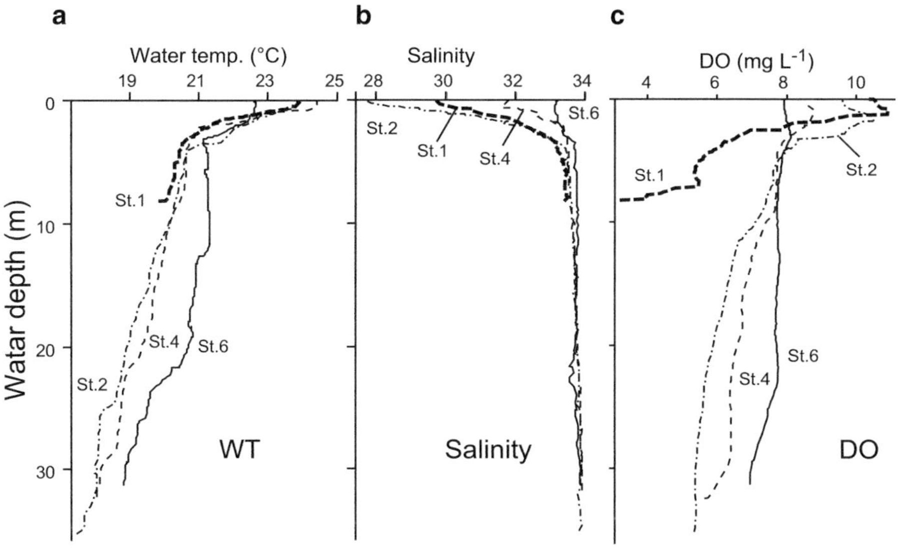

Fig. 15.3 Vertical profiles of water temperature (WT) (**a**), salinity (**b**), and dissolved oxygen (DO) (**c**) at St. 1, 2, 4, and 6 in August 2014

Table 15.3 Summary of the canonical correspondence analysis (CCA) for three environmental variables and macrozoobenthic abundances from total 19 samples collected on four occasions. Inter-set correlations between axes 1–3 and the measured variables are provided

	Axis 1	Axis 2	Axis 3
Eigenvalue	0.157	0.120	0.048
Species–environment correlation	0.872	0.809	0.590
Cumulative % variance			
Species data	10.5	18.5	21.8
Species–environment relation	48.1	85.1	100
Correlation with environmental variables [a]			
Water temperature (WT)	–0.84***	–0.22	0.05
Silt–clay (SC)	0.26	–0.60**	–0.36
Salinity (Sal)	–0.23	0.27	–0.53*

[a]During the forward selection procedure, 11 environmental parameters (bottom dissolved oxygen [DO], water depth, total polycyclic aromatic hydrocarbons [TPAHs], Σalkylated-PAHs, pyrolytic PAHs, Eh, total nitrogen [TN], total organic carbon [TOC], isotopic signatures [$\delta^{13}C$, $\delta^{15}N$], and C/N ratio) were excluded from the analysis because of low significance level ($p>0.1$)
Significant correlations are indicated by asterisks: $^{*}p<0.05$, $^{**}p<0.01$, $^{***}p<0.001$

stations (55–86 %). PAH contents in sediments at every sampling station often exceeded the effects range-low (ERL) levels in US Sediment Quality Guidelines (SQG; NOAA 1999). Furthermore, PAHs, including fluoranthene, pyrene, benz[a]anthracene, chrysene, benzo[a]pyrene, and dibenz[a,h]anthracene and TPAHs, exceeded effects range-median (ERM) levels at St. 1–5.

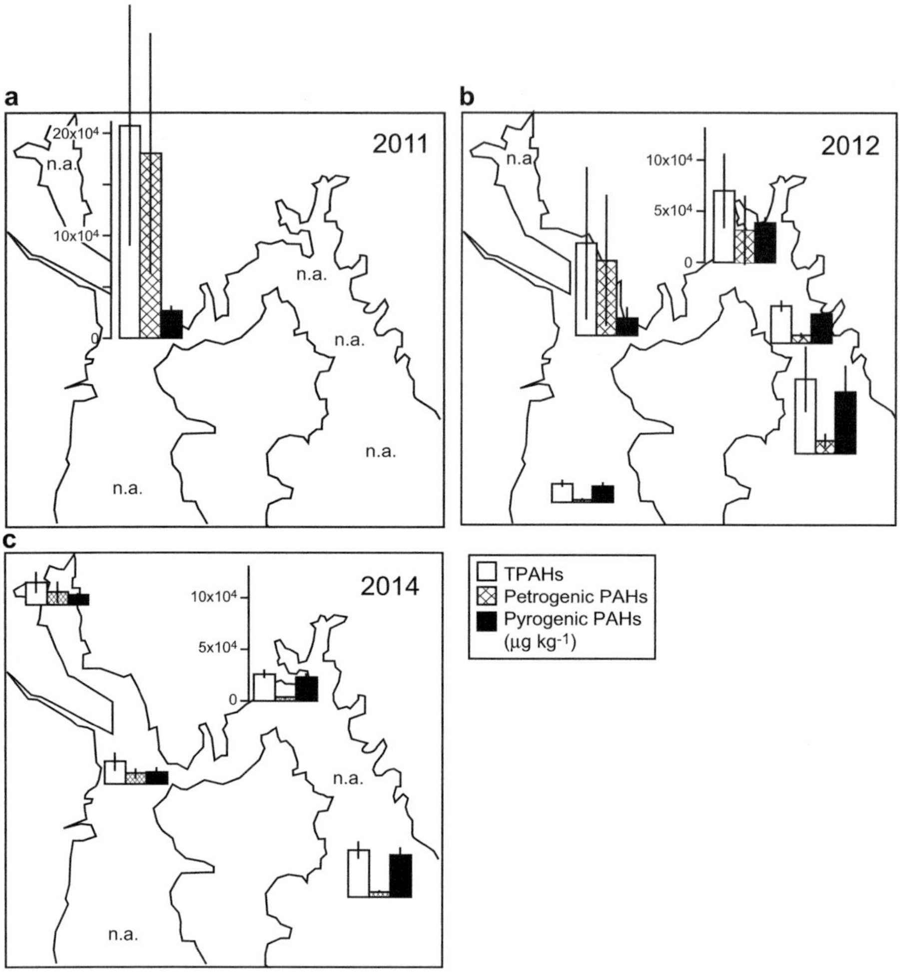

Fig. 15.4 Spatial variations in the contents of total polycyclic aromatic hydrocarbons (TPAHs), petrogenic PAHs, and pyrogenic PAHs in 2011 (**a**), 2012 (**b**), and 2014 (**c**). Error bars refer to SDs ($n=3$). *n.a* not available. Values in 2011 and 2012 are means of three sampling occasions

The TPAH content amounted to 200,000 μg kg^{-1} at St. 2 in 2011 (Fig. 15.4) and largely comprised petrogenic PAHs (87 %). In 2012, the TPAH content was still highest at St. 2 (89,197 μg kg^{-1}), followed by St. 6 and 4 (69,363–71,605 μg kg^{-1}), St. 5 (36,201 μg kg^{-1}), and St. 3 (18,023 μg kg^{-1}). At St. 2, the PAH pool was largely petrogenic (82 %), while more pyrogenic PAHs occurred at the other four stations (55–86 %). In 2014, the TPAH content was highest at St. 6 (45,949 μg kg^{-1}), followed by St. 4 (26,180 μg kg^{-1}) and St. 1 and 2 (21,677–21,968 μg kg^{-1}). At St. 1 and 2,

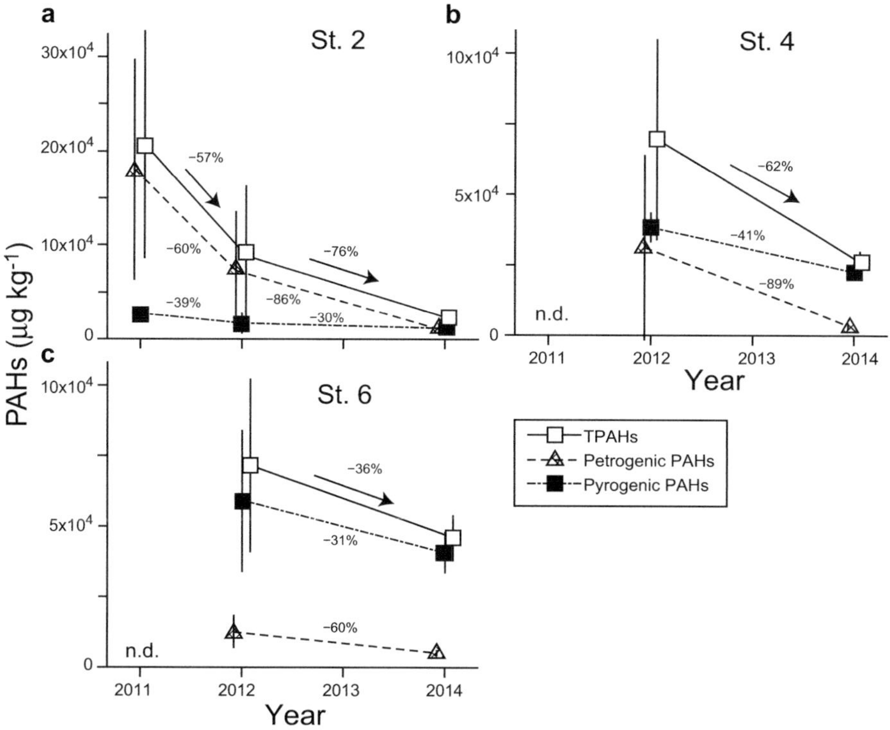

Fig. 15.5 Temporal changes in total polycyclic aromatic hydrocarbons (TPAHs), petrogenic PAHs, and pyrogenic PAHs in the sediments in the period 2011–2014 at St. 2 (**a**), St. 4 (**b**), and St. 6 (**c**). Error bars refer to SDs ($n=3$). *n.a.* not available. Values in 2011 and 2012 are means of three sampling occasions

more than half of the PAH pool was still petrogenic in 2014 (53–57 %), but was dominated by pyrogenic PAHs at the other two stations (87–89 %).

Within 3 years, the TPAH content had decreased considerably and had fallen below the ERM level (44,792 μg kg^{-1}) at St. 2 and 4 (Fig. 15.5). Decreased rate of the concentration was much steeper among petrogenic PAHs (30–45 % year^{-1}) than among pyrogenic PAHs (15–39 % year^{-1}). The decreased rate of TPAHs (18 % year^{-1}) was much lower at Stn 6 than at the other two stations (28–38 % year^{-1}), which we attribute to the high fractional contribution of recalcitrant pyrogenic PAHs at St. 6. TPAHs in the sediments peaked at ca. 8–15 cm below the sediment surface; they were much lower near the surface (less than the ERM levels in the US SQG) (Fig. 15.6).

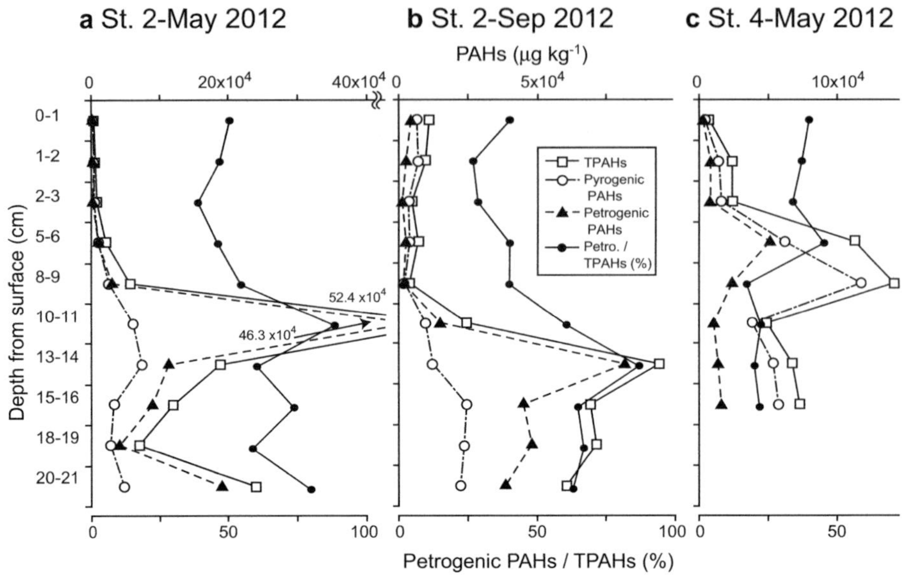

Fig. 15.6 Vertical distribution of polycyclic aromatic hydrocarbons (PAHs) in the sediment at St. 2 in May 2012 (a) and September 2012 (b), and at St. 4 in May 2012 (c)

15.3.4 Macrozoobenthos

We found no macrofauna at St. 2 in 2011. On the four sampling occasions in 2012–2014 we encountered a total of 96 taxa (see Appendix Table 15.4). The taxon number was highest in September 2012 (55 taxa), followed by May 2012 (49 taxa), January 2013 (45 taxa), and August 2014 (37 taxa). The low taxon number in August 2014 may bc attributed to the lack of sampling at St. 3 and 5 in that month. Polychaetes, including *Euchone* sp., *Chaetozone* sp., and *Asabellides* sp., and the bivalve *Alvenius ojianus* dominated in May 2012 (overall mean density: 131–216 ind. m^{-2}). The densities of the polychaetes *Asabellides* sp. and *Nephtys polybranchia* and of the bivalve *Theora fragilis* had increased considerably by September 2012 (151–1288 ind. m^{-2}). *Chaetozone* sp. was a dominant taxon in September 2012 (200 ind. m^{-2}), as it had been in May of the same year. In January 2013, *T. fragilis*, *Euchone* sp., *N. polybranchia*, and *Scoletoma longifolia* were the dominant taxa (90–174 ind. m^{-2}). In August 2014, macrozoobenthic density and biomass decreased to much lower levels than those in the period 2012–2013. *Asabellides* sp. (279 ind. m^{-2}), *Capitella* sp., *S. longifolia*, *Nephtys oligobranchia*, and *T. fragilis* (33–92 ind. m^{-2}) were the dominant taxa in August 2014.

Spatiotemporal changes in densities of all macrozoobenthos and polychaetes are shown in Fig. 15.7a. Overall values were high in September 2012 (632–7605 ind. m^{-2}), intermediate in May 2012 (211–2342 ind. m^{-2}) and January 2013 (395–2281

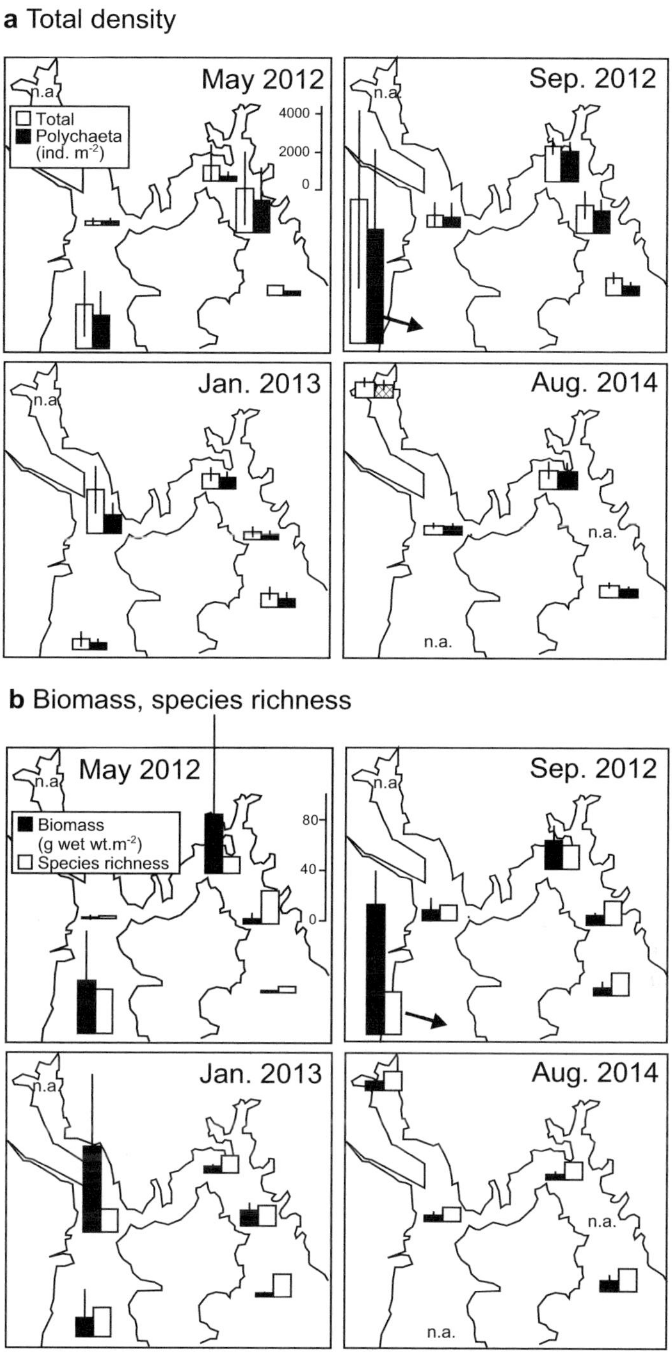

Fig. 15.7 Densities (ind. m^{-2}) of total macrozoobenthos and polychaetes (**a**), and biomass (g wet wt m^{-2}) and species richness of macrozoobenthos (**b**) in May 2012, September 2012, January 2013, and August 2014. Error bars refer to SDs ($n = 3$). *n.a.* not available

ind. m^{-2}), and low in August 2014 (447–982 ind. m^{-2}). In most cases, the communities were dominated by polychaetes (31–100 %) and bivalves (0–42 %); amphipods were scarce or completely absent (0–9 %). Compared with other stations, total biomass of macrozoobenthos was much higher at St. 3 and 4 in May 2012 (42.0–46.4 g m^{-2}) and September 2012 (22.5–103.2 g m^{-2}), at St. 2 in January 2013 (67.9 g m^{-2}), and at St. 1 and 6 in August 2014 (7.0–8.7 g m^{-2}) (Fig. 15.7b). Species richness were lowest at St. 2 in May and September 2012 (2–12 taxa), at St. 4 in January 2013 (14 taxa), and St. 2 in August 2014 (11 taxa).

15.3.5 Macrozoobenthos–Environment Relations

No significant relationships were detected between sediment TPAH content and densities of (i) total macrozoobenthos and (ii) five dominant taxa, including *Asabellides* sp., *T. fragilis*, *Chaetozone* sp., *Euchone* sp., and *N. polybranchia* (Pearson's correlation coefficient; $p > 0.05$, Fig. 15.8). The densities of macrozoobenthos were rather high in grossly polluted sediments (TPAHs > ERM).

In the forward selection procedure of the CCA, bottom DO, water depth, TPAHs, petrogenic PAHs, pyrogenic PAHs, Eh, TN, TOC, $\delta^{13}C$, $\delta^{15}N$, and C/N ratio were discarded from the analyses at a significant level of $p > 0.1$ (Table 15.3). Thus, we used WT, silt–clay content, and salinity as significant explanatory variables for further analysis. These three environmental variables explained 85.1 % of the variance in the first two CCA axes. The first CCA axis had the highest positive correlation with WT ($r = -0.84$, $p < 0.001$), likely reflecting seasonality. The second axis had the highest correlation with silt–clay content ($r = -0.60$, $p < 0.01$), likely reflecting conditions on the seafloor.

Samples collected in the summer to fall period clustered to the left side of the CCA plot, and those collected in spring and winter clustered to the right side (Fig. 15.9a), reflecting changes in WT across the sampling period. Samples taken on the same sampling occasion were distributed along axis 2, reflecting the differences in sediment granulometry across stations.

The CCA also displays relationships between 22 common taxa and each of the environmental variables (Fig. 15.9b). Of the ten dominant taxa in the samples (black triangles), the polychaetes *Chaetozone* sp. and *Phyllodoce* sp. and the bivalve *Axinopsida subquadrata* clustered in the upper sector of coordinate space. Thus, they occurred abundantly in relatively coarse sediments (silt–clay: 63–94 %). In contrast, the symbols for the cumacean *Iphinoe sagamiensis* clustered in the lower sector of the plot, indicating their preferential occurrence in muddy sediments (silt–clay: 93–100 %). Symbols for the polychaete *Asabellides* sp. clustered in the left sector of the plot, whereas symbols for *Euchone* sp. and the gastropod *Philine argentata* clustered in the right sector, indicating that they were more abundant in warmer (September 2012 and/or August 2014) and colder months (May 2012 and/or January 2013), respectively. Symbols for other dominant taxa, such as

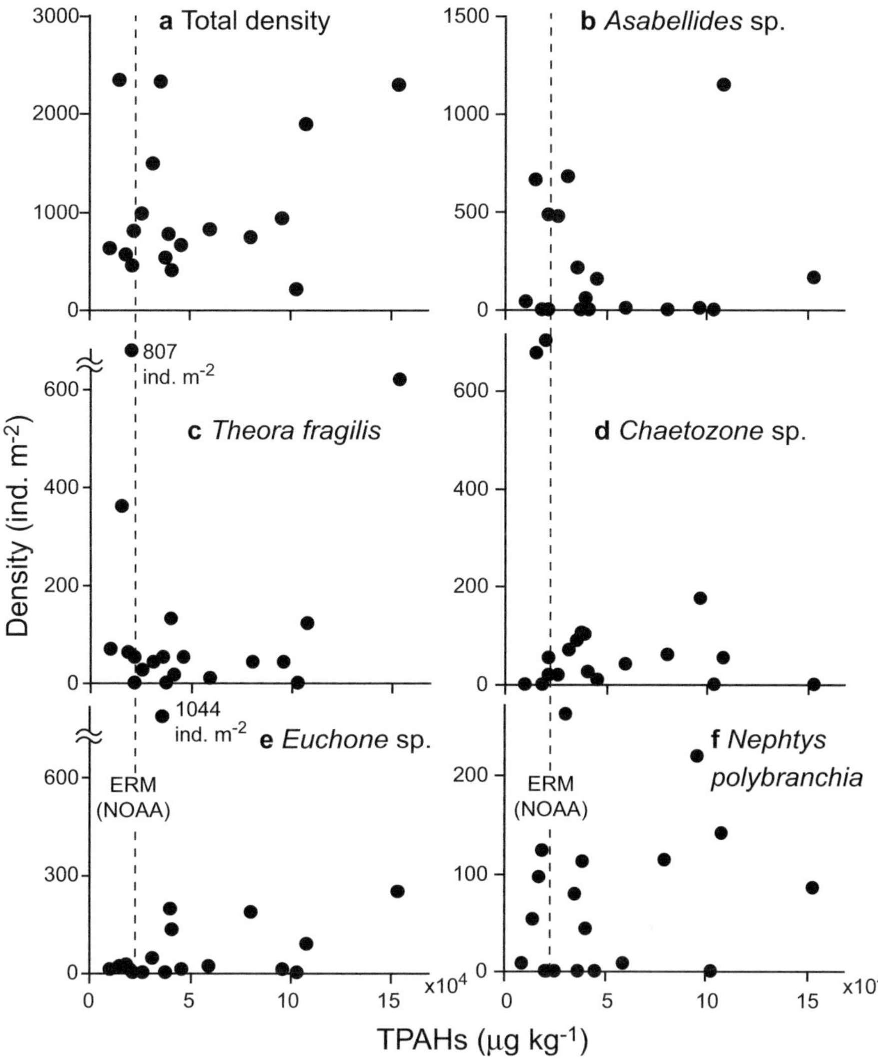

Fig. 15.8 Relationships between total polycyclic aromatic hydrocarbons (TPAHs) in the sediments (μg kg^{-1}) and the densities of total macrozoobenthos (**a**), *Asabellides* sp. (**b**), *Theora fragilis* (**c**), *Chaetozone* sp. (**d**), *Euchone* sp. (**e**), or *Nephtys polybranchia* (**f**). The effects range median (ERM) of US Sediment Quality Guidelines (SQG; NOAA 1999) is indicated by a *dashed line*

S. longifolia, *N. polybranchia*, *Capitella* sp., *Alvenius ojianus*, *T. fragilis*, and Nemertinea sp., were plotted close to the origin of the coordinate plot, indicating that their abundances were not well explained by either sediment granular structure or temperature.

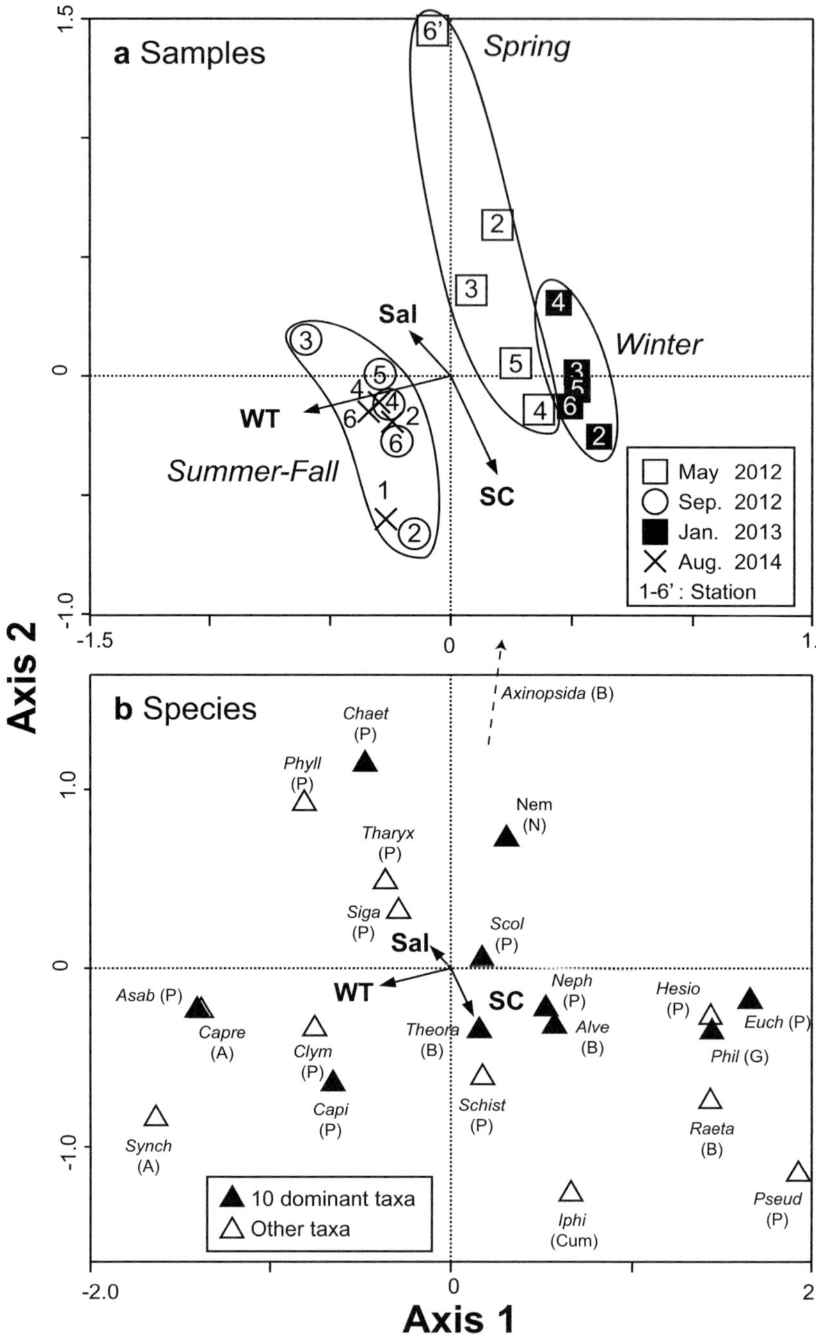

Fig. 15.9 Biplots of canonical correspondence analysis (CCA) showing sample scores (**a**) and species scores (**b**). *Arrows* indicate correlations of the environmental variables with the two canonical axes. Samples in plot (**a**) are grouped by sampling occasions. The first three to six characters of each taxon name (See Appendix Table 15.4) are provided in plot (**b**). *N* Nemartinea, *G* Gastropoda, *B* Bivalvia, *P* Polychaeta, *Cum* Cumacea. See also Table 15.3 for summary of the CCA

15.4 Discussion

The bottom sediments of Kesennuma Bay were grossly polluted by petrogenic and pyrogenic PAHs released by the fuel spill and conflagration that followed the 2011 earthquake and tsunami. Fuel-derived petrogenic PAHs accumulated mainly in the inner bay stations that were close to the fuel storage tanks (St. 1–2); conflagration-derived pyrogenic PAHs were dispersed and deposited at stations in outer portion of the bay (St. 3–6). The high levels of sediment PAHs may have had toxic effects on benthic invertebrates.

We compared measured PAH contents in the sediments with the ERL and ERM levels of the US SQG (NOAA 1999) to assess ecotoxicological risks for marine organisms. The TPAH contents and those of six individual PAHs (fluoranthene, pyrene, benz [a] anthracene, chrysene, benzo [a] pyrene, dibenz [a,h] anthracene) exceeded the ERM level at several or most of the stations. TPAHs at St. 1, 3, and 5 were four- to eightfold higher than the ERL (4022 $\mu g\ kg^{-1}$) and exceeded the ERM (44,792 $\mu g\ kg^{-1}$) at other stations. Fluoranthene and chrysene exceeded the ERM at St. 6; pyrene and benz [a] anthracene exceeded the ERM at St. 4–6. Benzo [a] pyrene and dibenz [a,h] anthracene exceeded the ERM at five of the six stations. Page et al. (2002) calculated the TPAH low threshold (LT) value, LC10 value (lethal concentration for 10% of a population), and the LC50 value (median lethal concentration) of 2600 $\mu g\ kg^{-1}$, 4100 $\mu g\ kg^{-1}$, and 10,750 $\mu g\ kg^{-1}$, respectively, using a standard amphipod bioassay with the *Rhepoxynius abronius*. The TPAH values we measured far exceeded the LC50 concentration at all stations, suggesting that there was a high ecotoxicological risk for benthic invertebrates on the seafloor across all of Kesennuma Bay.

Amphipods, including the genera *Ampelisca* and *Harpinia* and the family Isaeidae, are reported to be highly sensitive to PAH pollution and have often declined or almost disappeared after oil spills (Gómez Gesteira and Dauvin 2000; Nikitik and Robinson 2003; Joydas et al. 2011, 2012). On the other hand, polychaetes are generally more tolerant of toxic chemicals, including PAHs (Forbes et al. 2001; Lenihan et al. 2003). Thus, benthic communities in oil-polluted areas are often dominated by opportunistic polychaetes (Nikitik and Robinson 2003; Joydas et al. 2011, 2012), which was the case in the fuel-polluted areas we studied, where small polychaetes (e.g., *Chaetozone* sp., *Asabellides* sp., *Euchone* sp., and *Capitella* sp.) were predominant and amphipod densities and diversity were low.

The absence of macrozoobenthos at St. 2 in 2011 may have been the result of PAH toxicity in the sediments. Despite the extraordinarily high PAH levels in the stations (e.g., >60,000 $\mu g\ kg^{-1}$ TPAHs), however, we did not detect clear negative effects on dominant macrozoobenthos in the following years (Fig. 15.8). Even in grossly polluted areas (TPAHs >> ERM), members of the benthos sometimes occurred densely in the sediments. Our CCA also showed that PAHs in the sediments (TPAHs, petrogenic PAHs, and pyrogenic PAHs) were not significant explanatory factors for observed variations in macrozoobenthic density. The dominant taxa in our study site, such as the bivalve *T. fragilis* and the polychaetes *Asabellides* sp., *Capitella* sp., *Chaetozone* sp., *Euchone* sp., and *N. polybranchia*, may have physiologically tolerated the PAHs. *Capitella* sp. is known to dominate soft-bottom communities shortly after oil spills (Grassle and Grassle 1974) and is able to metabolize and degrade PAHs in the sediment (Forbes et al. 2001).

Another potential explanation for the lack of a PAH effect on the animals is that they colonized a thin surface sediment layer where the PAH content was much lower than in subsurface sediments. Similar sediment surface distributions are common among infauna inhabiting organically enriched mud with high levels of toxic sulfides in subsurface layers (Nilsson and Rosenberg 2000; Kanaya et al. 2016). For surface-dwelling macrozoobenthos, the region we studied would not have been toxic because the PAH pool was accumulated in the deeper sediment layer.

We detected major changes in the dominant macrozoobenthic taxa among sampling occasions (Appendix Table 15.4). Macrozoobenthic composition, abundance, and biomass clearly differed between September 2012 and August 2014, although bottom water quality (DO, WT, and salinity) were largely similar. Factors that may explain the large temporal shift in the biota at the scale of the whole bay are not readily identifiable. It is possible that the changes were part of the intrinsic sequence of community succession initiated by the huge pulsed disturbance events in 2011 (see Seike et al. 2013; Kanaya et al. in press). Environmental factors other than those we examined such as primary productivity, ocean current, and local climate may have triggered the temporal shifts in macrozoobenthic assemblage structure on the seafloor (see Nakaoka 1992; Sirabella et al. 2001; Cusson and Bourget 2005). All of these postulates require testing in future research efforts conducted in the bay.

We showed that the subtidal soft-bottom habitats in Kesennuma Bay were grossly polluted by petrogenic and pyrogenic PAHs derived from the fuel storage tanks and the subsequent conflagration. However, most of the pollutants became buried in the deep sediment layer (see Fig. 15.6); the fraction in the upper sediment layer declined over 4 years. The rapid decline in superficial concentrations may have been a result of continuing surface sediment accretion and PAH degradation through weathering, microbial decomposition, and/or metabolic and bio-diffusion processes of the macroinvertebrates (Forbes et al. 2001; Page et al. 2002; Johnsen et al. 2005; Boopathy et al. 2012). We also found that water quality improved, i.e., severe hypoxia did not occur even in summer, whereas oxygen depletion had been a serious environmental problem in the bay before the 2011 tsunami (Ito 2008). Our knowledge of the macrozoobenthic community structure before the tsunami was quite limited (Igarashi et al. 1978; Ito 2008). Thus, predictions on the reestablishment of the former community configuration were difficult to formulate. Nevertheless, decreasing PAH levels in the surface sediment layer and improving DO condition at across the bay may have allowed the recolonization of pollution-sensitive benthic invertebrate taxa in the future.

Borja et al. (2010) suggested that full recovery of coastal marine and estuarine ecosystems to former configurations of biotic composition and diversity requires a minimum of 15–25 years after an extreme disturbance. Recovery of marine soft-bottom community from oil spills often requires in excess of 10 years (Gómez Gesteira and Dauvin 2000; Nikitik and Robinson 2003; Joydas et al. 2011, 2012).

Bottom sediments in Kesennuma Bay still contain some toxic PAHs in the subsurface layer. Whether the benthic assemblage in Kesennuma Bay is reached to a steady state or still underway of the community succession is not clear. Long-term monitoring (e.g., >10 years) is required to more fully comprehend the ecological consequences of the 2011 fuel spill on the soft-bottom communities of the Sanriku coast.

Acknowledgments We thank H. Misui, J. Nakano, A. Oishi, S. Kimura, the Japan Fisheries cooperative in Miyagi Prefecture (Kesennuma, Hajikami), the Econavigate Co., Ltd., and the Marine Biological Research Institute of Japan Co., Ltd. for their assistance with field and laboratory work.

Appendix

Appendix Table 15.4 Density (ind. m^{-2}), biomass (wet weight; g m^{-2}), and species richness of macrozoobenthos at the stations on 22 May 2012. Data are mean value. $n=1$ for St. 6′ and $n=3$ for other stations. The first to fourth dominant taxa for each station were bolded

Taxa		St. 2	St. 3	St. 4	St. 5	St. 6′	All
N	Nemertinea sp.	–	–	–	9	**52**	12
G	*Voorwindia paludinoides*	–	44	**127**	105	–	55
	Turbonilla sp.	–	9	–	–	–	2
	Cylichnatys angustus	–	9	–	–	–	2
	Liloa porcellana	–	–	–	9	–	2
	Philine argentata	–	26	**84**	132	–	48
	Aglajidae sp.	–	9	–	9	–	4
B	*Alvenius ojianus*	–	44	**338**	**272**	–	**131**
	Axinopsida subquadrata	–	9	–	9	**209**	45
	Raeta pulchellus	–	26	–	9	–	7
	Theora fragilis	–	**360**	9	53	–	84
P	*Eteone longa*	–	9	–	–	–	2
	Eumida sp.	–	9	–	–	–	2
	Phyllodoce sp.	**42**	9	–	9	–	12
	Harmothoe sp.	–	35	–	–	–	7
	Polynoidae sp.	–	–	9	**149**	–	32
	Sigambra hanaokai	–	26	–	9	–	7
	Hesionidae sp.	–	9	–	–	–	2
	Eusyllinae sp.	–	9	–	–	–	2
	Nectoneanthes latipoda	–	–	42	–	–	8
	Platynereis bicanaliculata	–	9	–	–	–	2
	Glycera macintoshi	–	–	–	9	–	2
	Glycera sp.	–	–	–	18	–	4
	Nephtys caeca	–	18	–	–	–	4
	N. californiensis	–	9	–	–	–	2

(continued)

Appendix Table 15.4 (continued)

Taxa		St. 2	St. 3	St. 4	St. 5	St. 6′	All
	N. polybranchia	–	53	9	79	–	28
	Diopatra sugokai	–	9	–	–	–	2
	Scoletoma longifolia	–	**114**	–	–	–	23
	Schistomeringos rudolphi	**169**	–	42	–	–	42
	Prionospio ehlersi	–	9	–	–	–	2
	Chaetozone sp.	–	**675**	42	88	**105**	**182**
	Tharyx sp.	–	9	–	9	–	4
	Capitella sp.	–	–	**84**	35	**105**	45
	Clymenella collaris	–	–	–	9	–	2
	Maldane cristata	–	35	–	–	–	7
	Ampharete sp.	–	18	–	–	–	4
	Asabellides sp.	–	**667**	9	**211**	–	**177**
	Chone sp.	–	9	–	–	–	2
	Euchone sp.	–	18	18	**1044**	–	**216**
	Serpulidae sp.	–	–	–	9	–	2
C	*Iphinoe sagamiensis*	–	–	–	9	–	2
	Dimorphostylis sp.	–	9	–	9	–	4
I	*Natatolana japonensis*	–	–	–	–	**52**	10
Am	*Cerapus tubularis*	–	18	–	–	–	4
	Synchelidium sp.	–	–	9	–	–	2
	Caprella sp.	–	9	–	9	–	4
As	*Asterias amurensis*	–	9	–	–	–	2
O	Amphiuridae sp.	–	–	–	18	–	4
	Ophiura kinbergi	–	9	–	–	–	2
Total	Density	211	2342	821	2325	524	1244
	Biomass	0.8	42.0	46.4	4.0	1.6	19.0
	Taxa	2	35	13	26	5	49
	Gastropoda	–	96	211	254	–	112
	Bivalvia	–	439	346	342	209	267
	Polychaeta	211	1754	255	1675	209	821
	Amphipoda	–	26	9	9	–	9
Sampling apparatus		H	E	H/E	E	K	

T Turbellaria, *N* Nematoda, *Sca* Scaphopoda, *G* Gastropoda, *B* Bivalvia, *S* Sipuncula, *E* Echiura, *P* Polychaeta, *M* Myodocopa, *Py* Pycnogonida, *L* Leptostraca, *C* Cumacea, *I* Isopoda, *Am* Amphipoda, *D* Decapoda, *As* Asteroidea, *O* Ophiuroidea, *Ho* Holothuroidea, *H* HR type core sampler, *E* Ekman-Birge sampler, *K* KS bottom sampler

Appendix Table 15.4 (Continued) Data for macrozoobenthos collected on 6 September 2012. $n=3$

Taxa		St. 2	St. 3	St. 4	St. 5	St. 6	All
N	Nemertinea sp.	9	53	35	35	9	28
G	*Voorwindia paludinoides*	9	–	–	–	–	2
	Ringicula doliaris	–	–	–	–	61	12
	Philine argentata	18	–	9	26	35	18
B	*Musculus cupreus*	–	9	–	–	–	2
	Cardiidae sp.	–	–	–	9	–	2
	Alvenius ojianus	–	44	9	53	**70**	35
	Raeta pulchellus	–	–	–	–	**70**	14
	Macoma incongrua	–	9	–	–	–	2
	Theora fragilis	**70**	**807**	**123**	44	44	**218**
S	*Golfingia* sp.	–	26	–	–	–	5
E	Echiuroidea sp.	–	–	–	–	9	2
P	*Phyllodoce* sp.	–	26	–	–	–	5
	Harmothoe sp.	–	18	–	–	–	4
	Sigambra hanaokai	–	26	–	9	–	7
	Hesionidae sp.	–	–	26	–	–	5
	Glycera alba	–	–	–	–	18	4
	Glycera sp.	–	–	9	–	–	2
	Glycinde wireni	–	9	–	18	–	5
	Nephtys caeca	9	–	–	–	–	2
	N. polybranchia	9	123	**140**	**263**	**219**	**151**
	Scoletoma longifolia	–	**272**	61	18	–	70
	Schistomeringos rudolphi	**61**	–	–	–	–	12
	Leitposcoloplos pugettensi	–	9	–	–	–	2
	Paradoneis nipponica	–	9	–	–	–	2
	Laonice sp.	–	–	9	9	–	4
	Prionospio ehlersi	–	–	18	18	–	7
	Pseudopolydora sp.	18	–	–	18	–	7
	Magelona japonica	–	9	–	–	–	2
	Chaetozone sp.	–	**702**	53	**70**	**175**	**200**
	Tharyx sp.	–	114	–	–	–	23
	Cossura sp.	–	–	–	–	18	4
	Capitella sp.	**368**	9	–	–	35	82
	Mediomastus sp.	–	18	–	–	–	4
	Clymenella collaris	–	70	44	26	–	28
	Maldane cristata	–	26	–	–	–	5
	Praxillella pacifica	–	9	–	–	–	2
	Ammotrypane aulogaster	–	–	9	–	–	2
	Scalibregma inflatum	–	–	9	–	–	2
	Asabellides sp.	**44**	**4553**	**1149**	**684**	9	**1288**
	Euchone sp.	9	9	**88**	44	9	32

(continued)

Taxa		St. 2	St. 3	St. 4	St. 5	St. 6	All
M	*Xenoleberis yamadai*	–	9	–	–	–	2
C	*Iphinoe sagamiensis*	–	–	–	–	**<u>123</u>**	25
Am	*Aoroides* sp.	–	9	–	–	–	2
	Grandidierella japonica	–	18	–	–	–	4
	Monocorophium acherusicum	–	96	–	–	–	19
	Cerapus tubularis	–	377	–	–	9	77
	Ericthonius pugnax	–	44	–	9	–	11
	Orchomene sp.	–	18	–	–	–	4
	Synchelidium sp.	9	18	35	**<u>88</u>**	–	30
	Caprella sp.	–	53	44	44	–	28
D	*Paradorippe granulata*	–	–	–	–	9	2
	Pinnixa rathbuni	–	–	–	–	9	2
	Brachyura sp. larva	–	–	18	–	–	4
Ho	*Labidoplax dubia*	–	9	–	–	–	2
Total	Density	632	7605	1886	1482	930	2507
	Biomass	8.5	103	22.5	7.3	6.0	29.5
	Taxa	12	34	19	19	18	55
	Gastropoda	26	–	9	26	96	32
	Bivalvia	70	868	132	105	184	272
	Polychaeta	518	6009	1614	1175	482	1960
	Amphipoda	9	632	79	140	9	174
Sampling apparatus		E	E	E	E	E	

Appendix Table 15.4 (Continued) Data for macrozoobenthos collected on 24 January 2013. $n=3$

Taxa		St. 2	St. 3	St. 4	St. 5	St. 6	All
T	Polycladida sp.	–	9	–	–	18	5
N	Nemertinea sp.	51	**<u>35</u>**	–	9	44	28
G	*Alvania* aff. *coninna*	9	–	–	–	–	2
	Ringicula doliaris	–	9	–	–	–	2
	Liloa porcellana	–	9	–	–	–	2
	Philine argentata	9	18	51	**<u>35</u>**	35	29
B	*Alvenius ojianus*	51	9	–	–	–	12
	Axinopsida subquadrata	–	–	–	9	26	7
	Raeta pulchellus	84	**<u>35</u>**	–	–	26	29
	Theora fragilis	**<u>619</u>**	**<u>61</u>**	**<u>130</u>**	18	44	**<u>174</u>**
S	*Golfingia* sp.	–	9	–	–	–	2

Taxa		St. 2	St. 3	St. 4	St. 5	St. 6	All
P	*Labiosthenolepis laevis*	–	–	–	–	9	2
	Sigambra hanaokai	–	18	–	9	–	5
	Hesionidae sp.	9	9	–	–	9	5
	Platynereis bicanaliculata	–	9	–	–	–	2
	Glycera alba	–	–	–	–	9	2
	G. macintoshi	–	–	–	9	–	2
	Glycera sp.	51	–	–	–	–	10
	Nephtys californiensis	–	9	–	9	–	4
	N. polybranchia	84	**96**	**112**	**44**	**114**	**90**
	Scoletoma longifolia	**313**	**88**	53	9	–	**92**
	Ninoe palmata	–	–	–	–	9	2
	Schistomeringos rudolphi	–	–	9	–	–	2
	Paradoneis nipponica	–	9	–	–	–	2
	Prionospio ehlersi	–	–	9	–	–	2
	Prionospio membranacea	18	–	–	–	–	4
	Pseudopolydora sp.	69	9	18	–	9	21
	Paraprionospio patiens	–	–	9	–	9	4
	Magelona japonica	–	–	–	–	18	4
	Chaetozone sp.	–	–	**102**	**26**	**61**	38
	Tharyx sp.	9	**35**	–	–	–	9
	Cossura sp.	–	–	9	–	–	2
	Capitella sp.	–	–	9	–	–	2
	Mediomastus sp.	–	9	–	18	–	5
	Clymenella collaris	–	18	–	9	26	11
	Maldane cristata	–	9	–	–	–	2
	Asabellides sp.	162	–	60	–	–	44
	Euchone sp.	**246**	26	**197**	**132**	**184**	**157**
Py	*Propallene* aff. *longiceps*	42	18	–	–	–	12
C	*Iphinoe sagamiensis*	–	–	–	18	**88**	21
Am	*Cerapus tubularis*	186	–	–	–	–	37
	Caprella sp.	–	–	9	–	–	2
D	*Alpheus japonicus*	–	–	–	**35**	–	7
	Labidoplax dubia	–	–	–	9	–	2
Ho	*Corella japonica*	**271**	–	–	–	–	54
Total	Density	2281	553	774	395	737	948
	Biomass	67.9	14.7	5.1	12.4	3.2	20.6
	Taxa	18	23	14	16	18	45
	Bivalvia	754	105	130	26	96	222
	Polychaeta	960	342	585	263	456	521
	Amphipoda	186	–	9	–	–	39
Sampling apparatus		H/E	E	H/E	E	E	

Appendix Table 15.4 (Continued) Data for macrozoobenthos collected on 20 August 2014. $n = 3$

Taxa		St. 1	St. 2	St. 4	St. 6	All
N	Nemertinea sp.	–	–	9	–	2
Sca	Dentaliidae sp.	–	–	–	18	4
G	Unid. Gastropod sp.	–	9	–	–	2
	Yokoyamaia ornatissima	–	–	–	26	7
B	*Alvenius ojianus*	–	–	–	26	7
	Raeta pulchellus	–	–	–	18	4
	Ruditapes philippinarum	9	–	–	–	2
	Macoma incongrua	9	–	–	–	2
	Theora fragilis	**53**	–	26	**53**	**33**
P	*Eteone longa*	–	9	–	–	2
	Phyllodoce sp.	18	–	26	–	11
	Harmothoe sp.	9	–	18	–	7
	Sigambra hanaokai	–	18	–	18	9
	Podarkeopsis sp.	9	9	–	–	4
	Glycera sp.	–	–	9	–	2
	Glycinde wireni	–	–	–	9	2
	Nephtys oligobranchia	–	–	18	**114**	**33**
	Scoletoma longifolia	–	**44**	**184**	–	**57**
	Schistomeringos rudolphi	**79**	35	–	–	29
	Paraprionospio coora	–	–	–	9	2
	Magelona japonica	–	–	–	18	4
	Chaetozone sp.	18	**53**	18	9	24
	Tharyx sp.	–	–	18	–	4
	Capitella sp.	**44**	**202**	**123**	–	**92**
	Clymenella collaris	–	–	18	**105**	31
	Scalibregma inflatum	9	**53**	–	–	15
	Asabellides sp.	**482**	–	**474**	**158**	**279**
	Euchone sp.	–	–	–	9	2
L	*Nebalia japonensis*	9	–	–	–	2
C	*Iphinoe sagamiensis*	–	–	–	26	7
Am	*Monocorophium acherusicum*	18	–	–	–	4
	Synchelidium sp.	–	–	–	26	7
	Caprella sp.	18	9	**35**	–	15
D	*Paradorippe granulata*	–	–	–	9	2
O	Amphiuridae sp.	–	–	–	9	2
	Labidoplax dubia	–	9	9	–	4
Ho	Molgulidae sp.	26	–	–	–	7
Total	Density	807	447	982	658	724
	Biomass	7.0	5.0	4.1	8.7	6.2
	Taxa	15	11	14	18	37
	Gastropoda	–	9	–	44	13
	Bivalvia	70	–	26	96	48
	Polychaeta	667	421	904	447	610
	Amphipoda	35	9	35	26	26
Sampling apparatus		E	E	E	E	

References

Bacosa HP, Inoue C (2015) Polycyclic aromatic hydrocarbons (PAHs) biodegradation potential and diversity of microbial consortia enriched from tsunami sediments in Miyagi, Japan. J Hazard Mater 283:689–697

Belan TA (2003) Benthos abundance pattern and species composition in conditions of pollution in Amursky Bay (the Peter the Great Bay, the Sea of Japan). Mar Pollut Bull 46:1111–1119

Boopathy R, Shields S, Nunna S (2012) Biodegradation of crude oil from the BP oil spill in the marsh sediments of southeast Louisiana, USA. Appl Biochem Biotechnol 167:1560–1568

Borja Á, Dauer DM, Elliott M, Simenstad CA (2010) Medium- and long-term recovery of estuarine and coastal ecosystems: patterns, rates and restoration effectiveness. Estuar Coast Shelf Sci 33:1249–1260

Colombo JC, Pelletier E, Brochu C, Khalllt M (1989) Determination of hydrocarbon sources using *n*-alkane and polyaromatic hydrocarbon distribution indexes. Case study: Rio de La Plata Estuary, Argentina. Environ Sci Technol 23:888–894

Cusson M, Bourget E (2005) Global patterns of macroinvertebrate production in marine benthic habitats. Mar Ecol Prog Ser 297:1–14

Engraff M, Solere C, Smith KEC, Mayer P, Dahllöf I (2011) Aquatic toxicity of PAHs and PAH mixtures at saturation to benthic amphipods: linking toxic effects to chemical activity. Aquat Toxicol 102:142–149

Forbes V, Andreassen M, Christensen L (2001) Metabolism of the polycyclic aromatic hydrocarbon fluoranthene by the polychaete *Capitella capitata* species I. Environ Toxicol Chem 20:1012–1021

Gómez Gesteira GJL, Dauvin J-C (2000) Amphipods are good bioindicators of the impact of oil spills on soft-bottom macrobenthic communities. Mar Pollut Bull 40:1017–1027

Grassle JF, Grassle JP (1974) Opportunistic life histories and genetic systems in marine benthic polychaetes. J Mar Res 32:253–284

Haraguchi T, Goto K, Sato M, Yoshinaga Y, Yamaguchi N, Takahashi T (2013) Large bedform generated by the 2011 Tohoku-oki tsunami at Kesennuma Bay, Japan. Mar Geol 335:200–205

Hayasaka D, Shimada N, Konno H, Sudayama H, Kawanishi M, Uchida T, Goka K (2012) Floristic variation of beach vegetation caused by the 2011 Tohoku-oki tsunami in northern Tohoku, Japan. Ecol Eng 44:227–232

Igarashi T, Watanabe M, Iwai T, Fujita N (1978) Changes in sediment quality and macrozoobenthic composition by removal of organic rich mud. Bull Kesennuma Miyagi Prefectural Fish Exp Station 4:38–58 (in Japanese)

Ito H (2008) The occurrence of oxygen-deficient water and yearly changes in environmental condition of bottom sediments in Kesennuma Bay. Miyagi Pref Rep Fish Sci 8:41–47

Johnsen AR, Wick LY, Harms H (2005) Principles of microbial PAH-degradation in soil. Environ Pollut 133:71–84

Joydas TV, Krishnakumar PK, Qurban M, Ali SM, Al-Suwailem A, Al-Abdulkhader K (2011) Status of macrobenthic community of Manifa-Tanajib Bay System of Saudi Arabia based on a once-off sampling event. Mar Pollut Bull 62:1249–1260

Joydas TV, Qurban MA, Al-Suwailem A, Krishnakumar PK, Nazeer Z, Cali NA (2012) Macrobenthic community structure in the northern Saudi waters of the Gulf, 14 years after the 1991 oil spill. Mar Pollut Bull 64:325–335

Kanaya G, Maki H, Suzuki T, Sato-Okoshi W, Kikuchi E (2014) Tsunami-induced changes in a shallow brackish lagoon ecosystem (Gamo Lagoon) in Sendai Bay, Japan. Glob Environ Res 18:35–46

Kanaya G, Suzuki T, Kanou K, Kondoh T, Sato-Okoshi W, Kikuchi E (2016) Ecological consequences of the tsunamis Caused by the Great East Japan Earthquake and subsequent disturbance events in a shallow brackish lagoon in Sendai Bay, Japan. In: Urabe J,Nakashizuka T (eds) Ecological impacts of Tsunamis on coastal ecosystems: lessons from the Great East Japan Earthquake. Springer, pp 85–104

Kanaya G, Uehara T, Kikuchi E (2016) Effects of sedimentary sulfide on community structure, population dynamics, and colonization depth of macrozoobenthos in organic-rich estuarine sediments. Mar Pollut Bull 109:393–401

Lenihan HS, Peterson CH, Kim SL, Conlan KE, Fairey R, McDonald C, Grabowski JH, Oliver JS (2003) Variation in marine benthic community composition allows discrimination of multiple stressors. Mar Ecol Prog Ser 261:63–73

Masuda A (2011) Report for the fuel spills from collapsed tanks in Kesennuma Port (Translated title). RISTEX CT J 11:1–6 (in Japanese) http://www.ristex.jp/aboutus/enterprize/security/pdf/ind_01.pdf. Last accessed on 9 Jul 2015

Ministry of the Environment (2003) Manual for analytical methods on the chemicals to be monitored. Ministry of the Environment, Tokyo (in Japanese) https://www.env.go.jp/water/chosa/h15-03.pdf. Last accessed on 9 Jul 2015

Nakaoka M (1992) Spatial and seasonal variation in growth rate and secondary production of *Yoldia notabilis* in Otsuchi Bay, Japan, with reference to the influence of food supply from the water column. Mar Ecol Prog Ser 88:215–223

Nikitik CCS, Robinson AW (2003) Patterns in benthic populations in the Milford Haven waterway following the "Sea Empress" oil spill with special reference to amphipods. Mar Pollut Bull 46:1125–1141

Nilsson HC, Rosenberg R (2000) Succession in marine benthic habitats and fauna in response to oxygen deficiency: analysed by sediment profile-imaging and by grab samples. Mar Ecol Prog Ser 197:139–149

NOAA (1999) Sediment quality guidelines developed for the National Status and Trends Program. NOAA, Silver Spring. http://ccma.nos.noaa.gov/publications/sqg.pdf. Last accessed on 9 Jul 2015

Oug E, Næs K, Rygg B (1998) Relationship between soft bottom macrofauna and polycyclic aromatic hydrocarbons (PAH) from smelter discharge in Norwegian fjords and coastal waters. Mar Ecol Prog Ser 173:39–52

Page DS, Boehm PD, Stubblefield WA, Parker KR, Gilfillan ES, Neff JM, Maki AW (2002) Hydrocarbon composition and toxicity of sediments following the Exxon Valdez oil spill in Prince William Sound, Alaska, USA. Environ Toxicol Chem 21:1438–1450

Samanta SK, Singh OV, Jain RK (2002) Polycyclic aromatic hydrocarbons: environmental pollution and bioremediation. Trends Biotechnol 20:243–248

Seike K, Shirai K, Kogure Y (2013) Disturbance of shallow marine soft-bottom environments and megabenthos assemblages by a huge tsunami induced by the 2011 M9.0 Tohoku-Oki earthquake. PLoS One 8:e65417

Sera K, Baba F, Goto S, Takahashi C, Saitoh Y, Matsumasa M (2012) Analysis of plants and sediments from the tidelands of the coastal regions of the Tohoku District following the 2011 tsunami. Int J PIXE 22:139–147

Sirabella P, Giuliani A, Colosimo A, Dippner JW (2001) Breaking down the climate effects on cod recruitment by principal component analysis and canonical correlation. Mar Ecol Prog Ser 216:213–222

ter Braak CJF, Smilauer P (1998) CANOCO reference manual and user's guide to Canoco for Windows: software for canonical community ordination (version 4). Microcomputer Power, Ithaca, 351 pp

Yamaoka S, Nakaie H, Chiba A (2011) Secular changes in environmental condition of bottom sediment in Kesennuma Bay. Miyagi Pref Rep Fish Sci 11:31–41 (in Japanese) http://agriknowledge.affrc.go.jp/RN/2010832737.pdf. Last accessed on 9 Jul 2015

Part II
Terrestrial Ecosystems

Chapter 16
Impact of the Great East Japan Earthquake and Tsunami on Coastal Vegetation and Landscapes in Northeast Japan: Findings Based on Remotely Sensed Data Analysis

Keitarou Hara, Yi Zhao, Mizuki Tomita, Noritoshi Kamagata, and Yunqing Li

Abstract The Great East Japan Earthquake of 11th March 2011 and subsequent huge tsunami caused widespread damage along the Pacific Ocean coast of eastern Honshu, Japan. This research utilizes field surveys and analyses of multi-resolution remote sensing images to clarify the impact on landscapes caused by this disaster and also to monitor the subsequent survival and recovery process in the Sendai Bay region. MODIS data (250 m resolution) were employed to map the overall extent of inundation and damage on the regional landscape scale. The major damage caused by the tsunami, destruction of coastal pine forests and inundation of rice paddies on the plain, was identified at this level. Progressively finer-scale analyses were then implemented using SPOT/HRG-2 (10 m resolution) data, GeoEye-1 fine-resolution data (0.5 m), very fine-resolution aerial photographs (10 cm) and LiDAR. These results demonstrated the minute details of the damage and recovery process down to the level of individual groves and trees. The results of the research show that on the wide-scale regional level, plant communities on rocky cliffs along the rias-style Sanriku Coast survived relatively intact, but coastal forests and other vegetations on sandy beaches and other low-lying coasts were severely damaged. Finer-scale analyses of a coastal pine forest in the Sendai Bay area showed that local topography influenced the patterns of damage and survival for individual trees. At the finest level, damaged pine forests and sandy beach vegetation could be seen recovering

K. Hara (✉) • M. Tomita
Tokyo University of Information Sciences, Chiba, Japan
e-mail: hara@rsch.tuis.ac.jp

Y. Zhao
Pasco Corporation, Higashiyama, Tokyo, Japan

N. Kamagata
Kokusai Kogyo Co., Ltd., Fuchu, Japan

Y. Li
Japan Space Imaging Corporation, Tokyo, Japan

J. Urabe, T. Nakashizuka (eds.), *Ecological Impacts of Tsunamis on Coastal Ecosystems*, Ecological Research Monographs,
DOI 10.1007/978-4-431-56448-5_16

only a little more than a year after the disaster. Continuous monitoring using fieldwork and remote sensing is required for balanced regional strategies that provide for economic and social recovery as well as restoration of vegetation, biodiversity and vital ecotones.

Keywords Disturbance • Great East Japan Earthquake • Remote sensing • Tsunami • Vegetation

16.1 Introduction

The Great Eastern Japan Earthquake and subsequent tsunamis, which struck on the 11th of March, 2011, caused enormous damage to vegetation and landscapes along the Pacific Ocean side of the Tohoku Region (Hara and Higuchi 2013; Zhao et al. 2013; Hara 2014; Himiyama et al. 2014). The tsunami raced several kilometres inland across the coastal plains and along the river valleys, especially along the Kitakami River in Miyagi Prefecture, where the water reached as far as 49 km from the sea (The 2011 Tohoku Earthquake Tsunami Joint Survey Group 2011). In addition, the earthquake caused severe ground subsidence in some coastal areas, which remained underwater even after the tsunami had receded (Okada 2012). Earthquakes and tsunami of this scale are relatively rare on the Sendai Plain, occurring on a cycle of several hundred to a thousand years. These sort of ‘large and infrequent disturbances’ (Tomita et al. 2014) have a special impact on regional ecosystems. Previous researches have shown that remote sensing is a useful tool for evaluating the damage caused by a large and infrequent disturbance such as a tsunami (Dharanirajan et al. 2007; Olwig et al. 2007; Sirikulchayanon et al. 2008; Roemer et al. 2010; Römer et al. 2012).

Despite the severity of the tsunami and the enormous scale of the disturbance, many plant communities in the stricken area survived (Hara M 2012; Hara K 2014), and others have been recovering remarkably (Kanno et al. 2014; Oka and Hirabuki 2014). Future recovery efforts will be enhanced by a clear understanding of which areas and which types of coastal habitat suffered the severest damage and which experienced a high degree of survival and recovery. This research employs various types of remote sensing data to analyse the patterns of destruction, survival and recovery in disturbed coastal ecosystems and landscapes.

16.2 Study Area

The Pacific Ocean coast of the Tohoku District, in northern Honshu, suffered the greatest tsunami damage. This region, comprising Iwate, Miyagi and Fukushima Prefectures, consists of two ancient mountain ranges, the Kitakami Mountains in the north and the Abukuma Mountains in the south, with the Sendai Plain located in

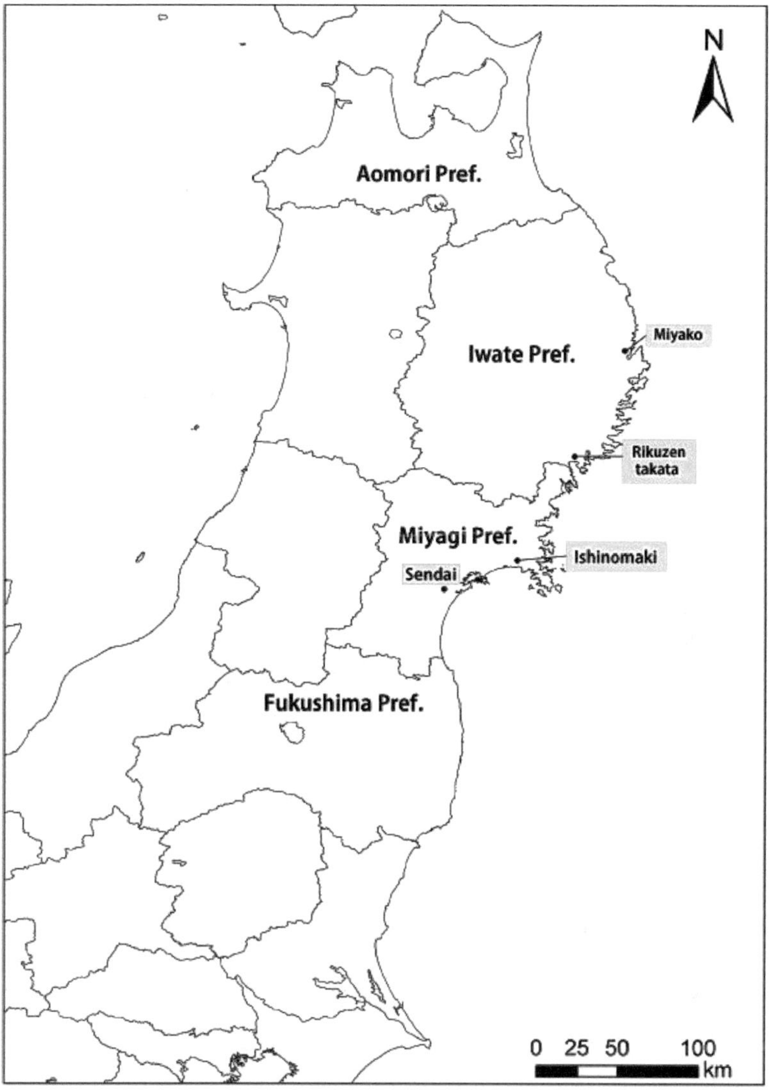

Fig. 16.1 Map of the eastern Tohoku District

between (Figs. 16.1 and 16.2). North of Sendai Bay, the Kitakami Mountains extend all the way to the Sanriku Coast, which is subsiding. As the land subsides, the former river valleys are inundated, forming a series of long narrow inlets, known as rias coast, which are separated by steep rocky promontories. Narrow sections of alluvial plain, which are used as agricultural and residential land, as well as small tidal flats and sandy beaches, form around the river mouths at the heads of the inlets. The narrow inlets are also dotted with numerous islands of various sizes, as well as many small fishing ports.

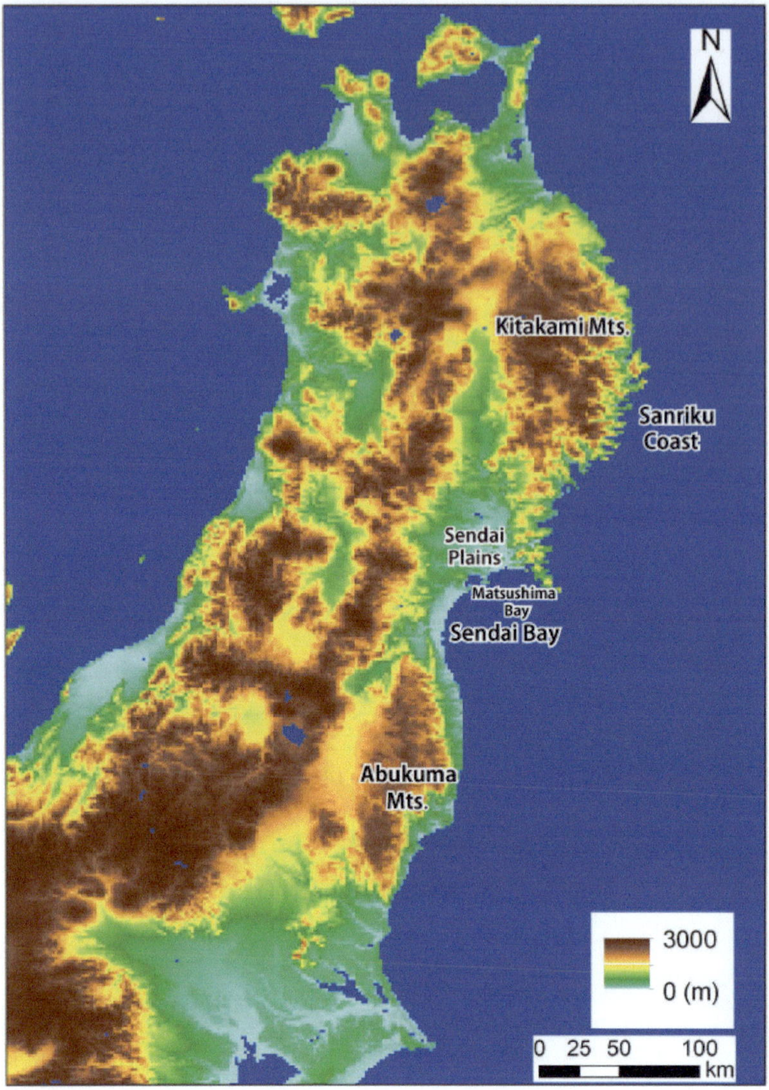

Fig. 16.2 Topography of the Tohoku District

Both the Kitakami and Abukuma Rivers, as well as many smaller rivers, empty into the Pacific Ocean at Sendai Bay. Sendai itself is the largest and most important city in the region. Outer coastlines along Sendai Bay are comprised primarily of sandy beaches and sand barriers, with extensive coastal wetlands developing at the river mouths and behind the barriers. Slightly inland from the coast, there is a levee, beyond which the low-lying marshes have been drained and dyked and converted to rice paddies, vegetable fields and residential land (Miura 1987).

16.3 Data and Methods

Remotely sensed satellite data are an effective tool for analysing changes caused by large and infrequent disasters. Data for the study region were collected periodically from before the disaster, and intensive satellite monitoring was implemented afterwards. As a result, a large amount of data is now available. The remote sensing data employed in this research included aerial photographs, wide-scale 250 m terrestrial resolution MODIS from NASA's Terra and Aqua satellite, 10 m resolution SPOT/HRG-2 from France, fine-scale 1.0 m resolution IKONOS data and 0.5 m resolution GeoEye-1 data from the United States and 10 cm resolution LiDAR. This data was combined with the results of various field research and research projects carried out after the disaster and was employed to evaluate changes in vegetation and land cover caused by the earthquake and tsunami, as well as monitoring the process of vegetation recovery and the ongoing effects of reconstruction projects. The level of analyses starts with an entire regional landscape and then is step by step fine-tuned to the level of individual habitats and tree groves.

16.4 Results and Discussion

16.4.1 Inundation Area

Figure 16.3, based on Terra/MODIS data acquired and archived at Tokyo University of Information Sciences on 14th March 2011, immediately after the disaster, shows the extent of inundation in the study area. As can be seen, the greatest amount of inundation occurred on the low-lying coasts of Sendai Bay, but the sandy beaches and tidal flats at the heads of the rias inlets were also covered.

16.4.2 Rocky Cliffs on the Sanriku Coast

The rias inlets on the submerging Sanriku Coast are bounded by steep rocky cliffs. According to Takayama (1987a,b, 1989), vegetation on this type of coastal landform consists of several zones. The areas closest to the splash zone are dominated by species in common with tidal wetlands, as well as annual and biennial herbaceous plants such as *Juncus gracillimus*, *Puccinellia nipponica*, *Sagina maxima* and *Lysimachia mauritiana*. Next is a zone characterized by perennials such as *Chrysanthemum japonica*, *Boehmeria biloba*, *Phedimus aizoon* var. *floribundus* (Fig. 16.4a) and *Lilium maculatum* followed by *Miscanthus sinensis* and other grasses near the top of the cliff. Above that is a zone of hardy shrubs such as *Pittosporum tobira* (Fig. 16.4b), *Juniperus chinensis* and *Viburnum dilatatum*. The hills beyond the cliff are forested with *Pinus densiflora*, and the islets in the bays support vegetation communities characterized by *Machilus thunbergii*.

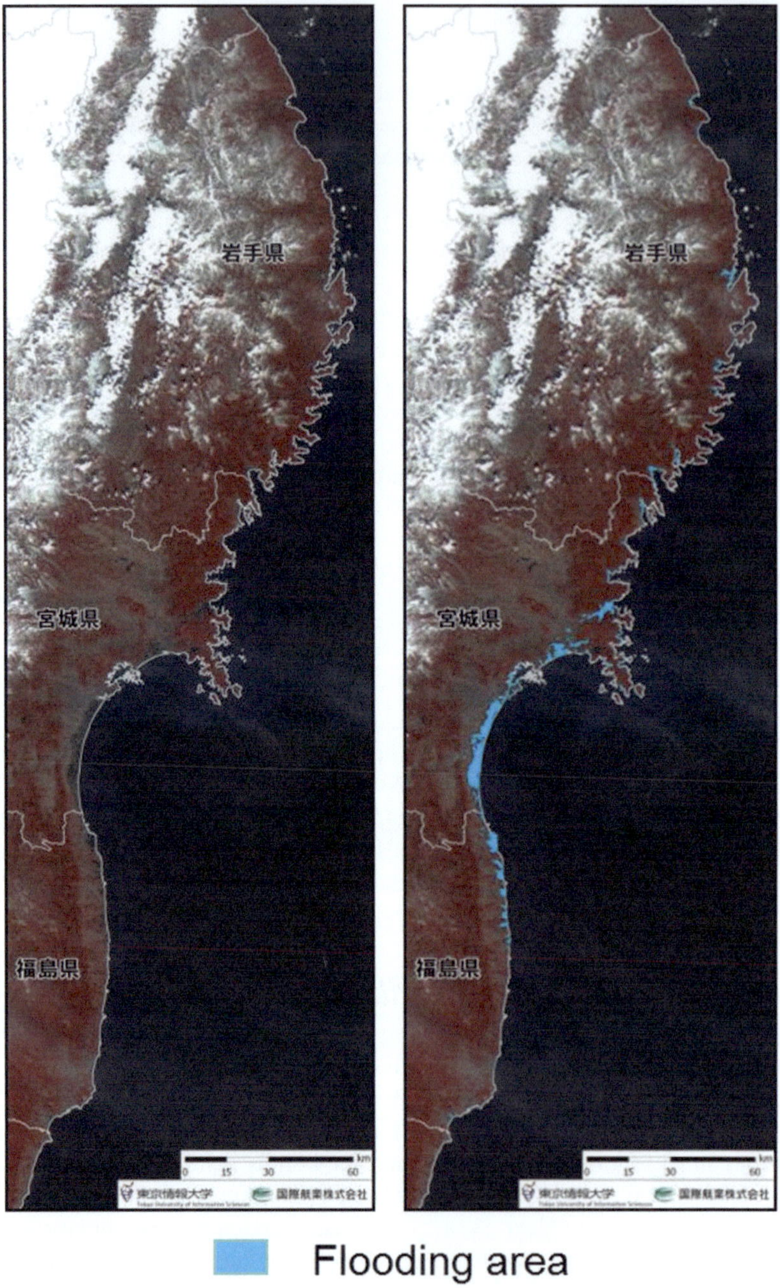

Fig. 16.3 Terra/MODIS image of Tohoku District after disaster (After Hara 2014). Analysis results for 14th March 2011 image, with submerged areas shown in *sky blue*

Fig. 16.4 Typical plants of rocky cliff. **a** *Phedimus aizoon* var. *floribundus* blooming during the following summer. **b** *Pittosporum tobira* surviving in the shrub zone (Photographed at Arajima Island, Minamisanriku Town, 15th September 2011)

The lower two vegetation zones along this coast, which develop in the areas hit directly by the tsunami, are accustomed to periodic inundation by waves and salt spray. The plants in these zones have thus acquired a tolerance to salt water and as such were not severely damaged by inundation by the tsunami. In addition, the disaster occurred during the winter dormant period, which also helped to limit the damage to plants. Figure 16.4 shows photographs of surviving vegetation taken only about 8 months after the disaster. With the exception of some areas where the rocky substrate itself was physically destroyed, damage to ecosystems along these rocky cliffs was minimal (Hara 2012; Hara et al. 2012; Ooue 2012).

16.4.3 Sandy Beaches and Back Marshes

As can be seen in Fig. 16.3, large areas of coastline along Sendai Bay, as well as the shallow wetlands at the head of the rias inlets, were completely inundated. The lower zones of sandy beaches in this region are dominated by perennials such as *Carex kobomugi*, *Calystegia soldanella*, *Lathyrus japonicus*, *Glehnia littoralis* and *Ixeris repens* (Fig. 16.5a). Landwards of these, where the substrate is more stable and more moisture is available, are found low shrubs such as *Rosa rugosa* (Fig. 16.5b) and *Juniperus conferta* (Suzuki 1987). The sandy beach habitats also support a number of bees, beetles and other insects unique to these plant communities (Nagahata 2012).

Even before the disaster, coastal engineering works such as jetties and wave breaks had disrupted the sandy coast environments in many areas. The tsunami struck all of the sandy beaches in the region, but the degree of damage fluctuated from place to place. Along the Sendai Bay shoreline, for example, damage to the beaches that fronted the seawall appears to have been minimal. The tsunami swept over these areas, swirling the top sediment but leaving the substrate intact. In these

Fig. 16.5 Typical vegetation of sandy coast ecosystem. **a** *Ixeris repens* in bloom. **b** A shrub of *Rosa rugosa* (Photographed near Arahama, Sendai City, 20th September 2012)

habitats vegetation was already recovering during the summer of 2011 (Hirabuki et al. 2011). Figure 16.5 shows recovering sandy beach vegetation in the autumn of the year following the disaster.

In contrast, in areas that experienced substantial ground subsidence, the entire substrate was washed away. Sandy shorelines at the heads of the rias inlets also suffered heavy physical damage (Hayasaka et al. 2012).

16.4.4 Coastal Pine Forests

Before the disaster, groves of *Pinus thunbergii* had been planted along extensive areas of the eastern Tohoku coastline. These groves were planted to stabilize the sandy substrate and to mitigate damage from high waves and salt spray. The well-known pine forest at Rikuzentakata, for example, has been shown to have mitigated damage from past tsunamis (Ishikawa 1992).

These pine forests also supported a rich ecosystem, but were severely damaged by the earthquake and tsunami. A total of 3660 ha of coastal forest were inundated by the tsunami. About a third of this area, totalling 1072 ha, suffered more than 75 % damage. The greatest damage occurred in Miyagi Prefecture (750 ha) (Committee for restoration of disaster prevention coastal forests damaged by the Great East Japan Earthquake 2012).

The extent of damage suffered by pine forests varied according to local topography and height of the tsunami (Tanaka 2012; Tomita et al. 2013). Those on the sandy shorelines at the heads of inlets along the rias coast were devastated by powerful tsunami, with most of the trees broken or washed away. As an example, Figs. 16.6 and 16.7 show the tremendous extent of damage to the pines at Rikuzentakata.

On the Sendai Plain, pines were first planted along the Teizan-bori, a banked canal dug parallel to the coast, about 400 years ago. Later the plantings were

Fig. 16.6 Damage to coastal vegetation at Rikuzentakata (After Hara 2014). Near-infrared band applied to false colour composite images with vegetated areas shown in *red*. The dense pine forest along the coast at the head of the inlet (left: 20th May 2005) has completely disappeared (right: 13th March 2011). (© DigitalGlobe/JSI)

Fig. 16.7 Photographs showing damage to pine forest at Rikuzentakata. Disappeared pine forest (Photographed 5th July 2012)

expanded to include both the landward and seaward sides of the canal. The pre-disaster extent of this magnificent pine forest can be seen in the left-side images of Fig. 16.8. According to research conducted by Hirabuki et al. (2011) and Tomita et al. (2013, 2014), many of the shorter pines seawards of the Teizan-bori were bent over or broken, while taller trees on the high banks along the canal survived much better. Landwards of the canal, segments of pine forest survived in narrow comb-

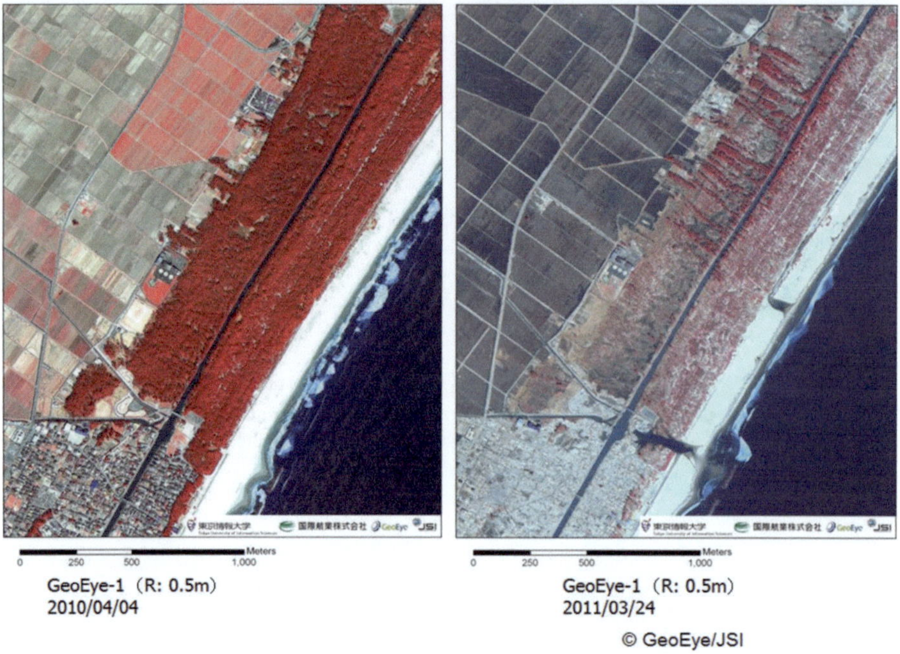

Fig. 16.8 GeoEye-1 images showing damage to coastal pine forest along the coast of Sendai Bay (After Hara 2014: left: 4th April 2010, right: 24th March 2011). Near-infrared band applied to false colour composite images with vegetated areas shown in *red*. The strip of pine forest running parallel to the coast has almost completely disappeared. Dense pine forest along Teizan-bori Canal has been decimated by the tsunami, but some trees remain along the raised banks of the canal, as well as in narrow strips to the landward side (© DigitalGlobe)

like strips running perpendicular to the coast and direction of the wave. This pattern of destruction and survival can be clearly seen in the GeoEye-1 images acquired for 4th April 2010 and 24th March 2011 (Fig. 16.8) and in the photographs of Fig. 16.9.

The area landwards of the canal was originally a low-lying back marsh that had been filled in and planted with pines. Along the coast pine trees naturally prefer to sink their roots deep into dry sandy soils. In this thick marshy soil, however, the roots were unable to penetrate downwards and were forced to spread outwards. This reduced the stability of the trees and their ability to withstand the force of the wave. The high water table in this area is attested to by the appearance of reeds in spots where the trees were destroyed. The precise reasons for the comb-like survival pattern, however, have yet to be deciphered.

Kamagata et al. (2013) employed very fine 10 cm resolution aerial photographs and airborne LiDAR to analyse changes in vegetation and clarify the process of recovery in this same area. Figure 16.10a shows a close-up image of the damage soon after the disaster (31st March 2011), while Fig. 16.10b–d shows the same area at intervals following the disaster. These images indicate that most of the pine forest patches that survived the tsunami were still viable more than a year later, on 10th July 2012. In

Fig. 16.9 Damage to coastal pine forest along the coast of Sendai Bay. (a) Remaining pine trees along the raised banks of Teizan-bori Canal. (b) Bent and broken trees in foreground, with narrow strip of surviving trees behind. (c) Reeds appearing in inundated area landwards of canal (Photographed 14th May 2011)

addition, the density of red colour indicates that by this time the understory layers, dominated by *Robinia pseudoacacia*, had already started to recover. A vegetated patch on the sandy shoreline is expanded in Fig. 16.10e. This patch, dominated by *Rosa rugosa*, represents a sandy beach plant community that is beginning to recover.

The extent of destruction, survival and recovery of the pine grove can also be monitored in fine detail using shadow maps compiled from LiDAR data. In Fig. 16.11, for example, a LiDAR image is compared with a photograph of the same area. The photo shows the tree, shrub and herbaceous layers all together (Fig. 16.11a), but the LiDAR image is able to pick out just the tree layer (Fig. 16.11b) (Kamagata et al. 2013).

16.4.5 Landscape-Level Changes Caused by Tsunami

Zhao et al. (2013) employed land use classifications of SPOT/HRG-2 satellite images to analyse landscape changes in the Sendai Bay coastal landscape caused by the earthquake and tsunami. Ten-metre resolution images obtained in October of

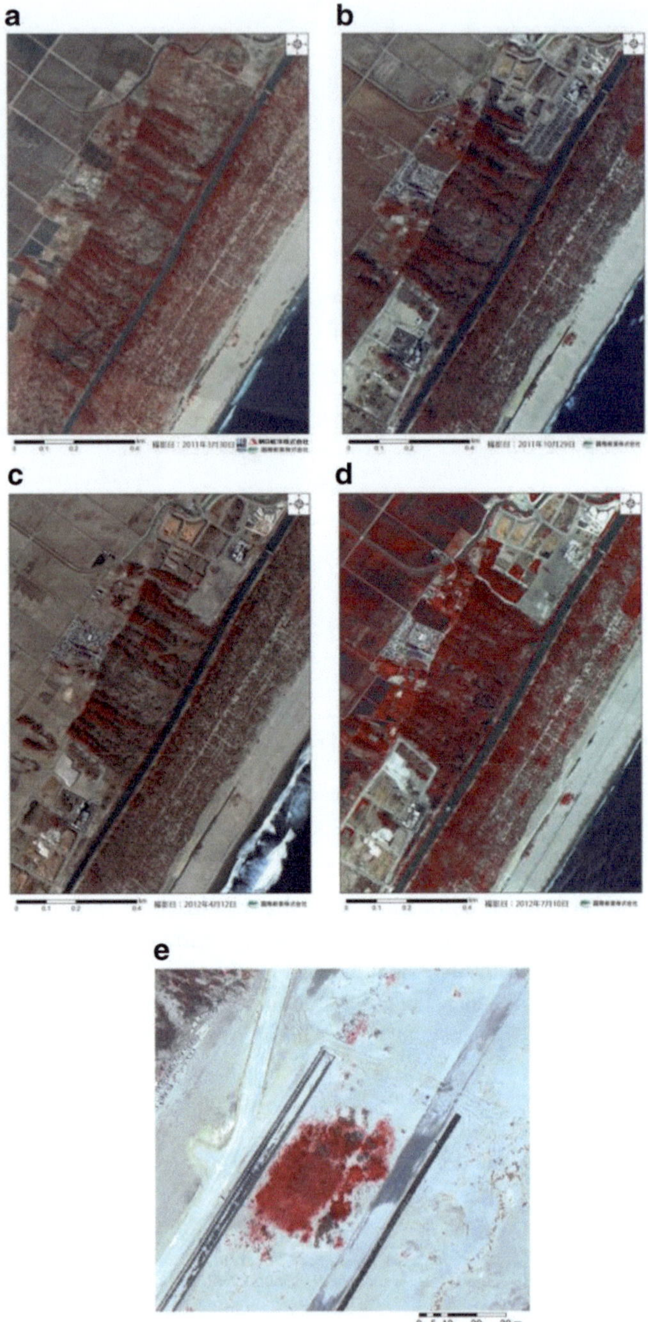

Fig. 16.10 Damage and survival of pine forest along the coast of Sendai Bay analysed using digital aerial photographs. (**a**) 30th March 2011. (**b**) 29th October 2011. (**c**) 12th April 2012. (**d**) 10th July 2012. (**e**) Zoom in view of **d**, a shrub of *Rosa rugosa*

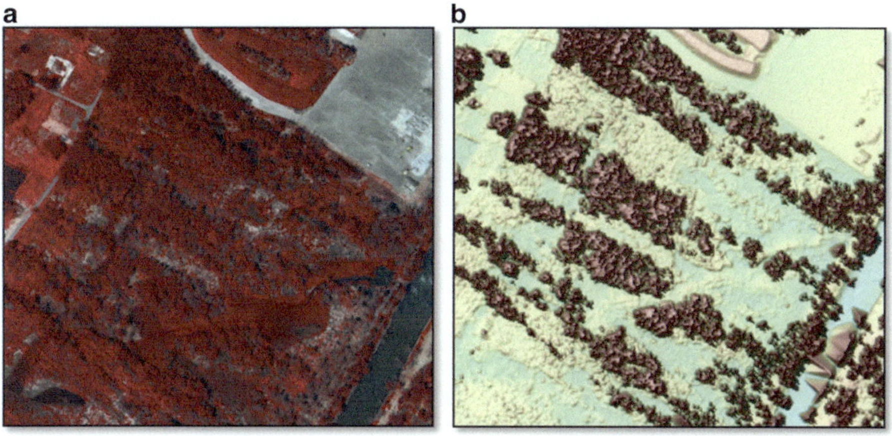

Fig. 16.11 (**a**, **b**) Comparison of aerial photographs (**a**: *left*) with LiDAR image (**b**: *right*). Both images acquired for 10th July 2012

2010 and November of 2011 were classified and the results compared. The images covered a 10×20 km area, from which the parts relevant to the research were selected and subjected to supervised classification employing the maximum likelihood classifier. Six land use classes were identified: forest, grassland/scrub-forest, paddy field, water, open land and urban area. The results of the classifications are shown in Fig. 16.12, with pre-disaster on the left and post-disaster on the right. Before the disaster, large sections of back marsh and valley bottom habitats on the Sendai Plain, as well as smaller sections of flatland at the heads of inlets along the rias coast, had been converted to rice paddies. Many of these paddies were inundated by the tsunami and in areas where ground subsidence was an issue remained submerged for months afterwards. Even after the water retreated, the drainage system was disrupted, and many paddies were completely taken over by *Echinochloa crus-galli* wild barnyard millet. Landscape-level changes were not limited to rice paddies. Of the 4.2 km^2 of pre-disaster forest, 3.6 km^2 (85 %) wound up in the damaged forest class; and of the original 4.8 km^2 of urban area, 1.4 km^2 (29 %) was classified as damaged urban area (Zhao et al. 2013).

16.5 Conclusions

In terms of geographic extent and intensity, the Great Eastern Japan Earthquake and Tsunami was a disaster of the worst sort. The entire regional landscape, which can be thought of as the composite of various ecosystems, was radically altered. Despite this, the results of this research showed that damage to coastal ecosystems varied according to landforms and local topography. Severe damage, for example, was suffered by sandy beach vegetation, forests along low-lying coasts and marshes. Rice

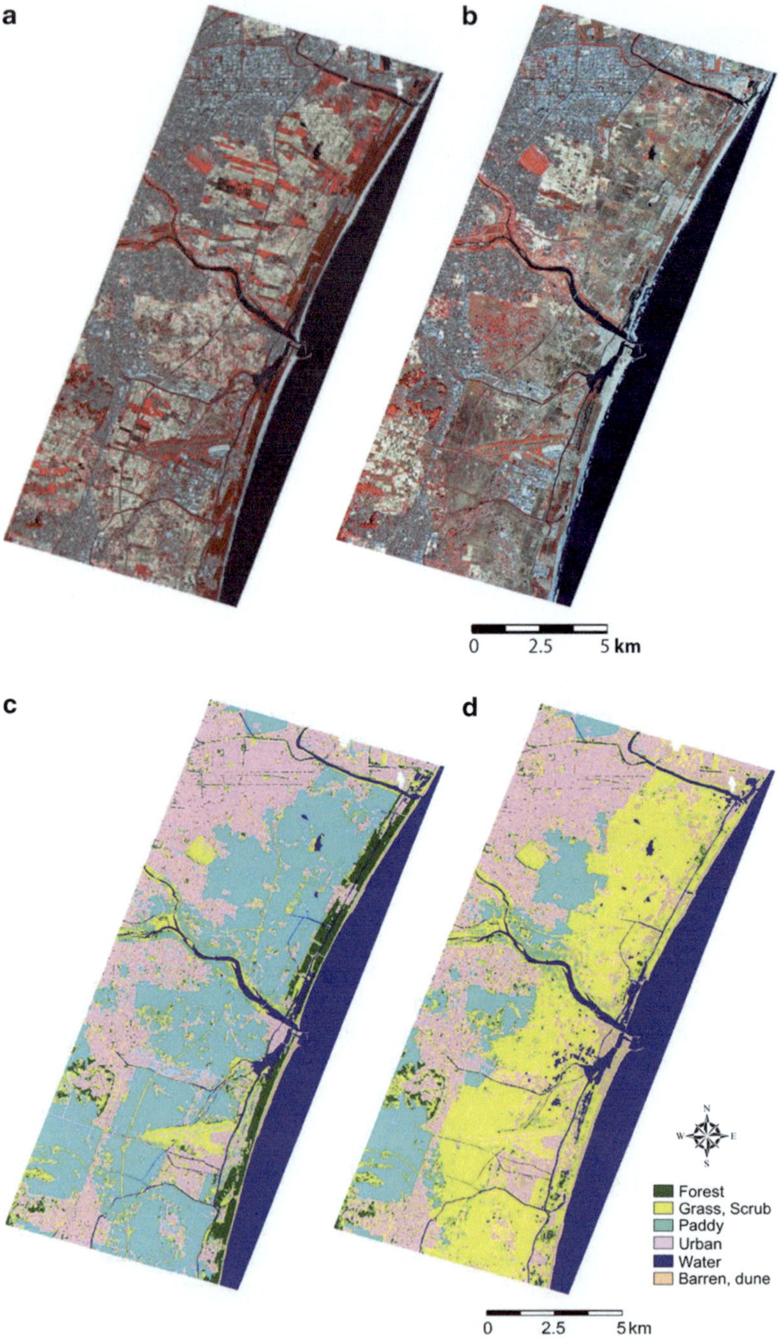

Fig. 16.12 (**a**) SPOT/HRG-2 images for the Sendai Bay area, before (*left*: 2nd September 2010) and after (*right*: 2nd November 2011) the disaster. R:G:B=3:2:1 © CNES 2010–2011, Distribution Astrium Service/Spot Image S.A. (**b**) Results of land cover classification of SPOT/HRG-2 images for the Sendai Bay area, before (*left*: 2nd September 2010) and after (*right*: 2nd November 2011) the disaster

paddies that had been developed in the low-lying areas behind coastal barriers were also heavily damaged. Rocky cliff vegetation, in contrast, sustained far less damage. Ecological damage to coastal and terrestrial ecosystems can be classified into two basic types: direct physical damage to habitats and life forms and subsequent damage caused by inundation in salt water (Hara and Higuchi 2013). The research also showed that despite the severity of the original damage, some coastal plant communities are already beginning to recover.

As the current international usage of this Japanese word indicates, Japan is especially susceptible to large tsunami. Research indicates that the study region has been struck by a similar scale tsunami on a cycle of about several hundreds to 1000 years (Matsumoto 2011; Matsumoto and Kumagai 2011; Hirakawa 2012). All of the past tsunami must have wreaked great havoc on the regional ecosystems. Research on previous disasters focused on disaster mitigation and the damage to human lives and lifestyles. This time, however, there is also a strong emphasis on the damage incurred by natural ecosystems.

Currently, various recovery plans and programmes are being implemented in the disaster region. This research shows clearly that multi-scale remote sensing data is useful not only for evaluating the extent of ecological damage but also for monitoring the subsequent processes of survival and recovery. As such remote sensing can play a vital role in designing regional plans that balance socio-economic reconstruction with recovery of vital coastal habitats and conservation of biodiversity.

Acknowledgements The authors wish to thank Kevin Short for editorial assistance. This research was supported by JSPS KAKENHI 24510332, 24810024, 25830153 and the Environment Research and Technology Development Fund (1-1405) of the Ministry of Environment, Japan.

References

Committee for restoration of disaster prevention coastal forests damaged by the Great East Japan Earthquake (2012) Restoration of disaster prevention coastal forests. Forestry Agency (In Japanese)

Dharanirajan K, Pandian PK, Gurugnanam B, Narayanan R, Ramachandran S (2007) An integrated study for the assessment of tsunami impacts: a case study of south Andaman Island, India using remote sensing and GIS. Coast Eng J 49(3):229–266

Hara M (2012) Situation of vegetation in damaged areas in Miyagi and Iwate Prefectures by the Great East Japan Earthquake. Veg Sci News 16:32–39 (In Japanese)

Hara K (2014) Damage to coastal vegetation due to the 2011 tsunami in Northeast Japan and subsequent restoration process: analyses using remotely sensed data. Glob Environ Res 18(1):27–34

Hara K, Higuchi H (2013) Effects of the 2011 Great East Japan Earthquake and tsunami disaster on ecosystems. Glob Environ (Chikyuu Kankyo) 18(1):23–33 (In Japanese with English abstract)

Hara M, Asami K, Tomita M (2012) A research report of Minami-Sanriku area damaged by the Great East Japan Earthquake. Veg Sci News 16:40–48 (In Japanese)

Hayasaka D, Shimada N, Konno H, Sudayama H, Kawanishi M, Uchida T, Goka K (2012) Floristic variation of beach vegetation caused by the 2011 Tohoku-oki tsunami in northern Tohoku. Jpn Ecol Eng 44:227–232

Himiyama Y, Watanabe T, Hara K (2014) The impact of the Great East Japan Tsunami of 2011 on land use. Glob Environ Res 18(2):191–206

Hirabuki Y, Tomita M, Kanno H, Hara K (2011) Impact of Great East Japan earthquake and subsequent tsunami on vegetation in the sand-dune coastal ecotone along the shores of Sendai Bay, in the Tohoku District of northern Japan. Med Plants Res (Yakuyo-shokubutu-kenkyu) 33(2):45–57 (In Japanese)

Hirakawa K (2012) Outsize tsunami sediment since last 6000 years along the Japan- and Kuril-Trench: a tentative idea on sources and supercycle. Kagaku 82(2):172–181 (In Japanese)

Ishikawa M (1992) Protection of high tide and tsunami. In: Murai H, Ishikawa M, Endo J, Tadaki R (eds) The coastal forest in Japan. Soft Science, Tokyo, pp 284–299 (In Japanese)

Kamagata N, Akamatsu Y, Hara K, Tomita M, Hirabuki Y (2013) Environmental monitoring of areas damaged by Great East Japan Earthquake and subsequent tsunami using digital aerial photographs and airborne LiDAR: a case study of the Sendai Bay coastal forests. Landsc Ecol Manag 18(1):29–34 (In Japanese)

Kanno H, Hirabuki Y, Sugiyama T, Tomita M, Hara K (2014) Vegetation change in various coastal forest habitats after a huge tsunami: a three-year study. Jpn J Conserv Ecol 99:201–220 (in Japanese with English abstract)

Matsumoto H (2011) The three-times huge tsunami reached into the Sendai Plains. Q J Tohoku Stud 28:114–126 (In Japanese)

Matsumoto H, Kumagai M (2011) Inundation distance of paleo-tsunami deposits in 2,000 yr BP and 1,000 yr BP on the Sendai coastal area. Northeast Jpn Q J Geogr 63(3):172–173 (In Japanese)

Miura O (1987) Topography and geology. In: Miyawaki A (ed) Vegetation of Japan, vol 8. Tohoku. Shibundo, Tokyo, pp 59–64 (In Japanese)

Nagahata Y (2012) Changes in ecosystems affected by the huge tsunami. Kodansya, Tokyo (In Japanese)

Oka K, Hirabuki Y (2014) Revegetation of coastal plants damaged by the 2011 Tohoku tsunami. Jpn J Conserv Ecol 99:189–199 (in Japanese with English abstract)

Okada Y (2012) Outline of the 2011 off the pacific coast of Tohoku earthquake. Natur Disaster Res Rep 48:1–14 (In Japanese with English abstract)

Olwig MF, Sørensen MK, Rasmussen MS, Danielsen F, Selvam V, Hansen LB, Nyborg L, Vestergaard KB, Parish F, Karunagaran VM (2007) Using remote sensing to assess the protective role of coastal woody vegetation against tsunami waves. Int J Remote Sens 28(13&14):3153–3169

Ooue M (2012) Current situation of coastal vegetation damaged by tsunami in the middle and northern part of Kitakami Mountains. Veg Sci News 16:49–58 (In Japanese)

Roemer H, Kaiser G, Sterr H, Ludwig R (2010) Using remote sensing to assess tsunami-induced impacts on coastal forest ecosystems at the Andaman Sea coast of Thailand. Nat Hazards Earth Syst Sci 10(4):729–745

Römer H, Jeewarongkakul J, Kaiser G, Ludwig R, Sterr H (2012) Monitoring post-tsunami vegetation recovery in Phang-Nga province, Thailand, based on IKONOS imagery and field investigations–a contribution to the analysis of tsunami vulnerability of coastal ecosystems. Int J Remote Sens 33(10):3090–3121

Sirikulchayanon P, Sun W, Oyana TJ (2008) Assessing the impact of the 2004 tsunami on mangroves using remote sensing and GIS techniques. Int J Remote Sens 29(12):3553–3576

Suzuki K (1987) Sand dune dwarf vegetation, Sand dune herbaceous community, Sea cliff herbaceous community. In: Miyawaki A (ed) Vegetation of Japan, vol 8. Tohoku. Shibundo, Tokyo, pp 222–242 (In Japanese)

Takayama H (1987a) Coastal cliff vegetation in the Tohoku District, Japan. I. Zonation. Ecol Rev 21:67–75

Takayama H (1987b) Sanriku coast. In: Miyawaki A (ed) Vegetation of Japan, vol 8. Tohoku. Shibundo, Tokyo, pp 414–419 (In Japanese)
Takayama H (1989) Plants of Shizugawa. In: Editing committee of History of Shizugawa (ed) History of Shizugawa, vol 1. Nature, Shizugawa-Town, pp 313–463 (In Japanese)
Tanaka N (2012) Effectiveness and limitation of coastal forest in large tsunami: conditions of Japanese pine trees on coastal sand dunes in tsunami caused by Great East Japan Earthquake. J Jpn Soc Civ Eng Ser B1 68(4):II_7–II_15
The 2011 Tohoku Earthquake Tsunami Joint Survey Group (2011) Nationwide field survey of the 2011 off the pacific coast of Tohoku earthquake tsunami. J Jpn Soc Civ Eng Ser B 67(1):63–66
Tomita M, Hirabuki Y, Kanno K, Hara K (2013) Spatial distribution of biological legacies and depth of accumulated sand in a coastal forest disturbed by the 2011 tsunami. J Nat Restor Conserv 6(1):51–60 (In Japanese with English abstract)
Tomita M, Hirabuki Y, Kanno K, Hara K (2014) Influence of tsunamis as large, infrequent disturbances on tree communities of coastal forest. Jpn J Conserv Ecol 99:163–176 (in Japanese with English abstract)
Zhao Y, Tomita M, Hara K (2013) Landscape change analysis before and after the earthquake disaster in Sendai coastal area by using SPOT satellite data. J Nat Restor Conserv 6(1):43–49 (in Japanese with English abstract)

Chapter 17
Susceptibility of Sandy Beach Flora to the Great East Japan Earthquake and Tsunami in Northern Tohoku, Japan

Daisuke Hayasaka, Katsumasa Yamada, and Taizo Uchida

Abstract Coastal ecosystems in northern Japan have been undergoing succession and recovery since the Great East Japan Earthquake and Tsunami. In order to elucidate the mechanisms of vegetation responses to such catastrophe, we assessed the differences in susceptibility of sandy beach flora based on species composition and trait group composition as indicated by differences in ecomorphological traits of species, namely, Raunkiaer life form and clonality, to the tsunami. From the field observation and aerial photo interpretation, we found the clear landscape change of the four studied beaches before and after the tsunami. Osuka beach (Aomori Prefecture), which had no beach protection structures, including revetments or wave-dissipating blocks, and remained in a natural condition before the tsunami, experienced a relatively small wave impact (about 10 m height) associated with the disturbance. Tofugaura and Aketo beaches (Iwate Prefecture), which were exposed to a wave height of more than 20 m, were severely damaged, with fracturing and collapse of protective structures after the tsunami. Revetments at Kuji (Iwate Prefecture), which was subjected to a wave height of approximately 14 m, were not obviously damaged by the tsunami. Among the 103 plant species recorded in the four studied beaches, the occurrence of 14 species (13.6 %) changed significantly after the tsunami. In the case of the Raunkiaer life forms, the number of therophytes (mainly non-beach species) and chamaephytes increased after the tsunami on Aketo and Kuji beaches, which had high anthropogenic disturbance before the tsunami, and many beach geophytes decreased on Osuka beach. Impacts of the tsunami on

D. Hayasaka (✉)
Faculty of Agriculture, KINDAI University,
3327-204 Nakamachi, Nara 631-8505, Japan
e-mail: hayasaka@nara.kindai.ac.jp; awayotou@hotmail.com

K. Yamada
Seikai National Fisheries Research Institute, Fisheries Research Agency,
1551-8 Taira-machi, Nagasaki, Nagasaki 851-2213, Japan

T. Uchida
Faculty of Engineering, Kyushu Sangyo University (KSU),
2-3-1 Higashi-ku, Matsukadai, Fukuoka, Fukuoka 813-8503, Japan

J. Urabe, T. Nakashizuka (eds.), *Ecological Impacts of Tsunamis on Coastal Ecosystems*, Ecological Research Monographs,
DOI 10.1007/978-4-431-56448-5_17

woody species were smaller than on herbaceous species. Except Aketo beach, the proportions of clonality among the beaches did not clearly change before and after the disturbance. Rhizomatous plants showed relatively good tolerance to the tsunami, whereas monophytes and stoloniferous species underwent damages by sand erosion and accretion, perhaps due to their shallow roots, and could not protect beaches from coastal disturbances. Thus, beaches such as Aketo and Tofugaura, with a high proportion of ephemeral monophytes and stoloniferous species, may be more vulnerable to tsunamis.

Keywords Catastrophe • Clonality • Comparative study • Disturbance • Ecological impacts • Plant functional traits • Raunkiaer life form • Species' sensitivity • Vegetation

17.1 Introduction

Effects of disturbances to biological communities vary according to their intensity, frequency, magnitude (Pickett and White 1985), and timing (Sousa 1985; Barrat-Segretian and Amoros 1995). In a frequently disturbed ecosystem, species are adapted to the disturbance regime and post-disturbance recovery is fast, whereas the recovery process can be very long in rarely disturbed ecosystems (Brock et al. 1987; Lamberti et al. 1991). Although small-scale disturbances tend to leave survivors, which determine the successional sere (Platt and Connell 2003), when a disturbance removes all plant species, the direction of ecological succession after the disturbance usually changes. During the recovery and succession of plant communities on bare sites cleared by threshold disturbances such as volcanic eruptions (del Moral and Bliss 1993) and landslides (White 1979), both environmental changes and biological effects (viz., residence effects and inter- and intraspecies competition) are important for determination of community structure (e.g., Connell and Slatyer 1977; Begon et al. 1996; Garnier et al. 2004).

The tsunami waves followed the Great East Japan Earthquake with magnitude 9.0 on the Richter scale on 11 March 2011, the strongest ever recorded in Japan (Mimura et al. 2011). This catastrophic disaster caused threshold disturbances in coastal areas (Seike et al. 2013; Mukai et al. 2014; Shimada et al. 2014). Many coastal biotic communities were destroyed by wave impact pressures and the extensive beach erosion and sediment deposition caused by the tsunami (Hayasaka et al. 2012c; Tanaka et al. 2012, 2013; Urabe et al. 2013; Ishida et al. 2015), and even the ground sunk due to the huge earthquake. Previous research reported a significant interaction between the tsunami impact and both the number of typical beach species and Shannon–Wiener diversity index (H') values (Hayasaka et al. 2012c); that is, the tsunami significantly changed both biotic factors. Following this destruction, surviving species would have been forced to compete with invasive species in a restructuring of coastal ecosystems. In such a case, species' functional traits, including ecomorphological traits, can be used to evaluate and predict the succession and recovery of communities after catastrophes (Hayasaka et al. 2009, 2012b).

Plants must be able to tolerate changes in their surroundings and to make use of this heterogeneity in both space and time (Grime 1994). The life history traits of surviving and resident species may determine community recovery processes after disturbances (Weiher et al. 1999). For example, it has been suggested that growth stage and growth rate (Lambers and Poorter 1992), lateral spread capacity (Lavorel and Garnier 2002; Ackerly 2004), environmental adaptivity (Grime 1977; Chapin 1991), and ability for competitive exclusion (Tilman 1986) by species can be used to predict the resilience of coastal communities after tsunami. Following the 2004 Indian Ocean tsunami, many non-beach ephemeral monophytes (particularly in the Poaceae and Asteraceae) rapidly invaded vegetation gaps and bare sites, biasing the succession of coastal sand dune vegetation (Hayasaka et al. 2009, 2012b). However, it is not clear whether this phenomenon would apply to the case of the 2011 Tohoku-oki tsunami as well. Following volcanic eruptions, rapid colonizers in the primary succession usually have long subterranean rhizomes that can tolerate disturbance better than plants with short rhizomes (Tsuyuzaki 1989). Species groups with similar ecological traits to these (e.g., Symstad 2000; Bellwood et al. 2004; Yamada et al. 2007, 2010, 2014) should lead the succession and recovery of communities after other disturbances such as landslides and land excavations (White 1979; Burke 2008). Therefore, the evaluation of community responses based on species' life history traits would help to elucidate the mechanisms by which vegetation responds to environmental change.

In this chapter, we report the differences in susceptibility of sandy beach flora to the tsunamis caused by the Great East Japan Earthquake. We focused on both species composition and trait group composition indicated by differences in ecomorphological traits, namely, Raunkiaer life form (Raunkiaer 1934) and clonality (Numata and Asano 1969).

17.2 Materials and Methods

17.2.1 Study Sites

Surveys of beach vegetation were conducted on four sites of northern Tohoku regions, Japan, because these beaches had similar topographical characteristics (i.e., flat and sandy beaches in ria coast facing the Pacific Ocean) and were located within a relatively small geographical scale (Fig. 17.1), but exposed to different heights of tsunamis (10–23 m height) (Table 17.1). Large variations of beach impacts by the 2011 tsunami among the four studied beaches were found (Fig. 17.2). Aketo (39°56′ N and 141°56′ E), Tofugaura (40°05′ N and 141° 49′ E), and Kuji (40°12′ N and 141°47′ E) beaches were protected by revetments or wave-dissipating blocks, but the structures of the former two beaches were severely damaged (i.e., were fractured

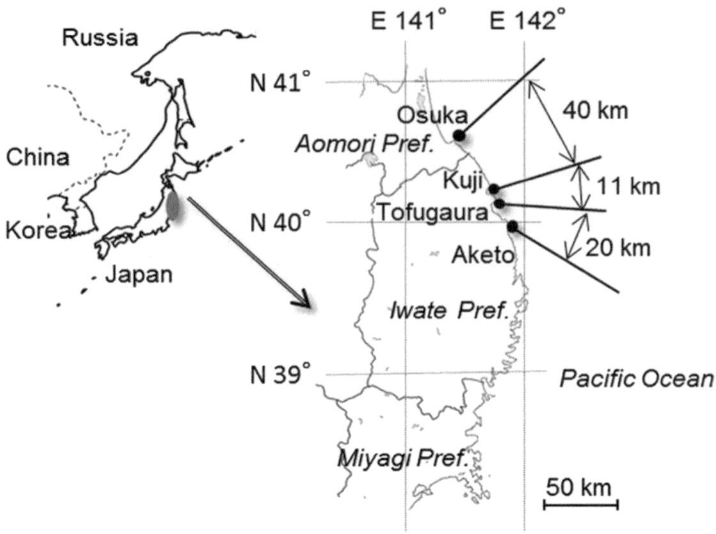

Fig. 17.1 Locations of studied beaches. The beaches are described in Table 17.1

Table 17.1 Beach characteristics and impacts of the 2011 tsunami on each beach

	Osuka	Kuji	Tofugaura	Aketo
Matrix types†	Coarse sand	Coarse sand	Coarse sand	Fine sand
Dominant landscapes around the studied beach†	Forest	Residential district	Paddy fields, forest	Forest
Beach protection structures†	–	Revetments	Revetments, wave- dissipating brocks	Revetments
Damage to the structures by the tsunami†	–	No damage	Fracturing	Collapse
Disappearance rate (%) of each studied beach before and after the tsunami‡	7.3 (16.7/18.0)	0.0 (9.0/9.0)	0.0 (10.0/10.0)	6.3 (4.5/4.8)
Run-up wave height (m)†	10.3	14.1	22.8	20.8

†Data from Hayasaka et al. (2012c)
‡Data from Ishida et al. (2015). Parenthesis shows beach areas (ha) in this study before and after the tsunami

and collapsed) by the tsunami. Osuka beach (40°31′ N and 141°34′ E) was in a natural condition before the tsunami. All beaches except Osuka experienced an elevation change approximately from 10 to 30 cm due to the Great East Japan Earthquake (see Geospatial Information Authority of Japan (http://www.gsi.go.jp/common/000062924.pdf) for detailed information on elevation changes of each beach due to the earthquake). The study region lies entirely in the cool-temperate

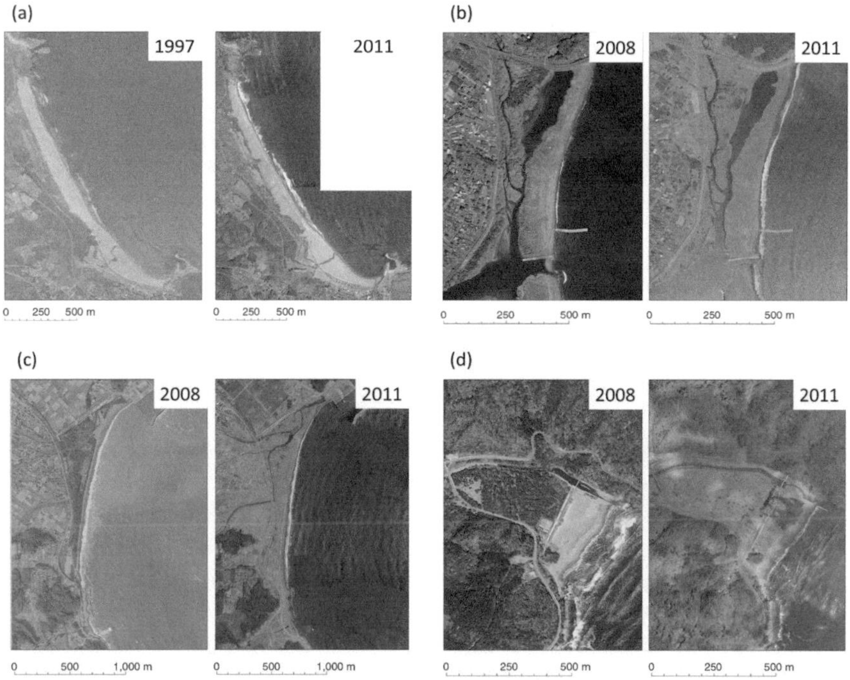

Fig. 17.2 Landscape changes in the four studied beaches before and after the tsunami based on aerial photos. (**a**) Osuka, (**b**) Kuji, (**c**) Tofugaura, and (**d**) Aketo

zone, with values of Kira's (1977) warmth index (WI) ranging from 45 to 85 °C. The climatic conditions of the four study sites, based on meteorological data collected from 2000 to 2010 (http://www.data.jma.go.jp/obd/stats/etrn/index.php), were relatively similar, with mean annual temperatures of 9.5–10.4 °C, WI values of 72.3–81.3 °C, and mean annual rainfall of 1,078–1,494 mm (Hayasaka et al. 2012c).

17.2.2 Field Survey

A previous vegetation survey (Braun-Blanquet 1964) was conducted before the tsunami, in August 2003 (8 years before the disturbance), and a second survey was conducted in August 2011, 5 months after the tsunami, using the same quadrats as in the first one (Hayasaka et al. 2012c). The study plots were selected based on vegetation zones (i.e., herbaceous vegetation and shrub vegetation) from the shoreline to inland areas. The author surveyed a total of 96 plots before the tsunami (including 17 plots at Aketo that later disappeared) and 83 plots afterward. The field relevés (i.e., separate sampling of the tree, shrub, and herbaceous layers in spatially homogenous vegetation) investigated at the studied beaches range in size from 1 m^2 (herbaceous vegetation) to 6 m^2 (shrub vegetation).

17.2.3 Classification of Species' Ecomorphological Traits

Attributes of species that influence recovery patterns after disturbances are determined not only by aboveground vegetation structures but also by underground root systems (Tsuyuzaki and del Moral 1995). Underground organs are important at sites with unstable surfaces (Platt and Connell 2003). Beaches, particularly foredunes, are unstable habitats that are subjected to constant fluxes such as sand movement, plant burial, and drought (Hesp 2004). So we selected two ecomorphological traits, namely, Raunkiaer life form and clonality, as bioindicators of species' sensitivity to wave impact pressures and the extensive beach erosion and accretion caused by the tsunami (e.g., Choowong et al. 2009; Scheucher and Vortisch 2011).

We followed the method of Raunkiaer (1934) for categorization of the Raunkiaer life form (dormancy form) of the species. The Raunkiaer life form falls into one of four major groups: (1) annual (therophyte, a species that produces seeds and dies within one growing season), (2) perennial (a plant that grows vegetatively for more than one growing season and flowers more than once before dying), (3) shrub (nanophanerophyte, dormant buds are located between 0.5 and 3 m above ground), and (4) tree (phanerophyte, >50 cm in height, and shoots do not die back). Plant populations with a long life span may persist for long periods, even when recruitment is no longer occurring (Eriksson 1996), and may have great significance for the vegetation response to environmental disturbances. Perennials can be subdivided into four categories according to the location of their dormant buds: (5) chamaephyte (bears dormant buds on persistent shoots near the ground), (6) hemicryptophyte (dormant buds sit at soil level, remnant shoot system), (7) geophyte (stem dies, leaving a remnant shoot system with storage organs in the soil), and (8) helophyte (dormant buds are buried in saturated soil or water). The Raunkiaer life form allows classification of the way plants are adapted morphologically for surviving stresses and unfavorable environmental conditions.

Clonality, mediated by subterranean organs, is most simply delineated into clonal and non-clonal (monophyte) growth. The capacity for lateral spread can be estimated as the number of connections from one ramet to other ramets. Clonal species are subdivided into two categories: (1) with rhizomatous growth and (2) with stoloniferous growth.

17.2.4 Data Analysis

To test whether the frequency of occurrence of each species differed between before and after the tsunami, we conducted Fisher's exact test; the null hypothesis was that the species' occurrence was equivalent before and after the tsunami. We used a generalized linear mixed model (GLMM), including the total number of species in each Raunkiaer life form as an offset variable, to investigate whether the Raunkiaer life-form composition patterns differed between pre- and post-tsunami; the null hypothesis was that the proportions of life forms did not change after the tsunami. We modeled patterns of Raunkiaer life-form composition with a Poisson error distribution and a log link. Differences in the proportions of clonality before and after the tsunami were also analyzed by GLMM. These analyses were conducted using the package "lme4" of the statistical software R ver. 3.0.3 (R Development Core Team 2014).

17.3 Results and Discussion

In this study, of 103 plant species recorded, 14 species (13.6 %) showed significant differences ($P<0.05$, Fisher's exact test) in their frequency of occurrence between before and after the tsunami; it included eight beach species and six non-beach species (Table 17.2). Species that increased in occurrence after the tsunami were the non-beach species *Artemisia indica* var. *maximowiczii* (Aketo), *Digitaria ciliaris* (Aketo), *Setaria* sp. (Tofugaura, Aketo), *Trifolium repens* (Aketo), and *Zoysia japonica* (Aketo) and the beach species *Chenopodium glaucum* (Kuji), *Rosa rugosa* (Kuji), and *Salsola komarovii* (Aketo). Beaches where these species increased were under strong anthropogenic disturbances before the tsunami. All of these species except *S. komarovii* frequently occupied vegetation gaps created by the tsunami. Invasion of vegetation gaps by non-beach species also followed the 2004 Indian Ocean tsunami in Thailand (Hayasaka et al. 2012b). Undisturbed patches act as refuges for surviving species, from where recolonization of the cleared patches can take place. It is necessary to assess the invasiveness of non-beach species that invade beaches after a tsunami and their ecological risks to beach communities for a realistic prediction of the resilience of beach flora. Following the tsunami, the herbaceous species composition on almost all beaches was characterized by a few non-beach species, which occupied disturbed sites such as urban land, developed land, and wasteland (Miyawaki and Okuda 1990), as reported in a previous study (Hayasaka et al. 2012a). Although we found ten introduced species designated as invasive species, due to their potential to invade ecosystems (https://www.env.go.jp/nature/intro/1outline/caution/list_sho.html), there was not an increase of these species after the tsunami on any beach. In addition, we found no species regulated by the Invasive Alien Species Act of Japan (http://www.env.go.jp/nature/intro/1outline/files/siteisyu_list_e.pdf). The beach species *Artemisia stelleriana*, *Atriplex subcordata*, *Carex kobomugi*, *Elymus mollis*, and *Ixeris repens* decreased significantly after the tsunami mainly on Osuka beach, which experienced little anthropogenic disturbance and on Kuji. Among these species, *E. mollis* plays an important role in protecting embryonic dunes from wave impacts, allowing *C. kobomugi* and *I. repens* to establish (Ishizuka 1962). These species are essentially maritime plants (Ridley 1930), and their disseminules are often transported by ocean currents (Nakanishi 1988), so their reductions would be temporary.

The number of vascular plant species on Osuka beach decreased after the tsunami, but that on the other beaches, particularly Aketo, generally increased (Fig. 17.3).

Table 17.2 Frequency of occurrence of species in study plots before (8 years before the tsunami) and after (5 months after the tsunami) the 2011 tsunami. Species are listed in decreasing order of occurrence

	Life-form characteristics			Occurrence frequency							
				Osuka		Kuji		Tofugaura		Aketo	
Species name	Dormancy forms	Radicoid forms	Flowering period	Before ($n=31$)	After ($n=31$)	Before ($n=25$)	After ($n=26$)	Before ($n=13$)	After ($n=16$)	Before ($n=27$)	After ($n=10$)
Ixeris repens†	G	R	Apr–Jul	28a	−13b	10a	−8b	0	±0	21a	−17b
Carex kobomugi†	G	R	Apr–Jun	25a	−13b	1	±0	0	+1	14	−10
Calystegia soldanella†	G	R	May–Jun	23	+1	8	+7	5	+6	8	−2
Elymus mollis†	G	R	May–Jul	19a	−9b	19a	−8b	0	±0	13	−11
Zoysia macrostachya†	G	R	Jul–Aug	12	−6	0	±0	1	−1	0	±0
Ischaemum anthephoroides†	G	R	Jul–Aug	10	−4	0	±0	0	±0	2	−2
Linaria japonica†	G	R	Jul–Aug	9	−6	9	−2	0	±0	4	−3
Bromus japonicus	Th	M	Jun–Jul	9a	−8b	0	±0	0	±0	0	±0
Artemisia stelleriana†	H	R	Aug–Oct	8	−5	10a	−8b	0	±0	0	±0
Rosa rugosa†	N	M	Jun–Aug	8	+1	1a	+6b	2	+4	1	−1
Lathyrus japonicus†	G	R	May–Jul	7	−6	14	+2	10	+2	8	−7
Carex pumila†	G	R	Apr–May	5	−3	2	±0	5	±0	6	−3
Sonchus brachyotus†	H	R	Jul–Oct	5	+2	0	±0	0	+2	0	±0
Glehnia littoralis†	Ch	M	Jun–Jul	4	+2	1	+1	0	±0	19	−16
Pinus thunbergii†	P	M	Apr–May	4	−3	0	±0	1	−1	1	−1
Lactuca indica	Th	M	Sep–Nov	4	−4	0	+2	0	±0	0	±0
Dendranthema arcticum spp. *maekawanum*†	H	R	Sep–Nov	4	−2	0	±0	0	±0	0	+1
Ligusticum hultenii†	Ch	M	Jul–Aug	4	−2	0	±0	0	±0	0	±0
Calamagrostis epigejos	G	R	Jul–Aug	3	−1	0	±0	0	±0	0	±0

Oenothera erythrosepala	Th	M	Jun–Aug	2	±0	5	−1	1	−1	11	−7
Cakile edentula†	Th	M	Jun–Aug	2	+2	0	±0	0	±0	3	+1
Dactylis glomerata‡	H	M	Jun–Jul	1	+1	1	−1	0	+1	5	−4
Artemisia indica var. *maximowiczii*	Ch	R	Aug–Oct	1	±0	3	+4	2	+1	1a	+4b
Taraxacum officinale‡	H	R	Oct–Nov	1	+3	1	−1	1	−1	2	−2
Plantago lanceolata‡	H	R	May–Jun	1	+5	3	+2	0	±0	0	±0
Ampelopsis glandulosa var. *heterophylla*	N	R	Jun–Jul	0	+3	0	+3	1	−1	0	±0
Carex fibrillose†	H	R	Apr–Jun	0	+4	0	+1	0	±0	0	+1
Zoysia japonica	Ch	S	Jun–Jul	0	+2	0	±0	0	±0	0a	+3b
Salsola komarovii†	Th	M	Jun–Aug	0	+1	1	+2	0	±0	0a	+6b
Scutellaria strigillosa†	H	R	May–Jun	0	+1	0	+1	0	+1	2	−2
Senecio vulgaris	Th	M	Feb–Sep	0	+1	3	−2	3	−3	0	±0
Atriplex subcordata†	Th	M	Jul–Aug	0	±0	6a	−5b	4	−1	3	−1
Mertensia maritima subsp. *asiatica*†	H	S	Jul–Aug	0	±0	4	−4	5	−2	0	±0
Phragmites australis	HH	R	Aug–Oct	0	±0	3	−3	0	+1	1	−1
Sonchus oleraceus	Th	M	Apr–Jul	0	±0	2	−1	2	−2	0	+1
Chenopodium glaucum†	Th	M	Jul–Oct	0	±0	0a	+6b	0	+2	0	+2
Setaria sp.	Th	M	–	0	±0	0	+3	0a	+7b	0a	+4b
Digitaria ciliaris	Th	S	Jul–Sep	0	±0	0	+2	3	−1	0a	+4b
Rumex japonicus	H	M	May–Aug	2	−2	1	±0	0	±0	0	±0
Honckenya peploides var. *major*†	G	R	May–Jun	2	−2	1	−1	0	±0	0	±0

(continued)

Table 17.2 (continued)

Species name	Life-form characteristics			Occurrence frequency							
				Osuka		Kuji		Tofugaura		Aketo	
	Dormancy forms	Radicoid forms	Flowering period	Before ($n=31$)	After ($n=31$)	Before ($n=25$)	After ($n=26$)	Before ($n=13$)	After ($n=16$)	Before ($n=27$)	After ($n=10$)
Juniperus conferta†	N	M	Apr–May	2	±0	0	±0	0	±0	0	±0
Artemisia annua	Th	M	Aug–Oct	1	−1	0	±0	0	±0	0	±0
Rumex acetosa	H	M	May–Jul	1	−1	0	±0	0	±0	0	±0
Toxicodendron orientale	P	M	May–Jun	1	+2	0	±0	0	±0	0	±0
Alnus sp.	–	M	–	1	−1	0	±0	0	±0	0	±0
Cocculus orbiculatus	N	R	Jul–Aug	0	+2	0	±0	0	±0	0	±0
Liriope minor	G	R	Aug–Oct	0	+2	0	±0	0	±0	0	±0
Rubia argyi	G	R	Aug–Oct	0	+1	0	±0	0	±0	0	±0
Ligustrum ovalifolium†	N	M	May–Jun	0	+1	0	±0	0	±0	0	±0
Messerschmidia sibirica†	G	R	May–Aug	0	+1	0	±0	0	±0	0	±0
Celastrus orbiculatus var. *orbiculatus*	P	M	May–Jun	0	+1	0	+1	0	±0	0	±0
Rubus parvifolius	N	M	May–Jun	0	+1	0	±0	0	±0	0	±0
Aster microcephalus var. *ovatus*	H	R	Aug–Nov	0	+1	0	+1	0	±0	0	±0
Vitis coignetiae	N	R	Jun–Jul	0	+1	0	±0	0	±0	0	±0
Setaria pallide-fusca	Th	M	Aug–Oct	0	±0	3	−3	0	±0	0	±0
Juglans mandshurica var. *sachalinensis*	P	M	Jan–Jun	0	±0	1	−1	0	±0	0	±0
Commelina communis	Th	M,S	Jun–Oct	0	±0	1	−1	0	+1	1	−1
Humulus japonicus	Th	M,S	Sep–Oct	0	±0	1	−1	0	±0	0	±0

Agropyron tsukushiensis var. *transiens*	H	M,S	May–Jun	0	±0	1	−1	0	±0	0	±0
Atriplex gmelinii†	Th	M	Jul–Aug	0	±0	1	−1	1	−1	0	+1
Stachys aspera var. *hispidula*	G	R	Jul–Aug	0	±0	1	−1	0	±0	0	±0
Polygonum polyneuron†	Th	M	Sep–Oct	0	±0	1	−1	0	±0	0	±0
Solidago altissima‡	H	R	Oct–Nov	0	±0	1	−1	0	±0	0	±0
Robinia pseudoacacia‡	P	M	May–Jun	0	±0	1	−1	0	±0	0	±0
Dianthus superbus var. *longicalycinus*	Ch	R	Jun–Sep	0	±0	1	−1	0	±0	0	±0
Lotus corniculatus var. *japonicus*	H	M	Apr–Oct	0	±0	1	±0	0	±0	0	±0
Saponaria officinalis	G	R	Jun–Aug	0	±0	0	+3	0	±0	0	±0
Phragmites japonica	Ch	S	Jul–Aug	0	±0	0	+3	0	±0	0	±0
Portulaca oleracea	Th	M	Jun–Sep	0	±0	0	+2	0	±0	0	±0
Bidens frondosa‡	Th	M	Sep–Oct	0	±0	0	+1	0	±0	0	±0
Solidago virgaurea var. *asiatica*	H	R	Sep–Oct	0	±0	0	+1	0	±0	0	±0
Festuca arundinacea‡	H	R	Jan–Aug	0	±0	0	+1	0	+1	0	±0
Euphorbia supine	Th	M	Jul–Oct	0	±0	0	+1	0	±0	0	±0
Lepidium virginicum	Th	M	May–Jul	0	±0	0	+1	0	±0	0	±0
Bromus carinatus	G	R	Apr–Jun	0	±0	0	±0	2	−2	0	±0
Trifolium pretense	H	M	May–Sep	0	±0	0	±0	1	−1	2	−2
Oxalis corniculata	Ch	S	May–Sep	0	±0	0	±0	1	−1	0	±0
Equisetum arvense	G	R	–	0	±0	0	±0	1	−1	0	±0
Poa annua	Th	M	Jan–Feb	0	±0	0	±0	1	−1	0	±0

(continued)

Table 17.2 (continued)

Species name	Life-form characteristics			Occurrence frequency							
				Osuka		Kuji		Tofugaura		Aketo	
	Dormancy forms	Radicoid forms	Flowering period	Before ($n=31$)	After ($n=31$)	Before ($n=25$)	After ($n=26$)	Before ($n=13$)	After ($n=16$)	Before ($n=27$)	After ($n=10$)
Sagina maxima[†]	H	M	Jun–Aug	0	±0	0	±0	1	−1	0	±0
Lolium perenne[‡]	Th	M	May–Aug	0	±0	0	±0	1	−1	0	±0
Plantago camtschatica[†]	H	R	May–Jul	0	±0	0	±0	0	+3	0	±0
Calystegia japonica	G	R	Jun–Sep	0	±0	0	±0	0	+2	0	±0
Rumex obtusifolius var. *agrestis*[‡]	H	M	May–Jul	0	±0	0	±0	0	+1	0	±0
Festuca rubra var. *rubra*	H	R	Jun–Jul	0	±0	0	±0	0	+1	0	±0
Lycopus lucidus var. *lucidus*	HH	R	Aug–Sep	0	±0	0	±0	0	+1	0	±0
Pueraria lobata	P	M	Jul–Aug	0	±0	0	±0	0	±0	3	−3
Lonicera japonica	N	R	May–Jun	0	±0	0	±0	0	±0	2	−2
Agropyron sp.	Th	M	–	0	±0	0	±0	0	±0	2	−2
Arabis stelleri var. *japonica*[†]	Th	M	Apr–Jun	0	±0	0	±0	0	±0	2	−2
Silene armeria	Th	M	Jun–Sep	0	±0	0	±0	0	±0	2	−2
Agropyron ciliare var. *minus*	Th	M	May–Jul	0	±0	0	±0	0	±0	1	−1
Pinus densiflora	P	M	Jan–Apr	0	±0	0	±0	0	±0	1	±0
Xanthium canadense[‡]	Th	M	Aug–Sep	0	±0	0	±0	0	±0	1	−1
Miscanthus sinensis	H	R	Jan–Sep	0	±0	0	±0	0	±0	1	−1

Plantago asiatica	Th	R	May–Jul	0	±0	0	±0	0	±0	1	−1
Setaria viridis var. *pachystachys*†	Th	M	Jul–Aug	0	±0	0	±0	0	±0	1	−1
Lysimachia mauritiana†	Th	R	May–Jul	0	±0	0	±0	0	±0	1	−1
Trifolium repens	Ch	S	Apr–Jul	0	±0	0	±0	0	±0	0a	+3b
Festuca ovina var. *ovina*†	H	R	Jun–Jul	0	±0	0	±0	0	±0	0	+1
Polygonaceae sp.	–	–	–	0	±0	0	±0	0	±0	0	+1
Agrostis clavata var. *nukabo*	H	R	May–Jun	0	±0	0	±0	0	±0	0	+1
Dioscorea japonica	G	M	Jul–Aug	0	±0	0	±0	0	±0	0	+1

Abbreviations of Raunkiaer life form (dormancy form) are as follows: *Ch*, chamaephyte, *G* geophytes, *H* hemicryptophyte, *HH* helophyte, *N* nanophanerophyte, *P* phanerophyte, *Th* throphyte. Clonality (subterranean organ form) was classified as *R* (rhizomatous growth), *S* (stoloniferous growth), and *M* (non-clonal: monophyte). Different letters indicate significant differences between before and after the tsunami by Fisher's exact test ($P<0.05$)

†Indicates beach species

‡Shows species designated as introduced invasive species with potential to invade ecosystems. *n* indicates the total number of study plots

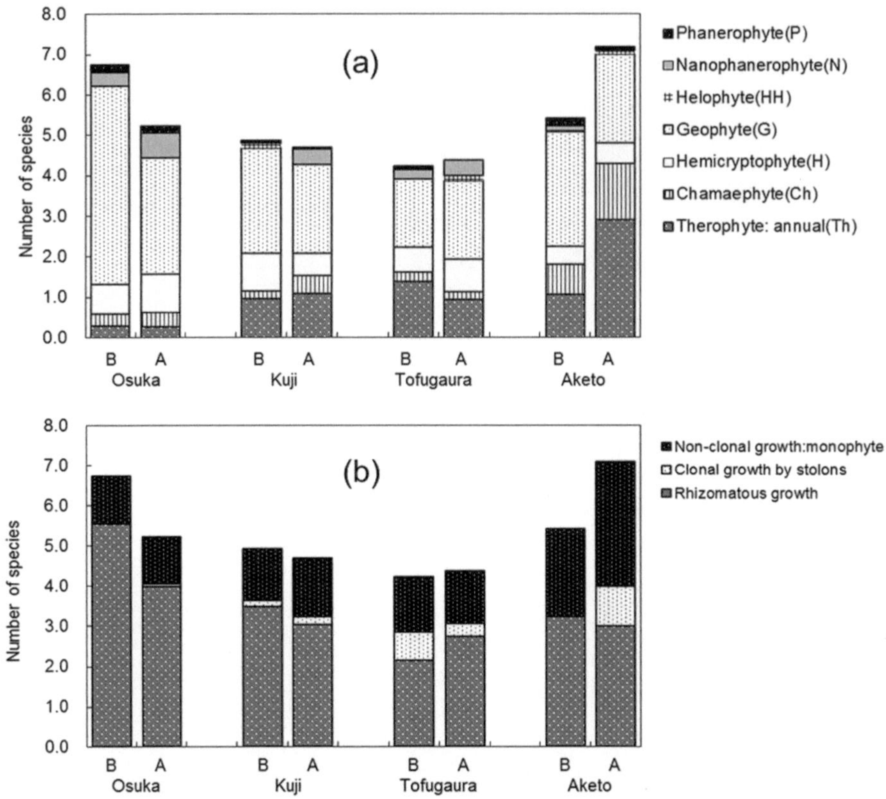

Fig. 17.3 General linear mixed model results of differences in the proportions of species of each ecomorphological type on the studied beaches before and after the 2011 tsunami. (**a**) Raunkiaer life form (dormancy form). (**b**) Clonality (subterranean organ form). Symbols: *B* indicates 8 years before the tsunami and *A* shows 5 months after the tsunami

In the case of the Raunkiaer life forms (Fig. 17.3a), the number of therophytes (mainly non-beach species) and chamaephytes increased after the tsunami on Aketo ($df=6$, $x^2=16.96$, $P=0.009$, GLMM) and Kuji ($df=6$, $x^2=19.06$, $P=0.004$), and the number of geophytes decreased on Osuka ($df=6$, $x^2=14.81$, $P=0.022$). There were no significant differences on Tofugaura ($df=6$, $x^2=6.38$, $P=0.381$). After disturbances, ephemerals or short-lived taxa (therophytes) invade bare sites much more rapidly than long-lived species regardless of disturbance (Hayasaka et al. 2009, 2011). As we reported previously (Hayasaka et al. 2012c), we found smaller impacts on woody species (nanophanerophytes and phanerophytes) than on herbaceous species.

Plants occurring in habitats frequently disturbed by floods can rapidly regenerate and recolonize disturbed sites through vegetative organs such as turions (winter buds) and tubers, whenever the disturbance occurs (Barrat-Segretain and Bornette 2000). Vegetative reproduction has an important role in plant community recovery in harsh environments (Tsuyuzaki 1989). In this study, the proportions of clonality among the beaches studied did not significantly differ between before and after the tsunami (*df*=2, $x^2=25.69$, $P<0.001$), except at Aketo beach (Fig. 17.3b). These results were similar to those following the 2004 Indian Ocean tsunami in Thailand (Hayasaka et al. 2009). Species with rhizomatous growth have a high lateral spread ability (Lavorel and Garnier 2002) regardless of the length of their rhizomes (Tsuyuzaki 1989), and the distance among ramets suggests that rhizomatous perennials have a good tolerance of disturbances, including tsunamis (Weiher et al. 1999; Hayasaka et al. 2009). By contrast, monophytes and stoloniferous species are prone to heavy damages by sand movement due to the tsunamis, perhaps on account of their shallow roots, and would be less able to protect beaches from coastal disturbances. Thus, beaches such as Aketo and Tofugaura, which had a higher proportion of ephemeral monophytes and stoloniferous species than Osuka and were exposed to particularly strong wave impacts during the tsunami (>20 m height), would be more vulnerable to coastal disturbances. In addition, seashore plants have a low success rate of sexual reproduction compared with vegetative reproduction in their dispersal and colonization (Bouzillé et al. 1997; Rautiainen et al. 2004).

These results show that species that share ecomorphological traits respond with similar sensitivities to tsunamis. Trait-based evaluation of ecosystem status (e.g., succession, recovery, and susceptibility of flora) is a useful tool for the elucidation of the mechanisms of community resilience after disturbances.

Acknowledgments We would like to thank Dr. Naoaki Shimada, Dr. Motohiro Kawanishi, and Miho Kazahaya for helping with fieldwork. The authors wish to thank Dr. Gen Kanaya and Dr. Francisco Sánchez-Bayo for their valuable technical advice. In this paper, we used aerial photos of the four studied beaches before and after the tsunami with permission from the Geospatial Information Authority of Japan and the Land Improvement Association of Iwate Prefecture, Japan (see Fig. 17.2). Part of this research was financially supported by a research grant from the Mitsui & Co., Ltd. Environment Fund (R11-F1-015-1, Leader: D. Hayasaka), the Tohoku Ecosystem-Associated Marine Sciences, the Research Institute of Marine Invertebrate Foundation, the Sasagawa Scientific Research Grant from the Japan Science Society (No. 21-507, Leader: K. Yamada), and the Center Project in National Institute for Environmental Studies, Japan (No.1112AF001, Leader: K. Yamada; No. 1213AR001, Leader: D. Hayasaka).

References

Ackerly D (2004) Functional strategies of chaparral shrubs in relation to seasonal water deficit and disturbance. Ecol Monogr 74:25–44

Barrat-Segretain MH, Amoros C (1995) Influence of flood timing on the recovery of macrophytes in a former river channel. Hydrobiologia 316:91–101

Barrat-Segretain MH, Bornette G (2000) Regeneration and colonization abilities of aquatic plant fragments: effect of disturbance seasonality. Hydrobiologia 421:31–39

Begon M, Harper JL, Townsend CR (1996) Ecology: individuals, populations and communities, 3rd edn. Blackwell Science Ltd., Oxford

Bellwood DR, Hughes TP, Folke C et al (2004) Confronting the coral reef crisis. Nature 429:827–833

Bouzillé JB, Bonis A, Clément B et al (1997) Growth patterns of *Juncus gerardi* clonal populations in a coastal habitat. Plant Ecol 132:39–48
Braun-Blanquet J (1964) Pflanzensoziologie, 3rd edn. Springer, Vienna
Brock TCM, van der Velde G, van de Steeg HM (1987) The effects of extreme water level fluctuations on the wetland vegetation of a Nymphaeid-dominated oxbow lake in the Netherlands. Arch Hydrobiol 27:57–73
Burke A (2008) The effect of topsoil treatment on the recovery of rocky plain and outcrop plant communities in Namibia. J Arid Environ 72:1531–1536
Chapin FS III (1991) Integrated responses of plants to stress. Bioscience 41:29–36
Choowong M, Phantuwongraj S, Charoentitirat T et al (2009) Beach recovery after 2004 Indian Ocean tsunami from Phang-nga, Thailand. Geomorphology 104:134–142
Connell JH, Slatyer RO (1977) Mechanisms of succession in natural communities and their role in community stability and organization. Am Nat 111:1119–1144
del Moral R, Bliss LC (1993) Mechanisms of primary succession: insights resulting from the eruption of Mount St. Helens. Adv Ecol Res 24:1–66
Eriksson O (1996) Regional dynamics of plants: a review of evidence for remnant, source-sink and metapopulations. Oikos 77:248–258
Garnier E, Cortez J, Billés G et al (2004) Plant functional markers capture ecosystem properties during secondary succession. Ecology 85:2630–2637
Grime JP (1977) Evidence for the existence of three primary strategies in plants and its relevance to ecological and evolutionary theory. Am Nat 111:1169–1194
Grime JP (1994) The role of plasticity in exploring environmental heterogeneity. In: Caldwell MM, Pearcy RW (eds) Exploitation of environmental heterogeneity by plants: ecophysiological processes above- and below-ground. Academic, New York, pp 1–19
Hayasaka D, Fujiwara K, Box EO (2009) Recovery of sandy beach and maritime forest vegetation on Phuket Island (Thailand) after the major Indian Ocean tsunami of 2004. Appl Veg Sci 12:211–224
Hayasaka D, Akasaka M, Miyauchi D et al (2011) Classification of roadside weeds along two highways in different climatic zones according to ecomorphological traits. Weed Technol 25:411–421
Hayasaka D, Akasaka M, Miyauchi D et al (2012a) Qualitative variation in roadside weed vegetation along an urban–rural road gradient. Flora 207:126–132
Hayasaka D, Goka K, Thawatchai W et al (2012b) Ecological impacts of the 2004 Indian Ocean tsunami on coastal sand-dune species on Phuket Island, Thailand. Biodivers Conserv 21:1971–1985
Hayasaka D, Shimada N, Konno H et al (2012c) Floristic variation of beach vegetation caused by the 2011 Tohoku-oki tsunami in northern Tohoku, Japan. Ecol Eng 44:227–232
Hesp PA (2004) Coastal dunes in the tropics and temperate regions: location, formation, morphology and vegetation processes. In: Martínez ML, Psuty NP (eds) Coastal dunes, ecology and conservation. Springer, Berlin, pp 29–49
Ishida T, Sawahata T, Kanaya G, Hayasaka D (2015) Population dynamics of two sympatric sandhoppers (*Trinorchestia* species) in the Pacific coast of northern Tohoku after the 2011 Tohoku-oki tsunami. Crustaceana 88:511–521
Ishizuka K (1962) Ecological studies on the vegetation of coastal sand bars. II Succession in vegetation and developmental processes of dunes. Ann Gakugei Fac Iwate Univ 20:139–171
Kira T (1977) A climatological interpretation of Japanese vegetation zones. In: Miyawaki A, Tüxen R (eds) Vegetation science and environmental protection. Maruzen Co. Ltd., Tokyo, pp 21–30
Lambers H, Poorter H (1992) Inherent variation in growth rate between higher plants: a search for physiological causes and ecological consequences. Adv Ecol Res 23:187–261
Lamberti GA, Gregory SV, Ashkenas LR et al (1991) Stream ecosystem recovery following a catastrophic debris flow. Can J Fish Aquat Sci 48:196–208

Lavorel S, Garnier E (2002) Predicting changes in community composition and ecosystem functioning from plant traits: revisiting the Holy Grail. Funct Ecol 16:545–556
Mimura N, Yasuhara K, Kawagoe S et al (2011) Damage from the Great East Japan earthquake and tsunami – a quick report. Mitig Adapt Strateg Glob Chang 16:803–818
Miyawaki A, Okuda S (1990) Vegetation of Japan illustrated. Shibundo Co., Ltd., Tokyo (in Japanese)
Mukai Y, Suzuki T, Makino W, Iwabuchi T, So M, Urabe J (2014) Ecological impacts of the 2011 Tohoku Earthquake Tsunami on aquatic animals in rice paddies. Limnology 15:201–211
Nakanishi H (1988) Dispersal ecology of the maritime plants in the Ryukyu islands, Japan. Ecol Res 3:163–173
Numata M, Asano S (1969) Biological flora of Japan, 2nd edn. Tsukiji Shokan, Tokyo (in Japanese)
Pickett STA, White PS (1985) The Ecology of natural disturbance and patch dynamics. Academic, New York
Platt WJ, Connell JH (2003) Natural disturbances and directional replacement of species. Ecol Monogr 73:507–522
Raunkiaer C (1934) Life forms of plants and plant geography. Oxford University Press, London
Rautiainen P, Laine AL, Aikio S et al (2004) Seashore disturbance and management of the clonal Arctophila fulva: modeling patch dynamics. Appl Veg Sci 7:221–228
R-Development-Core-Team (2014) R: a language and environment for statistical computing. R Foundation for Statistical Computing, Vienna (http://cran.r-project.org/)
Ridley HN (1930) The dispersal of plants throughout the world. L. Reeve & Co., Ltd., Ashford
Scheucher LEA, Vortisch W (2011) Sedimentological and geomorphological effects of the Sumatra-Andaman Tsunami in the area of Khao Lak, southern Thailand. Environ Earth Sci 63:785–796
Seike K, Shirai K, Kogure Y (2013) Disturbance of shallow marine soft-bottom environments and megabenthos assemblages by a huge tsunami induced by the 2011 M 9.0 Tohoku-oki Earthquake. PLoS One 8:e65417
Shimada N, Kawanishi M, Hayasaka D (2014) Ecological evaluation of beach vegetation recovery after the Tohoku Region Pacific Coast Earthquake, and a proposal for environmental conservation measures in Iwate Prefecture. J Policy Stud 16:19–34 (in Japanese with English summary)
Sousa WP (1985) Disturbance and patch dynamics on rocky intertidal shores. In: Pickett STA, White PS (eds) The ecology of natural disturbance and patch dynamics. Academic, New York, pp 101–124
Symstad AJ (2000) A test of the effects of functional group richness and composition on grassland invasibility. Ecology 81:99–109
Tanaka G, Naruse H, Yamashita S et al (2012) Ostracodes reveal the sea-bed origin of tsunami deposits. Geophys Res Lett 39:L05406
Tanaka N, Yagisawa J, Yasuda S (2013) Breaking pattern and critical breaking condition of Japanese pine trees on coastal sand dunes in huge tsunami caused by Great East Japan Earthquake. Nat Hazards 65:423–442
Tilman D (1986) Resources, competition and the dynamics of plant communities. In: Crawley MJ (ed) Plant ecology. Black-well Scientific, Oxford, pp 51–75
Tsuyuzaki S (1989) Analysis of revegetation dynamics on the volcano Usu, northern Japan, deforested by 1977–78 eruptions. Am J Bot 76:1468–1477
Tsuyuzaki S, der Moral R (1995) Species attributes in the early stages of volcanic succession. J Veg Sci 6:517–522
Urabe J, Suzuki T, Nishita T, Makino W (2013) Immediate ecological impacts of the 2011 Earthquake Tsunami on intertidal flat communities. PLoS One 8:e62779
Weiher E, van der Werf A, Thompson K et al (1999) Challenging theophrastus: a common core list of plant traits for functional ecology. J Veg Sci 10:609–620
White PS (1979) Pattern, process and natural disturbance in vegetation. Bot Rev 45:229–299

Yamada K, Hori M, Tanaka Y et al (2007) Temporal and spatial macrofaunal community changes along a salinity gradient in seagrass meadows of Akkeshi-ko estuary and Akkeshi Bay, northern Japan. Hydrobiologia 592:345–358

Yamada K, Hori M, Tanaka Y et al (2010) Contribution of different functional groups to the diet of major predatory fishes at a seagrass meadow in northeastern Japan. Estuar Coast Shelf Sci 86:71–82

Yamada K, Tanaka Y, Era T et al (2014) Environmental and spatial controls of macroinvertebrate functional assemblages in seagrass ecosystems along the Pacific coast of northern Japan. Glob Ecol Conserv 2:47–61

Chapter 18
The Species Composition of Buried Seeds of Seashore Vegetation Disturbed by the Great East Japan Earthquake and Tsunami in Northern Tohoku, Japan

Motohiro Kawanishi, Daisuke Hayasaka, and Naoaki Shimada

Abstract The 2011 Tohoku-oki tsunami damaged most sandy coastlines in Tohoku, Japan, and sandy beach vegetation was extensively disturbed. Buried seeds contribute greatly to vegetation recovery and ecological succession in general. Knowing the impact of tsunamis on seed banks and drift seeds is important in estimating the recovery of destroyed vegetation, but such information is lacking. Therefore, we examined buried seed and drift seed populations to elucidate the potential for the recovery of species diversity on sandy beaches. We collected samples of sandy sediment and debris in Iwate and Aomori prefectures, northern Tohoku, Japan. We estimated the seed populations in these samples using both the germination and seed-floating methods. In December 2011, the species composition of buried seeds in sandy sediment was very simple; ruderal plants comprising *Chenopodium*, Compositae, and Gramineae were remarkable. Although we found no coastal plant seeds in December 2011 sediment, eight coastal plants, *Atriplex subcordata*, *Salsola komarovii*, *Setaria viridis* var. *pachystachys*, *Boehmeria splitgerbera*, *Leymus mollis*, *Linaria japonica*, *Carex kobomugi*, and *Glehnia littoralis*, emerged in November 2012. Concerning alien plants, four species were observed in 2011—*Bidens frondosa*, *Erigeron canadensis*, *Erigeron annuus*, and *Atriplex prostrata*—and *Plantago lanceolata* was seen in 2012. The coastal species *S. komarovii* and the alien species *Cakile edentula* occurred in a debris sample in 2011. In the following year, we iden-

M. Kawanishi (✉)
Faculty of Education, Kagoshima University,
Korimoto 1-20-6, Kagoshima, Kagoshima 890-0065, Japan
e-mail: kawanishi@edu.kagoshima-u.ac.jp

D. Hayasaka
Faculty of Agriculture, KINDAI University, 3327-204 Nakamachi, Nara 631-8505, Japan

N. Shimada
Faculty of Policy Studies, Iwate Prefectural University,
152-52, Sugo, Takizawa, Iwate 020-0693, Japan

J. Urabe, T. Nakashizuka (eds.), *Ecological Impacts of Tsunamis on Coastal Ecosystems*, Ecological Research Monographs,
DOI 10.1007/978-4-431-56448-5_18

tified many types of coastal plants, alien plants, and other species. The seed bank that existed before the tsunami appears to have been mostly lost, and its contribution to vegetation recovery in 2011 would therefore have been small. The species composition and diversity of the seed bank after the tsunami will depend on the introduction of dispersed seeds directly from living vegetation and/or via the ocean.

Keywords Alien plants • Buried seeds • Coastal plants • Drift seeds • Sandy coast • Seed bank • Resilience of vegetation

18.1 Introduction

Most sandy coastlines in northern Tohoku, Japan, were deleteriously affected by the tsunami caused by the Great East Japan Earthquake, with coastal vegetation suffering extensive damage. Some reports have clarified the impact of the tsunami on coastal vegetation (e.g., Hara et al. 2012; Ooue 2012; Hara 2014; Shimada 2014) and described the floristic changes in beach vegetation since the tsunami (Hayasaka et al. 2012). Ecological evaluations have been reported and conservation measures presented (Shibuya et al. 2014; Shimada 2014). Determining the resilience mechanisms of vegetation after disturbance by a tsunami is important and will contribute to the conservation of coastal ecosystems.

Buried seeds play an important role in vegetation restoration (Tsuyuzaki 1989; Bakker et al. 1996; Baskin and Baskin 2014) and ecological succession in general (Oosting and Humphreys 1940; Ishizuka 1962; Numata et al. 1964). The existing research on seed banks will be useful for estimating the resilience mechanisms of destroyed vegetation. However, most studies on seed banks have been conducted in grasslands and forests, and few have examined sandy coastal habitats. Furthermore, information regarding the impact of tsunamis on the seed banks of sandy beach vegetation is lacking due to the infrequency of these events.

Various seeds dispersed by seawater drift ashore (Bakker et al. 1996; Nakanishi 1988, 2013; Tokura et al. 1988). In addition, some studies have reported that seedlings of coastal plants, such as *Carex kobomugi*, become established in drift lines where much debris is deposited (Ishizuka 1962; Oka et al. 2009b). Ishizuka (1962) suggested that the establishment of coastal plants from seeds often starts in drift lines because the sediment includes many drifted seeds and provides a good germination bed. Although the relationship between drift seeds and seedlings in drift lines was unclear in these studies, drifted seeds may have the potential to restore damaged populations. Some species may be on the verge of local extinction due to the catastrophic disturbance resulting from a tsunami. In such cases, the introduction of seeds from the sea may contribute to the recovery of populations and increase species diversity.

Therefore, we examined buried seed populations in sandy sediment and in debris using germination tests to clarify the potential capacity for the recovery of species diversity on sandy beaches.

18.2 Study Sites

We sampled beach sediments on six sandy beaches—Yoshihama, Nebama, Akedo, Tofugaura, Kuji in Iwate Prefecture, and Oosuga in Aomori Prefecture, northern Tohoku—in December 2011 and November 2012 (Fig. 18.1). In Nebama (including the Murohama area) and Oosuga, sampling was conducted only in 2012.

All study sites were severely damaged by the tsunami on 11 March 2011 (Table 18.1). Wave heights reached over 10 m at every site. In Tofugaura and Akedo, waves over 20 m were observed. Seawalls and other structures were completely destroyed at Yoshihama. At Nebama, Akedo, and Tofugaura, seawalls were severely damaged as a result of the tsunami, with fracturing and collapse of protective structures. No artificial structures existed beyond the sandy beaches at Kuji (sandspit) and Oosuga (natural coast). At Yoshihama and Nebama, all areas of the sandy beach became bare; all vegetation vanished due to the tsunami and ground subsidence. At the other sites, coastal vegetation partially or mostly survived. The dominant species were *Leymus mollis*, *C. kobomugi*, and *Rosa rugosa*, among others. Coastal forests dominated by *Pinus* were not included in this investigation.

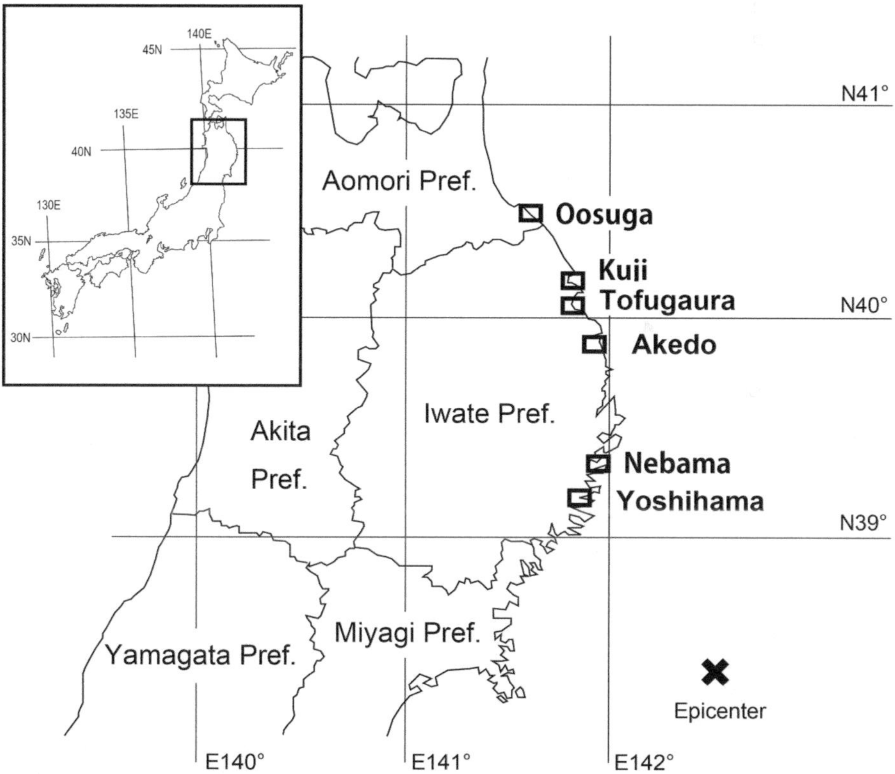

Fig. 18.1 Locations of the study sites. The *cross* indicates the epicenter of the Great East Japan Earthquake

Table 18.1 Impact of the tsunami caused by the Great East Japan Earthquake on six study beaches

	Yoshihama	Nebama	Akedo	Tofugaura	Kuji	Osuga
Wave height (height from msl [m])[a]	17.8	15.0	20.8	22.8	14.1	10.3
Structures for beach protection	Seawall	Seawall	Seawall	Seawall wave-dissipating blocks	Sandspit	Natural coast
Damage of the structures	Vanished	Crack	Collapse	Fracturing	–	–
Vegetation	Vanished	Vanished	Partially remained	Partially remained	Partially remained	Remained

[a]The 2011 Tohoku Earthquake Tsunami Joint Survey (TTJS) Group, http://www.coastal.jp/ttjt/index.php (2011/06/01)

18.3 Methods

18.3.1 Sampling and Analysis

We performed a detailed analysis of buried seed communities in December 2011 with the view that seeds present at that time might strongly affect subsequent recovery of vegetation. We collected sediment samples from transects on each beach. To obtain buried seeds from various coastal habitats, we collected samples from several types of plant communities, as well as from bare ground.

To discriminate tsunami deposits from the collection samples, we tried digging a ~50–70-cm-deep pit, but we were unable to identify whether each sample corresponded to tsunami deposits. Thus, sand samples were collected from depths of 0–5 cm (surface) and 30–35 cm (underground) at each sampling point. These sampling units were 20×20 cm$^2 \times 5$ cm thick in each sampling layer; i.e., one sample was 2 liters (L). These 2-L samples were divided into two 1-L samples, one of which was used in the germination test and the other in the environmental analysis. In the 2012 survey, we obtained sediment only from 0 to 5 cm depth because few seeds were buried in the deeper sediment (30–35 cm depth), as described below. We collected samples from five to eight points per beach at Yoshihama, Akedo, Tofugaura, and Kuji on 3–4 December 2011. In November 2012, we collected samples at the same points as in the 2011 investigation, plus some additional points. Furthermore, we collected some samples at Nebama and Oosuga. In total, we obtained 8–12 samples on six coastlines: Yoshihama, Nebama, Akedo, Tofugaura, Kuji, and Oosuga.

Samples of debris were also collected from drift lines (linear structures formed by the deposition of drifted materials) along sand dunes (Fig. 18.2). We collected two samples per beach at Akedo and Kuji on 3–4 December 2011 and in November 2012; then we collected three to eight samples from each coastline (Yoshihama,

Fig. 18.2 Debris along a beach

Nebama, Akedo, Tofugaura, Kuji, and Oosuga) on 3–4 December 2011. The sampling units were again 20 cm^2. All debris within the square (except anthropogenic garbage and large algae) was collected. One liter (1 L) of the samples was used for the germination test after removing large organic materials.

SPSS Statistics (IBM) was used for statistical analysis.

18.3.2 Germination Method

Samples were placed on seed-free vermiculite in containers and set under lights in a greenhouse at the Korimoto Experimental Field of the Nature Education and Research Department, Faculty of Education, Kagoshima University, Kagoshima, Japan. These samples in containers were covered by nonwoven cloth (white in color, translucency 90 %, mesh size about 0.2–0.5 mm) to prevent outside seeds from entering the sample. Samples were watered twice daily (early morning and evening) using a water sprinkler. The test was conducted for about 1 year, from 26 December 2011 to 2 December 2012 for the samples collected in 2011 and 30 December 2012 to 29 November 2013 for those collected in 2012. If we could not identify a species from the morphological characteristics of the stem and leaves (e.g., Gramineae, Cyperaceae), the seedlings were transplanted and cultivated until they flowered.

18.3.3 Floating Seed Method

To detect seeds that did not emerge in the germination test, the sample from the container used in the germination test was immersed in a solution of 50 % K_2CO_3 (Tsuyuzaki and Goto 2001). This test was performed only for the 2011 samples and followed the germination test, i.e., from December 2012 to June 2013.

The samples were immersed in 50 % K_2CO_3 solution and agitated; organic deposits that floated to the surface were then scooped off and washed with water. This operation was repeated about five times, until no further deposits floated to the surface. Seeds were identified based on their morphological characteristics and size, according to Nakayama et al. (2000), Asano (2005), Shimizu et al. (2001), Suzuki et al. (2012), and Uemura et al. (2010). When species determination was difficult, identification was completed to the genus or family level. After identification, we estimated seed viability using the crushing technique (Naka and Yoda 1984) and TTC (2,3,5-triphenyltetrazolium chloride) methods (Weed Science Society of Japan 2001). If the embryo and albumen were empty or brown, the seeds were considered dead (Naka and Yoda 1984); if white, viability was validated using the 0.5 % TTC solution, and if pink the seed was considered alive.

18.3.4 Ecological Character of Identified Plants

We categorized species identified in the germination and flotation tests into three types: coastal plants, alien plants, and other species. The list of coastal plants in Japan was derived from Sawada et al. (2007) and that of alien species was based on YList (Yonekura and Kajita 2003).

18.4 Results and Discussion

18.4.1 Species Composition of the Seed Bank in December 2011

18.4.1.1 Buried Seeds in the Sandy Surface Sediment (0–5 cm Depth)

We identified 18 species in the germination test for all study sites, not counting unknown species: six species at Yoshihama and Akedo, four at Tofugaura, and nine at Kuji. Seed density was highest at Akedo, at $6.4 \pm 11.8\ L^{-1}$ (Table 18.2). The densities at other sites were about 3 L^{-1} (Yoshihama and Kuji) or 1 L^{-1} (Tofugaura). The overall seed density in these samples was low. *Chenopodium album* emerged at every study site. Other *Chenopodium* species (*C. glaucum* and *C. ambrosioides*) were also observed. In addition, Compositae (e.g., *Artemisia indica* var.

Table 18.2 Species composition of buried seeds using the germination method. The samples comprise 1 liter (L) of sandy sediment from beaches and debris collected in 2011 and 2012. Mean individual density (no. of L-1 of soil) is shown. The standard deviations are omitted

	Sandy sediment														Debris							
	Dec. 2011																					
	0–5cm				30–35cm				Nov. 2012						Dec. 2011		Nov. 2012					
	Y	A	T	K	Y	A	T	K	Y	N	A	T	K	O	A	K	Y	N	A	T	K	O
Number of samples	5	8	5	6	3	8	5	4	10	11	9	8	11	12	2	2	4	8	4	4	3	4
Total number of species	6	6	4	9	3	1	0	1	48	19	25	4	21	4	9	3	2	13	0	2	11	19
Coastal plants																						
Atriplex subcordata	–	–	–	–	–	–	–	–	0.5	0.1	–	–	0.1	–	–	–	0.3	–	–	–	–	–
Setaria viridis var. *pachystachys*	–	–	–	–	–	–	–	–	0.1	–	0.1	–	3.1	–	–	–	–	–	–	–	–	–
Boehmeria biloba	–	–	–	–	–	–	–	–	0.1	–	–	–	–	–	–	–	–	–	–	–	–	–
Carex kobomugi	–	–	–	–	–	–	–	–	–	0.4	–	–	0.1	–	–	–	–	–	–	–	–	0.3
Salsola komarovii	–	–	–	–	–	–	–	–	–	–	0.1	–	–	–	6.5	2.0	–	–	–	–	0.3	–
Glehnia littoralis	–	–	–	–	–	–	–	–	–	–	0.1	–	–	–	–	–	–	–	–	–	–	1.0
Linaria japonica	–	–	–	–	–	–	–	–	–	–	–	–	0.1	0.3	–	–	–	–	–	–	–	–
Leymus mollis	–	–	–	–	–	–	–	–	–	–	–	–	0.2	–	–	–	–	–	–	–	1.0	1.0
Atriplex patens	–	–	–	–	–	–	–	–	–	–	–	–	–	–	–	–	–	0.3	–	–	–	0.3
Artemisia stelleriana	–	–	–	–	–	–	–	–	–	–	–	–	–	–	–	–	–	–	–	–	1.0	0.3
Ischaemum anthephoroides	–	–	–	–	–	–	–	–	–	–	–	–	–	–	–	–	–	–	–	–	–	0.3
Ixeris repens	–	–	–	–	–	–	–	–	–	–	–	–	–	–	–	–	–	–	–	–	–	0.3
Alien species																						
Chenopodium album	1.4	0.4	0.4	0.3	–	–	–	–	1.0	0.3	3.2	0.3	16.5	–	–	–	–	–	–	–	0.3	–

(continued)

Table 18.2 (continued)

	Sandy sediment														Debris							
	Dec. 2011																					
	0–5cm				30–35cm				Nov. 2012						Dec. 2011		Nov. 2012					
	Y	A	T	K	Y	A	T	K	Y	N	A	T	K	O	A	K	Y	N	A	T	K	O
Number of samples	5	8	5	6	3	8	5	4	10	11	9	8	11	12	2	2	4	8	4	4	3	4
Total number of species	6	6	4	9	3	1	0	1	48	19	25	4	21	4	9	3	2	13	0	2	11	19
Chenopodium ambrosioides	0.2	–	–	–	–	–	–	–	16.0	–	–	–	–	–	–	–	–	–	–	–	–	–
Oenothera spp.	–	0.1	–	0.3	–	–	–	–	0.4	0.6	2.4	–	1.4	1.5	–	–	–	0.4	–	–	0.3	0.5
Chenopodium glaucum	–	–	–	1.0	–	–	–	–	23.4	0.1	0.2	–	0.5	0.1	–	–	5.5	–	–	–	–	–
Bidens frondosa	–	–	–	–	–	–	–	–	13.9	0.1	0.2	–	1.0	–	–	–	–	1.6	–	–	0.7	0.5
Erigeron canadensis	–	–	–	–	–	–	–	–	0.5	–	–	0.1	0.2	–	–	–	–	0.3	–	–	0.7	–
Erigeron annuus	–	–	–	–	–	–	–	–	0.4	–	0.4	–	–	–	–	–	–	–	–	–	–	0.3
Plantago lanceolata	–	–	–	–	–	–	–	–	1.2	–	0.1	–	–	–	–	–	–	–	–	–	–	–
Atriplex prostrata	–	–	–	–	–	–	–	–	–	0.1	0.2	–	–	–	–	–	–	7.5	–	–	–	–
Asparagus sp.	–	–	–	–	–	–	–	–	–	0.1	–	–	–	–	–	–	–	–	–	–	–	–
Trifolium repens	–	–	–	–	–	–	–	–	–	–	0.1	–	–	–	–	–	–	–	–	–	–	–
Cakile edentula	–	–	–	–	–	–	–	–	–	–	–	–	–	–	0.5	–	–	–	–	–	–	0.3
Momordica charantia	–	–	–	–	–	–	–	–	–	–	–	–	–	–	–	–	–	0.1	–	–	–	–
Senecio vulgaris	–	–	–	–	–	–	–	–	–	–	–	–	–	–	–	–	–	–	–	0.3	–	–
Zea mays	–	–	–	–	–	–	–	–	–	–	–	–	–	–	–	–	–	–	–	–	0.3	–

Other species																						
Artemisia indica var. maximowiczii	0.6	0.9	–	0.3	–	0.1	–	–	0.2	0.1	4.0	0.1	3.6	–	–	–	b–	–	–	–	–	1.8
Digitaria ciliaris	0.6	3.0	–	–	0.3	–	–	–	1.6	2.0	1.7	–	1.5	–	–	–	–	3.5	–	0.3	–	–
Panicum dichotomiflorum	0.2	–	–	–	–	–	–	–	0.4	–	–	–	–	–	–	–	–	–	–	–	–	–
Digitaria violascens	–	1.5	–	–	–	–	–	–	0.1	0.3	1.6	–	0.2	–	–	–	–	–	–	–	–	–
Zoysia japonica	–	0.5	–	–	–	–	–	–	–	–	–	–	0.1	–	–	–	–	–	–	–	–	–
Sonchus brachyotus	–	–	0.2	–	–	–	–	–	0.1	–	0.1	–	–	0.2	–	–	–	0.3	–	–	–	–
Sagina maxima	–	–	0.2	–	–	–	–	–	–	–	–	–	–	–	–	–	–	–	–	–	–	0.3
Gramineae sp. 1	–	–	–	0.3	–	–	–	–	0.3	0.1	–	–	0.4	–	0.5	–	–	–	–	–	0.7	–
Agrostis clavata var. *nukabo*	–	–	–	0.2	–	–	–	–	0.1	–	0.2	–	–	–	–	–	–	–	–	–	–	–
Youngia japonica	–	–	–	0.2	–	–	–	–	–	–	–	–	–	–	–	0.5	–	–	–	–	–	–
Gramineae sp. 2	–	–	–	0.2	–	–	–	–	–	–	–	–	–	–	–	0.5	–	–	–	–	–	–
Eragrostis multicaulis	–	–	–	–	0.3	–	–	–	–	–	–	–	–	–	–	–	–	–	–	–	–	–
Poa annua	–	–	–	–	0.3	–	–	–	–	–	–	–	–	–	–	–	–	–	–	–	–	–
Eclipta thermalis	–	–	–	–	–	–	–	–	14.2	–	0.1	–	–	–	–	–	–	–	–	–	–	–
Rorippa palustris	–	–	–	–	–	–	–	–	10.8	0.4	–	–	1.1	–	–	–	–	–	–	–	–	–
Echinochloa crus-galli var. *crus-galli*	–	–	–	–	–	–	–	–	0.4	–	–	–	0.1	–	–	–	–	–	–	–	–	–
Setaria pumila	–	–	–	–	–	–	–	–	0.1	0.1	4.7	–	–	–	–	–	–	–	–	–	–	–
Carex sp.	–	–	–	–	–	–	–	–	0.2	0.2	–	–	–	–	–	–	–	0.1	–	–	–	–
Galium gracilens	–	–	–	–	–	–	–	–	0.2	–	–	–	0.1	–	–	–	–	–	–	–	–	–

(continued)

Table 18.2 (continued)

	Sandy sediment														Debris							
	Dec. 2011																					
	0–5cm				30–35cm				Nov. 2012						Dec. 2011		Nov. 2012					
	Y	A	T	K	Y	A	T	K	Y	N	A	T	K	O	A	K	Y	N	A	T	K	O
Number of samples	5	8	5	6	3	8	5	4	10	11	9	8	11	12	2	2	4	8	4	4	3	4
Total number of species	6	6	4	9	3	1	0	1	48	19	25	4	21	4	9	3	2	13	0	2	11	19
Achyranthes bidentata var. *japonica*	–	–	–	–	–	–	–	–	0.1	–	–	–	–	–	–	–	–	–	–	–	–	–
Boehmeria japonica var. *longispica*	–	–	–	–	–	–	–	–	0.1	–	–	–	–	–	–	–	–	–	–	–	–	–
Centipeda minima	–	–	–	–	–	–	–	–	0.9	–	–	–	–	–	–	–	–	–	–	–	–	–
Cyperus brevifolius var. *leiolepis*	–	–	–	–	–	–	–	–	4.0	–	–	–	–	–	–	–	–	–	–	–	–	–
Cyperus difformis	–	–	–	–	–	–	–	–	0.2	–	–	–	–	–	–	–	–	–	–	–	–	–
Cyperus microiria	–	–	–	–	–	–	–	–	0.1	–	–	–	–	–	–	–	–	–	–	–	–	–
Penthorum chinense	–	–	–	–	–	–	–	–	0.4	–	–	–	–	–	–	–	–	–	–	–	–	–
Persicaria longiseta	–	–	–	–	–	–	–	–	0.5	–	–	–	–	–	–	–	–	–	–	–	–	–
Phyllanthus lepidocarpus	–	–	–	–	–	–	–	–	0.1	–	–	–	–	–	–	–	–	–	–	–	–	–
Plantago asiatica	–	–	–	–	–	–	–	–	0.3	–	–	–	–	–	–	–	–	–	–	–	–	–
Poa annua	–	–	–	–	–	–	–	–	0.1	–	–	–	–	–	–	–	–	–	–	–	–	–
Ranunculus silerifolius	–	–	–	–	–	–	–	–	0.2	–	–	–	–	–	–	–	–	–	–	–	–	–
Rumex spp.	–	–	–	–	–	–	–	–	0.2	–	–	–	–	–	0.5	–	–	0.5	–	–	–	2.0
Sagina japonica	–	–	–	–	–	–	–	–	0.2	–	–	–	–	–	–	–	–	–	–	–	–	–
Sedum bulbiferum	–	–	–	–	–	–	–	–	0.1	–	–	–	–	–	–	–	–	–	–	–	–	–
Setaria viridis	–	–	–	–	–	–	–	–	0.1	–	–	–	–	–	–	–	–	–	–	–	–	–
Portulaca oleracea	–	–	–	–	–	–	–	–	–	1.7	1.6	–	–	–	–	–	–	–	–	–	–	–

Acalypha australis	–	–	–	–	–	–	–	–	–	–	1.6	–	–	–	–	–	–	–	–	–	–	–
Arthraxon hispidus	–	–	–	–	–	–	–	–	–	–	0.2	–	–	–	–	–	–	–	–	–	–	–
Elsholtzia ciliata	–	–	–	–	–	–	–	–	–	–	0.2	–	–	–	–	–	–	–	–	–	–	–
Picris hieracioides ssp. *japonica*	–	–	–	–	–	–	–	–	–	–	0.1	–	–	–	–	–	–	0.1	–	–	–	–
Sonchus asper	–	–	–	–	–	–	–	–	–	–	–	–	0.1	–	–	–	–	–	–	–	–	0.3
Aeschynomene indica	–	–	–	–	–	–	–	–	–	–	–	–	–	–	0.5	–	–	–	–	–	–	–
Zelkova serrata	–	–	–	–	–	–	–	–	–	–	–	–	–	–	0.5	–	–	–	–	–	–	–
Pterocarya rhoifolia	–	–	–	–	–	–	–	–	–	–	–	–	–	–	0.5	–	–	–	–	–	–	–
Phalaris arundinacea	–	–	–	–	–	–	–	–	–	–	–	–	–	–	–	–	–	–	–	–	0.3	–
Allium grayi	–	–	–	–	–	–	–	–	–	–	–	–	–	–	–	–	–	–	–	–	–	0.5

* Unknown species were omitted

Abbreviations for the study sites: *Y* Yoshihama, *N* Nebama, *A* Akedo, *T* Tofugaura, *K* Kuji, and *O* Oosuga

maximowiczii, *Youngia japonica*), Gramineae (e.g., *Digitaria ciliaris*, *Digitaria violascens*, *Panicum dichotomiflorum*, *Zoysia japonica*), and *Oenothera* (perhaps including *O. biennis*, *O. parviflora*, *O. glazioviana*) were remarkable.

The floating seed method detected live seeds of *Chenopodium* spp. and some freshly dead seeds of *Chenopodium* spp., *Cyperus* sp., Compositae sp., *Eleusine indica*, and *Salsola komarovii*. Some of these freshly dead seeds may have died during the germination or floating tests, as they still contained white, apparently living tissue. Although the cause of death of these seeds was unclear, they might have survived if they had been preserved in good condition on the beach. Because the tests in our study took a long time, seed survivorship might have been underestimated. Some studies have reported that the seed burial depth affects seedling emergence in some annual plants (Yamanaka et al. 2000; Kondo et al. 2002; Tobe et al. 2007). Although we observed some species that comprise the vegetation of sandy coasts (e.g., *Sagina maxima* and *Sonchus brachyotus*), coastal species (Sawada et al. 2007) were lacking. Also, we observed many species as dead seeds, e.g., members of *Rosa* (perhaps the dominant species in living vegetation, *R. rugosa*), common species of the Polygonaceae (*Persicaria*, *Rumex*), and deciduous trees distributed on mountain slopes—*Acer*, *Actinidia*, *Zelkova*, and *Tilia*, among others.

As a general trend at these study sites, alien species and other species consisting of ruderal plants belonging to the Chenopodiaceae, Compositae, and Gramineae dominated the buried seed community. Coastal plants were absent from the seed bank in the surface layer of sandy sediment.

In this study, we could not determine whether the sandy sediment on the ground surface was deposited by the tsunami. Szczuciński et al. (2012a) reported the existence of tsunami deposits (a few cm thick) on a beach on the Sendai Plain. These tsunami deposits comprised mainly of medium and coarse sand. This tendency was also reported following the 2004 Indian Ocean tsunami (Szczuciński et al. 2012b). In general, medium and coarse sand were found along the coastlines examined. In addition, coastal sand is frequently moved by wind, and thus distinguishing tsunami deposits is difficult.

18.4.1.2 Buried Seeds in Underground Sediment (30–35 cm Depth)

We found very few seeds in samples collected in sediment from 30 to 35 cm depth (Table 18.2). The total seed density at each site was below 1.0 L^{-1} (Table 18.2), significantly lower than on the surface (Mann–Whitney *U* test, $p<0.01$). We found four species (*A. indica* var. *maximowiczii*, *D. ciliaris*, *Eragrostis multicaulis*, and *Poa annua*), but the number of species in each sample was also smaller than on the surface ($p<0.01$).

Floating revealed a few seeds in samples from 30 to 35 cm (Table 18.3). Freshly dead *Chenopodium* spp. and *Cyperus* spp. seeds were confirmed, but the other species' seeds were dead. We found no living seeds.

Table 18.3 Mean numbers of seeds using the floating seed method in 1-liter (L) sand samples of beach sediment in December 2011. See Table 18.2 for abbreviations of the study sites

			0–5 cm				30–35 cm			
			Y	A	T	K	Y	A	T	K
Total			2.4±4.3	4.6±5.9	1.4±2.1	3.0±9.9	0.7±1.2	2.1±4.9	0.4±0.9	0.8±0.5
Live			0	0.3±0.7	0	0	0	0	0	0
Fresh			0.6±0.9	1.3±3.5	0.2±0.4	0.3±1.4	0	0.8±2.1	0	0.3±0.4
Dry			1.8±3.5	3.1±4.7	1.2±1.6	2.7±11.3	0.7±1.2	1.4±2.9	0.4±0.9	0.5±0.5
	Acer sp.	Dry	–	–	–	–	0.3±0.6	–	–	–
	Actinidia arguta	Dry	–	0.1±0.4	–	–	–	–	–	–
	Actinidia polygama	Dry	–	0.3±0.7	–	–	–	–	–	–
	Aralia elata	Dry	0.1±0.4	0.1±0.4	–	–	–	–	–	–
a	*Carex kobomugi*	Dry	–	0.1±0.4	–	–	–	–	–	–
	Carex sp. 1	Dry	–	–	–	–	–	0.1±0.4	–	–
	Carex sp. 2	Dry	–	0.1±0.4	–	0.2±0.2	–	0.1±0.4	–	–
b	*Chenopodium* sp.	Live	–	0.3±0.7	–	c	–	–	–	–
		Fresh	0.3±0.9	1.0±2.8	–	c	–	0.8±2.1	–	–
		Dry	1.6±3.0	0.5±1.1	–	c	0.3±0.6	1.0±2.8	–	–
b	*Compositae* sp. 1	Fresh	–	–	0.2±0.4	–	–	–	–	–
b	*Compositae* sp. 2	Dry	–	–	–	–	–	–	0.2±0.4	–
	Cyperus sp.	Fresh	0.1±0.4	–	–	–	–	–	–	0.3±0.3
	Digitaria ciliaris	Dry	–	–	–	0.2±0.2	–	–	–	–
	Eleusine indica	Fresh	–	–	–	–	–	–	–	0.3±0.3
b	*Gramineae* sp.	Dry	–	0.1±0.4	–	–	–	–	–	–
b	*Oenothera* sp.	Fresh	–	0.1±0.4	–	–	–	–	–	–
b	*Oryza sativa*	Dry	–	0.1±0.4	–	–	–	–	–	–
	Persicaria longiseta	Dry	–	0.1±0.4	–	–	–	–	–	–

(continued)

Table 18.3 (continued)

			0–5 cm				30–35 cm			
			Y	A	T	K	Y	A	T	K
[b]	*Polygonaceae* sp.	Dry	–	0.1±0.4	–	–	–	–	–	–
	Polygonum sp.	Dry	–	0.1±0.4	–	–	–	–	–	–
[a]	*Rosa* sp.	Dry	–	–	1.0±1.7	2.2±2.2	–	–	0.2±0.4	–
	Rubus sp.	Dry	–	0.3±0.7	–	–	–	–	–	–
[b]	*Rumex* sp.	Dry	–	0.1±0.4	0.2±0.4	0.2±0.2	–	–	–	–
[a]	*Salsola komarovii*	Fresh	–	0.1±0.4	–	–	–	–	–	–
	Schoenoplectiella sp.	Dry	–	0.1±0.4	–	–	–	–	–	–
	Tilia japonica	Dry	–	0.1±0.4	–	–	–	–	–	–
	Umbelliferae sp.	Dry	–	0.1±0.4	–	–	–	–	–	–
	Zelkova serrata	Dry	–	0.1±0.4	–	–	–	–	–	–
	Unknown spp.	Dry	–	0.4±0.7	–	–	–	–	0.1±0.4	0.3±0.3

[a]Coastal plants in Japan (Sawada et al. 2007)
[b]Taxa including alien plants or crops
[c]No data, because new seeds were dispersed from germinated seedling

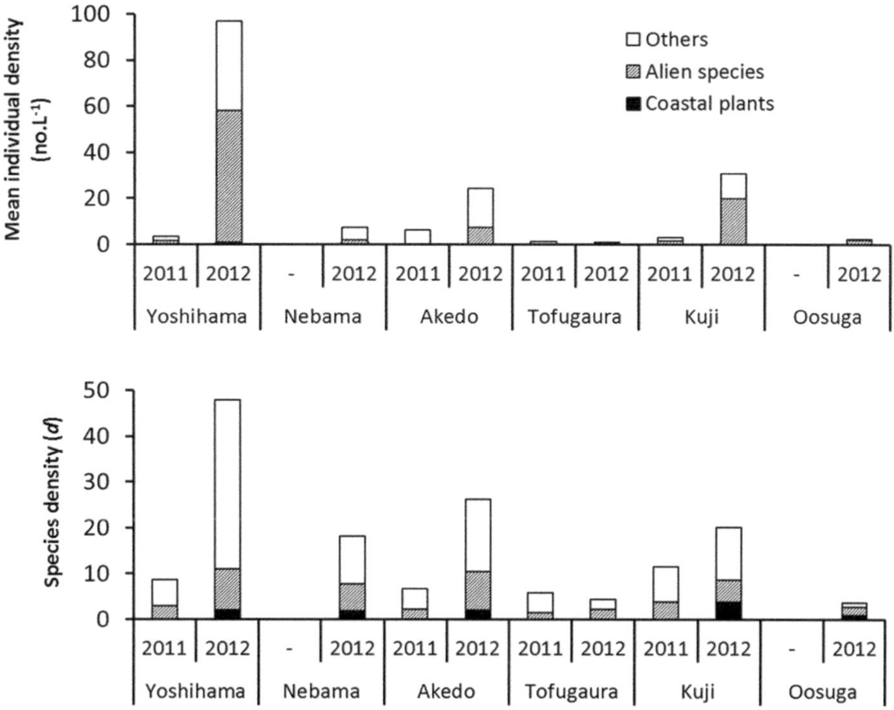

Fig. 18.3 Species density (*d*) and individual density (no. of L^{-1} of soil) of plants, including coastal sand sediment. $d = S/\log_{10}A$, *A* total volume of sample (litter soil), *S* number of species

As a result, the number of living seeds (germinated seeds + living seeds) from the 30 to 35-cm layer was significantly less than on the surface ($p < 0.01$). Although buried seeds of some non-coastal species (e.g., *Chenopodium*, *Artemisia*, *Digitaria*, and *Poa*) were found in the deeper layers, their regeneration would have been unlikely as the seeds were very few in number and almost dead.

18.4.2 Changes in Buried Seed Diversity and Arrival of Drifted Seeds

We examined changes in buried seeds' composition and the arrival of drifted seeds after 1 year, as these factors will greatly affect vegetation recovery. At that time, the seeds buried in the surface sandy sediment and the drifted seeds in the debris as determined by the germination test in 2012 will also be present (Table 18.2, Figs. 18.3 and 18.4).

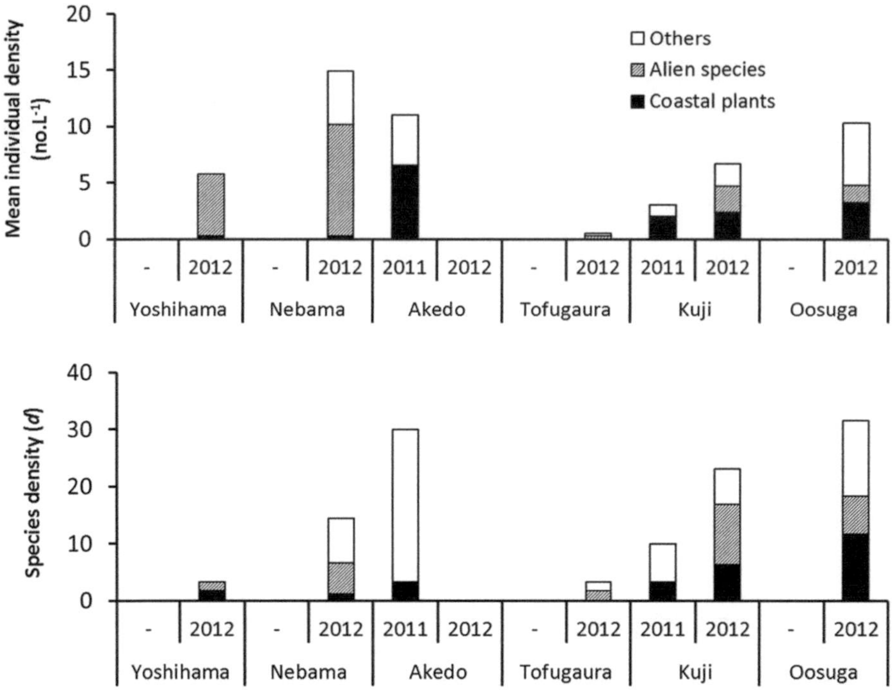

Fig. 18.4 Species density (*d*) and individual density (no. of L^{-1} of soil) of plants from debris along the beach. $d = S/\log_{10}A$

18.4.2.1 Changes in Buried Seed Composition in Sandy Sediments

The mean density of seeds at Yoshihama, Akedo, and Kuji increased greatly in 2012, to 96.6 ± 177.3, 24.3 ± 32.1, and 30.7 ± 75.3 L^{-1}, respectively (Fig. 18.3). Increases in the density of alien species and other species were responsible for this result. We found almost no change at Tofugaura. Species density (*d*; $d = S/\log_{10}A$) had the same trend. Although *d* values were <10 at all sites, the values increased at Yoshihama (48.0), Akedo (26.2), Tofugaura (4.4), and Kuji (20.2). The *d* values at Nebama and Oosuga were 18.2 and 3.7, respectively.

We found no coastal plants' seeds in December 2011 sediment, as noted above. However, eight coastal plants (*Atriplex subcordata*, *S. komarovii*, *Setaria viridis* var. *pachystachys*, *Boehmeria splitgerbera*, *L. mollis*, *Linaria japonica*, *C. kobomugi*, and *Glehnia littoralis*) emerged in November 2012 (Table 18.2). Four alien plant species were observed in 2011, comprising only *Chenopodium* and *Oenothera* species. In 2012, Compositae (*Bidens frondosa*, *Erigeron canadensis*, *Erigeron annuus*), *A. prostrata*, and *Plantago lanceolata* were also observed. Although the "other species" groups in 2011 consisted mainly of weeds from an upland field that belonged to Compositae, Gramineae, and *Oenothera*, many species were added to the buried seed communities in 2012, e.g., paddy field weeds (*Rorippa palustris*,

Echinochloa crus-galli var. *crus-galli*, and *Eclipta thermalis*), *Rumex* spp., and *Setaria pumila*, among others.

The number of species observed in each species group increased, probably because of seeds provided from surrounding vegetation and/or other shore plants.

18.4.2.2 Seeds That Drifted Ashore

In 2011, the mean densities of drifted seeds (mixed in debris) at Akedo and Kuji were 11.0 ± 11.3 and 3.0 ± 2.8 (no. $L-^{1}$), respectively (Fig. 18.4). Coastal plants accounted for more than half of the seeds at both sites. The coastal species *S. komarovii* and the alien species *Cakile edentula* occurred at Akedo (Table 18.2). Additionally, we also observed some ruderal herbs (*Rumex* spp., *Y. japonica*) and riparian trees (e.g., *Zelkova serrata* and *Pterocarya rhoifolia*).

During 2012, we obtained samples at the other four sites; the mean density of drifted seeds was below 15 L^{-1} at every site (Fig. 18.4). We could not confirm the seeds in the Akedo debris samples. Many kinds of coastal plants, alien plants, and species were confirmed at five sites, except Akedo (Table 18.2). Species density, *d*, was higher at Nebama, Kuji, and Oosuga than at the other sites.

Coastal plants observed in debris included *C. kobomugi*, *S. komarovii*, *L. mollis*, *Atriplex patens*, and *Artemisia stelleriana*. Among alien species, various Compositae (e.g., *B. frondosa* and *E. canadensis*) and Chenopodiaceae (e.g., *C. glaucum* and *A. prostrata*) had relatively high densities. Additionally, ruderal plants (e.g., *A. indica* var. *maximowiczii*, *D. ciliaris*, and *Rumex* spp.) and some crop seeds (e.g., *Momordica charantia* and *Zea mays*) were also observed. At Akedo, we found no buried seeds in debris in 2012, despite finding many seeds in the 2011 samples, perhaps because local people had cleaned up the beach in the interim.

18.4.3 Contribution of the Seed Bank to Vegetation Recovery After the Tsunami Disturbance

In this study, we detected buried seeds that had been dispersed before December 2011. However, as it is not possible to know the distribution of buried seeds immediately after and before the tsunami, the origin of seeds buried in sandy sediments has been estimated in some reports.

The existence of seed banks on sandy coastlines has been demonstrated previously (e.g., Pierce and Cowling 1991; Looney and Gibson 1995; Aziz and Khan 1996). In Japan, Fujiki et al. (2001) reported the seed bank species composition at Tottori Sand Dunes. That study showed that the *C. kobomugi* community seed bank consisted of *C. kobomugi*, *Ischaemum anthephoroides*, and *E. indica* (mean seed density: 6330.0 ± 8385.5 $m^{-2} \times 10$ cm in November). Sawada and Tsuda (2005b) clarified the formation potential of a persistent seed bank of *Lathyrus japonicus*, *G.*

littoralis, *Calystegia soldanella*, *C. kobomugi*, *Carex pumila*, *Oenothera laciniata*, *Ixeris repens*, *I. anthephoroides*, and others. *C. kobomugi* had a seed bank and viable seeds concentrated in the surface layer (0–4 cm depth; Yamanaka et al. 2000). The above-cited species were widely distributed at our study sites. Consequently, coastal vegetation may have had seed banks in the surface sand before the tsunami.

According to vegetation survey data collected by Hayasaka et al. (2012) before the tsunami, major coastal plants, such as *L. mollis*, *C. kobomugi*, *C. pumila*, *L. japonicas*, *I. repens*, and *C. soldanella*, were widely distributed at these study sites. *Chenopodium album* (the dominant species among the buried seeds in our study) was not found at Yoshihama, Akedo, Tofugaura, or Kuji before the tsunami. At our study sites, however, *C. album* emerged at every site, and coastal plants were absent from the seed bank in the surface layer of sandy sediment.

In addition, some reports have shown erosion of sandy beaches and dunes in Miyagi Prefecture (Nishi et al. 2012; Richmond et al. 2012). According to aerial photographs and reports by Ooue (2012) and Shimada (2014), coastal *Pinus* forests at Yoshihama, Nebama, and Akedo vanished, probably due to erosion and ground subsidence.

Given these observations, it is likely that the sandy beach vegetation and topsoil at our study sites were severely eroded by the tsunami. Therefore, the seed bank that existed before the tsunami was mostly lost, and the contribution of a seed bank to vegetation recovery in 2011 would have been small. The species composition and diversity of the seed bank after the tsunami will depend on the introduction of dispersed seeds directly from living vegetation and/or through the ocean.

The dispersal route of seeds shows three general patterns: (1) dispersal from coastal vegetation on the same beach, (2) dispersal from inland vegetation in the same region, and (3) dispersal from coastal vegetation on another beach. At Oosuga, Kuji, Tofugaura, and Akedo, all three dispersal routes might be in effect because coastal vegetation remained only partially after the tsunami disturbance. If some perennial plants survive, recovery of vegetation after a tsunami may be relatively rapid. The vegetation used to occupy a broader area in these regions (Kawanishi, unpublished data). Many surviving perennial plants seem to be growing by vegetative reproduction. These plants will also produce seeds, and some dispersed seeds will become dormant and comprise the seed bank. However, the regeneration of plants that have vanished from a beach, especially in the case of bare beaches such as Yoshihama and Nebama, will depend largely on routes (2) and (3). In this case, from the viewpoint of the recovery of coastal plants, route (3) (i.e., drifted seeds) will be very important.

In this study, we observed many drifted seeds of *S. komarovii* (a coastal plant) at Akedo and Kuji in 2011. This species can disperse continuously via the sea and form communities on drift lines (Nakanishi 2013). Additionally, we observed *A. subcordata*, *C. kobomugi*, *L. mollis*, *G. littoralis*, *A. patens*, *A. stelleriana*, *I. anthephoroides*, and *I. repens* in 2012. Sawada and Tsuda (2005a) clarified the potential sea dispersal of *A. subcordata*, *A. patens*, *L. japonicas*, *G. littoralis*, *C. soldanella*, *I. repens*, *C. kobomugi*, and *C. pumila* due to their ability to float on seawater for extended periods. Additionally, *C. pumila*, *C. kobomugi*, *C. soldanella*, and *Tetragonia expansa* can disperse to remote beaches (Oka et al. 2009a). These species (distributed at our study sites) may disperse widely on sandy coastlines. However, some coastal plants cannot float for long, and dispersal to distant locations will be difficult for them (Sawada and Tsuda 2005a).

Therefore, if local vegetation has survived a tsunami, it should be preserved because it will have a very important role as a source of seeds and contribute to the efficient recovery of damaged populations.

Additionally, attention should be given to the behavior of alien species. Ajima (2001) clarified the seed accumulation of alien species in seed banks in many types of plant communities. After extensive disturbance by the tsunami, some alien species, such as the Chenopodiaceae, Compositae, and Cruciferae, formed a community (Hayasaka et al. 2012; Kanno et al. 2014; Oka and Hirabuki 2014). Although alien species' communities have declined over the years since the tsunami (Kanno et al. 2014), we found alien seeds, especially Chenopodiaceae (*C. album*, *C. ambrosioides*, *C. glaucum*, and *A. prostrata*), Compositae (*B. frondosa*, *E. canadensis*, and *E. annuus*), and other taxa (*Oenothera* spp., *P. lanceolata*, and *Trifolium repens*, Table 18.2, Fig. 18.4), buried in sediment and debris. Hence, these alien species have the potential to accumulate in the seed bank. The species composition of buried and drifted seeds will be important to conservation. Our results indicate the importance of the remaining living vegetation and seed inputs from the sea. Considering the recovery of coastal vegetation, we must estimate the ecological succession at each site. To recognize the process and mechanism of succession, we have to continue to monitor, survey, and analyze the relationships between seed banks and vegetation.

Acknowledgments We thank Mr. Mitsuru Ikeda and Mrs. Miyo Ryuuno of the Experimental Field of the Nature Education and Research Department, Faculty of Education, Kagoshima University, for their technical support and helpful suggestions on the germination test. We also thank Ms. Mio Hidaka, Mr. Takuma Maeda, Mr. Teppei Hashimoto, Mr. Ryou Yamakita, and members of the Laboratory of Botany, Faculty of Education, Kagoshima University, for their assistance in the field survey. This study was supported by a research grant from the Mitsui & Co., Ltd. Environment Fund (R11-F1-015-1, Leader: Daisuke Hayasaka).

References

Ajima M (2001) Accumulation of seeds of exotic species in the soil seed bank in Japan. Jpn J Conserv Ecol 6:155–177 (in Japanese)

Asano S (2005) Seeds/fruits and seedling of plants in Japan. Zenkoku Noson Kyoiku Kyokai, Tokyo, 279 pp (in Japanese)

Aziz S, Khan MA (1996) Seed bank dynamics of a semi-arid coastal shrub community in Pakistan. J Arid Environ 34:81–87

Bakker JP, Poschlod P, Strykstra RJ, Bakker RM, Thompson K (1996) Seed banks and seed dispersal: important topics in restoration ecology. Acta Bot Neerlandica 45(4):461–490

Baskin C, Baskin J (2014) Seeds: ecology, biogeography, and evolution of dormancy and germination, 2nd edn. Academic, San Diego, 1586 pp

Fujiki D, Yamanaka N, Tamai S (2001) Relationship between vegetation types and seed banks on Tottori Sand Dune. J Jpn Soc Revegetation Technol 26(3):209–222 (in Japanese with English summary)

Hara M (2014) Results of a survey of tsunami effects. Veg Sci News 18:21–40 (in Japanese)

Hara M, Asami K, Tomita M (2012) The society of vegetation science report on the great east Japan earthquake in the south sanriku region. Veg Sci News 16:40–48 (in Japanese)

Hayasaka D, Shimada N, Konno H, Sudayama H, Kawanishi M, Uchida T, Goka K (2012) Floristic variation of beach vegetation caused by the 2011 Tohoku-oki tsunami in northern Tohoku, Japan. Ecol Eng 44:227–232

Ishizuka K (1962) Ecological studies on the vegetation of coastal sand bars II. Succession in vegetation and developmental processes of dunes. Ann Rep Gakugei Fac Iwate Univ 20:140–168

Kanno H, Hirabuki Y, Sugiyama T, Tomita M, Hara K (2014) Vegetation change in various coastal forest habitats after a huge tsunami: a three-year study. In: Disturbance and recovery of the coastal-ecotone vegetation following the Great East Japan Earthquake and Tsunami. Jpn J Conserv Ecol 19:201–220

Kondo T, Sakai A, Sasaki S (2002) Effects of sowing dates and burial depth on seedling emergence and survival of *Calystegia soldanella* Roem. et Schult. and *Lathyrus japonicus* Willd. J Jpn Soc Revegetation Technol 28(2):330–341 (in Japanese with English abstract)

Looney PB, Gibson DJ (1995) The relationship between the soil seed bank and above-ground vegetation of a coastal barrier island. J Veg Sci 6:825–836

Naka K, Yoda K (1984) Community dynamics of evergreen broadleaf forests in south western Japan–2–Species composition and density of seeds buried in the soil of a climax evergreen oak forest. Bot Mag Tokyo 97:61–79

Nakanishi H (1988) Dispersal ecology of the maritime plants in the Ryukyu Islands, Japan. Ecol Res 3:163–173

Nakanishi H (2013) Seedling populations from sea-borne seeds on drift debris in northern and western Kyushu, Japan. Veg Sci 30:17–24 (in Japanese with English abstract)

Nakayama S, Marehide I, Tadashi M (2000) Seeds of wild plants in Japan. Tohoku University Press, Sendai, 678 pp (in Japanese)

Nishi R, Manu J, Jansen T, Hayashi K (2012) Scarp generation behind a dune and shore protection structure by tsunami on March 11, 2011. Proc Civ Eng 68(2):198–203 (in Japanese with English summary)

Numata M, Hayashi I, Komura T, Oki K (1964) Ecological studies on the buried-seed population in the soil as related to plant succession I. Jpn J Ecol 14:207–215 (in Japanese with English synopsis)

Oka K, Hirabuki Y (2014) Revegetation of coastal plants damaged by the 2011 Tohoku tsunami. In: Disturbance and recovery of the coastal-ecotone vegetation following the Great East Japan Earthquake and Tsunami. Jpn J Conserv Ecol 19:189–199 (in Japanese with English abstract)

Oka K, Okuma M, Yoshizaki S (2009a) Estimation of the source of seed supply by thalassochory for the Higashimachi district of the Shonan coast, Kanagawa Prefecture. Bull Int Assoc Landsc Ecol Jpn 13:45–54 (in Japanese with English abstract)

Oka K, Yoshizaki S, Kobori H (2009b) The emergence and establishment of seedlings of five coastal dune plants on the Enshu-nada coast, Shizuoka Prefecture, central Japan. Veg Sci 26:9–20 (in Japanese with English abstract)

Oosting HJ, Humphreys ME (1940) Buried viable seeds in a successional series of old field and forest soils. Bull Torrey Bot Club 67:253–273

Ooue M (2012) Circumstance of coastal vegetation affected by tsunami in northern and mid area of Kitakami Mountains. Veg Sci News 16:49–58 (in Japanese)

Pierce SM, Cowling RM (1991) Disturbance regimes as determinants of seed banks in coastal dune vegetation of the southeastern Cape. J Veg Sci 2:403–412

Richmond B, Szczuciński W, Chagué-Goff C, Goto K, Sugawara D, Witter R, Tappin DR, Bruce J, Fujino S, Nishimura Y, Goff J (2012) Erosion, deposition and landscape change on the Sendai coastal plain, Japan, resulting from the March 11, 2011 Tohoku-oki tsunami. Sediment Geol 282:27–39

Sawada Y, Tsuda S (2005a) Thalassochory potential in 14 species of coastal plants in the warm temperate zone of Japan. Veg Sci 22:53–61 (in Japanese with English abstract)

Sawada Y, Tsuda S (2005b) Potential for persistent seed bank formation in 14 coastal dune plants in the warm temperate zone in Japan. Veg Sci 22:135–146 (in Japanese with English abstract)

Sawada Y, Nakanishi H, Oshida K, Hattori T (2007) A check list of coastal plants in Japan. Hum Nat 17:85–101 (in Japanese with English abstract)

Shibuya K, Shimada N, Suzuki M (2014) Study of the present condition of the Sanriku coastal eco-tone and environmental conservation measures in Iwate Prefecture. J Policy Stud 15:181–199 (in Japanese with English abstract)

Shimada N (2014) Effect of tsunami with the Great East Japan earthquake to coastal vegetation and circumstance of conservation, in Iwate Pref. Veg Sci News 18:44–54 (in Japanese)

Shimizu N, Morita H, Hirota S (2001) Picture book of alien plants in Japan 1. Zenkoku Noson Kyoiku Kyokai, Tokyo, 554 pp (in Japanese)

Suzuki I, Takahashi F, Yasunobu N (2012) Seeds and fruits illustrated. Seibundo-Shinkosha, Tokyo, 272 pp (in Japanese)

Szczuciński W, Kokociński M, Rzeszewski M, Chagué-Goff C, Cachão M, Goto K, Sugawara D (2012a) Sediment sources and sedimentation processes of 2011 Tohoku-oki tsunami deposits on the Sendai Plain, Japan—insights from diatoms, nannoliths and grain size distribution. Sedimentol Geol 282:40–56

Szczuciński W, Rachlewicz G, Chaimanee N, Saisuttichai D, Tepsuwan T, Lorenc S (2012b) 26 December 2004 tsunami deposits left in areas of various tsunami runup in coastal zone of Thailand. Earth, Planets and Space 64:843–858

Tobe K, Liping Z, Omasa K (2007) Seed germination and seedling emergence of three annuals growing on desert sand dunes in China. Ann Bot 95:649–659

Tokura R, Tanaka T, Yanobu N (1988) Current-dispersed seeds from Kumihama Sand Dune. Bulletin of Kyoto University of Education. Series B. Math Nat Sci 73:25–30 (in Japanese with English Abstract)

Tsuyuzaki S (1989) Contribution of buried seeds to revegetation after eruptions of the volcano Usu, northern Japan. Bot Mag Tokyo 102:511–520

Tsuyuzaki S, Goto M (2001) Persistence of seed bank under thick volcanic deposits twenty years after eruptions of Mount Usu, Hokkaido Island, Japan. Am J Bot 88:1813–1817

Uemura S, Katsuyama T, Shimizu N, Mizuta M, Morita H, Hirota S, Ikehara N (2010) Picture book of alien plants in Japan, vol 2. Zenkoku Noson Kyoiku Kyokai, Tokyo, 579 pp (in Japanese)

Weed Science Society of Japan (ed) (2001) Method in weed science. Weed Science Society of Japan, Tokyo, 463 pp (in Japanese)

Yamanaka N, Imada M, Tamai S (2000) Effects of sand movement on the seed bank formation and seedling dynamics of *Carex kobomugi* Ohwi. Sand Dune Res 47(1):1–11 (in Japanese with English abstract)

Yonekura K, Kajita T 2003 BG plants (YList). http://ylist.info/. Last accessed 22 July 2015 (in Japanese)

Chapter 19
Plant Diversity and Considerations for Conservation of It in Infrastructure Reconstruction Planning After the Great East Japan Earthquake and Tsunami of 2011

Takahide Kurosawa

Abstract This review examines the impacts of the Great East Japan Earthquake and Tsunami of 2011 and subsequent reconstruction activities on coastal plant diversity. Arable lands and coastal forests in lowlands areas were seriously and extensively damaged by the tsunami and by land subsidence. Although most salt marshes and back marshes were probably seriously damaged, new marshes have emerged in the lowlands. The vegetation on sandy beaches disappeared or was reduced in many areas. Sand dune vegetation at some sandy beaches survived the disaster and showed remarkable recovery. The influence of the tsunami and land subsidence on the plant communities of the coastal cliffs and on weed communities along roadsides and in vacant lots was limited. Threatened species, especially *Potamogeton pusillus* and *Eleocharis parvula*, appeared in inundated paddy fields and residential areas, in ponds and in wet areas within surviving coastal forest. The origin of these newly appeared plants has not been confirmed, but the possibilities include germination from the seed banks, transport from another location by the tsunami and seed dispersal from another area. Alternatively, the presence of the plants before the disaster might simply have been overlooked. Many infrastructure reconstruction projects were planned with little consideration for biodiversity. Most of the new tidal flats and salt marshes containing some threatened plants have already been lost due to infrastructure reconstruction and the resumption of cultivation on inundated arable land. Some attempts have been made to include biodiversity conservation in plans for infrastructure reconstruction.

Keywords Back marsh • Biodiversity conservation • Coastal cliff • Coastal forest • Infrastructure reconstruction • Paddy field • Salt marsh • Sandy beach • Threatened plant

---

T. Kurosawa (✉)
Faculty of Symbiotic Systems Science, Fukushima University, Fukushima, Japan
e-mail: kurosawa@sss.fukushima-u.ac.jp

J. Urabe, T. Nakashizuka (eds.), *Ecological Impacts of Tsunamis on Coastal Ecosystems*, Ecological Research Monographs,
DOI 10.1007/978-4-431-56448-5_19

19.1 Introduction

The Great East Japan Earthquake and Tsunami of 2011 hit the Pacific coast of Japan which is a temperate coast area that had long been impacted by human activities, such as afforestation and land reclamation. In general, ecologists, taxonomists and especially amateur botanists have actively studied the local vegetation and flora in Japan (e.g. Kadoya et al. 2014). Therefore, many studies have been published that document the impacts of the disaster on plant diversity in specific areas. Most of these reports were written in Japanese. For this reason, the changes in plant diversity caused by the disaster are not yet known in foreign countries. In this chapter, first, the state of the vegetation and land use in coastal areas on the Pacific side of the Tōhoku district before the tsunami are introduced. Next, literatures reporting changes in plant diversity after the disaster are reviewed. Last, considerations for biodiversity conservation in infrastructure reconstruction planning are introduced.

19.2 Vegetation Type and Land Use in Coastal Areas Before the Great East Japan Earthquake and Tsunami of 2011

The plant diversity on the Pacific side of the Tōhoku district, Japan, was dramatically changed by the tsunami and land subsidence that accompanied the Great East Japan Earthquake and Tsunami of 2011 (Prec Institute 2011; BCJ 2013, 2014, 2015; Tomita et al. 2014; see also Chap. 16). Most of the plant communities in this region had been previously modified by human activities, however. The Nature Conservation Bureau, the Environment Agency (1998a, b) had reported that artificial seacoast (e.g. landfill and ports) and seminatural seacoast (e.g. breakwaters, roads and embankments) accounted for 41 % and 15 %, respectively, of the seacoast of the Japanese mainland in 1996–1998. In the Fukushima (southern Tōhoku district) and Ibaraki (northern Kantō district) Prefectures, only a small proportion of the seacoast had been natural, while the majority had consisted of artificial landfills and seminatural sandy coastlines (The Nature Conservation Bureau, the Environment Agency 1998a, b; BCJ 2013; Table 19.1). On the other hand, the Sanriku-kaigan Coast from the Iwate Prefecture to north of the Sendai Bay in Miyagi Prefecture—a well-known rias (deeply indented coastlines with cliffs)—had always been dominated by natural seacoast (BCJ 2013; Hara and Higuchi 2013; Hara 2014a; Takiguchi et al. 2014). It is noteworthy that the tsunami hit one of relatively highly modified coastal areas in Japan (from Fukushima Prefecture to Ibaraki Prefecture) and one of the most remarkable natural sea cliffs in the country (from Iwate Prefecture to northern Miyagi Prefecture).

Well before the disaster, the Pacific coast of the Kantō and Tōhoku districts had been significantly affected by human activities, which included afforestation, land reclamation, drainage and construction of revetments (BCJ 2013). Coastal

Table 19.1 Coast type of Pacific coast in Tōhoku district and eastern Kantō district in 1996–1998. Data from Agency of the Environment (1998)

		Total seacoast length	Artificial seacoast[a]	Seminatural seacoast[b]	Natural seacoast[c]	Estuary	Main shore type
Tōhoku district							
	Aomori Prefecture (east coast of Shimokita Peninsula)	161.27 km	46.59 km (28.89 %)	13.47 km (8.35 %)	99.15 km (61.48 %)	2.06 km (1.28 %)	Natural sandy coastlines (75.29 km, 46.69 %), artificial landfills (36.79 km, 22.81 %)
	Iwate Prefecture	671.25 km	135.74 km (20.22 %)	13.47 km (2.01 %)	516.87 km (77.00 %)	5.17 km (0.77 %)	Natural sea cliffs (405.17 km, 60.36 %)
	Miyagi Prefecture	838.54 km	244.28 km (29.13 %)	48.70 km (5.81 %)	542.07 km (64.64 %)	3.49 km (0.42 %)	Natural sea cliffs (488.09 km, 58.21 %)
	(Miyagi-sanriku Coast)	(421.84 km)	(97.38 km, 23.08 %)	(16.82 km, 3.99 %)	(306.58 km, 72.68 %)	(1.06 km, 0.25 %)	(Natural sea cliffs (282.68 km, 67.01 %))
	(Sendai Bay)	(416.70 km)	(146.90 km, 35.25 %)	(31.88 km, 7.65 %)	(235.49 km, 56.51 %)	(2.43 km, 0.58 %)	(Natural sea cliffs (205.41 km, 49.29 %))
	Fukushima Prefecture	187.38 km	78.63 km (41.96 %)	69.68 km (37.19 %)	36.88 km (19.68 %)	2.19 km (1.17 %)	Artificial landfills (58.58 km, 31.26 %), seminatural sandy coastlines (44.11 km, 23.54 %),
Kantō district							
	Ibaraki Prefecture	199.64 km	155.21 km (77.74 %)	15.54 km (7.78 %)	26.75 km (13.40 %)	2.14 km (1.07 %)	Artificial landfills (54.09 km, 27.06 %)
	Chiba Prefecture (Boso area)	248.74 km	74.17 km (29.82 %)	48.24 km (19.39 %)	123.79 km (49.77 %)	2.54 km (1.02 %)	Sandy coastlines (50.79 km, 20.42 %)
Total		2306.82 km	734.62 km (31.84 %)	209.1 km (9.06 %)	1345.51 km (58.32 %)	17.59 km (0.76 %)	

[a]Seacoast remarkably artificially modified (e.g. landfill, port and reclaimed land)
[b]Seacoast partly artificially modified (e.g. breakwater bank, road and embankment)
[c]Seacoast not artificially modified

afforestation with Japanese black pine, *Pinus thunbergii* Parl., had been carried out on coastal dunes or reclaimed wetlands during and after the Edo era (1603–1868). This had been done to reduce the damages caused by coastal sand blows, strong winds, flood tides and tsunamis to villages, cities, rice paddies and other inland crops (Hirabuki et al. 2011; Hara and Higuchi 2013; see also Chap. 20). During and after the Meiji era (1868–1912), most of the back marshes—wetlands that lie behind sea dunes or sandbars on alluvial plains—and lagoons in lowland areas near coasts had been drained and reclaimed to increase the area available for paddy fields (BCJ 2013; Hara and Higuchi 2013; Himiyama et al. 2014). Afforestation with *Pinus thunbergii* in coastal areas had resumed after the Second World War as forestry techniques improved. As such, most existing Japanese coastal forests were originally planted: natural, lowland forests are quite rare. This had been true of the Pacific coast of the Kantō and Tōhoku districts (BCJ 2013, 2014; Hara and Higuchi 2013). Concrete structures, such as revetments and wave-dissipating blocks—meant to protect against tsunamis and flood tides—had been built along the Pacific coast of the Kantō and Tōhoku districts after the Second World War (BCJ 2013). Changes to the coastal areas in Sōma, Fukushima Prefecture, resulting from these activities are shown in Fig. 19.1. Two lagoons had been drained during and after the Meiji era to create new ricefields, and coastal forests had been planted on coastal dunes or reclaimed wetlands (Sugiyama and Koseki 2005). Concrete revetments of T.P. (Tokyo Peil) +6.2 m in height had been constructed along nearly all of the lowland shores. A large harbour and its facilities had been built after the Second World War.

Before the disaster, paddy fields in reclaimed or drained lands often had flooded. Most rivers in these areas had been below sea level, and the water had been pumped upwards by drainage pump stations and then had drained off to the sea or a lagoon. Planted pine trees in coastal forests had been often weakened or had been damaged by high groundwater levels. Furthermore, pine wilt disease had been spreading to the Tōhoku district in and after the 1970s (Hirabuki et al. 2011). Coastal pine forests at Ōsu, the sandbar of the Matsukawa-ura Lagoon, northern Fukushima Prefecture, had been maintained by continuous planting and aerial application of pesticides to control the beetle vector of pine wilt disease. Paddy fields and planted pine forests on lowlands in coastal areas tended to be vulnerable to natural disasters, and people had spent considerable labour and money to maintain them before the earthquake.

Originally, natural habitats—salt marshes, back marshes, sandy dunes and sea cliffs—had covered the coastal areas. These had consisted of plants adapted to those environments. The extent of these types of habitat, especially salt and back marshes, had decreased due to human activity (BCJ 2013; Hara and Higuchi 2013; Hara 2014a). Coastal areas are home to many endangered plant species in Japan (Sawada et al. 2007; Kaneko et al. 2009).

Fig. 19.1 Changes of Matsukawa-ura Lagoon and around area in Fukushima Prefecture, Japan. (**a**) In Meiji era. (**b**) Before the Great East Japan Earthquake and Tsunami of 2011. (**c**) Just after the disaster. (**d**) Three years after the disaster. *Dc* destroyed coastal forest, *Dr* destroyed residential area, *Ip* inundated paddy field. (*a*) 1/50,000 Geographical Map in 1908 by Geospatial Authority of Japan; (*b–d*) Google Earth satellite photographs of Sep. 21, 2010, Mar. 14, 2011, and Jun. 1, 2015, respectively

19.3 Vegetation Changes Caused by the Great East Japan Earthquake and Tsunami of 2011

Coastal topography was changed by the tsunami and land subsidence associated with the Great East Japan Earthquake and Tsunami of 2011 (Richmond et al. 2012; Tanaka et al. 2012; Udo et al. 2012; Kanaya et al. 2012, 2014). Moreover, coastal protection structures and vegetation were severely and widely damaged (Prec Institute 2011; Kanaya et al. 2012, 2014; BCJ 2013, 2014, 2015; Hara and Higuchi 2013; Zhao et al. 2013; Hara 2014a; Himiyama et al. 2014). The influence of the earthquakes, fire and other disasters on coastal vegetation and plant diversity was relatively small (Yoshino 2014). The estimated maximum height of the tsunami was 2.9–17 m in Iwate, Miyagi and Fukushima Prefectures (e.g. 16.7 m at Ōfunato, southern Iwate Prefecture; 7.2 m at Sendai, Miyagi Prefecture; and 8.9 m at Sōma, northern Fukushima Prefecture) (Japan Meteorological Agency 2011; BCJ 2013; Yoshino 2014). Especially, the tsunami inundation was 10–15 m high in the southern Sanriku region (Mori et al. 2012). Land subsidence associated with the earthquake measured −20 to −116 cm in Iwate, Miyagi and Fukushima Prefectures (e.g. −77 cm in Ōfunato, Iwate Prefecture; −28 cm in Natori, Miyagi Prefecture; and −32 cm in Iwaki, Fukushima Prefecture) (Geospatial Information Authority of Japan 2011; BCJ 2013). The tsunami and land subsidence seriously disturbed plant communities and affected plant species diversity in coastal areas. The tsunami caused erosion and sometimes washed away plants along with the soil (Tanaka et al. 2012) or buried the plants under the sand and mud (Richmond et al. 2012). The tsunami and land subsidence resulted in a flood of seawater or brackish water that extended several kilometres inland (Mori et al. 2012; BCJ 2013; Une and Koarai 2014; see also Chap. 16).

Were all kinds of vegetation type equally damaged by the tsunami and land subsidence associated with the earthquake? What kinds of vegetation were resistant to damage? Unfortunately, we cannot answer these questions adequately nor scientifically, as the effects of the tsunami and land subsidence on vegetation type have not been well studied, except in coastal forests and on sandy beaches (Table 19.2). On the other hand, many ecologists and amateur botanists have recorded their observations of vegetation changes after the disaster. These limited studies and a number of records show that the extent of the damage after the earthquake depended on the type of vegetation or land use. Arable lands and coastal forests in lowlands were seriously and extensively damaged. Although it is likely that most salt marshes and back marshes were seriously damaged, new marshes emerged in the lowlands. The vegetation on sandy beaches disappeared or was reduced in many areas. Sand dune vegetation at some sandy beaches survived the disaster and showed remarkable recovery. The influence of the tsunami and land subsidence on the plant communities of the coastal cliffs and on weed communities along roadsides and in vacant lots was apparently limited.

Vast areas of arable land in the Sendai Plains of Miyagi Prefecture and the coastal plains of northern Fukushima Prefecture and many small areas of arable land along the deeply indented coastline in the Iwate and the northern Miyagi Prefectures were flooded because of the tsunami and land subsidence (Hirabuki et al. 2011; Mineta

Table 19.2 Studies and observations on plant diversity after the Great East Japan Earthquake and Tsunami of 2011

Vegetation or land use before the earthquake	Vegetational study	Floristic study	Observation on vegetation or remarkable plants	Observation on threatened plants
Arable land	Mineta and Tomosyo (2012), Naito (2012), BCJ (2013, 2014, 2015)	BCJ (2014)	Hirabuki et al. (2011), Oyama (2012), Hara and Higuchi (2013), Kasai (2013), Nemoto et al. (2013), Sugiyama and Kasai (2013), Sugiyama et al. (2013), Yuzawa (2013), Hara (2014a), Hiyama et al. (2014), Shibuya et al. (2014), Shimada (2014), Yoshikawa and Ayukawa (2014)	Kasai (2013), Nemoto et al. (2013), Sugiyama and Kasai (2013)[a], Sugiyama et al. (2013), Yuzawa (2013)[b], Shibuya et al. (2014), Shimada (2014)
Coastal forests	BCJ (2013, 2014, 2015), Tomita et al. (2013, 2014), Kanno et al. (2014), Oka and Hirabuki (2014), Onza et al. (2014)	BCJ (2014), Onza et al. (2014), Watanabe and Kurosawa (2015)	Hirabuki et al. (2011), Prec Institute (2011), Hara (2012, 2014a, b), Hara et al. (2012), Kanaya et al. (2012), Nagahata (2012), BCJ (2013), Hara and Higuchi (2013), Kasai (2013), Miyagi et al. (2013), Yuzawa (2013), Hiyama et al. (2014), Shimada (2014), Takiguchi et al. (2014), Watanabe and Kurosawa (2015)	Kasai (2013), Yuzawa (2013)[b], Onza et al. (2014)[b], Takiguchi et al. (2014), Watanabe and Kurosawa (2015)[b]
Salt marshes	BCJ (2013, 2014, 2015), Shibuya et al. (2014)	BCJ (2014), Oyamada et al. (2012)	Hara et al. (2012), Nagahata (2012), Kasai (2013), Shimada (2014), Takiguchi et al. (2014), Yoshikawa and Ayukawa (2014)	Oyamada et al. (2012)[b], Takiguchi et al. (2014)
Back marshes	BCJ (2013, 2014, 2015), Kanno et al. (2014)	BCJ (2014)	Hara (2012), Kanaya et al. (2012), Nagahata (2012), Hara and Higuchi (2013), Takiguchi et al. (2014)	Takiguchi et al. (2014)

(continued)

Table 19.2 (continued)

Vegetation or land use before the earthquake	Vegetational study	Floristic study	Observation on vegetation or remarkable plants	Observation on threatened plants
Sandy beaches	Prec Institute (2011), Hayasaka et al. (2012)[a], BCJ (2013, 2014, 2015), Yuzawa (2013)[a], Oka and Hirabuki (2014), Shibuya et al. (2014)	Hayasaka et al. (2012)[a], Oyamada et al. (2012), BCJ (2014), Shimada et al. (2014)	Hirabuki et al. (2011), Prec Institute (2011), Hara et al. (2012), Kanaya et al. (2012), Oyama (2012), Nagahata (2012), Hara and Higuchi (2013), Nemoto et al. (2013), Shimoyama (2013), Shimada (2014), Takiguchi et al. (2014), Yoshikawa and Ayukawa (2014)	Oyamada et al. (2012)[b], Kasai (2013), Nemoto et al. (2013)[b], Yuzawa (2013)[b], Shimada et al. (2014), Takiguchi et al. (2014)
Coastal cliffs	Prec Institute (2011)	Oyamada et al. (2012)	Prec Institute (2011), Hara (2012), Hara et al. (2012), Hara and Higuchi (2013), Nemoto et al. (2013), Yuzawa (2013), Shimada (2014), Takiguchi et al. (2014), Yoshikawa and Ayukawa (2014)	Oyamada et al. (2012)[b], Nemoto et al. (2013)[b], Yuzawa (2013)[b]
Other habitats	Watanabe and Kurosawa (2015, weed community)	Sugiyama et al. (2013, pond), Watanabe and Kurosawa (2015, weed community), Yonekura and Maki (2015, island including various vegetation and land use)	Hara et al. (2012, riverside), Sugiyama (2012, regulating reservoir), BCJ (2013, town area), Watanabe and Kurosawa (2015, weed community)	Sugiyama (2012, regulating reservoir), Sugiyama et al. (2013, pond), Shibuya et al. (2014, residential area), Watanabe and Kurosawa (2015, weed community)

[a]Comparative study before and after the earthquake
[b]Observation with number of individuals of endangered plants

and Tomosyo 2012; Hara and Higuchi 2013; Himiyama 2013; Matsumoto 2013, 2014; Yuzawa 2013; BCJ 2014, 2015; Himiyama et al. 2014). Satellite remote sensing data showed that total paddy field area along the Sendai coast decreased from 4.2 to 0.5 km^2 because of immersion (Zhao et al. 2013). Large wetlands and numerous small wetland areas then emerged, and paddy weeds, such as *Echinochloa crus-galli* (L.) P. Beauv. var. *crus-galli*, frequently came to dominate the community (Hirabuki et al. 2011; Mineta and Tomosyo 2012; Naito 2012; Hara and Higuchi 2013; Kasai 2013; Yuzawa 2013; Himiyama et al. 2014; Kurosawa 2014). Coastal plants, such as *Scirpus maritimus* L., *Cyperus polystachyos* Rottb. and *Suaeda australis* (R.Br.) Moq., occasionally dominated, and threatened species, such as *Nymphoides coreana* (H.Lév.) H.Hara and *Potamogeton pusillus* L., sometimes appeared (Mineta and Tomosyo 2012; Nagahata 2012; Kasai 2013; Nemoto et al. 2013; Sugiyama and Kasai 2013; Sugiyama et al. 2013; Yuzawa 2013; BCJ 2014; Kurosawa 2014; Shibuya et al. 2014; Shimada 2014; Yonekura and Maki 2015; see also Chap. 21). Previously, these arable lands had been predominantly paddy fields, often created by reclaiming marshes or draining lagoons. For example, of three lagoons in Sōma that had been drained during 1906–1966, two were then used as paddy fields, and all were flooded after the earthquake and tsunami. Thus, many arable lands in the lowlands, especially paddy fields, returned to their natural wetland or lagoon state after the disaster (BCJ 2013; Hara 2014a). It is noteworthy that many threatened species emerged in ponds within arable lands (Kasai 2013; Sugiyama et al. 2013).

Coastal forests in lowlands were also heavily damaged by the tsunami and land subsidence (Hirabuki et al. 2011; Prec Institute 2011; Kanaya et al. 2012, 2014; Hara 2012, 2014a; Hara et al. 2012; Oyama 2012; Nagahata 2012; Tanaka 2012; Udo et al. 2012; Hara and Higuchi 2013; Miyagi et al. 2013; Tomita et al. 2013, 2014; Yuzawa 2013; BCJ 2014, 2015; Kanno et al. 2014; Kurosawa 2014; Oka and Hirabuki 2014; Onza et al. 2014; Takiguchi et al. 2014; Watanabe and Kurosawa 2015; see also Table 19.2). Zhao et al. (2013) used satellite remote sensing data to estimate that about 90 % of coastal forests in the Sendai coastal region collapsed. It was widely reported that most of the planted pines in coastal lowlands from Iwate Prefecture to Fukushima Prefecture were uprooted or damaged by seawater inundation after the tsunami (Prec Institute 2011; Tanaka et al. 2012; Udo et al. 2012; BCJ 2013; Hara and Higuchi 2013; Miyagi et al. 2013; Tomita et al. 2013, 2014). Planted pines that survived the tsunami were so rare in Takata-matsubara (Rikuzentakata, Iwate Prefecture) (Shimada 2014), Matsukawa-ura (Sōma, Fukushima Prefecture) and coastal areas of Minamisōma (Fukushima Prefecture) that they were often called 'kiseki-no-matsu' (the 'miracle lone pine tree' or 'pine tree of hope'). Although information about understory plants is still limited, some observers reported the disappearance of forest floor vegetation and the emergence or invasion of weedy plants, such as *Chenopodium album* L., *Oenothera laciniata* Hill and *Erigeron canadensis* L., in the remaining coastal forests (Hirabuki et al. 2011; Hara et al. 2012; Kasai 2013; Yuzawa 2013; Hara 2014a; Kurosawa 2014; Oka and Hirabuki 2014; Takiguchi et al. 2014). Wetlands and salt marshes, along with several endangered plant species, appeared at some lowland sites where coastal forests had been destroyed (Kasai 2013; Kanno et al. 2014; Kurosawa 2014; Onza et al. 2014; Takiguchi et al. 2014; Watanabe and Kurosawa 2015).

Before the earthquake and tsunami of 2011, salt marshes had been restricted to small areas because of reclamations and draining operations, and many of the species in them had been threatened. It seems that these rare salt marsh plants were seriously damaged by erosion due to the tsunami (Table 19.3; see also Hara et al. 2012; Nagahata 2012; Hara and Higuchi 2013; Kasai 2013; Hara 2014a; Kurosawa 2014; Takiguchi et al. 2014), although this damage has been little studied. On the other hand, new salt marshes emerged in seawater-inundated areas in lowlands that had been paddy fields or coastal forests before the earthquake (Mineta and Tomosyo 2012; Nagahata 2012; Kasai 2013; Kurosawa 2014; Watanabe and Kurosawa 2015).

Back marshes that had been reduced in size by human activity before the earthquakes were also damaged by land subsidence and erosion, seawater inundation and sediment deposition by the tsunami (Hirabuki et al. 2011; Kanaya et al. 2012, 2014; Nagahata 2012; Hara and Higuchi 2013; Hara 2014a; Kanno et al. 2014; Kurosawa 2014; Takiguchi et al. 2014; see also Table 19.3). Although many of these were destroyed, new swamps and marshes emerged in lowland areas (Mineta and Tomosyo 2012; Nagahata 2012; BCJ 2013; Hara 2014a; Shibuya et al. 2014; Takiguchi et al. 2014; see also Chap. 21).

The vegetation on sandy beaches disappeared or was reduced in many areas because of erosion or sedimentation caused by the tsunami and land subsidence (Prec Institute 2011; Hara et al. 2012; Hayasaka et al. 2012; Kanaya et al. 2012; Oyama 2012; Nagahata 2012; BCJ 2013, 2014, 2015; Yuzawa 2013; Shimada et al. 2014; Takiguchi et al. 2014; see also Table 19.3). Some disturbed sandy beaches were invaded by non-beach weeds (Hayasaka et al. 2012). Sand dune vegetation at some sandy beaches survived the disaster (Prec Institute 2011; Hara 2012; Hayasaka et al. 2012; Nagahata 2012; Nemoto et al. 2013; BCJ 2014, 2015) and showed remarkable recovery (Prec Institute 2011; Nagahata 2012; Yuzawa 2013; Shimada 2014). New sandy beaches and sand dune vegetation emerged in some coastal areas (Shimada 2014; Shimada et al. 2014).

Coastal cliffs were hit directly by the tsunami. Some coastal cliff communities disappeared entirely because of landslides caused by the shock of the earthquake and tsunami, and others were washed away by the tsunami (Prec Institute 2011; Yuzawa 2013). Damage to coastal cliff communities was, however, relatively minor (Prec Institute 2011; Hara 2012, 2014a; Nemoto et al. 2013; Hara and Higuchi 2013). This may be partly because the communities had been accustomed to wave disturbance and seawater inundation, and partly because the disaster occurred at the end of winter, when most of the plant species were dormant (Hara and Higuchi 2013; Hara 2014a).

Although some roadsides and vacant lots were flooded, occasionally becoming marshes or tidal flats, damage to the weed communities was minimal in general (Watanabe and Kurosawa 2015; see also Table 19.3). Large, new communities of weeds emerged in town and hamlet areas and on arable lands (Mineta and Tomosyo 2012; BCJ 2013).

Table 19.3 Vegetational changes at 82 quadrates around Matsukawa-ura Lagoon, Fukushima Prefecture, Japan, after the Great East Japan Earthquake and Tsunami of 2011. Data from Kurosawa et al. (unpublished)

Vegetation in 2003–2004			Vegetation in 2011			
					With vegetation	
Vegetation type	Number of quadrates	Dominant species (number of quadrates)	Number of quadrates with no vegetation because of remarkably eroded	Number of quadrates with no vegetation because of land subsidence, being buried by sand and mud and so on	Number of quadrates	Dominant species (number of quadrates)
Coastal forests	11	*Pinus thunbergii* (11)	1	2	8	*Pinus thunbergii* (1); *Pinus thunbergii* and *Eurya japonica* (1); *Eurya japonica* (1); *Phragmites australis* (2); *Hemarthria sibirica* (1); *Cyperus polystachyos*, *Echinochloa crus-galli* and *Digitaria ciliaris* (1); *Amorpha fruticosa* (1)
Salt marshes	25	*Carex scabrifolia* (4); *Carex scabrifolia* and *Phragmites australis* (2); *Triglochin* spp. (3); *Suaeda maritima* ssp. *asiatica* (4); *Puccinellia nipponica* (2); *Zoysia macrostachya* (2); *Limonium tetragonum* (2); *Limonium tetragonum* and *Suaeda maritima* ssp. *asiatica* (1); *Limonium tetragonum*, *Suaeda maritima* ssp. *asiatica* and *Phacelurus latifolius* (1); *Polygonum polyneuron*, *Suaeda maritima* ssp. *asiatica* and *Phragmites australis* (1); *Polygonum polyneuron* and *Phragmites australis* (1); *Ruppia maritima* (2)	9	12	4	*Phragmites australis* (3), *Zoysia sinica* var. *nipponica* and *Phragmites australis* (1)

(continued)

Table 19.3 (continued)

Vegetation in 2003–2004			Vegetation in 2011			
			Number of quadrates with no vegetation because of remarkably eroded	Number of quadrates with no vegetation because of land subsidence, being buried by sand and mud and so on	With vegetation	
Vegetation type	Number of quadrates	Dominant species (number of quadrates)			Number of quadrates	Dominant species (number of quadrates)
Back marshes	30	*Phragmites australis* (10); *Phragmites australis* and *Carex rugulosa* (1); *Phragmites australis* and *Phacelurus latifolius* (1); *Phragmites australis* and *Persicaria japonica* (1); *Phragmites australis*, *Rubia jesoensis*, *Solidago altissima* and *Imperata cylindrica* (1); *Ischaemum aristatum* var. *crassipes* (4); *Ischaemum aristatum* var. *crassipes*, *Juncus fauriei* and *Phacelurus latifolius* (1); *Phacelurus latifolius* (3); *Carex* sp. (1); *Calamagrostis epigejos* (1); *Calamagrostis epigejos* and *Phragmites australis* (1); *Calystegia soldanella* (1); *Schoenoplectus tabernaemontani* (1); *Bolboschoenus koshevnikovii* (1); *Bolboschoenus koshevnikovii* and *Typha domingensis* (1); *Carex dispalata* (1)	5	17	8	*Phragmites australis* (6), *Bolboschoenus maritimus* and *Phragmites australis* (1), *Carex disparata* (1)
Sandy beaches	9	*Calystegia soldanella* (2), *Zoysia macrostachya* (1), *Artemisia stelleriana* (1), *Carex kobomugi* (1), *Leymus mollis* (1), *Imperata cylindrica* (1), *Lathyrus japonicus* (1), *Carex pumila* (1)	2	5	2	*Phragmites australis* (1), *Calystegia soldanella* (1)
Weedy habitats	7	*Miscanthus sinensis* (1), *Rosa multiflora* (1), *Solidago altissima* (1), *Eragrostis curvula* (1), *Imperata cylindrica* and *Phragmites australis* (1), *Oenothera laciniata* (1), *Digitaria violascens* and *Trifolium repens* (1)	2	0	5	*Miscanthus sinensis* (1), *Rosa luciae* (1), *Phragmites australis* (1), *Lespedeza cuneata* (1), *Digitaria ciliaris* (1)
Total	82		19	36	27	

19.4 The Changes in Diversity of Plant Species Caused by the Great East Japan Earthquake and Tsunami of 2011

Several floristic studies and many observations on remarkable plants and threatened plants after the Great East Japan Earthquake and Tsunami of 2011 have been published (Table 19.2). Our knowledge of the changes in diversity of plant species caused by the disaster is, however, quite limited because of a lack of our knowledge about prior conditions. There are no comparative studies of flora before and after the earthquake and tsunami except for those by Hayasaka et al. (2012) and Watanabe and Kurosawa (2015). Information on threatened plants before and after the disaster is also limited (Table 19.2). Quantitative studies are even rarer. Therefore, it should be kept in mind that the following was written based on only scant information.

Three threatened plants inhabiting the coastal forest understory—*Epipactis papillosa* var. *sayekiana*, *Cephalanthera longifolia* and *Inula linariifolia*—disappeared from all observed locations in the disaster-stricken prefectures, namely, Iwate, Miyagi and Fukushima (Oyamada et al. 2012; Kasai 2013; Yuzawa 2013; Table 19.4). Since most of the *Pinus thunbergii* forests where these plants inhabited were lost or experienced changes in their understory environment, it is to be expected that these threatened plants would not survive the disaster. A small perennial herb growing in salt marshes, *Halerpestes kawakamii*, also disappeared from two localities (Oyamada et al. 2012; Kasai 2013; Takiguchi et al. 2014). Other threatened coastal plants reportedly survived in at least one locality.

The most striking aspect of plant species diversity after the disaster is that threatened species appeared in many inundated areas (Table 19.4). They were found in inundated paddy fields and residential areas, in ponds and in wet areas within surviving coastal forest. This was especially remarkable in the case of *Potamogeton pusillus*, a submerged aquatic plant found in coastal areas, and *Eleocharis parvula*, a tiny annual herb found in coastal wetlands. Eight near-threatened plants, *Tillaea aquatica* L. (Crassulaceae), *Najas gracillima* (A.Braun ex Engelm.) Magnus (Hydrocharitaceae), *Penthorum chinense* Pursh (Penthoraceae), *Veronica undulata* Wall. (Plantaginaceae), *Monochoria korsakowii* Regel et Maack (Pontederiaceae), *Potamogeton pectinatus* L. (Potamogetonaceae), *Ruppia maritima* L. (Ruppiaceae) and *Sparganium erectum* L. (Sparganiaceae), were also reported to have appeared after the disaster (Kasai 2013; Nemoto et al. 2013; Sugiyama et al. 2013; Yonekura and Maki 2015). The origin of these newly appeared plants has not yet been confirmed, but there are three possibilities: (1) they germinated from seed banks after the disturbance caused by the tsunami or land subsidence (Kasai 2013; Nemoto et al. 2013; Sugiyama et al. 2013); (2) they were carried there from another location by the tsunami (Kasai 2013; Nemoto et al. 2013; Sugiyama et al. 2013); or (3) they were dispersed from another area and established in the new location after the tsunami. Alternatively, they originally could have occurred in those locations, but had been overlooked before the disaster. Afterwards, they could have increased in number and become more conspicuous (Kasai 2013). It is very likely that newly appeared plants, even within a species in the same location, have different origins.

Table 19.4 Habitat and number of reported locality of threatened plant disappeared by, survived in, re-emerged after and newly appeared after the Great East Japan Earthquake of 2011 in the disaster-stricken prefectures of Iwate, Miyagi and Fukushima

Scientific name	Family	Red list category in Japan (Ministry of the Environment 2015)	Habitat before the disaster	Habitat after the disaster	Number of reported locality						References
					Disap-peared	Survived	Revived[a]	Newly appeared	Present (survived, revived or newly appeared, but not sure)	Total	
Sium suave Walter var. *nipponicum* (Maxim.) H.Hara	Apiaceae	VU	Pond side and ridge between ricefields	Ridge between ricefields	1	1				2	Kasai (2013)
Achillea ptarmica L. subsp. *macrocephala* (Rupr.) Heimerl var. *yezoensis* Kitam.	Asteraceae	EN	Sea cliff	Sea cliff		1				1	Oyamada et al. (2012)
Inula linariifolia Turcz.	Asteraceae	VU	Coastal forest		1					1	Kasai (2013)
Carex capricornis Meinsh. ex Maxim.	Cyperaceae	VU	Pond side and river side	Wet place in coastal forest	2			1		3	Kasai (2013), Takiguchi et al. (2014)
Eleocharis parvula (Roem. et Schult.) Link ex Bluff, Nees et Schauer	Cyperaceae	VU		Inundated paddy field			2	9		11	Kasai (2013), Takiguchi et al. (2014), Yonekura and Maki (2015)
Lespedeza tomentosa (Thunb.) Siebold ex Maxim.	Fabaceae	VU	River bank	River bank		1				1	Kasai (2013)

Swertia tosaensis Makino	Gentianaceae	VU	Pond side and wet place in coastal forest	Wet place in coastal forest	2	1			1	4	Kasai (2013), Onza et al. (2014)
Najas chinensis N.Z.Wang	Hydrocharitaceae	VU	Fallow ricefield	Pond and inundated paddy field		2	1	2		5	Kasai (2013), Sugiyama and Kasai (2013), Sugiyama et al. (2013), Yonekura and Maki (2015)
Ottelia alismoides (L.) Pers.	Hydrocharitaceae	VU	Wet place on side of vegetable or paddy field and fallow ricefield	Pond and inundated paddy field		1	1	2		4	Kasai (2013)
Rotala mexicana Cham. et Schltdl.	Lythraceae	VU	Paddy field and fallow ricefield	Pond		1				1	Sugiyama et al. (2013)
Nymphoides coreana (H.Lév.) H.Hara	Menyanthaceae	VU	Paddy field, fallow ricefield and pond	Pond and inundated paddy field		2				2	Kasai (2013), Sugiyama and Kasai (2013)
Cephalanthera longifolia (L.) Fritsch	Orchidaceae	VU	Coastal forest		1					1	Kasai (2013)

(continued)

Table 19.4 (continued)

Scientific name	Family	Red list category in Japan (Ministry of the Environment 2015)	Habitat before the disaster	Habitat after the disaster	Number of reported locality						References
					Disappeared	Survived	Revived[a]	Newly appeared	Present (survived, revived or newly appeared, but not sure)	Total	
Epipactis papillosa Franch. et Sav. var. *sayekiana* (Makino) T.Koyama et Asai	Orchidaceae	VU	Coastal forest		7					7	Oyamada et al. (2012), Kasai (2013), Yuzawa (2013)
Gratiola japonica Miq.	Plantaginaceae	VU	Paddy field	Inundated paddy field		2				2	Kasai (2013)
Persicaria foliosa (H.Lindb.) Kitag. var. *paludicola* (Makino) H.Hara	Polygonaceae	VU	Wet place on side of vegetable or paddy field		1					1	Kasai (2013)
Potamogeton cristatus Regel et Maack	Potamogetonaceae	VU		Inundated paddy field		1				1	Yonekura and Maki (2015)
Potamogeton pusillus L.	Potamogetonaceae	VU	Irrigation or drainage ditch between paddy fields	Pond and inundated paddy field		2		14		16	Sugiyama (2012), Kasai (2013), Takiguchi et al. (2014), Yonekura and Maki (2015)

Zannichellia palustris L.	Potamogetonaceae	VU	Irrigation or drainage ditch between paddy fields	Pond and inundated paddy field or residential area	1	1	2	3		7	Kasai (2013), Takiguchi et al. (2014), Yonekura and Maki (2015)
Halerpestes kawakamii (Makino) Tamura	Ranunculaceae	VU	Salt marsh and estuary		2					2	Oyamada et al. (2012), Kasai (2013), Takiguchi et al. (2014)
Ranunculus kazusensis Makino	Ranunculaceae	EN		Inundated road		1				1	Kasai (2013), Yonekura and Maki (2015)
Galium tokyoense Makino	Rubiaceae	VU	River terrace		1					1	Kasai (2013)
Scrophularia buergeriana Miq.	Scrophulariaceae	VU	Canal bank	Canal bank		1				1	Kasai (2013)
Zostera caulescens Miki	Zosteraceae	VU	Bay with sandy mud		1					1	Yonekura and Maki (2015)
Total					20	18	6	31	1	76	

[a]Plant disappeared before 2011 and appeared after the earthquake and tsunami

Only a few attempts at quantitative estimation of the effect of the Great East Japan Earthquake and Tsunami in 2011 on plant diversity have been made, largely because there are few places where flora had been studied before the disaster. In a sandbar habitat, which included a coastal forest of planted *Pinus thunbergii*, sandy beach, coastal grassland, bog, weedy habitat and a small salt marsh, the number of vascular plant taxa decreased from ca. 360 before the disaster to ca. 130 afterwards (Kurosawa et al. unpublished data). Hayasaka et al. (2012) compared various indices of biodiversity on four sandy beaches before and after the tsunami. They reported a decrease in the number of beach species, Pielou evenness index values and Shannon-Wiener diversity index values after the tsunami and an increase in the number of non-beach plants tolerant of human disturbances. There was no significant relationship between the number of all vascular plant species or non-beach plants species and the magnitude of the tsunami disturbance. Since these discussions are written based on only one or a few cases, generalizations should be made carefully.

19.5 The Changes of Plant Diversity Caused by Infrastructure Reconstruction After the Disaster

Vegetation type and diversity of plant species have changed not only because of the Great East Japan Earthquake and Tsunami of 2011 but also because of post-disaster reconstruction in coastal areas (Nishihiro et al. 2014). This includes the reconstruction of protective structures and the construction of new green spaces as preventive measures against future disasters. These efforts have been carried out by the Ministry of Land, Infrastructure, Transport and Tourism or the Civil Works departments of local governments, while the reconstruction of coastal forest has been conducted by the Forestry Agency or the Agriculture and Forestry departments of local governments. Revetments (7.2 m in height at Sendai Bay), wave-dissipating blocks and *Pinus thunbergii* afforestation on moundings of ca. 3 m in height and more than 200 m in width are now being reconstructed over the lowlands of the Pacific coast of Tōhoku (Kurosawa 2014; Nishihiro et al. 2014; Takiguchi et al. 2014), except around the Fukushima Daiichi Nuclear Power Plant, which was damaged in the disaster and caused serious nuclear accidents. The changes can be easily seen in Google Earth satellite photographs (Fig. 19.1d). Most of the new tidal flats and salt marshes containing threatened plants have already been lost to infrastructure reconstruction and the resumption of agricultural activities on inundated arable land (Shimada 2014). Moreover, wetlands known to have contained threatened plants before the disaster have also disappeared because of draining and reclamation efforts aimed at creating temporary areas for the deposition of debris or the reconstruction of coastal forests (Kasai 2013; Kurosawa 2014; Takiguchi et al. 2014).

19.6 Biodiversity Conservation and Post-disaster Reconstruction

Many infrastructure reconstruction projects on the Pacific coast of the Tōhoku district were planned with little consideration for biodiversity. The damage to human life caused by the Great East Japan Earthquake and Tsunami of 2011 was so severe that there has been little opposition from either citizens or nature conservation organizations to reconstruction efforts that are destructive to nature. Some Japanese societies for ecology or biodiversity submitted their written opinions to the Ministry of the Environment, the Forestry Agency and local governments (Nakamura et al. 2012; Nemoto et al. 2013; Nishihiro et al. 2014). These opinions stated that infrastructure reconstruction should consider the conservation of biodiversity. The Science Council of Japan recommended that the planned uniform approach to infrastructure reconstruction should be modified to a diverse approach, taking local ecosystems into account (Science Council of Japan 2014a, b).

Some attempts have been made to include biodiversity conservation concerns in plans. There are too few instances, however, to introduce them systematically. The following are examples of reconstruction plans of seawall in the Pacific coast of Tōhoku district that include noteworthy conservation efforts:

1. The Local Government of Rikuzentakata City planned for a revetment to be set back seawall by about 150 m to conserve a tidal flat of 200 × 500 m, which had newly appeared on previously drained land in Otomoura, Rikuzentakata City, Iwate Prefecture (Shibuya et al. 2014; Shimada 2014).
2. The Government of Miyagi Prefecture planned for a revetment to be set back seawall by 30–40 m to conserve a tidal flat in Gamō-higata Lagoon, Sendai City, Miyagi Prefecture.
3. The Ministry of Land, Infrastructure, Transport and Tourism planned for a revetment to be set back seawall by 50 m to conserve sand dune vegetation in Okadashinhama, Sendai City, Miyagi Prefecture.
4. The Government of Fukushima Prefecture planned three conservation areas totalling ca. 2 ha of newly appeared bog and salt marsh on arable lands. These form part of a new green space for preventing disasters in Shinchi Town, Fukushima Prefecture (Kurosawa 2014). These conservation areas are currently being used for the deposition of debris and construction materials and for the construction's offices. The bog and salt marsh will be restored after construction is complete.
5. The Government of Fukushima Prefecture and the Forestry Agency designated two protection areas of ca. 1.4 and 1.8 ha and a conservation area of ca. 9.5 ha that contain new tidal flats and salt marshes in inundated coastal forests and on the grounds of an outdoor learning centre (Kurosawa 2014; Watanabe and Kurosawa 2015). These areas are at a planned coastal forest reconstruction site on the sandbar of Matsukawa-ura Lagoon in Ōsu, Sōma City, Fukushima Prefecture (Fig. 19.2). The conservation area is currently used for the deposition

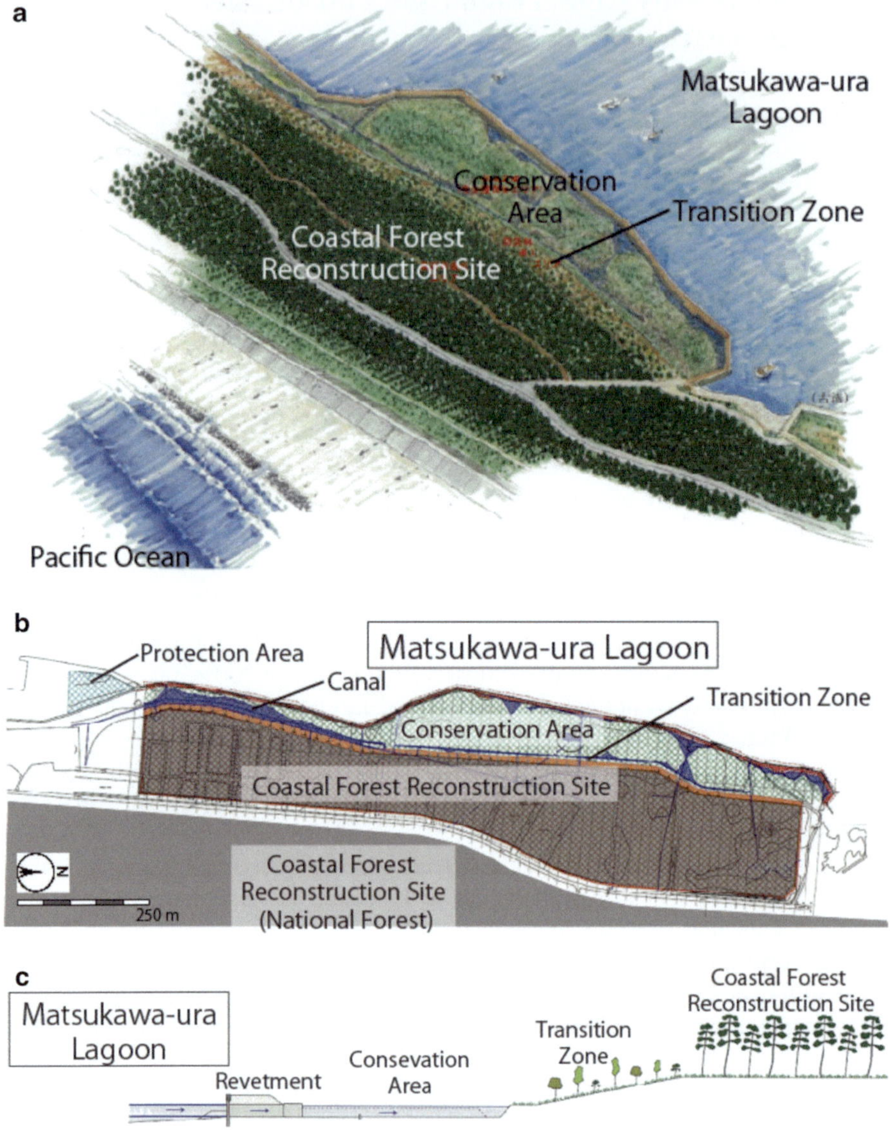

Fig. 19.2 The protection area, conservation area and transition zone planned at the coastal forest reconstruction site on the sandbar of Matsukawa-ura Lagoon, Fukushima Prefecture, Japan, after the Great East Japan Earthquake and Tsunami of 2011 (Created from the document released from the Government of Fukushima Prefecture)

of debris and construction materials and for the construction's offices. The tidal flats and salt marshes will be restored after construction is complete. The Government of Fukushima Prefecture and the Forestry Agency designated several other protection areas in Sōma-shi and Minamisōma-shi (Kurosawa 2014).

Revetment setbacks and designation of protection or conservation areas seem to be effective means to conserve sand dune, salt marsh and tidal flat vegetation. However, it remains to be seen whether the attempts will be successful or not. It is necessary to monitor biodiversity and the areas should be adaptively managed.

19.7 Conclusions

The Great East Japan Earthquake and Tsunami of 2011 changed plant diversity along the Pacific coast of the Tōhoku district. The appearance of new bogs and salt marshes and the re-emergence of populations of threatened plants species indicate that, in some instances, biodiversity has been given the opportunity to recover from human disturbance predating the disaster. Most of the new bogs and salt marshes hosting populations of threatened plants are doomed to be destroyed by infrastructure reconstruction projects, however. The revetments, coastal forests and green spaces will be restored in large scale, are easily observed by satellite, and will affect the whole Pacific coast ecosystem in the Tōhoku district. Will attempts to conserve biodiversity during reconstruction go well? Can threatened plants survive? The Pacific coast of the Tōhoku district is just like an experimental field in relation to conservation for coastal plant diversity.

References

BCJ (Biodiversity Center of Japan, Nature Conservation Bureau, Ministry of the Environment) (2013) 2012 survey of the natural environment in the Pacific Coast of Tohoku district. Biodiversity Center of Japan, Nature Conservation Bureau, Ministry of the Environment, Fujiyoshida

BCJ (Biodiversity Center of Japan, Nature Conservation Bureau, Ministry of the Environment) (2014) 2013 report on the survey project for changes in vegetation and wetlands on the Pacific Coastal area of the Tohoku region. Biodiversity Center of Japan, Nature Conservation Bureau, Ministry of the Environment, Fujiyoshida

BCJ (Biodiversity Center of Japan, Nature Conservation Bureau, Ministry of the Environment) (2015) Fiscal year 2014 survey of the vegetation and sea area in the Pacific Coastal area of the Tohoku region, survey report. Biodiversity Center of Japan, Nature Conservation Bureau, Ministry of the Environment, Fujiyoshida

Geospatial Information Authority of Japan (2011) No title. http://www.gsi.go.jp/common/000059961.pdf. Accessed 4 Jul 2015

Hara M (2012) Miyagi-Iwate-no-hisaichi-no-shokusei-wo-mite (Observations on vegetation at the disaster stricken area in Miyagi Prefecture and Iwate Prefecture). Veg Sci News 16:32–39

Hara K (2014a) Damage to coastal vegetation due to the 2011 tsunami in northeast Japan and subsequent restoration process: analyses using remotely sensed data. Glob Environ Res 18:27–34

Hara M (2014b) Tsunami-eikyou-chousa-no-kekka-ni-tsuite (A report of research on impacts of the Great East Japan Tsunami of 2011). Veg Sci News 18:21–40

Hara K, Higuchi H (2013) Effects of the 2011 Great East Japan Earthquake and Tsunami disasters on ecosystems. Chikyu Kankyo 18:23–33

Hara M, Asami K, Tomita M (2012) Shokusei-gakkai Higashinihon-daishinsai-Minami-sanriku-chiiki-chousa-houkoku (A report on impacts of the East Japan Earthquake of 2011 in Minami-sanriku region). Veg Sci News 16:40–48

Hayasaka D, Shimada N, Konno H, Sudayama H, Kawanishi M, Uchida T, Goka K (2012) Floristic variation of beach vegetation caused by the 2011 Tohoku-oki tsunami in northern Tohoku, Japan. Ecol Eng 44:227–232. doi:10.1016/j.ecoleng.2012.03.014

Himiyama Y (2013) The Great East Japan Earthquake and Tsunami of 2011 and land use. Chikyu Kankyo 18:35–44

Himiyama Y, Watanabe T, Hara K (2014) The impact of the Great East Japan Tsunami of 2011 on land use. Glob Environ Res 18:191–206

Hirabuki Y, Tomita M, Kanno H, Hara K (2011) Impact of Great East Japan Earthquake and subsequent tsunami on vegetation in the sand-dune coastal ecotone along the shores of Sendai Bay, in the Tohoku District of northern Japan. Jpn J Med Res 33:45–57

Japan Meteorological Agency (2011) Monthly report on earthquakes and volcanoes in Japan, March 2011. http://www.data.jma.go.jp/svd/eqev/data/gaikyo/monthly/201103/monthly201103.pdf. Accessed 4 Jul 2015

Kadoya T, Takenaka A, Ishihama F, Fujita T, Ogawa M, Katsuyama T, Kadono Y, Kawakubo N, Serizawa S, Takahashi H, Takamiya M, Fujii S, Matsuda H, Muneda K, Yokota M, Yonekura K, Yahara T (2014) Crisis of Japanese vascular flora shown by quantifying extinction risks for 1618 taxa. PLoS ONE 9(6), e98954. doi:10.1371/journal.pone.0098954

Kanaya G, Suzuki T, Maki H, Nakamura Y, Miyajima Y, Kikuchi E (2012) Effects of the 2011 tsunami on the topography, vegetation, and macrobenthic fauna in Gamo Lagoon, Japan. Jpn J Benthol 67:20–32

Kanaya G, Maki H, Suzuki T, Sato-Okoshi W, Kikuchi E (2014) Tsunami-induced changes in a shallow brackish lagoon ecosystem (Gamo Lagoon) in Sendai Bay, Japan. Glob Environ Res 18:35–46

Kaneko S, Ohta Y, Shirakawa K, Inoue M, Tsutsumi M, Watanabe S, Sakuma T, Takahashi Y (2009) An attempt to evaluate the habitat type of endangered plant species in the Chugoku region, western Japan. Jpn J Conserv Ecol 14:119–123

Kanno H, Hirabuki Y, Sugiyama T, Tomita M, Hara K (2014) Vegetation change in various coastal forest habitats after a huge tsunami: a three-year study. Jpn J Conserv Ecol 19:201–220

Kasai H (2013) Miyagi-ken-no-Higashi-nihon-daishinsai-tsunami-shinsui-iki-ni-okeru-kishou-shokubutsu-tou-no-2012-nen-no-genjou (Situation of rare plants and other plants in 2012 at inundation area in Miyagi Prefecture, Japan, by the Great East Japan Tsunami of 2011). Bull Bot Soc Tohoku 17:32–43

Kurosawa T (2014) Higashi-nihon-daishinsai-zengo-no-Fukushima-ken-no-kaigan-no-shokusei--to-shokubutsusou-no-henka-oyobi-shokusei-ya-shokubutsu-tayousei-no-hozen-no-joukyou (Changes of coastal vegetation and flora before and after the Great East Japan Earthquake of 2011 and situation of conservation of vegetation and plant diversity). Veg Sci News 18:70–80

Matsumoto S (2013) Invasion of seawater and marine sediments into arable lands. Chikyu Kankyo 18:55–62

Matsumoto S (2014) Invasion of seawater and marine sludge into arable land. Glob Environ Res 18:57–63

Mineta T, Tomosyo T (2012) Necessity of vegetation management of agricultural land flooded by tsunami after the 2011 off the Pacific coast of Tohoku Earthquake: report of vegetation investigation in the southern Sendai Plains. Tech Rep Nation Inst Rural Eng 213:297–303

Ministry of the Environment (2015) Red data book 2014: threatened wildlife of Japan, vol 8, vascular plants. Gyosei Co, Tokyo
Miyagi T, Yanagisawa H, Baba S (2013) The protective role of mangroves and other coastal forests against tsunami damage: lesson learned from case studies of two tsunamis. Glob Environ Res 17:247–254
Mori N, Takahashi T, The 2011 Tohoku Earthquake Tsunami Joint Survey Group (2012) Nationwide post event survey and analysis of the 2011 Tohoku Earthquake Tsunami. Coast Eng J 54:1250001. doi:10.1142/S0578563412500015
Nagahata Y (2012) Kyodai-tsunami-ha-seitaikei-wo-dou-kaetaka: ikimono-tachi-no-Higashi-nihon-daishinsai (What changes did occur in ecosystem due to the large tsunami? The Great East Japan Earthquake of 2011 for living things). Kodansha, Tokyo
Naito T (2012) Miyagi-ken-Watari-gun-Watari-machi-no-Higashi-nihon-daishinsai-no-tsunami-ni-yori-kaisui-no-shinnyushita-suiden-no-shokusei (Vegetation of paddy fields in seawater inundation area by the Great East Japan Tsunami of 2011 in Watari-machi, Watari-gun, Miyagi Prefecture, Japan). Miyagi-no-shokubutsu 37:15–18
Nakamura T, Kitazawa T, Sugita H (eds) (2012) A catastrophic disaster and biodiversity. Biodiversity Network Japan, Tokyo
Nature Conservation Bureau, the Environment Agency (1998a) Dai-5-kai-shizenkankyou-hozen--kiso-chousa, umibe-chousa, sougou-houkokusho (Report on the 5th basic survey for conservation of natural environments, coast survey). http://www.biodic.go.jp/reports/umibe/umibe.pdf. Accessed 3 Jul 2015
Nature Conservation Bureau, the Environment Agency (1998b) Dai-5-kai-shizenkankyou-hozen--kiso-chousa, umibe-chousa, dēta-hen (Data of the 5th basic survey for conservation of natural environments, coast survey). http://www.biodic.go.jp/reports/umibe/umibedata.pdf. Accessed 3 Jul 2015
Nemoto S, Kurosawa T, Fujiwara K (2013) Endangered plants on the coast in the southern part of Fukushima Prefecture, Japan, after the 2011 Tohoku earthquake and tsunami. J Center Reg Aff Fukushima Univ 24(2):81–95
Nishihiro J, Hara K, Hirabuki Y (2014) Biodiversity conservation and infrastructure reconstruction after a large-scale disaster: lessons from the coastal regions of southern Sendai Bay. Jpn J Conserv Ecol 19:221–226
Oka K, Hirabuki Y (2014) Revegetation of coastal plants damaged by the 2011 Tohoku tsunami. Jpn J Conserv Ecol 19:189–199
Onza N, Ishida I, Tomita M, Hara K, Hirabuki Y, Nishihiro J (2014) Environmental heterogeneity and plant diversity in a coastal forest affected by a severe tsunami. Jpn J Conserv Ecol 19:177–188
Oyama M (2012) Tsunami-no-eikyou-ni-yoru-Kitakami-sanchi-chu-hokubu-no-kaigan-shokusei-no-joukyou-ni-tsuite (Impacts of the Great East Japan Tsunami of 2011 on coastal vegetation in northern and central parts of Kitakami Mountains area, Japan). Veg Sci News 16:49–58
Oyamada T, Kurakake S, Arai R, Yamauchi K, Katayama C (2012) Disappearance of coastal plants by the tsunami of the East Japan great earthquake disaster in Iwate Prefecture. Jpn J Med Res 34:37–48
Prec Institute (2011) Heisei-23-nendo-Higashi-nihon-daishinsai-ni-yoru-shizenkouen-tou-heno--eikyou-chousa-gyoumu-houkokusho (Fiscal year 2011 survey of impacts on natural park and other areas by the Great East Japan Earthquake, survey report). http://www.shiokaze.biodic.go.jp/data/natural_park_survey23/natural_park_survey23-1.pdf, http://www.shiokaze.biodic.go.jp/data/natural_park_survey23/natural_park_survey23-2.pdf. Accessed 26 Jul 2015
Richmond B, Szczuciński W, Chagué-Goff C, Goto K, Sugawara D, Witter R, Tappin DR, Jaffe B, Fujino S, Nishimura Y, Goff J (2012) Erosion, deposition and landscape change on the Sendai coastal plain, Japan, resulting from the March 11, 2011 Tohoku-oki tsunami. Sediment Geol 282:27–39. doi:10.1016/j.sedgeo.2012.08.005
Sawada Y, Nakanishi H, Oshida K, Hattori T (2007) A check list of coastal plants in Japan. Hum Nat 17:85–101

Science Council of Japan (2014a) Inochi-wo-hagukumu-anzen-na-engan'iki-no-keisei-ni-muketa--kaiganrin-no-saisei-ni-kansuru-teigen (Recommendation for reconstruction of coastal forest to form safe coast area against a disaster in consideration of animal and plant life). http://www.scj.go.jp/ja/info/kohyo/pdf/kohyo-22-t140423.pdf. Accessed 28 Aug 2015

Science Council of Japan (2014b) Fukko-kokudo-kyoujinka-ni-okeru-seitaikei-infurasutorakuchā-katsuyou-no-susume (Recommendation for application of ecosystem infrastructure to reconstruction projects and building national resilience to disasters). http://www.scj.go.jp/ja/info/kohyo/pdf/kohyo-22-t199-2.pdf. Accessed 28 Aug 2015

Shibuya K, Shimada N, Suzuki M (2014) Study of the present condition of the Sanriku coastal eco-tone and environmental conservation measures in Iwate Prefecture. J Policy Stud 15:181–199

Shimada N (2014) Higashi-nihon-daishinsai-ni-yoru-tsunami-ga-Iwate-ken-no-kaigan-shokusei-ni-ataeta-eikyou-oyobi-sonogo-no-shokusei-saisei-to-hozen-joukyou-ni-tsuite (Impacts of the Great East Japan Earthquake and Tsunami of 2011 on coastal vegetation in Iwate Prefecture, Japan, and situation of subsequent vegetation recovery and conservation). Veg Sci News 18:44–54

Shimada N, Kawanishi M, Hayasaka D (2014) Ecological evaluation of beach vegetation recovery after the Tohoku Region Pacific Coast Earthquake, and a proposal for environmental conservation measures in Iwate Prefecture. J Policy Stud 16:19–34

Shimoyama Y (2013) Miyagi-ken-no-Sanriku-engan-ni-okeru-Higashi-nihon-daishinsai-go-no-sakyu-shokusei-no-bunpu-joukyou (Distribution of sand dune vegetation after the Great East Japan Earthquake of 2011 in Sanriku Coast in Miyagi Prefecture, Japan). Miyagi-no-shokubutsu 38:1–5

Sugiyama T (2012) Ōtsunami-go-no-Natori-shi-Mitazono-chouseiike-A-no-shokubutsu (sourui-fukumu) (Plants and algae after the Great East Japan Tsunami of 2011 in Mitazono Regulating Reservoir A, Natori-shi, Miyagi Prefecture, Japan). Miyagi-no-shokubutsu 37:19–23

Sugiyama T, Kasai H (2013) Ōtsunami-go-no-Natori-shi-Shimomasuda-Kitaharahigashi-no-shokubutsu (Plants of Kitaharahigashi, Shimomasuda, Natori-shi, Miyagi Prefecture, Japan, after the Great East Japan Tsunami of 2011). Miyagi-no-shokubutsu 38:9–15

Sugiyama H, Koseki S (2005) Shizen-kankyou-no-hensen (Changes in natural environment). In: Nature Conservation Group, Life and Environment Department, Fukushima Prefecture (ed) Juuyou-shicchi-Matsukawa-ura-sougou-chousa-houkokusho (Report on Survey Project of Matsukawa-ura, Fukushima Prefecture, Japan). Nature Conservation Group, Life and Environment Department, Fukushima Prefecture, Fukushima

Sugiyama T, Kasai H, Emi Y (2013) Higashi-nihon-daishinsai-ō-tsunami-go-no-Sendai-shi-Ōnuma-no-shokubutsusou-no-henka (Changes in flora of Ōnuma Pond, Sendai-shi, Miyagi Prefecture, Japan, after the Great East Japan Tsunami of 2011). Bull Bot Soc Tohoku 17:44–52

Takiguchi M, Hirabuki Y, Kanno H, Naitou T, Sugiyama T, Simoyama Y, Kasai H (2014) Drastic changes of vegetation in the huge tsunami-disturbed area, Pacific coast of Miyagi Prefecture. Veg Sci News 18:55–69

Tanaka N (2012) Effectiveness and limitations of coastal forest in large tsunami: conditions of Japanese pine trees on coastal sand dunes in tsunami caused by Great East Japan Earthquake. J Jpn Soc Civil Eng Ser B1 (Hydraul Eng) 68(II):7–15

Tanaka H, Tinh NX, Umeda M, Hirao R, Pradjoko E, Mano A, Udo K (2012) Coastal and estuarine morphology changes induces by the 2011 Great East Japan Earthquake Tsunami. Coast Eng J 54:1250010. doi:10.1142/S0578563412500106

Tomita M, Hirabuki Y, Kanno H, Hara K (2013) Spatial distribution of biological legacies and depth of accumulated sand in a coastal forest disturbed by the 2011 tsunami. J Nat Restor Conserv 6:51–60

Tomita M, Hirabuki Y, Kanno H, Hara K (2014) Influence of tsunamis as large, infrequent disturbances on tree communities of coastal forests. Jpn J Conserv Ecol 19:163–176

Udo K, Sugawara D, Tanaka H, Imai K, Mano A (2012) Impact of the 2011 Tohoku Earthquake and Tsunami on beach morphology along the northern Sendai Coast. Coast Eng J 54:1250009. doi:10.1142/S057856341250009X

Une H, Koarai M (2014) Disaster responses of the Geospatial Information Authority of Japan: special focus on tsunami inundation mapping. Glob Environ Res 18:9–18

Watanabe Y, Kurosawa T (2015) Flora and vegetation of conservation area established at tidal flat and salt marsh newly appeared due to the Great East Japan Earthquake and Tsunami of 2011 in Matsukawa-ura Lagoon, Fukushima Prefecture, Japan. J Center Reg Aff Fukushima Univ 27(1):67–92

Yonekura K, Maki M (2015) Miyatojima-no-shokubutsusou: Miyatojima-no-shokubutsusou-to-higashinihon-daishinsai-ni-yoru-tsunami-to-jibankouka-niyoru-sore-heno-eikyou (Flora of Miyatojima: impacts of the tsunami and land subsidence caused by the Great East Japan Earthquake of 2011 on flora of Miyatojima). The Historical Museum of Jomon Villege OkuMatsushima (ed) OkuMatsushima II, pp. 9–58. Miyato Nobiru-chiiki-no-bunkaisan-no--saisei katsuyou-kenntou-jikkou-iinnkai, Higashimatsushima

Yoshikawa M, Ayukawa E (2014) Aomori-ken-no-kaigan-shokusei-ni-taisuru-tsunami-no-eikyou (Impacts of the Great East Japan Tsunami of 2011 on coastal vegetation in Aomori Prefecture, Japan). Veg Sci News 18:41–43

Yoshino M (2014) Impact of tsunamis on human life and society. Glob Environ Res 18:3–8

Yuzawa Y (2013) Assessment of damages on coastal and seashore plants in Fukushima Prefecture caused by Japan Tsunami following the Great East Japan Earthquake on 2011. J Phytogeogr Taxon 61:1–14

Zhao Y, Tomita M, Hara K (2013) Landscape change analysis before and after the earthquake disaster in Sendai coastal area by using SPOT satellite data. J Nat Restor Conserv 6:43–49

Chapter 20
Regeneration of Damaged Coastal Forests Caused by the Great East Japan Earthquake and Tsunami

Tomoki Sakamoto

Abstract The Great East Japan Earthquake and subsequent tsunami damaged the coastal forests in the large area in Japan. Especially, almost all the coastal forests were lost from Iwate Prefecture to Fukushima Prefecture. The coastal forests had served as a form of disaster prevention against wind, tide, and blown sand as well as provided landscape amenities and places for recreation until their multifaceted functions were entirely lost. Considering these lost functions, regeneration of these coastal forests is essential in the post-earthquake recovery process. It is desirable that these forests be regenerated not only for restoring them to their condition before the tsunami but also for improving them to be stronger against tsunami than before and to work effectively for tsunami damage reduction. I review the damage this tsunami has caused to coastal forests and the ways by which coastal forests may have reduced tsunami damages from the disaster prevention point of view. I also summarize characteristics of coastal forests as disaster prevention facilities and, finally, describe my thoughts on regeneration and recovery of coastal forests.

Keywords Tsunami • Coastal forest • Disaster prevention • Tohoku

20.1 Introduction

On March 11, 2011, 14:46, a huge earthquake centered off the Sanriku coast occurred, resulting in a large tsunami and that led to the worst disaster of the post-war period on the Pacific coast in east Japan. Many lives and much property and social infrastructures were destroyed. Coastal forests were also considerably affected in a wide area between Aomori and Chiba Prefectures. In particular, forests located between Iwate and Fukushima Prefectures were severely damaged. The

---

T. Sakamoto (✉)
Tohoku Research Center, Forestry and Forest Products Research Institute,
Morioka, Iwate, Japan
e-mail: safe@ffpri.affrc.go.jp

J. Urabe, T. Nakashizuka (eds.), *Ecological Impacts of Tsunamis on Coastal Ecosystems*, Ecological Research Monographs,
DOI 10.1007/978-4-431-56448-5_20

coastal forests had served as a form of disaster prevention against wind, tide, and blown sand as well as provided landscape amenities and places for recreation until their multifaceted functions were entirely lost. Considering these lost functions, regeneration of these coastal forests is essential in the post-earthquake recovery process. It is desirable that these forests be regenerated not only for restoring them to their condition before the tsunami but also for improving them.

The Forestry and Forest Products Research Institute conducted "Fiscal 2011 Emergency Survey on Post-earthquake Recovery (Research on Reduction of Tsunami Damages by Coastal Disaster Protection Forests)" as a Forestry Agency-sponsored project, with the help of the Japanese Society of Coastal Forest. Leaving survey details to the survey report and publications by members of the Japanese Society of Coastal Forest (ex. Teramoto et al. 2012; Yamanaka et al. 2012; Yamanaka et al. 2013), I review herein, as a member of the project, the damage this tsunami has caused to coastal forests and the ways by which coastal forests may have reduced tsunami damages from the disaster prevention point of view. I also summarize characteristics of coastal forests as disaster prevention facilities and, finally, describe my thoughts on regeneration and recovery of coastal forests.

The damaged coastal forests were basically Japanese black pine (*Pinus thunbergii*) forests though Japanese red pine (*Pinus densiflora*) might mix, because coastal forests have been established with Japanese black pine in Japan historically and they were not an exception.

I thank the Conservation Division; Tohoku Regional Forest Office; Kanto Regional Forest Office of the Forestry Agency; Prefectures of Aomori, Iwate, Miyagi, Fukushima, Ibaraki, and Chiba; Japan Forest Technics Co.; Japan Conservation Engineers & Co., Ltd.; the Japanese Society of Coastal Forest; and many other entities for their support and cooperation in a series of surveys. The original of this article is published as a research achievement of the third midterm plan 29 (Sakamoto 2015). I made bare-bones finishing touches for this book.

20.2 Tsunami Damages to Coastal Forests

20.2.1 Huge Tsunami

The Great East Japan Earthquake (The 2011 off the Pacific coast of Tohoku Earthquake) that caused the tsunami of present interest was a huge earthquake of magnitude 9.0, with its epicenter located off the Sanriku coast. The tsunami that attacked coastal areas in the Tohoku region was never encountered before and was far larger than what people could have imagined (Fig. 20.1). The waves were particularly high in coastal areas in the northern part of Miyagi Prefecture and Iwate Prefecture, wherein the ria coastline had developed; seawalls were destroyed and forests were devastated in coastal areas affected by the huge tsunami. For example, in the scenic Takata-matsubara in Rikuzentakata city, because of pre-existing ground subsidence, the tsunami damage was not limited to trees but included the land, which was largely washed out and had become a sea. The wave height in the

Fig. 20.1 Catastrophically damaged coastal forest (Matsukawaura, Fukushima Prefecture; provided by ©Fukushima Prefecture Japan, 2011. All Rights Reserved). The coastal forest was lost with the ground, leaving only a part. The seawall along the shoreline was also destroyed

Fig. 20.2 Destroyed seawall and scouring behind it (Iwanuma City, Miyagi Prefecture; ©T. Sakamoto 2011. All Rights Reserved). The seawall was destroyed and the sea surface can be observed. The strip-shaped water surface on the right side is a depression considered to have been scoured by the tsunami that overflowed the seawall. Concrete pieces at the front and *white* objects in the center are constituent materials of the damaged seawall

Sendai Plain was not as high as that in the ria coastline, but some inundation heights exceeding 10 m were recorded.[1] In many places in the Sendai Plain, the tsunami affected seawalls and scoured the land to create depressions on the inland side of the seawalls. The seawalls themselves were also violently destroyed, presumably owing largely to the scouring (Fig. 20.2). A phenomenon similar to the scouring on the inland side of the seawalls also occurred in an artificial sand dune. Furthermore, there were some places that had been coastal forests but became seas, presumably because the topsoil was washed away mainly by the concentrated backrush.

20.2.2 Tsunami Damages to Coastal Forests

Types and degrees of damages of coastal forests were diverse, considering that the tsunami reached a wide range of areas and its scale, terrain conditions, affected backrushs, size and type (tree species, tree height, breast-height diameter, clear stem height, and stand density) of coastal forests, topsoil conditions, and disaster prevention facilities, such as seawalls, were different among locations. In the damage survey of coastal forests, the types of damage to trees were divided into stem breakage, uprooting, and leaning.

[1] Based on preliminary estimates by the 2011 Tohoku Earthquake Tsunami Joint Survey Group (http://www.coastal.jp/ttjt/) (seen November 11, 2011).

Fig. 20.3 Stem breakage (Soma City, Fukushima Prefecture; ©T. Sakamoto 2011. All Rights Reserved). The broken stem remained connected to the stump in some cases or was washed away in other cases. The state of the breakage edge also is not uniform

20.2.2.1 Stem Breakage, Uprooting, and Leaning

"Stem breakage" refers to a broken stem (Fig. 20.3). Although stems of some trees appeared to be just bent without being completely broken, cracking was confirmed inside the stems. Among trees with broken stems, the stem was completely separated from the stump in some trees or connected to the stump with bark or other tree parts. If only the stump was left without the stem being attached to it, it meant that the upper part (stem) had become driftwood.

"Uprooting" refers to a state wherein the stem tilted and the root mass was lifted. The degrees of uprooting widely varied, from the lodging state wherein the tree was fully turned over to the state wherein some roots appeared above the ground surface or roots were floating in the sediment layer.

Typical examples were observed on the inland side of the Teizan Canal in the Sendai Plain and lowlands such as Matsukawaura in Fukushima Prefecture (Satoh et al. 2012; Tamura 2012; Watanabe et al. 2014). For many uprooted trees, development of sinker roots (taproots and spindle roots) was not observed, and the roots revealed a thin board-like shape (Fig. 20.4). Further, many places wherein uprooting occurred were swampy lowlands (Fig. 20.5). These findings suggest that trees in these areas had no deep roots because of the high groundwater level and were

Fig. 20.4 Root system of drift trees (Soma City, Fukushima Prefecture; ©T. Sakamoto 2011. All Rights Reserved). Many drift trees with roots had board-shaped root systems and sinker roots had not been developed

Fig. 20.5 Uprooting (Sendai City, Miyagi Prefecture; ©T. Sakamoto 2011. All Rights Reserved). Uprooting continuously occurred from the seaside to the land side, leaving a strip-shaped forest zone such as after line logging. Water remained even 13 days after the tsunami, and the place was like wetland

Fig. 20.6 Pushed down coastal forest (Sendai City; ©T. Sakamoto 2011. All Rights Reserved). Trees with stems broken at the base and those that underwent uprooting are mixed. Although the damage was devastating, trees were not necessarily washed away (thus did not become driftwood)

readily pushed down by the tsunami with an additional effect of buoyant force from the tsunami rather than being pulled out by the tsunami. Root damage because of shaking by the earthquake and the loss of binding force of roots because of liquefaction are also believed to have facilitated uprooting (Committee on Regeneration of Coastal Disaster Prevention Forests related to the Great East Japan Earthquake 2012).

"Leaning" is a state wherein the stem is inclined (Fig. 20.6). Causes of leaning include stem breakage near the base and uprooting. If the cause could be identified, the event was classified into either of these causes. However, the cause could not be confirmed to be breakage or uprooting in some cases because the basal part was covered with sand brought by the tsunami or blown sand accumulated before the tsunami; these cases were classified as "leaning." Some trees were found to undergo both uprooting and stem breakage.

The association between stem and root strength (resistance to uprooting) against the force exerted by the tsunami determines whether the tree undergoes stem breakage or uprooting. In brief, uprooting occurs if the root is weaker than the stem against the force of the tsunami, and stem breakage occurs if the resistance of the root is high.

Compared with trees with a root system having a board-like shape of less than 1 m depth, those with a root system normally developed beyond a certain thickness

Fig. 20.7 Stem breakage (Rikuzentakata City, Iwate Prefecture; ©T. Sakamoto 2011. All Rights Reserved). In Takata-matsubara, most of the coastal forest has disappeared along with ground, but stem breakage was readily noticeable where the ground remained. Washed-out roots had well-developed sinking roots. The breakage edges were complex, suggesting that breakage was not completed in a single strike

were rarely uprooted or became driftwood with attached roots (described below). In contrast, stem breakage of trees with a relatively large diameter, leaving the stump, was observed in clusters (Fig. 20.7). It appeared that trees of a size beyond a certain level with developed sinking roots underwent stem breakage before uprooting occurred, unless the ground was scoured.

Watanabe et al. (2014) investigated damage coastal forest caused by the tsunami in relation to type of damage, tree size, and depth of groundwater in Sendai Plain Miyagi Prefecture (Fig. 20.8). According to their results, the stems with DBH over 20 cm were hard to be broken, but uprooting happened regardless of diameter. The trees are uprooted, and the cause is believed to be the weakness of the ground at this location and the undeveloped depth of the root. In other words, it is important that root systems are properly developed.

It is noteworthy that uprooting and stem breakage occurred not only from the power of the tsunami itself but also from impacts by tsunami-driven objects. For example, concrete blocks from broken seawalls did not stay in place but entered coastal forests on the inland side. These concrete blocks then directly strike coastal forests and pushed down or damaged trees therein (Fig. 20.9). In Ichikawa-cho in Hachinohe city, the coastal forest was possibly pushed down by such objects like fishing boats drifting into the forest (Fig. 20.10; Sakamoto et al. 2012).

"Driftwood" is a phenomenon wherein some part or the entire tree is moved from where it has grown to some other place. Some driftwood occurred with roots attached owing to uprooting (washout), and others occurred without roots owing to stem breakage. When such trees were found in the survey area, they were recorded as driftwood.

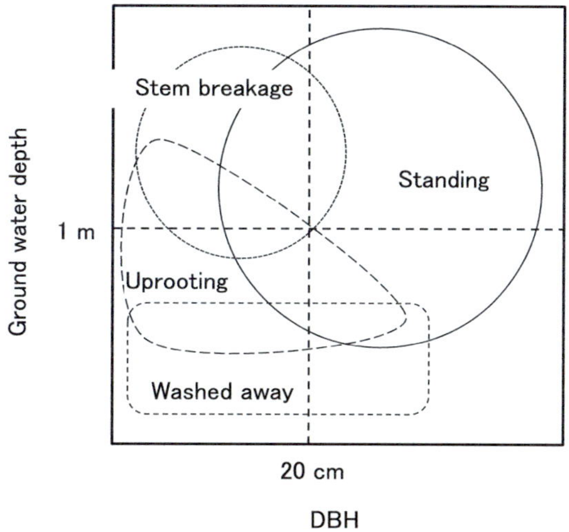

Fig. 20.8 Types of damage in relation to tree size and groundwater depth (Watanabe et al. 2014)

Fig. 20.9 Concrete blocks scattered in the coastal forest (Natori City, Miyagi Prefecture; ©T. Sakamoto 2011. All Rights Reserved). Constituent materials (such as concrete blocks) of seawalls destroyed by the tsunami became drifting objects and entered coastal forests. Coastal forests were severely damaged not only by the tsunami itself but also by these concrete blocks

20.2.2.2 Survival/Dieback

Although many coastal forests and trees constituting coastal forests were severely damaged, other coastal forests survived the tsunami. The Nagahama coastal forest in Ishinomaki city in Miyagi Prefecture is such an example (Fig. 20.11). The area

Fig. 20.10 Drifting objects transported into coastal forest (Hachinohe City, Aomori Prefecture; provided by ©Hachinohe city forest owners' cooperative, 2011. All Rights Reserved). Fishing boats and various other drifting objects entered coastal forests. From a different point of view, coastal forests captured objects drifting because of the tsunami and prevented them from being transported inland

Fig. 20.11 Surviving coastal forest (Ishinomaki City, Miyagi Prefecture: March 12, 2011; provided by ©Aero Asahi Corporation, 2011. All Rights Reserved). Surrounding areas, including those behind the coastal forest, were flooded, and many houses were washed away to the east of the coastal forest (at *right* in photo); however, damage to the forest was limited. However, many trees died later

behind this coastal forest was flooded to approximately 4 m high, and many houses on the west side (the right side facing the picture) of the coastal forest were washed away. Despite these serious damages in the surrounding area, the initial damage to the forest was limited to those at the front edge hit by constituent materials of seawalls. Plausible reasons for the survival of the forest include that the forest was constituted of relatively large-diameter trees with mean breast-height diameter of 29 cm[2] and that the clear stem height was relatively high compared with the inundation height. Furthermore, compared with the inundation depth of 4 m in the forest, the mean clear stem height was as high as 13 m; therefore, the tsunami presumably passed between the trees and did not strike the tree crowns (foliage layer). As a result, the trees received wave power weaker than the resistance of the stem, and this could be the reason why this forest survived.

Among coastal forests that appeared to have survived immediately after the tsunami, some subsequently showed browning leaves and were debilitated (Nakamura et al. 2012). Such trees were located where near base areas had been scoured or in relatively low places, wherein the water residence time was possibly longer. The possible cause of scouring is that the flow velocity increased in dunes inclining toward the inland side, intensifying the scouring force.

Debilitation of the surviving trees was also observed in the coastal areas in the Sendai Plain. Until 2 weeks after the disaster, a few isolated dead trees could be observed, but no cluster of dead trees was apparent in the coastal areas in the Sendai Plain. Moreover, even leaning trees had green leaves. However, when I revisited the Sendai Plain more than 2 months later, most leaning trees had brown leaves, and even some trees standing upright had browning leaves and seemed weakened.

The cause of debilitation in these trees is not necessarily clear, but many weakened trees were those with their roots partly exposed by scouring and those located in relatively low areas, wherein the flooding period was estimated to be longer than in the surrounding areas. Ono et al. (2013) reported that the incremental changes of soil pH(H_2O) in the 2A horizons caused by the tsunami may strongly affect the microenvironments of the rhizospheres and in turn cause severe damages, including needle discoloration, debilitation, droop, and finally death, to broad areas of pine forests along the eastern Pacific coastline.

Debilitation of the remaining trees has been confirmed in various regions, and the affected range and degree of damage to coastal forests may become wider and more severe than estimated immediately after the tsunami.

20.3 Functions of Coastal Forests Against Tsunami

Coastal forests, although they experienced a variety of damage, revealed tsunami damage-reducing functions in various places.

[2] Thirty-six centimeter, if only canopy-constituting trees (trees higher than 20 % in this coastal forest) were concerned.

Fig. 20.12 Leaning of coastal forest caused by the tsunami (Misawa City, Aomori Prefecture; ©T. Sakamoto 2011. All Rights Reserved). Although the coastal forest leaned toward the inland, leaning terminated in the middle of the coastal forest, with a clear boundary

20.3.1 Diminishing Wave Power

The wave power-reducing function of a coastal forest is produced by the forest acting as a resistor against running water for reducing the flow velocity and wave power of a tsunami. By this function, it is expected that structure damage by the tsunami will be reduced and that the tsunami arrival is delayed so that additional time for evacuation is acquired. During evacuation, it is possible for people to escape death by a hairbreadth; even a difference of a few seconds may matter at the last minute.

Although these functions have been clearly demonstrated in numerical simulations, it is difficult to clearly demonstrate them in the field, considering that multiple factors are involved in tsunami damage. Examples of wave power reduction by coastal forests have also been reported in the Indian Ocean Tsunami in 2004, but there has been an objection that the presented data were insufficient. A rebuttal to the objection has been published but has not adequately addressed the issue (Sakamoto and Noguchi 2009).

In the tsunami of present interest, in Aomori Prefecture, wherein the scale of the tsunami was smaller than in the area between Iwate and Fukushima Prefectures, there were some places in which leaning of trees because of the tsunami was clearly confirmed to have stopped in the middle of a coastal forest (Fig. 20.12). This fact suggests that the wave power was reduced while the tsunami was passing through the forest (Sato et al. 2012; Fig. 20.13, Noguchi et al. 2012).

20.3.2 Capturing Drifting Objects

This is a function to stop the transportation of drifting objects by tsunami. It includes a function of preventing drifting objects such as ships and rubble from colliding with conservation targets, such as houses, function of preventing additional tsunami

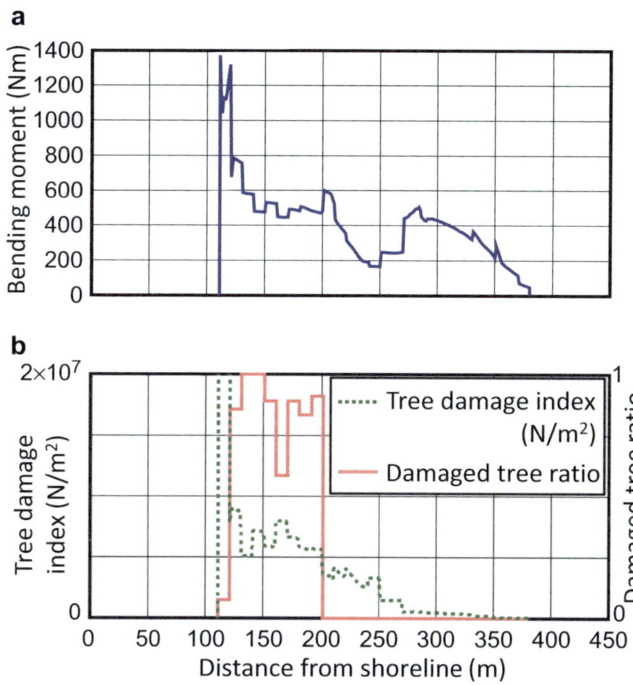

Fig. 20.13 Maximum bending moment every 10 m (**a**), tree damage index and damaged tree ratio every 10 m (**b**); Noguchi et al. (2012)

damages caused by houses and other objects that have become drifting objects, and function of preventing houses and other objects from flowing out to sea because of backrush. There is a certain level of uncertainty in these functions because drifting objects can pass between trees, but effects can be expected unless lodging or wash-out of the forest zone occurs.

Although drifting objects pass less readily through a wider forest zone, even a single tree can pose a barrier in some cases because even if it is a single row, a coconut grove has the potential to mitigate the damage caused by flotsam (Sakamoto et al. 2008), and there are important differences in these functions between places without and with trees. Unlike the wave power-reducing function, this function is easy to demonstrate, because floating objects are often left on site.

For this function, it has been highlighted that debris captured by the forest zone may be driven again by backrush and may cause additional damage (Laso Bayas et al. 2011). However, in the case of large-scale tsunami wherein rubble that becomes drifting objects increases damage, places attacked by rubble associated with backrush are possible to have been already damaged seriously by the spilling wave. Between additional damage caused by rubble being pulled by backrush and damage avoided by capturing rubble, the latter is expected to be larger. Therefore, even if the possibility that rubble captured by the coastal forest is driven again by backrush reflects the actual situation, underestimating the function of trees to capture floating objects is not justified.

In this survey, a case wherein ships had been stopped by a coastal forest (Fig. 20.10) and cases wherein concrete blocks from broken seawalls and other drifting objects stayed in the forest although they had pushed down trees in the forest (Figs. 20.2 and 20.9) were observed. Further, we observed a case wherein driftwood was trapped by the surviving trees.

20.3.3 Regulation of Land Use

This is a function for regulating land use and preventing conservation targets from danger by occupying a certain amount of coastal area with coastal forests. In this function, coastal forests do not directly act against tsunami, but their disaster prevention effect is reliable and large and deserves positive evaluation.

For example, in areas affected by the tsunami of present interest, concrete masses from seawalls were scattered in coastal forests (Fig. 20.9). If buildings had been there, these concrete masses could have directly struck them.

Moreover, in Nagahama in Ishinomaki city (Fig. 20.11), land development for housing was suppressed owing to the presence of coastal forest; thus, the forest prevented potential damage to houses that could have been built there and damaged by the tsunami had the coastal forest not been there. Furthermore, it has been evaluated that the coastal forest prevented potential additional damage from these houses that could have occurred if they had been washed away, broken into rubble, and driven inland by the tsunami (Fig. 20.14, Okada et al. 2012).

20.3.4 Means to Escape from Tsunami

Counter-tsunami functions of a coastal forest include providing means of escaping from the tsunami, such as climbing, holding, and soft landing. Specifically, these are functions that allow people to climb trees higher than the inundation depth and wait for the tsunami to cease, allow people attacked by the tsunami to hold on to trees in order to avoid being washed away, or to rescue people who have been washed away by the tsunami. Although these are somewhat primitive to be called functions, it is a fact that many lives were saved by trees during the Indian Ocean Tsunami (Sakamoto and Noguchi 2009).

In the case of the tsunami from the Great East Japan Earthquake, atmospheric temperature and water temperature were substantially lower than those in areas affected by the Indian Ocean Tsunami, and trees could not survive the tsunami in many locations because the scale was too large. Thus, benefits from this function might have been limited.

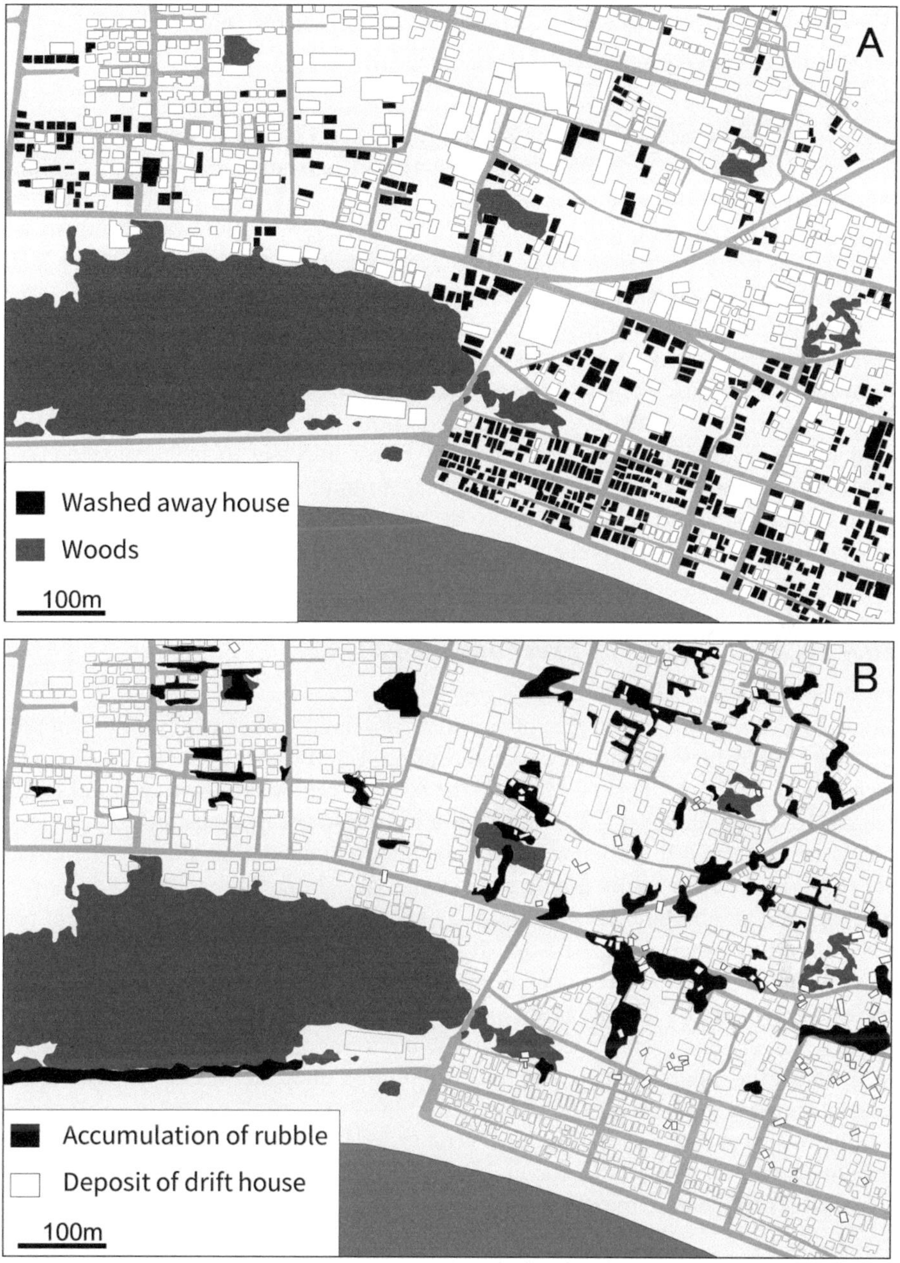

Fig. 20.14 Regulation of land use by the coastal forest (Ishinomaki City, Miyagi Prefecture; Okada et al. 2012). (**a**) Land development for housing was suppressed owing to the presence of coastal forest; thus, the coastal forest prevented potential damage to houses that could have been built there and damaged by the tsunami had the coastal forest not been there. (**b**) It has been evaluated that the coastal forest prevented potential additional damage from these houses that could have occurred if they had been washed away, broken into rubble, and driven inland by the tsunami

20.4 Characteristics of Coastal Forest as a Disaster-Preventing Facility During Tsunami

The basic disaster prevention plan issued by the Central Disaster Prevention Council (2008) in February 2008 recommends promoting afforestation for preventing mountain and avalanche disasters but does not include coastal forests as counter-tsunami measures. This omission presumably occurred because in Japan wherein advanced land use has progressed, it was difficult to assign locations for coastal forests, which are assumed to be invaded by tsunami, as disaster prevention facilities.

However, it is not realistic to construct seawalls sufficient for resisting tsunamis of this scale, owing to both cost and undesirable effects such as those on landscapes. Furthermore, because the risk of overreliance on seawalls has been widely recognized, comprehensive measures, including optimization of land use, will be considered in the future. In this process, coastal forest is likely to be recognized and actively incorporated for its characteristics as a multifaceted space responsible for the tsunami damage-reducing function. In this section, I describe characteristics of coastal forest as a disaster prevention facility.

20.4.1 Low Degree of Freedom

Coastal forests have a low degree of freedom compared with seawalls. For example, if a seawall has been constructed with a 5-m projected tsunami height, but the projected tsunami height increases to 8 m, the seawall can be modified to increase the height. However, increasing the height of a coastal forest is almost infeasible. Tree height has a certain limit depending on the environmental conditions in the target area and can be reduced by pruning but cannot be arbitrarily increased. Moreover, because tree height, tree density, breast-height diameter, and clear stem height are interrelated, for increasing tree density, breast-height diameter cannot exceed a certain limit. Therefore, tree density must be reduced as the tree height increases for creating a coastal forest comprising large-diameter trees. Furthermore, the tree species that can grow are dependent on the environmental conditions of the target area. Thus, the nature of a coastal forest is limited once the location, wherein the coastal forest is to be created has been determined.

The width of the forest zone and ground level can be artificially altered. To enhance functions of a coastal forest depending on the projected scale of tsunami, it is necessary to increase the width of the forest zone and elevate the ground.

20.4.2 Incompleteness

A seawall can completely stop seawater intrusion unless the tsunami height exceeds the seawall height. However, a coastal forest allows tsunami to pass through the forest zone even when the inundation depth does not exceed the tree height and thus cannot prevent flooding of the hinterland. Even small-scale tsunami pass through coastal forests.

However, a coastal forest acts as a barrier to seawater, reducing the wave power and delaying the arrival time. This function can be expected to work until the wave power exceeds the tolerance of trees and coastal forest lodges. Even after the coastal forest has fallen, the forest continues to resist flowing water, albeit to a lower degree. The larger the tsunami scale is, the less noticeable this function is. However, when the tsunami scale is larger, it would be reasonable to say that the wave power-reducing effect is still provided unless the forest trees become driftwood. In contrast, the function of a seawall is largely lost when the tsunami height exceeds the seawall height. As counter-tsunami facilities, coastal forests and seawalls should be evaluated on different criteria.

20.4.3 Uncertainty

Many actual cases are available for illustrating functions of drifting object blocking and climbing/holding/soft landing. However, it is largely uncertain to what extent these functions can be expected to work. For the drifting object-blocking function, drifting objects can pass between tree stems and break or push down trees, depending on the object type, irrespective of the tsunami scale. For climbing/holding/soft landing, the outcome largely depends on individual abilities and luck.

20.4.4 Time

A coastal forest cannot be created in a short time and requires longer than construction of a seawall. Although the required time depends on growth conditions, it is considered to take approximately 20 years until a coastal forest grows to the size required to execute its target functions.

In contrast, a longer life as a disaster-preventing facility can be expected compared with a seawall, which undergoes aging degradation. However, in the case of a pine forest, measures for preventing pine wilt disease spread by pine weevil (Mamiya 1983; Zhao et al. 2008) are required for this advantage to be realized.

20.4.5 Evaluation Under Normal Circumstances

Tsunami damage-mitigating functions of coastal forests involve many incomplete and uncertain parts compared with those of seawalls, and construction of coastal forests also takes time. However, coastal forests are superior in terms of fewer adverse effects and multifaceted utility during normal times. Examples include functions of reducing blown sand damage, serving as a windbreak, reducing wave damage, providing a place for walking, and providing a landscape of white sand and green pines. Coastal forests should be considered helpful in serving as useful spaces for a much longer period than serving as disaster prevention facilities against tsunami.

20.5 Toward Recovery

Although the exact nature of coastal forests to be regenerated/recovered varies under the influence of land use plans in the overall recovery plan, at least regeneration of the coastal forests is necessary regardless of the evaluation of their counter-tsunami functions. Here, I try to highlight how regenerated/recovered coastal forests should look while dividing the process into several stages.

20.5.1 Regeneration of Coastal Forest to the Pre-tsunami State

The first thing to do is regeneration of coastal forests. In regional reconstructions, it is necessary to restore the coastal forests for blown sand control and windbreak. This reforestation is intended to restore the conventional nature of coastal forests, centering on recovery of daily functions of the coastal forests, such as the blown sand defense and windbreak functions. It is noteworthy that wherever uprooting is believed to have occurred because of insufficient root system development because of a high groundwater level, measures, such as embankment, are necessary to ensure the availability of sufficient space for root spreading when we reconstruct a coastal forest in such a place. The need for embankments in low wetlands has been highlighted at the time of the Chile earthquake tsunami in 1960 (Izumi et al. 1961).

20.5.2 Healthy Coastal Forest

It is important that regenerated/recovered coastal forests comprise healthy trees because coastal forests in Japan are not necessarily in a healthy state; many of them are composed of Japanese black pine (*Pinus thunbergii*) or Japanese red pine trees (*Pinus densiflora*). Thus, countermeasures for pine wilt disease (spread by pine weevil) are essential but have not been adequately applied, and there are several almost completely devastated coastal forests. Moreover, in coastal pine forests, trees have been generally planted at a high density of 10,000 trees/ha; thus, it is necessary to adequately manage tree numbers as the planted trees grow. However, in many coastal forests, delayed tree number adjustment has resulted in overcrowded forests, wherein the trees are thin and have long clear stem height relative to their total height.

A coastal forest can provide the blown sand defense function as long as the sandy soil is covered with trees. It can provide the windbreak function without problems if branches on the seaward forest edge are not withered and a desirable tree height is ensured on the leeward edge. Therefore, maintaining forests healthy is the priority, and the stand structure is not specifically adjusted for functional enhancement. However, with the aim of tsunami resistance, the results of this survey suggest that a forest zone composed of large-diameter trees with a well-developed root system is desirable for reducing damages to the coastal forest itself. Thus, coastal forests composed of healthy trees are important for disaster prevention functions.

Concerning coastal forests as disaster prevention facilities, it is desirable to avoid trees turning into driftwood, which could increase damage inland. When trees can no longer withstand a tsunami, it is desirable that stem breakage should not lead to separation from the stump and that uprooting should result in lodging without roots being pulled out or cut off rather than driftwoods resulting from stem breakage and uprooting as shown in Fig. 20.6. If these aims are accomplished, the tsunami-derived force is substantially reduced, and trees are more likely to stay in their original locations. However, current knowledge is insufficient for defining how trees should be arranged and grown to achieve this goal. Our knowledge of associations between specific stand types and the wave power-reducing function of coastal forests has not advanced to a practical level allowing the design of stand types based on it. Thus, presently, a practical approach would be to prioritize establishment of healthy coastal forests and maintenance of existing coastal forests as healthy forests.

20.5.3 Functional Enhancement

As discussed above, it is not practical to alter the tree height, breast-height diameter, clear stem height, or tree density of black and red pines for enhancing tsunami damage-reducing functions, particularly the wave power-reducing function of healthy growing coastal forests. Practically, it is necessary to widen the forest zone to increase the wave power-reducing function. This function is expected to resist a tsunami of a greater magnitude as the width of the forest zone increases. In brief, the desired width of the forest zone varies with the projected tsunami scale and expected reducing effect.

A wider forest zone also helps secure a space to stop driftwood, if created, within the forest. A forest zone that is unlikely to become driftwood should be aimed as a disaster prevention facility, but it is equally important that the forest captures driftwood once it occurs. When the width of a forest zone is difficult to increase, it is desirable to consider creating another forest zone on the inland side to stop the washing out of trees from the coastal forest.

Cases wherein garden trees, hedges, or homestead woodlands reduced damages have also been reported after the Chile earthquake tsunami in 1960 (Izuni et al. 1961). Unlike in rural areas, establishing homestead woodlands is impractical in residential areas. Alternative approaches include establishing a row of trees with well-developed roots along the main road running parallel to the coast and increasing trees in public spaces such as parks. In particular, because the momentum of tsunami becomes weaker with distance from the coast, trees are likely to withstand the tsunami. The momentum of drifting objects is similarly reduced, and thus, an increased drifting object-capturing effect of trees can be expected.

If additional modifications to forest zones are to be considered, the functions may be enhanced by filling a lower space in a forest wherein trees have a large clear stem height as self-pruning of branches of forest trees is considered to be disadvantageous in terms of the wave power-reducing function. Specifically, a possible approach is to introduce evergreen broadleaf trees into the lower layer. When the drifting object-capturing function is specifically targeted, it is also effective to prune lower branches so that the clear stem height is high enough to reduce resistance and increase tolerance to tsunami.

A foredune is often used for constructing a coastal forest and can be expected to function like a seawall against tsunami. The seawall crown of the foredune is managed to prevent break and recess formation, wherein wind and blown sand can focus; this measure also effectively works for tsunami. In reality, it is not possible to eliminate breaks because of roads and other structures; thus, means of managing breaks must be prepared. It is noteworthy that although the coastal forest is aimed at avoiding concentration of a tsunami and receiving a tsunami as evenly as possible, as described above, effective approaches on the inland side near conservation targets include forest zone and embankment arrangement for protecting conservation targets in a focused manner.

20.5.4 Ideal Appearance

Appearance, particularly arrangement of regenerated/recovered coastal forests, is not necessarily possible to design from scratch because of interrelationships with other land uses and will often develop based on pre-tsunami conditions. Nevertheless, lastly, I have tried to outline how a regenerated/recovered coastal forest should ideally look.

If the tsunami damage-reducing effect of a coastal forest is intended to be substantially enhanced, it is required to construct the forest zone with an increased width and a raised ground level with an embankment.

If a high foredune or a high embankment can replace the function of a seawall, it would be possible to construct an area between the beach and coastal forest, similar to a natural coast, and if a coastal forest had enough space, it would be possible to leave places with a high groundwater level on the inland side partly as wetlands without forming a coastal forest.

If a seawall is to be constructed to prepare for tsunamis, possible approaches include placing the seawall inside or on the inland side of the coastal forest, similar to those on Fudai beach and old Takata-matsubara, both in Iwate Prefecture. The seawall constructed on the inland side of the coastal forest can also be used as a road.

A reason why a few ships were captured by coastal forests despite many ships being carried onshore by the 2011 tsunami is presumably that land use was advanced and no coastal forests were present in many locations where ships were moored. In future reforestation, I hope that coastal forests are created in places where there was no previous forest band. Although this operation requires coordination with land use in the surrounding regions, it is desirable that new space be secured for coastal forests.

It is noteworthy that pine trees constituting coastal forests are not necessarily growing steadily. Trees that previously underwent main shaft (stem) death and have the main shaft derived from branches are common and even dominant trees are no exception. Such trees have a stem, wherein the dead main shaft is contained and thus are structurally weak at that part. Moreover, in this tsunami, stem breakage occurred often near the ground level, but breakage at high positions occurred more in large-diameter trees. The same thing may also occur when dead branches are considered. A measure that may practically be difficult to implement but may prevent stem breakage at a high position would be cutting off dead main shafts and branches, if such detailed work is affordable.

One can assume that a larger-diameter tree is more resistant to wave power. However, aged large-diameter trees are not necessarily highly resistant to wave power, considering that they may have a stem, wherein a dead stem or branches are incorporated and have decayed from that part. In this regard, a report on a storm tsunami that attacked prefectures of Kochi, Tokushima, Osaka, and Wakayama in September 1934 also states that black pine (*Pinus thunbergii*) damage was least in trees with 30–60 cm diameter and that more breakage and lodging occurred in trees with a greater diameter (Bureau of Forestry, Ministry of Agriculture and Forestry 1935). Therefore, the forest zone should be renewed to maintain tolerance to tsunami.

20.5.5 Use of Broad-Leaved Trees

Japanese black pine has been planted as major and key species for the coastal forests in Japan because it has high ability to adapt to the dry and oligotrophic sand dune with salt spray. However, the broad-leaved coastal forests or use of broad-leaved trees partially in the pine coastal forests have being requested from the following points: measure to counter pine wilt disease spread by pine weevil and multiple/diverse coastal forests. In addition, inadequate supply of resistance seedling of Japanese black pine against the pine wilt disease is anticipated because planting for the regeneration of coastal forests must be finished in a short term (10 years) in large area fiscally. However, application technologies for broad-leaved trees have not been established yet and they are future tasks. For example, we have to know where we change broad-leaved tree species from seaward to inland because broad-leaved trees with higher ability to adapt to salt spray are not tall and those trees that can be tall do not have the ability. In addition, because broad-leaved trees in general require richer soil than pine trees, we need soil improvement or wait until soil becomes rich. Therefore, it is thought a realistic procedure to regenerate coastal forest with Japanese black pine first for recovering the disaster prevention function at the early stage and after that to change them to broad-leaved coastal forests using a natural invasion or planting from the inland side.

20.6 Epilogue

Coastal forests in many places were severely damaged by this tsunami. In regeneration of coastal forests, I want to take this opportunity to solve problems that coastal forests have faced pre-tsunami rather than engage in simple reforestation. In brief, I hope that a healthy stand type is established by appropriate tree number adjustment and that measures against pine wilt disease are properly conducted. It is noteworthy that seedlings are suspected to be unavailable in sufficient quantity because of the extremely large affected area. To address this concern, I propose to consider using seedling trees found in affected regions and reviewing the planting numbers. Further, functional strip roads must be available in coastal forests for detailed work and maintenance, and it is important to proceed with an image of the coastal forest to regenerate/recover from the stage of disposing damaged trees.

References

Bureau of Forestry, Ministry of Agriculture and Forestry [edited] (1935) Technical considerations on construction of tsunami disaster prevention forests (tide water control forests). Bureau of Forestry, Ministry of Agriculture and Forestry, 22pp. (in Japanese)

Central Disaster Prevention Council (2008) Basic disaster prevention plan, 399pp. http://www.bousai.go.jp/keikaku/090218_basic_plan.pdf. Confirmed November 28, 2011 (in Japanese)

Committee on Regeneration of Coastal Disaster Prevention Forests related to the Great East Japan Earthquake (2012) On regeneration of coastal disaster prevention forests in the future, 80pp (in Japanese)

Izuni T, Abe N, Yamauchi H, Doi Y (1961) Discussion on effects of tide-water control forests in the Chile earthquake tsunami. Miyagi Prefectural Agricultural Experiment Station Extraordinary Report, 5, 41pp (in Japanese)

Laso Bayas JC, Marohn C, Dercon G, Dewi S, Piepho HP, Joshi L, van Meine N, Cadisch G (2011) Influence of coastal vegetation on the 2004 tsunami wave impact in west Aceh. PNAS Early Ed 108(46):18612–18617

Mamiya Y (1983) Pathology of the pine wilt disease caused by Bursaphelenchus xylophilus. Annu Rev Phytopathol 21:201–220

Nakamura K, Kadani E, Ono K (2012) Debilitation and death of trees in tsunami-affected coastal forests. For Sci 66:7–12 (in Japanese)

Noguchi H, Sato H, Torita H, Masaka K, Abe T, Kimura K, Sakamoto T (2012) Numerical simulation of effect of inundation flow caused by the 2011 Tohoku earthquake tsunami on the *Pinus thunbergii* coastal forest – a case study of Misawa City of the Aomori Prefecture. J Jpn Soc Coast For 11(2):47–51 (in Japanese with English summary)

Okada M, Noguchi H, Okano M, Sakamoto T (2012) The evaluations of coastal forest by grade of the house damage caused by the 2011 off the pacific coast of Tohoku Earthquake Tsunami – a case study of Nagahama, Ishinomaki, Miyagi Prefecture. J Jpn Soc Coast For 11(2):59–64 (in Japanese with English summary)

Ono K, Nakamura K, Tanaka N, Furusawa H, Hirai K (2013) Soil conditions in coastal pine forests damaged by the Heisei-Sanriku Mega-tsunami following Tohoku Earthquake along the eastern Pacific coast of Japan 2011. Bull FFPRI 12(1):49–66 (in Japanese with English abstract)

Sakamoto T (2015) Regeneration of coastal forests affected by tsunami. Research achievement of the 3rd midterm plan 29. Forestry and Forest Products Research Institute, Tokyo, 24pp

Sakamoto T, Noguchi H (2009) For treating coastal forests as one of tsunami disaster countermeasures. 21st Ocean Engineering Symposium Aug 6–7, 2009. JFOES · JASNAOE, OES21-137(CD-ROM)

Sakamoto T, Inoue S, Okada M, Yanagihara A, Harada K, Hayashida M, Nakashima Y (2008) The collision mitigation function of coconut palm trees against marine debris transported by tsunami – a case study of Tangalla on the southern Sri Lanka coast. J Jpn Soc Coast For 7(2):1–6

Sakamoto T, Niiyama K, Nakamura K, Kodani K, Hirai K, Saito T, Kimura K, Kon J (2012) Effectiveness of the coastal forest in stopping drifts from entering the residential area -a case study of the *Pinus thunbergii* coastal forest in Hachinohe City Aomori Prefecture. J Jpn Soc Coast For 11(2):65–70 (in Japanese with English summary)

Sato H, Hiroyuki Torita H, Kazuhiko Masaka K, Tomoyuki Abe T, Hironori Noguchi H, Koki Kimura K, Tomoki Sakamoto T (2012) Relationship between tree fall damage and forest structure of *Pinus thunbergii* coastal forest by the 2011 Tohoku earthquake tsunami disaster – an example of Misawa City of Aomori Prefecture. J Jpn Soc Coast For 11(2):41–45 (in Japanese with English summary)

Satoh A, Tanaka S, Oono R, Iguci H, Sakamoto T (2012) Relationship between coastal forest damages by tsunami and vegetation foundation embankment. Collection of Scientific Lectures at the Annual Meeting of the Japan Forestry Society. 123, p D09- (in Japanese)

Tamura H (2012) Uprooting damages in coastal forests in the Sendai Plain. For Sci 66:3–6 (in Japanese)

Teramoto Y, Asano T, Hayashi K, Tada T, Imai K, Sakamoto T (2012) Coastal forest damage from the 2011 Tohoku tsunami and related mitigation by fill at Yuriage beach, Miyagi Prefecture, Natori City, Japan. J Jpn Soc Coast For 11(1):11–18 (in Japanese with English summary)

Watanabe K, Ebina H, Furukawa K, Horigome H, Otsuki K, Ueno M, Miyashita T, Sakamoto T (2014) Relationship of coastal forest damage in the Sendai Plain caused by the 2011 off the pacific coast of Tohoku Earthquake Tsunami between depth of groundwater and tree size. J Jpn Soc Coast For 13(1):7–14 (in Japanese with English summary)

Yamanaka K, Fujihara M, Hayashida M, Goto Y, Suzuki S, Miyamae T, Inoue S, Sakamoto T (2012) The influence of the tsunami caused by 2011 off the pacific coast of Tohoku earthquake in Ido, Sendai-city – relation between coastal embankment and damage to coastal forests. J Jpn Soc Coast For 11(1):19–25 (in Japanese with English summary)

Yamanaka K, Fujihara M, Hayashida M, Goto Y, Suzuki S, Miyamae T, Inoue S, Kodani E, Sakamoto T (2013) Impact of the tsunami caused by 2011 off the pacific coast of Tohoku earthquake in Ido, Miyagi Prefecture, Sendai-city, Japan (II) – Damage to coastal forests consisting of large-diameter *Pinus densiflora* Sieb. et Zucc and *Pinus thunbergii* Parl. J Jpn Soc Coast For 12(1):1–7 (in Japanese with English summary)

Zhao BG, Futai K, Sutherland JR, Takeuchi Y (eds) (2008) Pine wilt disease. Springer, Tokyo, 459p

Chapter 21
Flora of Freshwater Wetlands in the Tsunami-Affected Zone of the Tohoku Region

Suzuki Mahoro

Abstract The Great East Japan Earthquake and subsequent tsunami broke open seawalls, levees, and channels on the Pacific coast of the Tohoku region, creating temporary marshes and pools at sites that had been dry before the earthquake. In ponds, paddy fields, and canals that had existed before 2011, the tsunami disturbed sediments, deposited sand, and soil and changed the growth environments of wetland plants.

After the 2011 tsunami, plant researchers and lay botanists made observations in freshwater and brackish water wetlands, where the flora recovered for the first time in decades. Many threatened, extirpated, and newly recorded species were listed by them. In this study, literature on vegetation and floristic studies of the regions of Iwate, Miyagi, and Fukushima Prefectures that were inundated by the 2011 tsunami was reviewed, focusing on studies of inland waters.

Wetland plants that appeared after the tsunami likely germinated from seed banks in the soils of the flooded areas, rather than from seeds transported from surrounding regions by dispersal vectors. They were previously common weeds in the floodplain, especially in agricultural environments.

The reported appearance frequencies for rare species are considerably higher than those for common species because the aim of most taxonomic publications is to report first findings or rare species. Herbarium specimens and literature records focus disproportionately on rare species.

These wetlands are vanishing rapidly as restoration of housing and agriculture proceeds with the infilling of wetlands. Therefore, on-site conservation of threatened aquatic and wetland species is extremely difficult.

Keywords Floristic study • Tsunami • Freshwater wetlands • Seed bank • Propagule bank • Vegetation restoration • Red list • Endangered plants

---

The original version of this chapter was revised. An erratum to this chapter can be found at DOI 10.1007/978-4-431-56448-5_24

S. Mahoro (✉)
Iwate Prefectural Museum, 34 Matsuyashiki, Ueda, Morioka City 020-0202, Japan
e-mail: suzukim@iwapmus.jp

J. Urabe, T. Nakashizuka (eds.), *Ecological Impacts of Tsunamis on Coastal Ecosystems*, Ecological Research Monographs,
DOI 10.1007/978-4-431-56448-5_21

21.1 Introduction

The Great East Japan Earthquake, which was accompanied by subsidence, ground liquefaction, and tsunamis, demolished seawalls and levees, severed inland canals and pipelines, and severely damaged paddy fields and ponds along the Pacific coast of the Tohoku region (Suppasri et al. 2012; Tanaka 2012). The tsunami strewed sediments over ground-level surfaces, burying toppled trees and wrecking houses (Haraguchi et al. 2012; Sugawara et al. 2011). Water escaping from the affected canals flooded fields and pooled in pits and ledges created by subsidence, ground liquefaction, and erosion (Tanaka 2012; Nandasena et al. 2012). Shallow freshwater ponds and marshes subsequently formed in the inundated areas (Sugawara et al. 2011). The tsunami broke open seawalls and floodgates in river estuaries, allowing brackish water to flow inland.

The 2011 tsunami was the largest natural disaster in the history of Japan (EERI 2011). Reconstruction of agricultural lands and irrigation and drainage channels was delayed for several years by the immense scale of the damage. The work was ongoing in 2015. The delays ensured that ponds and marshes created by the tsunami remained untouched for several years, and the process of revegetation began.

Professional plant researchers and lay botanists in the Tohoku region began investigating the flora in the inundated lands during the summer of 2011. Attention focused mostly on coastal vegetation (Hayasaka et al. 2012; Kanaya et al. 2012; Shimoyama 2013), but plant species growing in inland freshwater ponds and marshes were also studied. Contrary to expectations, several threatened and rare plant species flourished in these waters (Kasai 2013; Takiguchi et al. 2014). Species that had disappeared from the records for the region decades previously reappeared (Arai and Nagahata 2013; Kasai 2013; Sugiyama et al. 2013). These findings have been reported in academic and less formal journals.

This studied compiled data from the available literature, added newly collected information, and constructed a floristic survey of freshwater and brackish wetlands in the tsunami inundation area.

21.2 Methods

Literature on vegetation and floristic studies of the regions of Iwate, Miyagi, and Fukushima Prefectures that were inundated by the 2011 tsunami was reviewed, focusing on studies of inland waters and excluding those that reported floristic elements of sand dunes, salt marshes, and beaches. The names of species that are characteristic of freshwater, brackish wetlands, and beaches were extracted from the assembled literature. Previously unpublished records for six sites in Iwate Prefecture were using my own data and those of the Society for Research on Flora of Iwate.

Using the compiled floristic list, the appearance frequency of each species was calculated. Most of the literature included in the study reported observations at several sites. Accordingly, the appearance frequency of a species was defined as the number of sites in which it had been recorded. When more than one published paper recorded a species at a location with a single site name, the species was assigned an appearance frequency of 1.0 for that site. In some cases, however, habitat descriptions and other relevant information indicated that study sites subsumed under a single name were in fact different locations, and the appearance frequencies were calculated based on occurrences at separate sites.

21.3 Results

This study located 14 published studies that included records of plant species in freshwater and brackish wetlands. Table 21.1 lists published literature, my own unpublished studies, and those of the Society for Research on Flora of Iwate. All

Table 21.1 Published and unpublished studies that included records of plant species in freshwater and brackish wetlands in the tsunami-affected zone of Tohoku region

No.	Authors	Year of publication	Study area	Year of observation
1	Arai R, Nagahata Y	2013	Iwate Pref., Miyagi Pref.	2011, 2012
2	Hoshino Y, Hoshino J	2014	Iwate Pref.	2013
3	Kanno H et al.	2014	Miyagi Pref.	2011–2013
4	Kasai H	2013	Miyagi Pref., Fukushima Pref.	2012
5	Mineta T, Tomosyo T	2012	Miyagi Pref.	2011
6	Nemoto S et al.	2013	Fukushima Pref.	2011
7	Shibuya K et al.	2014	Iwate Pref.	2013
8	Shimada N	2014	Iwate Pref.	2011–2013
9	Sugiyama T	2012	Miyagi Pref., Natori City	2011
10	Sugiyama T, Kasai H	2013	Miyagi Pref., Natori City	2011
11	Sugiyama T et al.	2013	Miyagi Pref., Sendai City	2011
12	Takano Y	2013	Iwate Pref.	2012
13	Takiguchi M et al.	2014	Miyagi Pref.	2011–2013
14	Usuba M	2013	Fukushima Pref.	2012
15	The Society for Research on Flora of Iwate	–	Iwate Pref., Kamaishi City	2012
16	The Society for Research on Flora of Iwate	–	Iwate Pref., Rikuzentakata City	2012
17	The Society for Research on Flora of Iwate	–	Iwate Pref., Rikuzentakata City	2013
18	Mahoro S	–	Iwate Pref., Rikuzentakata City	2012
19	Mahoro S	–	Iwate Pref., Rikuzentakata City	2012, 2013
20	Mahoro S	–	Iwate Pref., Rikuzentakata City	2013

studies were conducted in various seasons in the period 2011–2013 across 79 study sites.

The linear distance between study sites and the coastal shoreline ranged between 170 and 5000 m. The tsunami inundation of the Sendai Plain (Miyagi Prefecture) extended 5 km inland (EERI 2011; Sugawara et al. 2011); several of the widely separated study sites were >4000 m from the shoreline (Mineta and Tomosyo 2012; Sugiyama et al. 2013; Takiguchi et al. 2014).

Only six of the published works provided comprehensive floristic lists for all of the study sites (Hoshino and Hoshino 2014; Kanno et al. 2014; Mineta and Tomosyo 2012; Sugiyama 2012; Sugiyama and Kasai 2013; Sugiyama et al. 2013); the list of comprehensive works was increased to nine by including our unpublished data. The remaining published studies focused only on records of rare and threatened species and did not provide floristic lists. Therefore, the appearance records were strongly biased toward rare species.

Table 21.2 lists the appearance frequencies of species recorded in our compilations. *Monochoria korsakowii* Regel et Maack was recorded most frequently (at 40 study sites), followed in order by *Potamogeton pusillus* L. (at 21 sites), *Eleocharis parvula* (Roem. et Schult.) Link ex Bluff, Nees et Schauer (at 15 sites), *Sparganium erectum* L. var. *coreanum* (H. Lév) H. Hara (at 13 sites), *Rorippa palustris* (L.) Besser (at 11 sites), *Potamogeton pectinatus* L. (at 10 sites), *Penthorum chinense* Pursh (at 10 sites), *Bidens frondosa* L. (at 10 sites), *Ottelia alismoides* (L.) Pers. (at 9 sites), *Echinochloa crus-galli* (L.) P. Beauv. (at 9 sites), *Phragmites australis* (Cav.) Trin. ex Steud. (at 9 sites). Other than *R. palustris*, *B. frondosa*, *E. crus-galli*, and *P. australis*, all of these species were listed as “threatened” in the national Red Data Book (Ministry of the Environment 2015).

The ratios of threatened species to all recorded species in the previously unpublished data included in this study were 2:28 on the riverbank in Kamaishi, 2:20 on the riverbank in Rikuzentakata, 1:25 in a former rice field in Rikuzentakata, and 5:27 in a former residential area in Rikuzentakata. In published works, the ratios were 0:21 in paddy fields (Mineta and Tomosyo 2012), 2:24 in an artificial pond (Sugiyama 2012), 12:73 in a pond and in paddy fields (Sugiyama et al. 2013), 8:27 in paddy fields (Sugiyama and Kasai 2013), and 0:19 in a coastal pine forest (Kanno et al. 2014).

Few floristic studies of the Pacific coast of Tohoku district were undertaken before 2011. Only Sugiyama (Sugiyama 2012), Sugiyama et al. (Sugiyama et al. 2013), and Kanno et al. (Kanno et al. 2014) compared species compositions before and after the tsunami. However, the occurrences of rare species at 27 of the sites in 12 studies were clearly new because human infrastructure in place before the tsunami had precluded the formation of suitable habitats. For example, *Ottelia alismoides*, a threatened submergent species, was firstly found from ponds in Yamamoto Town, Miyagi Prefecture, after the tsunami (Takiguchi et al. 2014). *Potamogeton pusillus* and *Eleocharis parvula*, threatened coastal wetland species, were firstly recorded from Fukushima Prefecture after the disaster (Kasai 2013; Usuba 2013; Kurosawa 2016). In Iwate Prefecture, *Monochoria korsakowii*, which had disappeared from the records for the region, reappeared and flourished after

Table 21.2 Plant species recorded in freshwater and brackish wetlands in the tsunami-affected zone of the Tohoku region

Family	No	Species name	Life form	Habitat[a]	Conservation status[b]	Alien species	Apperance frequency	References[c]
Alismataceae	1	Alisma canaliculatum A.Braun et C.D.Bouché var. canaliculatum	Emergent				2	2, 4
	2	Alisma plantago-aquatica L. var. orientale Sam.	Emergent				1	4
	3	Sagittaria trifolia L. var. trifolia	Emergent				4	2, 9, 11, 19
Amaranthaceae	4	Atriplex patens (Litv.) Iljin	Terrestrial	Seashore			1	2
	5	Atriplex prostrata Boucher ex DC.	Terrestrial	Seashore		a	4	2, 15, 16, 17
	6	Atriplex subcordata Kitag.	Terrestrial	Seashore	NT (MY)		1	4
	7	Chenopodium acuminatum Willd. var. acuminatum	Terrestrial	Seashore			3	3, 5, 10
	8	Chenopodium glaucum L.	Terrestrial	Seashore		a	6	2, 4, 10, 15, 16, 17
	9	Suaeda glauca (Bunge) Bunge	Shoreline	Salt marsh	NT (MY)		2	4, 13
	10	Suaeda maritima (L.) Dumort.	Shoreline	Salt marsh	NT (MY), A (FK)		4	4, 13
Apiaceae	11	Oenanthe javanica (Blume) DC. subsp. javanica	Shoreline				3	2, 17, 19
	12	Sium suave Walter var. nipponicum (Maxim.) H.Hara	Shoreline		VU (ME), NT (MY)		1	4, 10
Araceae	13	Landoltia punctata (G.Mey.) D.H.Les et D.J.Crawford	Floating				1	2
	14	Lemna aoukikusa Beppu et Murata var. aoukikusa	Floating				4	2, 9, 11, 19

(continued)

Table 21.2 (continued)

Family	No	Species name	Life form	Habitat[a]	Conservation status[b]	Alien species	Apperance frequency	References[c]
Asteraceae	15	Ambrosia trifida L.	Shoreline			a	3	2, 15, 19
	16	Bidens frondosa L.	Shoreline			a	10	2, 3, 5, 11, 15, 16, 17, 19
	17	Bidens tripartita L. var. tripartita	Shoreline				2	2, 9
	18	Centipeda minima (L.) A. Braun et Asch.	Shoreline				4	2, 5, 11
	19	Eclipta alba (L.) Hassk.	Shoreline			a	2	9, 11
	20	Eclipta thermalis Bunge	Shoreline				2	3, 5
	21	Inula britannica L. subsp. japonica (Thunb.) Kitam.	Shoreline				2	2, 19
	22	Inula salicina L. var. asiatica Kitam.	Shoreline				1	2
	23	Senecio cannabifolius Less.	Shoreline				1	17
	24	Solidago altissima L.	Shoreline			a	4	2, 3, 15, 19
	25	Sonchus brachyotus DC.	Terrestrial	Seashore			1	2
	26	Symphyotrichum subulatum (Michx.) G.L.Nesom var. squamatum (Spreng.) S.D.Sundberg	Shoreline			a	4	2, 9, 16, 19
	27	Symphyotrichum subulatum (Michx.) G.L.Nesom var. subulatum	Shoreline			a	2	2, 3
	28	Turczaninovia fastigiatus Fisch.	Shoreline				1	4
	29	Xanthium orientale L. subsp. orientale	Shoreline			a	3	5, 11

Azollaceae	30	Azolla filiculoides Lam.	Floating			a	1	10
Betulaceae	31	Alnus hirsuta (Spach) Turcz. ex Rupr. var. sibirica (Spach) C.K.Schneid.	Shoreline				1	2
Brassicaceae	32	Nasturtium officinale R.Br.	Emergent			a	1	2
	33	Cardamine scutata Thunb. var. scutata	Shoreline				3	2, 3, 15
	34	Rorippa indica (L.) Hiern	Shoreline				3	2, 15, 16
	35	Rorippa palustris (L.) Besser	Shoreline				11	2, 3, 5, 9, 10, 11, 15, 16, 17, 19
	36	Rorippa sylvestris (L.) Besser	Shoreline			a	1	11
Caryophyllaceae	37	Sagina maxima A.Gray	Terrestrial	Seashore			1	2
	38	Spergularia bocconii (Scheele) Foucaud ex Merino	Shoreline	Salt marsh		a	1	2
	39	Spergularia marina (L.) Griseb.	Shoreline	Salt marsh			2	2, 4
Ceratophyllaceae	40	Ceratophyllum demersum L.	Submergent				1	4
Characeae	41	Chara braunii Gmelin	Submergent		II (ME)		1	10
	42	Nitella hyalina (DC.) C. Agardh	Submergent		I (ME)		1	9
Commelinaceae	43	Commelina communis L. var. communis	Shoreline				6	2, 5, 15, 16
	44	Murdannia keisak (Hassk.) Hand.-Mazz.	Shoreline				2	2, 11
Convolvulaceae	45	Calystegia soldanella (L.) R. Br.	Terrestrial	Seashore			1	11
Crassulaceae	46	Tillaea aquatica L.	Shoreline		NT (ME), II (MY)		2	4, 10

(continued)

Table 21.2 (continued)

Family	No	Species name	Life form	Habitat[a]	Conservation status[b]	Alien species	Apperance frequency	References[c]
Cyperaceae	47	Bolboschoenus fluviatilis (Torr.) Soják subsp. yagara (Ohwi) T.Koyama	Shoreline				3	1, 2, 11
	48	Bolboschoenus koshevnikovii (Litv. ex Zinger) A.E.Kozhevn.	Shoreline				4	5, 13
	49	Bulbostylis barbata (Rottb.) Kunth	Terrestrial	Seashore			1	10
	50	Carex capricornis Meinsh. ex Maxim.	Shoreline		VU (ME), I (MY)		3	4, 13
	51	Carex dimorpholepis Steud.	Shoreline				4	4
	52	Carex glabrescens (Kük.) Ohwi	Shoreline				1	16
	53	Carex maackii Maxim.	Shoreline				3	4
	54	Carex neurocarpa Maxim.	Shoreline				5	4
	55	Carex pumila Thunb.	Terrestrial	Seashore			2	2, 17
	56	Carex scabrifolia Steud.	Shoreline	Salt marsh			1	3
	57	Carex stipata Muhl. ex Willd.	Shoreline				2	4
	58	Carex thunbergii Steud. var. thunbergii	Shoreline				1	16
	59	Cyperus brevifolius (Rottb.) Hassk. var. leiolepis (Franch. et Sav.) T.Koyama	Shoreline				1	2
	60	Cyperus difformis L.	Shoreline				5	2, 9, 11, 19
	61	Cyperus flavidus Retz.	Shoreline				1	4
	62	Cyperus glomeratus L.	Shoreline				1	2
	63	Cyperus iria L.	Shoreline				3	2, 5, 11

	64	Cyperus microiria Steud.	Shoreline				2	2, 11
	65	Cyperus nipponicus Franch. et Sav. var. nipponicus	Shoreline				3	2, 9, 11
	66	Cyperus orthostachyus Franch. et Sav.	Shoreline				2	2, 11
	67	Cyperus pacificus (Ohwi) Ohwi	Shoreline				1	2
	68	Cyperus polystachyos Rottb.	Shoreline		NT (FK)		4	4, 6, 9, 10
	69	Cyperus sanguinolentus Vahl	Shoreline				2	2, 11
	70	Cyperus serotinus Rottb.	Shoreline				1	11
	71	Eleocharis acicularis (L.) Roem. et Schult. var. longiseta Svenson	Shoreline				5	2, 10, 11, 18, 20
	72	Eleocharis kamtschatica (C.A.Mey.) Kom. f. kamtschatica	Shoreline	Salt marsh			2	2, 16
	73	Eleocharis kuroguwai Ohwi	Emergent				1	12
	74	Eleocharis mamillata H.Lindb. var. cyclocarpa Kitag.	Emergent				1	15
	75	Eleocharis parvula (Roem. et Schult.) Link ex Bluff, Nees et Schauer	Shoreline	Salt marsh	VU (ME), C (IW), I (MY)		15	4, 12, 13, 16, 19
	76	Eleocharis pellucida J. et C. Presl	Shoreline				2	5, 11
	77	Fimbristylis dichotoma (L.) Vahl var. tentsuki T.Koyama	Shoreline				2	2, 10
	78	Fimbristylis littoralis Gaudich. var. littoralis	Shoreline				2	2, 11
	79	Fimbristylis longispica Steud.	Shoreline				1	13

(continued)

Table 21.2 (continued)

Family	No	Species name	Life form	Habitat[a]	Conservation status[b]	Alien species	Apperance frequency	References[c]
	80	Fimbristylis squarrosa Vahl	Shoreline				1	2
	81	Fimbristylis velata R.Br.	Shoreline				2	9, 11
	82	Lipocarpha microcephala (R.Br.) Kunth	Shoreline				1	11
	83	Schoenoplectus juncoides (Roxb.) Palla	Shoreline				5	2, 5, 10, 11
	84	Schoenoplectus mucronatus (L.) Palla var. tataranus (Honda) K.Kohno, Iokawa et Daigobo	Shoreline		II (MY)		3	10, 11
	85	Schoenoplectus tabernaemontani (C.C.Gmel.) Palla	Emergent				6	1, 2, 4, 9, 16, 17
	86	Schoenoplectus triangulatus (Roxb.) Soják	Emergent				1	9
	87	Schoenoplectus triqueter (L.) Palla	Emergent				5	11
	88	Schoenoplectus wallichii (Nees) T.Koyama	Shoreline				1	10
	89	Scirpus wichurae Boeck.	Shoreline				1	2
Elatinaceae	90	Elatine triandra Schkuhr var. pedicellata Krylov	Shoreline				4	9, 11, 19
Fabaceae	91	Aeschynomene indica L.	Shoreline				3	5, 9, 11
	92	Chamaecrista nomame (Siebold) H.Ohashi	Shoreline				1	10
	93	Lathyrus palustris L. var. pilosus (Cham.) Ledeb.	Shoreline		NT (FK)		1	6

Gentianaceae	94	Swertia tosaensis Makino	Shoreline		VU (ME), II (MY)		1	4
Haloragaceae	95	Myriophyllum spicatum L.	Submergent		II (FK)		2	4
Hydrocharitaceae	96	Najas chinensis N.Z.Wang	Submergent		VU (ME), II (MY)		7	4, 10, 11, 12
	97	Najas gracillima (A.Braun ex Engelm.) Magnus	Submergent		NT (ME), B (IW), NT (MY)		8	4, 10, 11, 12, 19
	98	Najas minor All.	Submergent		VU (ME), I (MY)		1	13
	99	Ottelia alismoides (L.) Pers.	Submergent		VU (ME), B (IW), NT (MY)		9	1, 4, 10, 11, 13, 18, 19
Iridaceae	100	Iris ensata Thunb. var. spontanea (Makino) Nakai ex Makino et Nemoto	Shoreline				1	4
	101	Iris pseudacorus L.	Emergent			a	1	2
Isoetaceae	102	Isoetes japonica A.Braun	Shoreline		NT (ME), NT (MY)		1	10
Juncaceae	103	Juncus alatus Franch. et Sav.	Shoreline		C (IW)		2	2, 11
	104	Juncus bufonius L.	Shoreline		II (MY)		4	2, 4, 15
	105	Juncus decipiens (Buchenau) Nakai	Shoreline				4	2, 3, 15, 19
	106	Juncus gracillimus (Buchenau) V.I.Krecz. et Gontsch.	Shoreline	Salt marsh			5	2, 4, 13
	107	Juncus papillosus Franch. et Sav.	Shoreline				1	2
	108	Juncus prismatocarpus R.Br. subsp. leschenaultii (J.Gay ex Laharpe) Kirschner	Shoreline				4	2, 3, 15, 16
	109	Juncus setchuensis Buchenau var. effusoides Buchenau	Shoreline				2	2, 15

(continued)

Table 21.2 (continued)

Family	No	Species name	Life form	Habitat[a]	Conservation status[b]	Alien species	Apperance frequency	References[c]
	110	Juncus tenuis Willd.	Shoreline				1	2
	111	Juncus wallichianus Laharpe	Shoreline				1	2
Juncaginaceae	112	Triglochin maritima L.	Shoreline	Salt marsh	NT (ME), I (MY)		2	4, 13
Lamiaceae	113	Lycopus cavaleriei H.Lév.	Shoreline				1	2
	114	Mentha japonica (Miq.) Makino	Shoreline		NT (ME), I (MY)		1	4
	115	Mosla dianthera (Buch.-Ham. ex Roxb.) Maxim.	Shoreline				3	3, 9, 11
Lentibulariaceae	116	Utricularia australis R.Br.	Submergent		NT (ME), D (IW), II (MY)		4	1, 4, 13, 18
Linderniaceae	117	Lindernia anagallidea (Michx.) Pennell	Shoreline			a	1	11
	118	Lindernia dubia (L.) Pennell subsp. dubia	Shoreline			a	2	9, 11
	119	Lindernia dubia (L.) Pennell subsp. major (Pursh) Pennell	Shoreline			a	1	2
	120	Lindernia procumbens (Krock.) Borbás	Shoreline				4	2, 10, 11, 19
Lythraceae	121	Ammannia multiflora Roxb.	Shoreline		I (MY)		1	10
	122	Lythrum anceps (Koehne) Makino	Shoreline				2	2, 15
	123	Lythrum hyssopifolia L.	Shoreline			a	1	2
	124	Lythrum salicaria L.	Shoreline				2	2, 19
	125	Rotala indica (Willd.) Koehne	Emergent				3	2, 11
	126	Rotala mexicana Cham. et Schltdl.	Shoreline		VU (ME), II (MY)		1	10
	127	Trapa japonica Flerow	Floating				4	11, 13

Mazaceae	128	Mazus pumilus (Burm.f.) Steenis	Shoreline				2	2, 16
Menispermaceae	129	Cocculus trilobus (Thunb.) DC.	Terrestrial	Seashore			1	2
Menyanthaceae	130	Nymphoides coreana (H.Lév.) H.Hara	Floating		VU (ME), II (MY)		5	4, 10, 11
	131	Nymphoides peltata (S.G.Gmel.) Kuntze	Floating		NT (ME), II (MY)		2	4, 9
Molluginaceae	132	Mollugo verticillata L.	Shoreline			a	1	10
Nelumbonaceae	133	Nelumbo nucifera Gaertn.	Emergent				3	4, 11, 13
Onagraceae	134	Epilobium pyrricholophum Franch. et Sav.	Shoreline				1	2
	135	Ludwigia epilobioides Maxim. subsp. epilobioides	Shoreline				3	2, 10, 11
	136	Oenothera laciniata Hill	Terrestrial	Seashore		a	1	11
Penthoraceae	137	Penthorum chinense Pursh	Shoreline		NT (ME), B (IW)		10	1, 2, 4, 7, 15, 20
Phrymaceae	138	Mimulus nepalensis Benth.	Shoreline				1	15
Plantaginaceae	139	Callitriche stagnalis Scop.	Emergent			a	1	10
	140	Deinostema adenocaulum (Maxim.) T.Yamaz.	Shoreline		VU (ME), II (MY)		1	10
	141	Deinostema violaceum (Maxim.) T.Yamaz.	Shoreline				1	10
	142	Gratiola japonica Miq.	Emergent		VU (ME)		3	4
	143	Limnophila sessiliflora (Vahl) Blume	Emergent				3	10, 11
	144	Plantago japonica Franch. et Sav.	Terrestrial	Seashore			3	2, 10, 16
	145	Veronica anagallis-aquatica L.	Shoreline			a	1	6

(continued)

Table 21.2 (continued)

Family	No	Species name	Life form	Habitat[a]	Conservation status[b]	Alien species	Apperance frequency	References[c]
	146	Veronica undulata Wall.	Shoreline		NT (ME), C (IW), NT (MY), C(FK)		5	1, 2, 4, 6, 15
Plumbaginaceae	147	Limonium tetragonum (Thunb.) A.A.Bullock	Shoreline	Salt marsh	NT (ME), I (MY)		3	4, 13
Poaceae	148	Arthraxon hispidus (Thunb.) Makino	Shoreline				1	2
	149	Echinochloa crus-galli (L.) P. Beauv. var. crus-galli	Shoreline				9	2, 3, 5, 11, 17, 19
	150	Echinochloa oryzicola (Vasing.) Vasing.	Shoreline				2	5, 16
	151	Ischaemum aristatum L. var. crassipes (Steud.) Yonek.	Terrestrial	Seashore			1	3
	152	Leersia japonica (Honda) Makino ex Honda	Shoreline				1	11
	153	Miscanthus sacchariflorus (Maxim.) Benth.	Shoreline				2	2, 15
	154	Panicum bisulcatum Thunb.	Shoreline				1	2
	155	Panicum dichotomiflorum Michx.	Shoreline			a	2	2, 11
	156	Phacelurus latifolius (Steud.) Ohwi	Shoreline	Salt marsh			1	3
	157	Phragmites australis (Cav.) Trin. ex Steud.	Emergent				9	2, 3, 7, 9, 11, 13, 15, 17, 19
	158	Phragmites japonicus Steud.	Shoreline				1	2
	159	Polypogon fugax Nees ex Steud.	Shoreline				1	2

	160	Puccinellia kurilensis (Takeda) Honda	Shoreline	Salt marsh	DT (IW)		1	2
	161	Puccinellia nipponica Ohwi	Shoreline	Salt marsh	I (FK)		3	4, 10
	162	Spodiopogon cotulifer (Thunb.) Hack. var. cotulifer	Shoreline				1	15
	163	Zizania latifolia (Griseb.) Turcz. ex Stapf	Shoreline				2	11, 13
Polygonaceae	164	Persicaria hydropiper (L.) Delarbre	Shoreline				5	2, 5, 11, 15
	165	Persicaria japonica (Meisn.) Nakai ex Ohki	Shoreline				1	3
	166	Persicaria lapathifolia (L.) Delarbre var. incana (Roth) H. Hara	Shoreline				2	5, 16
	167	Persicaria lapathifolia (L.) Delarbre var. lapathifolia	Shoreline				3	2, 5, 11
	168	Persicaria longiseta (Bruijn) Kitag.	Shoreline				4	2, 3, 5, 11, 19
	169	Persicaria muricata (Meisn.) Nemoto	Shoreline				1	2
	170	Persicaria odorata (Lour.) Soják subsp. conspicua (Nakai) Yonek.	Shoreline		C (IW)		1	2
	171	Persicaria perfoliata (L.) H. Gross	Shoreline				2	5, 15
	172	Persicaria posumbu (Buch.-Ham. ex D.Don) H.Gross var. posumbu	Shoreline				1	3
	173	Persicaria sagittata (L.) H. Gross var. sibirica (Meisn.) Miyabe	Shoreline				1	15

(continued)

Table 21.2 (continued)

Family	No	Species name	Life form	Habitat[a]	Conservation status[b]	Alien species	Apperance frequency	References[c]
	174	Persicaria thunbergii (Siebold et Zucc.) H.Gross var. thunbergii	Shoreline				3	2, 15, 19
	175	Persicaria viscosa (Buch.-Ham. ex D.Don) H. Gross ex T.Mori	Shoreline				1	10
Pontederiaceae	176	Monochoria korsakowii Regel et Maack	Emergent		NT (ME), A (IW), B (FK)		40	1, 2, 4, 6, 7, 8, 9, 10, 11, 13, 18, 19, 20
	177	Monochoria vaginalis (Burm.f.) C.Presl ex Kunth var. vaginata	Emergent				5	1, 2, 10, 11
Portulacaceae	178	Portulaca oleracea L.	Terrestrial	Seashore			4	2, 5, 11
Potamogetonaceae	179	Potamogeton berchtoldii Fieber	Submergent		NT (ME), B (IW)		6	8, 11, 13, 19
	180	Potamogeton cristatus Regel et Maack	Submergent		VU (ME), II (MY)		1	4
	181	Potamogeton distinctus A. Benn.	Floating				4	10, 11, 13, 19
	182	Potamogeton pectinatus L.	Submergent		NT (ME), I (MY)		10	4, 13
	183	Potamogeton pusillus L.	Submergent		VU (ME), I (MY)		21	4, 9, 10, 12, 13, 16
	184	Zannichellia palustris L.	Submergent	Brackish water	VU (ME), I (MY)		8	4, 12, 13
Primulaceae	185	Lysimachia fortunei Maxim.	Shoreline				1	13
	186	Lysimachia thyrsiflora L.	Shoreline		B (IW)		1	8

Ranunculaceae	187	Halerpestes kawakamii (Makino) Tamura	Shoreline		VU (ME), I (MY)		1	4
	188	Ranunculus sceleratus L.	Shoreline				3	2, 10, 16
	189	Ranunculus trichophyllus Chaix ex Vill. var. kazusensis (Makino) Wiegleb	Submergent		EN (ME), I (MY)		1	4
Ricciaceae	190	Ricciocarpus natans (L.) Corda	Floating		NT (ME)		1	2
Rosaceae	191	Potentilla anserina L. subsp. pacifica (Howell) Rousi	Shoreline	Salt marsh	II (MY)		1	4
	192	Rosa luciae Rochebr. et Franch. ex Crép.	Terrestrial	Seashore	AT (MY)		1	3
Ruppiaceae	193	Ruppia maritima L.	Submergent	Brackish water	NT (ME), A (IW), II (MY)		7	4, 13, 17
Salicaceae	194	Salix caprea L.	Shoreline				1	2
	195	Salix integra Thunb.	Shoreline				1	2
	196	Salix triandra L. subsp. nipponica (Franch. et Sav.) A.K.Skvortsov	Shoreline				1	2
	197	Salix udensis Trautv. et C.A.Mey.	Shoreline				1	2
Typhaceae	198	Sparganium erectum L. var. coreanum (H.Lév) H.Hara	Emergent		NT (ME), NT (MY)		13	1, 2, 4, 11, 15, 18
	199	Typha domingensis Pers.	Emergent				6	1, 2, 9, 11, 17, 19
	200	Typha latifolia L.	Emergent				8	1, 2, 7, 9, 11, 15, 17, 19
	201	Typha orientalis C.Presl	Emergent				3	2, 4, 16

[a]Usual habitat of the species

[b]Conservation status in the national Red Data Book (*ME* Ministry of the Environment 2015) or in each prefecture's red list (*IW* Iwate Prefecture 2014, *MY* Miyagi Prefecture 2013, *FK* Fukushima Prefecture 2002)

[c]Each number represents the literature number shown in Table 21.1

about 30 years (Arai and Nagahata 2013). On the other hand, the tsunami extirpated rare species from seven sites (Kasai 2013; Takiguchi et al. 2014).

The recorded species were assigned to life-forms; 139 species were shoreline plants, including 14 that are characteristic of salt marshes (Table 21.2). Twenty-two of the species were emergent, 16 were submergent, and 8 were free-floating or had floating leaves. Submergent taxa included 2 species that were specific to brackish waters. Sixteen species that are characteristic of seashores were recorded at inland sites 650–4300 m from the coastal shoreline.

21.4 Discussion

The Great East Japan Earthquake and tsunami broke open seawalls, levees, and channels on the Pacific coast of the Tohoku region, creating temporary marshes and pools at sites that had been dry before the earthquake. In ponds, paddy fields, and canals that had existed before 2011, the tsunami disturbed sediments, deposited sand, and soil and changed the growth environments of wetland plants.

Many threatened, extirpated, and newly recorded species were listed by professional and lay botanists studying these newly created freshwater and brackish water bodies. The threatened and extirpated species were previously common weeds in the floodplain, especially in agricultural environments; their abundance was reduced by infrastructure development and the use of toxic chemicals (Kohara et al. 2011; Yonemura et al. 2001; Washitani 2007).

Wetland plants that appeared after the tsunami likely germinated from seed banks in the soils of the flooded areas, rather than from seeds transported from surrounding regions by dispersal vectors. Most of the newly emerging species were gravity dispersed and were not observed outside the inundated areas (Kasai 2013; Mineta and Tomosyo 2012).

The sediments of freshwater lakes, ponds, and marshes hold persistent seed banks of diverse species, and water level fluctuations induce recruitment from these banks (Leck 1989). Vegetation restoration projects that make use of seed or propagule banks in the soils of wetlands have been undertaken since the 1990s (Nakamura et al. 2006; Nishihiro et al. 2006a, b). For example, Mizusawa et al. (2000), Nakamoto et al. (2000), Nishihiro et al. (2006a, b), Kushiro et al. (2009), Matsumoto et al. (2009), and Kitagawa and Shimano (2010) used soil seed banks obtained from abandoned Japanese paddy fields, lakeshores, and drained marshes to recover wetland vegetation. The recovered vegetation often included species that had disappeared from the aboveground flora decades previously (Kushiro et al. 2009; Matsumoto et al. 2009; Nishihiro et al. 2006a). Although the proportion of viable seeds in sediments declines with time, some seeds are able to germinate after 50 years of burial (Matsumoto et al. 2009; Momohara et al. 2001). The plant species composition of the wetlands that emerged after the 2011 tsunami (Table 21.2) was similar to that of wetland floras recovered experimentally from soil seed banks, especially those in Lake Kasumigaura (Nishihiro et al. 2006a).

Since the 1960s, the low-lying wetlands in Tohoku and elsewhere in Japan have decreased in area, largely through reclamation and modernization of paddy field agriculture by unification and dredging (Washitani 2007). Therefore, we suggest that seeds in the soil bank had been dormant for 30–50 years in the tsunami inundation zone. The tsunami itself and the debris swept with it excavated the ground and scattered the soil seed banks. After the tsunami, sediments in ponds, paddy fields, and canals were repeatedly excavated by human effort as rubble was removed and searched for victims of the disaster (2011–2012). Pooled freshwater water standing on the disturbed soil provided microtopographic and hydrological conditions that favored seed germination and seedling establishment of wetland plant species.

Fourteen species characteristics of salt marshes were found in the inland freshwater marshes (Table 21.2). Another 16 species characteristic of seashores were recorded at sites 650–4300 m inland from the shoreline. The coastal beaches in this area lost most of their vegetation in March 2011, and it is unlikely that plants revegetating the beaches were able to disperse their seeds far inland in the period 2011–2012 (Kawanishi et al. 2016). I propose that the tsunami beach swept up sand and mud containing seeds and propagules of seashore species and deposited it inland (Mineta and Tomosyo 2012).

The UNESCO-IOC International Tsunami Survey Team recorded the distribution of sand and mud deposited by the 2011 tsunami on the Sendai Plain (Sugawara et al. 2011). They found landward deposition over the length of their survey transect. Deposits were 15–20 cm thick 1 km from the shoreline, ca. 10 cm thick 2 km inland, and <5 cm thick 3 km inland. Tsunami deposits at sites <3 km from the shore comprised sand with overlying mud; only mud (except for a one grain-thick sand layer) was found at sites ≥3 km inland (Sugawara et al. 2011).

Comprehensive floristic lists for the study sites have been provided only by Mineta and Tomosyo (2012), Sugiyama (2012), Sugiyama and Kasai (2013), Sugiyama et al. (2013), Kanno et al. (2014), Hoshino and Hoshino (2014), and ourselves (unpublished data). These compilations show that the proportions of threatened species among all recorded species were in the range of 0–30 % and that common species did not decrease after the tsunami.

The reported appearance frequencies for rare species are considerably higher than those for common species because the aim of most taxonomic publications is to report first findings or rare species. Herbarium specimens and literature records focus disproportionately on rare species. Consequently, the threatened species *Monochoria korsakowii*, which has large, blue flowers and characteristic emergent, heart-shaped leaves, is the most frequently recorded, likely because of the higher probability of a recorder finding this species than any other (Rich and Woodruff 1992). The media reported the beautiful blue flowers of this threatened species blooming in land heavily damaged by the tsunami, which had a strong effect on the viewing public (Hiratsuka 2014; Nagahata 2012). On the other hand, *Potamogeton pusillus* and *Eleocharis parvula* are submergent plants that are inconspicuous. However, they were extremely rare in this region before the tsunami. Their occurrence was recognized by professional botanists as an uncommon phenomenon and reported in the scientific literature.

I suggest that across the entire tsunami inundation area, *Typha* spp. and *Phragmites australis* (Cav.) Trin. ex Steud. were the most frequent emergent taxa, and *Echinochloa crus-galli* (L.) P. Beauv. and its varieties the most frequent shoreline plants in the period 2011–2013. However, there have been few vegetation studies of inland wetlands in the region; some reports list dominant species in the vegetation (Kurosawa 2014; Mineta and Tomosyo 2012; Shibuya et al. 2014; Takiguchi et al. 2014). Naito (Naito 2012) investigated the vegetation of dry rice fields damaged by the tsunami at eight locations on the Sendai Plain and reported the dominance of *E. crus-galli* var. *aristata* and var. *crus-galli* and provided a list of 11 plant species, which contained no endangered taxa. During the floristic investigations of the Society for Research on Flora of Iwate at three dry land sites in tsunami-damaged housing developments, we found 77, 87, and 41 species at each, none of which was threatened (the Society for Research on Flora of Iwate, unpublished data). Overall, more threatened species have been recorded on wetlands than dry lands. The disparity is probably related to the process of soil seed bank formation in the lowlands of the coastal zone.

After the 2011 tsunami, plant researchers and lay botanists made observations in freshwater and brackish water wetlands, where the flora recovered for the first time in decades. However, these wetlands are vanishing rapidly as restoration of housing and agriculture proceeds with the infilling of wetlands. Housing and agriculture are urgent priorities in the disaster zone. Under these circumstances, on-site conservation of aquatic and wetland species is extremely difficult. Off-site conservation by transplantation and seed preservation has been restricted to a few species (Hiratsuka 2014; Shimada 2016).

The 2011 tsunami showed that soil seed banks in low-lying freshwater wetlands are large and persistent. However, technological and sociological developments will be required to utilize their potential in the restoration of local biodiversity after a huge natural disaster.

References

Arai R, Nagahata Y (2013) *Monochoria korsakowii* found in tsunami-affected area in Iwate Prefecture. Iwate-Syokubutsu-No-Kai-Kaihou 50:8–13 (In Japanese)

Earthquake Engineering Research Institute (2011) EERI Special earthquake report. Learning from earthquakes. The Japan Tohoku Tsunami of March 11, 2011. Available at: http://www.eqclearinghouse.org/2011-03-11-sendai/files/2011/11/Japan-eq-report-tsunami2.pdf

Haraguchi N, Kitagawa I, Tomosho T, Mineta T, Zukemura C, Wakasugi K (2012) Paddy fields damaged along the Coast Part of Miyagi Prefecture by the Tsunami originated from the 2011 off the Pacific Coast of Tohoku Earthquake. J Jpn Soc Soil Phys 121:11–17 (In Japanese)

Hayasaka D, Shimada N, Konno H, Sudayama H, Kawanishi M, Uchida T, Goka K (2012) Floristic variation of beach vegetation caused by the 2011 Tohoku-oki tsunami in northern Tohoku, Japan. Ecol Eng 44:227–232

Hiratsuka A (2014) Renaissance in the botanical world. Biotope 33:1 (In Japanese)

Hoshino Y, Hoshino J (2014) Flora and vegetation of the urban area in Otsuchi. In: Ohtsuchi Town (eds) The study report of spring water ecosystem in Otsuchi. Otsuchi Town, pp 50–73 (in Japanese). Available at: http://www.town.otsuchi.iwate.jp/docs/2014042200012/files/hokoku.pdf

Kanaya G, Suzuki T, Maki H, Nakamura Y, Miyajima Y, Kikuchi E (2012) Effects of the 2011 tsunami on the topography, vegetation, and macrobenthic fauna in Gamo Lagoon, Japan. Jpn J Benthology 67:20–32

Kanno H, Hirabuki Y, Sugiyama T, Tomita M, Hara K (2014) Vegetation change in various coastal forest habitats after a huge tsunami: a three-year study. Jpn J Conserv Ecol 19:201–220 (In Japanese with English summary)

Kasai H (2013) Situation of rare plants and other plants in 2012 at inundation area in Miyagi Prefecture, Japan, by the Great East Japan Tsunami of 2011. Bull Bot Soc Tohoku 17:32–43 (In Japanese)

Kawanishi M, Hayasaka D, Shimada N (2016) The species composition of buried-seeds in the seashore vegetation disturbed by 2011 Tohoku-oki tsunami in northern Tohoku, Japan. In: Urabe J, Nakashizuka T(eds) Ecological impacts of tsunamis on coastal ecosystems. Tokyo: Springer, pp 289–310.

Kitagawa K, Shimano K (2010) Evaluation of seed banks for plant community recovery in abandoned rice fields of Matsumoto Basin, Nagano, Japan. Jpn J Conserv Ecol 15:121–131 (In Japanese with English summary)

Kohara H, Wan X, Akai K, Wang G (2011) Weed Monograph. 7. *Monochoria korsakowii* Regel et Maack. J Weed Sci Technol 56:166–181 (In Japanese)

Kurosawa T (2014) Changes of coastal vegetation and flora before and after the Great East Japan Earthquake of 2011 and situation of conservation of vegetation and plant diversity. Veg Sci News 18:70–80 (In Japanese)

Kurosawa T (2016) Vegetation, plant diversity, and attempts on their conservation after the Great East Japan Earthquake and Tsunami of 2011. In: Urabe J, Nakashizuka T(eds) Ecological impacts of tsunamis on coastal ecosystems: lessons from the Great East Japan Earthquake, Ecological research monographs: Springer, pp 311–335.

Kushiro K, Hayashi N, Nishihiro J (2009) A restoration of submerged vegetations from propagule banks by making small ponds at the shore of Lake Imba-numa, Japan. Ecol Civil Eng 12:141–147 (In Japanese)

Leck MA (1989) Wetland seed banks. In: Leck MA, Parker VT, Simpson RL (eds) Ecology of soil seed banks. Academic, San Diego, pp 283–305

Matsumoto H, Imanishi A, Imanishi J, Susaki J, Morimoto Y, Natuhara Y (2009) Persistence and vertical distribution of diaspores of aquatic plants in the surface soil layer of Oguraike and Yokoojinuma drained lands. J Jpn Inst Landsc Archit 72:543–546 (In Japanese with English summary)

Mineta T, Tomosyo T (2012) Necessity of vegetation management of agricultural land flooded by tsunami after the 2011 off the Pacific coast of Tohoku Earthquake – Report of vegetation investigation in the southern Sendai plains. Tech Rep Natl Res Inst Agric Eng 213:297–304 (In Japanese with English summary)

Ministry of the Environment (2015) Red data book 2014 – threatened wildlife of Japan, vol 8. Vascular Plants (in Japanese).

Mizusawa T, Nakamoto M, Morimoto Y (2000) Restoration experiments of wet lowland vegetation using soil seed bank. J Jpn Soc Revegetation Technol 25:321–326 (In Japanese with English summary)

Momohara A, Uehara K, Fujiki T, Tanaka N (2001) Distribution and preservation of seed bank in lake sediment of Teganuma, Chiba, Central Japan. Ann Tsukuba Bot Gard 20:1–9 (In Japanese with English summary)

Nagahata Y (2012) What changes in ecosystems were generated by the huge tsunami following the Great East Japan Earthquake of 2011?. Kodansha, Tokyo (In Japanese)

Naito T (2012) Vegetation of paddy fields in the seawater inundation area created by the Great East Japan Tsunami of 2011 in Watari-machi, Watari-gun, Miyagi Prefecture, Japan. Miyagi-no-Shokubutsu 37:15–18 (In Japanese)

Nakamoto M, Natori S, Mizusawa T, Morimoto Y (2000) Seed banks of abandoned rice fields – a case study in Nakaikemi, Tsuruga City, Fukui Prefecture. Jpn J Jpn Soc Revegetation Technol 26:142–153 (In Japanese with English summary)

Nakamura K, Tockner K, Amano K (2006) River and wetland restoration: lessons from Japan. BioScience 56:419–429

Nandasena NAK, Sasaki Y, Tanaka N (2012) Modeling field observations of the 2011 Great East Japan tsunami: efficacy of artificial and natural structures on tsunami mitigation. Coast Eng 67:1–13

Nemoto S, Kurosawa T, Fujiwara K (2013) Endangered plants on the coast in the southern part of Fukushima Prefecture, Japan after the 2011 Tohoku earthquake and tsunami. J Center Reg Aff Fukushima Univ 24:81–95 (In Japanese)

Nishihiro J, Nishihiro MA, Washitani I (2006a) Assessing the potential for recovery of lakeshore vegetation: species richness of sediment propagule banks. Ecol Res 21:436–445

Nishihiro J, Nishihiro MA, Washitani I (2006b) Restoration of wetland vegetation using soil seed banks: lessons from a project in Lake Kasumigaura, Japan. Landsc Ecol Eng 2:171–176

Rich TCG, Woodruff ER (1992) Recording bias in botanical surveys. Watsonia 19:73–95

Shibuya K, Shimada N, Suzuki M (2014) Study of the present condition of the Sanriku coastal ecotone and environmental conservation measures in Iwate Prefecture. J Policy Stud 15:181–199 (In Japanese with English summary)

Shimada N (2014) Impacts of the Great East Japan Earthquake and Tsunami of 2011 on coastal vegetation in Iwate Prefecture, Japan, and situation of subsequent vegetation recovery and conservation. Veg Sci News 18:44–54 (In Japanese)

Shimada N (2016) Conservation activities for coastal plants of Tofugaura sand beach, Noda Village, Iwate Prefecture. In: Branch T (ed) The ecological impacts of the Great East Japan Earthquake and tsunami. Bun-ichi Sougou Shuppan, Tokyo, pp 177–182 (In Japanese)

Shimoyama Y (2013) Distribution of sand dune vegetation after the Great East Japan Earthquake of 2011 in Sanriku Coast in Miyagi Prefecture, Japan. Miyagi-no-Shokubutsu 38:1–5 (In Japanese)

Sugawara D, Goto K, Chagué-Goff C, Fujino S, Goff J, Jaffe B, Nishimura Y, Richmond B, Szczuciński W, Tappin DR, Witter R, Yulianto E (2011) Initial field survey report of the 2011 East Japan Tsunami in Sendai, Natori and Iwanuma Cities. Available at: https://walrus.wr.usgs.gov/news/UNESCO_ITST_FinalReport100711.pdf

Sugiyama T (2012) Plants and algae after the Great East Japan Tsunami of 2011 in Mitazono regulating reservoir A, Natori-shi, northern Japan. Miyagi-no-Shokubutsu 37:19–22 (In Japanese)

Sugiyama T, Kasai H (2013) Plants in Kitaharahigashi, Shimomasuda, Natori-shi, Japan, after the Great East Japan Tsunami of 2011. Miyagi-no-Shokubutsu 38:9–13 (In Japanese)

Sugiyama T, Kasai H, Emi Y (2013) Changes in the flora of Ônuma Pond, Sendai-shi, Japan, after the Great East Japan Tsunami of 2011. Bull Bot Soc Tohoku 17:44–52 (In Japanese)

Suppasri A, Koshimura S, Imai K, Mas E, Gokon H, Muhari A, Imamura F (2012) Damage characteristic and field survey of the 2011 Great East Japan Tsunami in Miyagi Prefecture. Coast Eng J 54:1250005. doi:10.1142/S0578563412500052

Takano Y (2013) Information on plant distribution in Iwate Prefecture. Newly recorded species in Iwate. 3. Iwate-Syokubutsu-no-kai-Kaihou 50:24–28, In Japanese

Takiguchi M, Hirabuki Y, Kanno H, Naitou T, Sugiyama T, Shimoyama Y, Kasai H (2014) Drastic changes in the vegetation of the huge tsunami-disturbed area on the Pacific coast of Miyagi Prefecture. Veget Sci News 18:55–69 (In Japanese)

Tanaka N (2012) Effectiveness and limitations of coastal forest in large tsunami: conditions of Japanese pine trees on coastal sand dunes in tsunami caused by Great East Japan Earthquake. J Jpn Soc Civil Eng Ser B1 (Hydraul Eng 68:II_7–II_15

Usuba M (2013) Newly recorded aquatic and wetland plants in Fukushima Prefecture. 6. Flora Fukushima 29:57–63 (In Japanese)

Washitani I (2007) Restoration of biologically-diverse floodplain wetlands including paddy fields. Global Environ Res 11:135–140

Yonemura S, Nasu M, Tazawa R, Matubara T, Kameyama A (2001) Restoration of a vulnerable species *Penthorum chinense* population in a regulating reservoir. J Jpn Soc Revegetation Technol 27:26–31 (in Japanese with English summary)

Chapter 22
Influences of Large, Infrequent Disturbance Caused by Tsunami on Coastal Forest Communities

Mizuki Tomita, Yoshihiko Hirabuki, Hiroshi Kanno, and Keitarou Hara

Abstract Physiographic environments, species composition, and population structure of coastal forests in Sendai City, Miyagi Prefecture, northeastern Japan, were investigated to clarify types of damage to trees and other impacts to the forests 3 months after the Great East Japan Earthquake and Tsunami of 2011. The coast at the study site was originally protected by a seawall, and a banked canal had been dug parallel to the coast. Sand dunes, dominated by young *P. thunbergii*, had developed seaward of the canal, while the landward side was characterized by back marsh dominated by mature *P. thunbergii* and *P. densiflora*.

A belt transect (540×40 m) was set perpendicular to the shoreline and transecting both sand dune and back marsh. In 2006, before the tsunami, elevations along the sand dune sections of the belt transect varied widely between 1.1 and 4.8 m, while those in the back marsh, except for the canal bank, measured between 0.5 and 1.2 m. After the tsunami, large decreases in elevation were recorded on the sand dunes, caused mainly by scouring directly behind the seawall and canal bank. Large decreases were also measured on the back marsh, due to uprooting of trees and stripping of soil by the tsunami.

The total number of both live and dead trees over 5 cm in DBH were 1,738 and 1,039 on the sand dune and back marsh, respectively. Healthy living trees of the two *Pinus* species were observed mainly on the back marsh, with the proportion of live trees increasing roughly with distance from shoreline. Various types of damage, such as leaning (2.0 trees/100 m^2), uprooted (1.4), floating (1.0), and stem breakage/bending (0.5), were observed on the back marsh, but leaning (7.9 trees/100 m^2) and

M. Tomita (✉) • K. Hara
Tokyo University of Information Sciences, Chiba, Japan
e-mail: tomita@rsch.tuis.ac.jp

Y. Hirabuki
Tohoku Gakuin University, Sendai, Japan

H. Kanno
Tohoku Afforestation and Environmental Protection Co., Ltd., Sendai, Japan

J. Urabe, T. Nakashizuka (eds.), *Ecological Impacts of Tsunamis on Coastal Ecosystems*, Ecological Research Monographs,
DOI 10.1007/978-4-431-56448-5_22

stem breakage/bending (6.4) were frequent on the sand dune. Seedlings and saplings of deciduous broad-leaved trees such as *Quercus serrata* were observed only on the back marsh. Those of *P. thunbergii* were also observed both in the back marsh and sand dune. These trees and/or other plants recruited after the disturbance may be taking advantage of new microhabitats, such as mounds and pits created by uprooted trees, created by the disturbance. Continuous monitoring is needed to clarify changes of coastal forest ecosystems after the disturbance.

Keywords Biological legacy • Disturbance • Tsunami • Sand dune • Back marsh • Types of damage • Coarse woody debris • Description

22.1 Introduction

On March 11, 2011, the Great East Japan Earthquake and subsequent tsunami caused extreme damage to coastal forest along Sendai Bay. This sort of ecological event can be thought of as a "large and infrequent disturbance." Despite the scale of the damage, various organisms and spatially heterogeneous habitats remained in the disturbed sites. Organisms, organically derived structures such as fallen trees, and organically produced patterns such as spatial distribution of remnant forest patches that remained from the pre-disturbance system are defined as biological legacies (Franklin et al. 2000). It is suggested that post-disturbance studies on these biological legacies can provide insights into the best management practices for restoring natural resources and ecosystems (Lindenmayer et al. 2004; Dale et al. 2005; Lindenmayer et al. 2010). Despite the importance of this kind of research, learning opportunities are often lost because of various difficulties involved in rapid research responses after large and infrequent disturbances (Lindenmayer et al. 2010). Recovery process of ecosystems after a major disturbance differs according to the type of disturbance, such as tsunami, hurricane, landslide, and volcanic eruption, as well as according to pre-disturbance characteristics of the local ecosystems, including species composition and structure of communities. Description of ecosystems immediately after the large and infrequent disturbance and subsequent long-term monitoring is thus important to clarify changes after the disturbance.

In this study, to clarify impacts of the tsunami on coastal forest and landforms, an initial field survey was implemented soon after the 2011 event. Objectives of this study were (1) to describe the population structure of trees comprising the coastal forest after the tsunami, (2) to elucidate both extent and forms of damage to trees, (3) to detect differences in forms of damages between coastal forests established on sand dunes along the coastline and those on marshlands that have formed in low-lying marshy environments behind the dunes, (4) to examine relationships between forms of damages and landforms, and (5) to describe number of surviving trees including seedlings and saplings after the extreme tsunami.

22.2 Study Site

The study was carried out in a coastal forest in Sendai City, Miyagi Prefecture, northeastern Japan. The Sendai Plain has been advancing seaward at a rate of approximately 1 m/year after the stabilization of the sea levels about 6,000 years ago (Matsumoto et al. 2011; Udo et al. 2012). Sediment is supplied from the Nanakitagawa, Natorigawa, and Abukumagawa Rivers, as well as the southern cliff coasts. The plain is about 10 km wide and is characterized by a low-lying terrain of 0–3 m above sea level. At present the coastal beach ridges (about 3 m above sea level) form the highest point of the plain (Miyagi et al. 2013). Freshwater back marshes (0–2 m above sea level) occur sparsely in the low-lying areas behind these beach ridges.

Before the Great East Japan Earthquake, the coastal forest in Sendai City was mainly dominated by *Pinus thunbergii* and *P. densiflora*. These pines were first planted along the Teizan canal, a banked canal dug parallel to the coast about 400 years ago, primarily to protect backlands from damage due to high tides and salt spray (Hara 2014). The coastal forest was separated into two sections by the canal (Fig. 22.1). The forest on the seaward side was established on sand dunes and dominated by young *P. thunbergii* (homogeneous-aged forest about 30 years old),

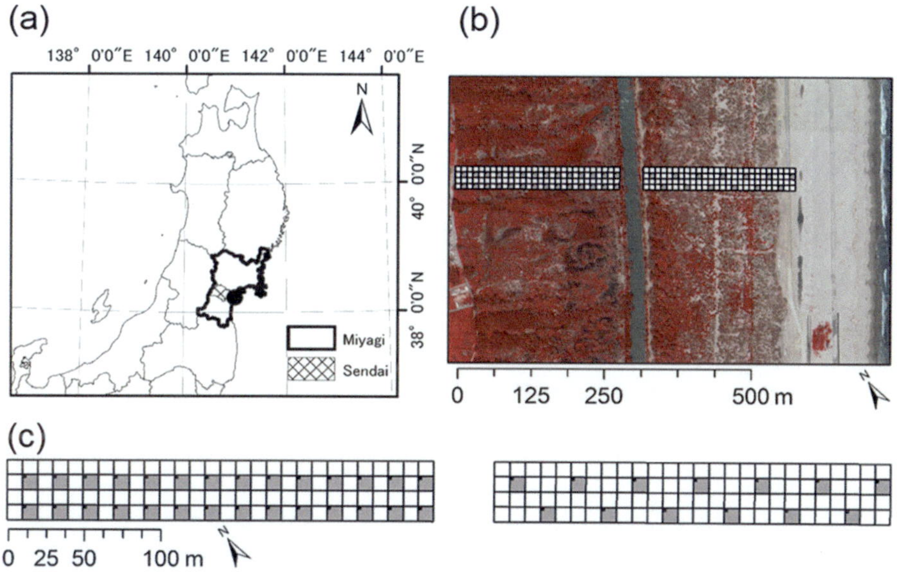

Fig. 22.1 (**a**) Location of the study site. *Black* point indicates the study site. (**b**) Location of the belt transect. Background image is an aerial photo obtained in summer of 2012 and shown by false color. The belt transect is separated into two parts by a canal. The *right*-side end of the image is the shoreline. (**c**) Extended indication of the belt transect. Trees in the S3 class were surveyed in all the quadrats (10 × 10 m). S2 class trees were surveyed only in the *gray* colored quadrats. S3 class trees were surveyed only in the smaller *black* color quadrats (2 × 2 m)

while the forest on the landward side was established on back marsh and dominated by mature *P. thunbergii* and *P. densiflora* (heterogeneous-aged forest about 70–110 years old). Some broad-leaved tree species, such as *Quercus myrsinifolia*, *Q. salicina*, *Q. serrata*, and *Prunus* spp., were also distributed in the landward forest, and small marshes were also interspersed among the trees. In a study conducted before the Great East Japan Earthquake, 759 species of vascular plant (40 % of the vascular plants in Sendai City) were confirmed in the coastal forest in and around the study site (Sugiyama et al. 2011). An especially high diversity of plants, including endangered and vulnerable species, were observed in the landward forest, where the back marshes are interspersed with areas in which forest management has been abandoned for the last several decades.

Based on the data obtained by the 2011 Tohoku Earthquake Tsunami Joint Survey (TTJS) Group (2011), the height of the tsunami reached 9–14 m on a sand dune about 1.9 km northeast from the study site and 300 m inland from the shoreline. The height, however, decreased to 4 m at 1 km inland. According to the Geospatial Information Authority of Japan (2011), after the earthquake land subsidence in the general area of the study site ranged between 0.2 and 0.5 m. Based on the land cover classification map from two satellite images (October 2010 and November 2011), Zhao et al. (2013) estimated that the total area of coastal forest in Sendai City decreased from 4.2 to 0.5 km^2 due to the tsunami. Although large areas of coastal forest were disturbed and destroyed by the tsunami, small patches remained in narrow comb-like stripes running perpendicular to the shoreline and direction of the tsunami at the study site and other nearby areas (Tomita et al. 2013; Hara 2014).

22.3 Methods

To clarify physiographic effects on population structures and types of damage to trees, a belt transect (540 × 40 m) was set perpendicular to the shoreline, transecting both sand dune and back marsh at Okada in Sendai City, June 2011 (Tomita et al. 2014). The belt transect was divided into 216 contiguous quadrats (10 × 10 m), 104 quadrats being located in seaward side, and 112 in landward side (Fig. 22.1). Distance between centroid of each quadrat and shoreline was measured by using ArcGIS 10.2.2. According to the TTJS Group (2011), height of tsunami was estimated to have reached about 10 m at beginning of the belt transect and 6 m at end of the belt transect.

In the belt transect, trees were classified into three size classes: S1 (trees less or equal to 5 cm in DBH and 2 m in height), S2 (trees less or equal to 5 cm in DBH and over 2 m in height), and S3 (trees over 5 cm in DBH). Species name and DBH for both live and dead trees in the S3 class were recorded. Major types of damages were classified as leaning (stem inclined physically with roots still in the ground), stem breakage/bending (stem broken or bent physically with roots still in the ground), uprooted (fallen trees with roots out of ground, found at or near their original spot), and floating (trees uprooted elsewhere and drifted to spot where found) and recorded for all trees of S3. Species name and height for living trees of the S1 and S2 classes

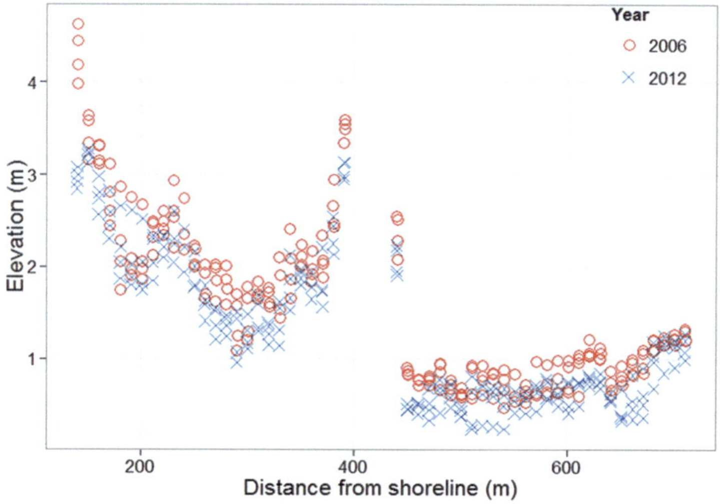

Fig. 22.2 Changes in elevation between 2006 and 2012. Left-side end of the scatter graph is the shoreline

were also recorded. Trees of the S3 class were surveyed in all quadrats. Trees of the S2 class were surveyed in 13 quadrats on seaward side and 28 on landward side of the belt transect. For survey of the S1 class trees, small-sized quadrats (2×2 m) were set in the same quadrats for S2 (Fig. 22.1).

Digital elevation model (DEM) of 1 m resolution, which was obtained by airborne laser scanning in December 2006 and July 2012, was used to describe changes in elevations in the belt transect. Median of DEM in each quadrat were calculated for each year and plotted for distance from shoreline (Fig. 22.2). DEM was corrected with reference to a ground control point so that differences of elevation between 2006 and 2012 could account for changes caused by land subsidence and sedimentation.

To clarify the extent of sand sedimentation carried by the tsunami, depth of accumulated sand to A layer was measured in September 2012. Locations of measurement points were also recorded using GPS.

22.4 Results and Discussion

22.4.1 Physiographic Environment

In 2006, elevations of the belt transect on seaward side varied widely between 1.1 and 4.8 m, although those in the landward side, except for canal bank, measured between 0.5 and 1.2 m (Fig. 22.2). After the tsunami, elevations decreased about 0.5–1.0 m in almost all quadrats. Particularly large decreases in elevation were recorded at the seawall around 130 m from the shoreline, at the canal bank about 400 m inland, and near small marshes around 500 m and 640 m inland. As a result,

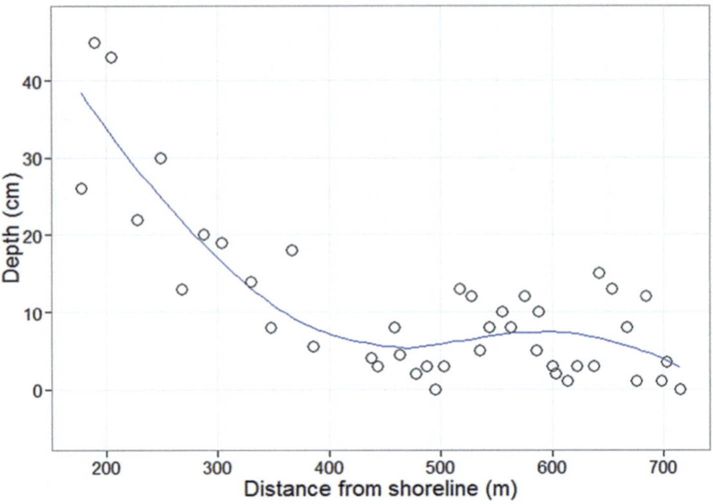

Fig. 22.3 Depth of sand sediment carried by tsunami. Left-side end of the scatter graph is the shoreline

in 2012, the range of elevation in the belt transect on the seaward side decreased, while that on the landward side increased. A decrease in elevation around 130 m and 400 m inland from the shoreline was caused mainly by scouring directly behind the seawall and canal bank. On the other hand, decrease in elevation around 500 m and 640 m inland occurred because of uprooting of trees and stripping of soil by the tsunami.

Erosion behind coastal engineering structures and deposition within the coastal forests were also reported by Udo et al. (2012) and Richmond et al. (2012). In the result of sand sedimentation measurement, depth of sand sediment was highest around 200 m inland and roughly decreased with distance (Fig. 22.3). Around 550 m and 660 m inland, however, deeper sand sediment compared to surrounding areas was observed. This was probably caused by accumulation of sand scoured from other areas, as well as sand from uprooting of trees and stripping of soil by the tsunami. Decrease and increase of elevation due to the tsunami disturbance provided heterogeneous habitats for plant species (Kanno et al. 2014; Onza et al. 2014), most likely leading to various successional pathways.

22.4.2 Physically Damaged Trees

After the tsunami, the total number of both live and dead trees in the S3 class was 1,738 in the belt transect on the seaward side and 1,039 on the landward side, respectively (Tables 22.1 and 22.2). *Pinus thunbergii* was most numerous, followed by *P. densiflora*. *Prunus* spp. and *Robinia pseudoacacia* were next numerous following the two *Pinus* species in each belt transect.

Table 22.1 Summary of trees in the belt transect on the seaward side

Species name	S1	S2	S3											
				Types of damages										
			Al	Ln		Sb		Ur		Fl		Sd	St	Total
				A	D	A	D	A	D	A	D			
Pinus thunbergii	38	8	21	19	791	9	637	2	33		5	150	2	1,669
P. densiflora			1	1	15		18		4			4		43
Robinia pseudoacacia	7	8	9	2	8	3			3			1		26
Total	45	16	31	22	814	12	655	2	40		5	155	2	1,738

After Tomita et al. (2014)

A number of healthy living trees for S1 and S2 and healthy and damaged trees for S3 are shown. A and D indicate live and dead trees at the time of survey, respectively. Al, Ln, Sb, Ur, Fl, Sd, and St indicate healthy living, leaning, stem breakage/bending, uprooted, floating, standing dead, and stumped trees, respectively. Total area of the quadrats in S1, S2, and S3 is 52 m^2, 1300 m^2, and 10,400 m^2, respectively

Table 22.2 Summary of trees in the belt transect on the landward side

Species name	S1	S2	S3											
				Types of damages										
			Al	Ln		Sb		Ur		Fl		Sd	St	Total
				A	D	A	D	A	D	A	D			
Pinus densiflora	22		134	13	67	2	15	5	68		56	36	22	418
P. thunbergii	31		70	24	117		35	8	75	5	50	25	8	417
Prunus spp.	28	38	56	29	11	6	11	6	10	8	14	5		156
Robinia pseudoacacia	23	5	5	9	2	1	1	1			1	1		21
Toxicodendron trichocarpum	6	14		2	1		1							4
Alnus japonica			1	1		1								3
Quercus serrata	18	5	1	1										2
Pinus spp.									1				1	2
Ilex crenata				1	1									2
Malus toringo		1		1										1
Fraxinus sieboldiana		5		1										1
Pourthiaea villosa var. *villosa*	10	1												
Morus australis	4													
Unidentified spp.					1		3		1		3	3	1	12
Total	142	69	267	82	200	10	66	20	155	13	124	70	32	1,039

After Tomita et al. (2014)

A number of healthy living trees for S1 and S2 and healthy and damaged trees for S3 are shown. For details, see Table 22.1

All types of damage were observed in the belt transect on the landward side, but uprooting and floating were almost absent on the seaward side (Figs. 22.4 and 22.5). On the seaward side, leaning and stem breakage/bending were frequent (7.9 and 6.4 trees/100 m^2) and accounted for 87.0 % of the *Pinus* species (Tables 22.1 and 22.2). On the other hand, in the belt transect on the landward side, leaning (2.0 trees/100 m^2) was most frequent, followed by uprooted (1.4), floating (1.0), and stem breakage/bending (0.5). Proportion of damage types also varied in broad-leaved trees, although the number of these trees was low (Fig. 22.6, Tables 22.1 and 22.2). A greater amount of leaning and stem breakage/bending damage was observed on the seaward side than on the landward side. Because the coastal forest on the seaward side was dominated by young *P. thunbergii* below about 8 m in height and 20 cm in DBH, almost all trees were flooded completely by the tsunami. The vertical root systems of these trees, however, were deep and well developed on the higher sand dunes, resulting in fewer uprooted trees compared to the landward side. On the other hand, trees on the landward side, growing at lower elevations with higher water tables around the marshes, were thus unable to develop deep roots and were thus more susceptible to uprooting. In addition, floating trees originated from scoured regions behind the seawall or on the downslopes of sand dunes (Tanaka 2012). Not only difference in susceptibility to uprooting due to physiographic features but also scouring by the tsunami should thus be taken into consideration in forest restoration projects.

22.4.3 Healthy Living Trees

After the tsunami, healthy living trees of both *Pinus* species were observed mainly in the belt transect on the landward side, with the proportion increasing roughly with distance from shoreline. Large-sized trees over 20 cm in DBH showed low susceptibility to physical damage by tsunami in the belt transect on the landward side (Figs. 22.4, 22.5, and 22.6), where the height of the tsunami had decreased to about 6 m. Consequently, crowns of large-sized trees over 20 cm in DBH escaped complete flooding, reducing physical damages. In the southern part of Sendai plain, similar situation was also reported by Miyagi et al. (2013). Some damage to these large-sized trees, however, including uprooting, floating, and leaning, occurred directly behind the canal bank and near the small marshes. As a result, not only smaller trees below about 20 cm in DBH but also the larger trees planted behind the canal bank or in unsuitable physiographic areas such as marshes suffered some physical damages due to the tsunami disturbance.

Numerous seedlings of *P. thunbergii* were observed in the both belt transects, and several seedlings and saplings of deciduous broad-leaved trees such as *Quercus serrata* in the belt transect on the landward side. The number of living trees in the S1 and S2 classes was 45 (86.5/100 m^2) and 16 (1.2/100 m^2), respectively, in the belt transect on the seaward side, and 142 (126.8/100 m^2) and 69 (2.5/100 m^2), respectively, on the landward side (Tables 22.1 and 22.2). In Japan, coastal forests designed

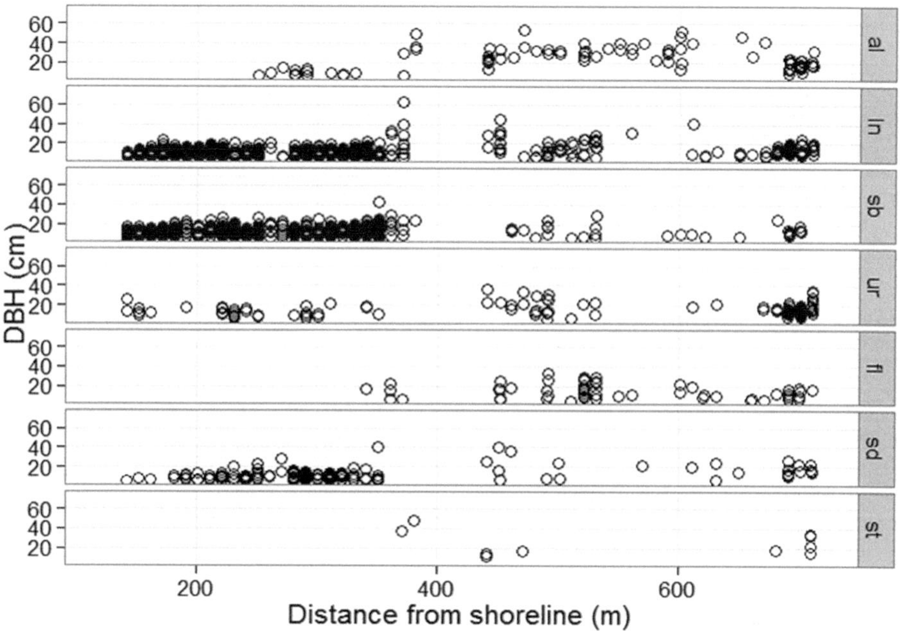

Fig. 22.4 DBH of *Pinus thunbergii* observed in the belt transect. For details of abbreviations, see Table 22.1 (After Tomita et al. 2014)

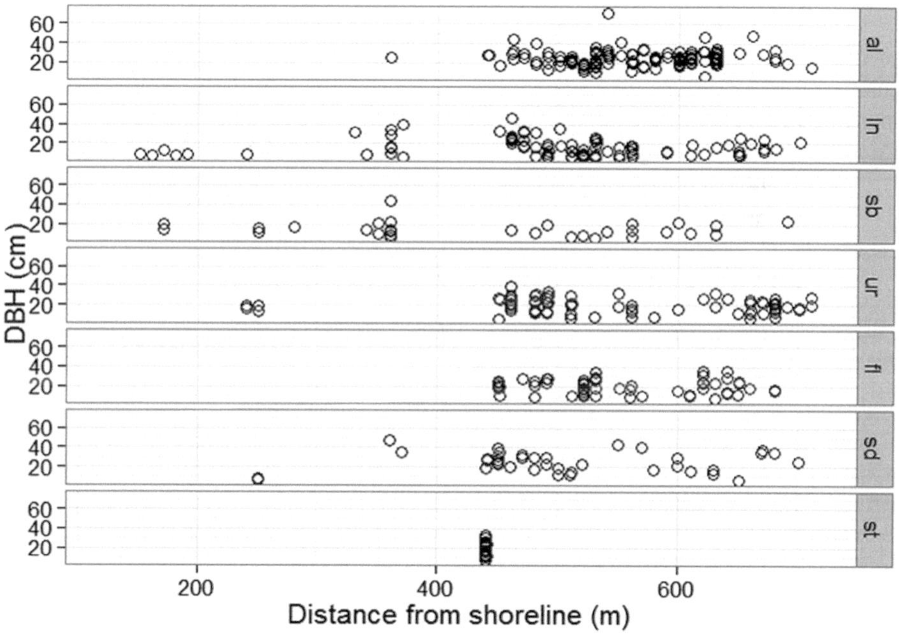

Fig. 22.5 DBH of *Pinus densiflora* observed in the belt transect. For details of abbreviations, see Table 22.1 (After Tomita et al. 2014)

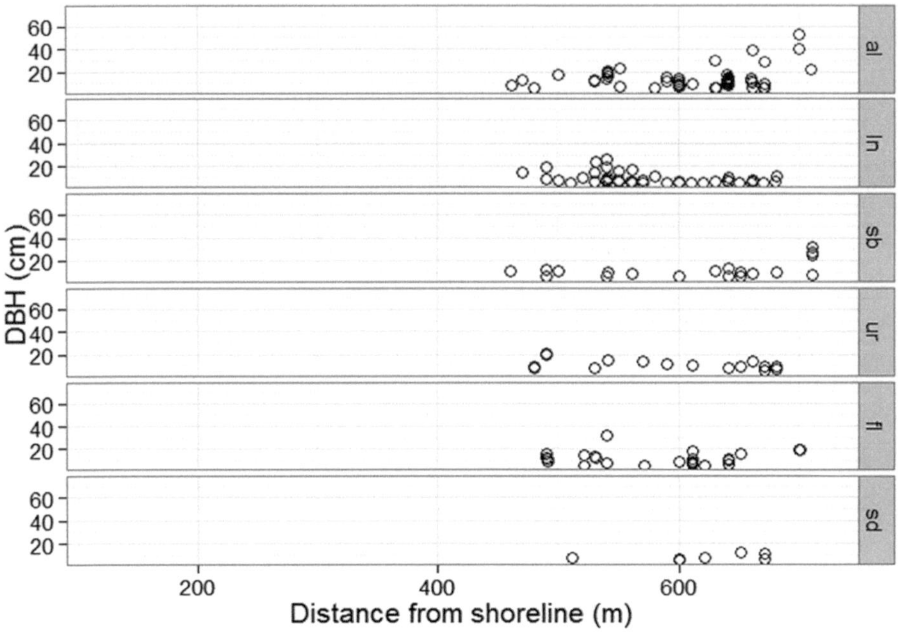

Fig. 22.6 DBH of broad-leaved trees observed in the belt transect. For details of abbreviations, see Table 22.1 (After Tomita et al. 2014)

as protection against salt spray and/or storm surge are generally planted at a density of 100 saplings/100 m^2. The density of surviving saplings in the S1 class, including some broad-leaved tree species, was close to this figure. A precise survey of the distribution of surviving saplings should be implemented, and the results used to formulate quick and efficient restoration plans.

22.5 Conclusion

Residual organic matters, organically derived structures, and organically generated spatial patterns, which persist through a disturbance and are incorporated into the recovering system, are defined as "biological legacies" and can drive the rate and pathway of ecosystem recovery following a disturbance (Franklin et al. 2000). The results of our survey showed that despite the scale and severity of the 2011 tsunami, live seedlings, saplings, and trees of *Pinus* and other species survived, and dead trees that suffered physical damages such as leaning, uprooting, and stem breakage remained in the disturbed coastal forest (see also, Tomita et al. 2013, 2014). In addition, the tsunami created heterogeneous environments for plant establishment, including pits, mounds, bare patches stripped by the tsunami, and forest patches consisting of remnant canopy trees (Tomita et al. 2013). Occurrence of various types of tree damage, as well as distribution of heterogeneous habitats for remnant and recruiting plants after the disturbance, seems to be depended on interactions between the pre-disturbance physiographic environment, height of tsunami, and size of trees. To understand the effect of large and infrequent disturbances such as an extreme tsunami on coastal forest ecosystems, it is important to continuously monitor how biological legacies and environmental heterogeneity influence regeneration.

Acknowledgments We wish to thank Kevin Short for editorial assistance. This research was supported by JSPS KAKENHI 24510332, 25830153.

References

Dale VH, Swanson FJ, Crisafulli CM (eds) (2005) Ecological responses to the eruption of Mount St Helens. Springer, New York

Franklin JF, Lindenmayer D, Macmahon JA et al (2000) Threads of continuity. Conserv Pract 1(1):8–16

Geospatial Information Authority of Japan (2011) http://www.gsi.go.jp/cais/topic110421-index-e.html. Accessed 5 June 2014

Hara K (2014) Damage to coastal vegetation due to the 2011 tsunami in northeast Japan and subsequent restoration process: analyses using remotely sensed data. Global Environ Res 18:27–34

Kanno H, Hirabuki Y, Sugiyama T et al (2014) Vegetation change in various coastal forest habitats after a huge tsunami: a three-year study. Jpn J Conserv Ecol 19:201–220

Lindenmayer DB, Foster DR, Franklin JF et al (2004) Salvage harvesting policies after natural disturbance. Science 303:1303–11303

Lindenmayer DB, Likens GE, Franklin JF (2010) Rapid responses to facilitate ecological discoveries from major disturbances. Front Ecol Environ 8(10):527–532

Matsumoto H, Kato A, Kumagai M (2011) Coastal erosion along the shoreline of Sendai coastal plain, northeastern part of Honshu-island, 1978–2010. J Hum Inf 16:103–110

Miyagi T, Yanagisawa H, Baba S (2013) The protective role of mangroves and other coastal forests against tsunami damage: lesson learned from case studies of two tsunamis. Global Environ Res 17:247–254

Onza N, Ishida I, Tomita M et al (2014) Environmental heterogeneity and plant diversity in a coastal forest affected by a severe tsunami. Jpn J Conserv Ecol 19:177–188

Richmond B, Szczuciński W, Chagué-Goff C et al (2012) Erosion, deposition and landscape change on the Sendai coastal plain, Japan, resulting from the March 11, 2011 Tohoku-oki tsunami. Sediment Geol 282:27–39. doi:10.1016/j.sedgeo.2012.08.005

Sugiyama T, Emi Y, Kasai H (2011) Flora of the coastal forest in Sendai-shi, Miyagi Pref. Bull Bot Soc Tohoku 16:59–68

Tanaka N (2012) Effectiveness and limitations of coastal forest in large tsunami: conditions of Japanese pine trees on coastal sand dunes in tsunami caused by Great East Japan Earthquake. J Jpn Soc Civil Eng Ser B1 (Hydraul Eng) 68:II:7–II:15

The 2011 Tohoku Earthquake Tsunami Joint Survey (TTJS) Group (2011) http://www.coastal.jp/ttjt/index.php. Accessed 5 June 2014

Tomita M, Hirabuki Y, Kanno H et al (2013) Spatial distribution of biological legacies and depth of accumulated sand in a coastal forest disturbed by the 2011 tsunami. J Nat Restor Conserv 6:51–60

Tomita M, Hirabuki Y, Kanno H et al (2014) Influence of tsunamis as large, infrequent disturbances on tree communities of coastal forests. Jpn J Conserv Ecol 19:163–176

Udo K, Sugawara D, Tanaka H et al (2012) Impact of the 2011 Tohoku Earthquake and Tsunami on beach morphology along the northern Sendai Coast. Coast Eng J 54:1250009. doi:10.1142/S057856341250009X

Zhao Y, Tomita M, Hara K (2013) Landscape change analysis before and after the earthquake disaster in Sendai coastal area by using SPOT satellite data. J Nat Restor Conserv 6:43–49

Chapter 23
Survey of Impact of the Great East Japan Earthquake on the Natural Environment in Tohoku Coastal Regions

Osamu Ichihashi, Daisuke Horii, Yoshio Tsukamoto, Takashi Someya, Hiroaki Terasawa, Shinji Iki, Maki Isoda, Kotaro Goto, Emiko Ariyasu, Takashi Inoue, and Shintaro Abe

Abstract The impacts of the Great East Japan Earthquake that occurred on March 11, 2011, to the natural environment in the coastal area were surveyed in April 2012 to March 2016 by Biodiversity Center of Japan, Ministry of the Environment. Sand dunes and coastal forests received direct impacts from the tsunami, while some new hygrophyte communities were generated from buried seeds in the remains of rice paddies. Although natural recoveries of vegetation (e.g., from open water to wet grassland) were observed at some areas, artificial impacts such as reconstruction projects turned wetland plant communities into artificial bare land. Some places (e.g., reclaimed lake) were turned back to original environment by the comparison with the land cover of 100 years before the earthquake. Many shorelines that had receded due to erosion accompanied with the tsunami had made early recovery except for some river mouth bars and pocket beaches. Distribution of seaweed (Phaeophyceae spp.) seemed to have little impacts. On the other hand, seagrass (*Zostera* spp.), which generally grow on the sediment of sand and silt, had heavy impact because the tsunami swiped the sediment away although many cases of recovery of seagrass beds were observed 4 years after the earthquake. Moreover, fauna and flora survey in key sites and monitoring on pre-surveyed sites of tidal flats, seagrass beds, seaweed beds, and seabird breeding sites have been conducted. According to the series of the surveys, the living organisms were seemed to be

O. Ichihashi (✉) • D. Horii • Y. Tsukamoto • T. Someya • H. Terasawa • S. Iki • M. Isoda
K. Goto • E. Ariyasu
Asia Air Survey Co., Ltd, 1-2-2 Manpukuji, Asao-ku, Kawasaki-shi,
Kanagawa 215-0004, Japan
e-mail: os.ichihashi@ajiko.co.jp

T. Inoue
Japan Wildlife Research Center, 3-3-7 Kotobashi, Sumida-ku, Tokyo 130-8606, Japan

S. Abe
Biodiversity Center of Japan, Ministry of the Environment,
5597-1, Kenmarubi, Kamiyoshida, Fujiyoshida-shi, Yamanashi 403-0005, Japan

J. Urabe, T. Nakashizuka (eds.), *Ecological Impacts of Tsunamis on Coastal Ecosystems*, Ecological Research Monographs,
DOI 10.1007/978-4-431-56448-5_23

adapted to the natural disturbance although some were still in the middle of recovery process. All of the data of these survey results were archived and are disclosed at "Shiokaze Natural Environment Log" <http://www.shiokaze.biodic.go.jp/>.

Keywords Great East Japan Earthquake • Ecosystem monitoring • Vegetation • Remote sensing • Land cover • Tohoku • Tsunami • Disturbance • Reconstruction • Biodiversity

23.1 Introduction

At 14:46 on March 11, 2011, the Great East Japan Earthquake (GEJE) of magnitude 9.0 occurred, with an epicenter off the Pacific coast of the Tohoku Region. When GEJE struck, a ground subsidence of approximately 1.2 m occurred at a point at the tip of the Ojika Peninsula of Miyagi Prefecture (GIAJ[1] 2011) and a huge tsunami struck the Pacific coasts from the Tohoku to Kanto Regions, causing unprecedented damage. The GEJE motion and tsunami have given significant impacts to the natural environment in the coastal area.

In accordance with the "Basic Guidelines for Recovery in Response to the Great East Japan Earthquake" released by Japanese Government in July 2011, which specifies to "conduct the current status survey on the natural environment affected by the tsunami and monitoring of secular changes," the Biodiversity Center of Japan Ministry of the Environment (BCMoE) has been monitoring the impact on and changes of the natural environment after the earthquake as well as collecting data before the earthquake in the Pacific coastal area of the Tohoku Region.

Series of surveys were conducted at the tsunami-flooded areas and their surroundings, covering a wide range of approx. 670 km in length and 578 km^2 in area, extending from Aomori to Chiba Prefectures (Fig. 23.1). The surveys started in 2012, including vegetation survey, fauna and flora survey, shorelines and land cover status survey on the coast, distribution survey of seagrass (*Zostera* spp.) and seaweed (Phaeophyceae spp.) beds, and monitoring surveys on tidal flats, seagrass beds, seaweed beds, and seabird breeding sites where the pre-earthquake records were available.

In this paper, the outline and the initial findings of the survey are introduced. Some of them are suggestive for recovery practices and management of coastal ecosystems, since such a huge disturbance seems so rare. It may also include helpful information of the future planning of the area where similar disasters are predicted to suffer.

---

[1] Geospatial Information Authority of Japan, Land, Infrastructure and Transportation Ministry.

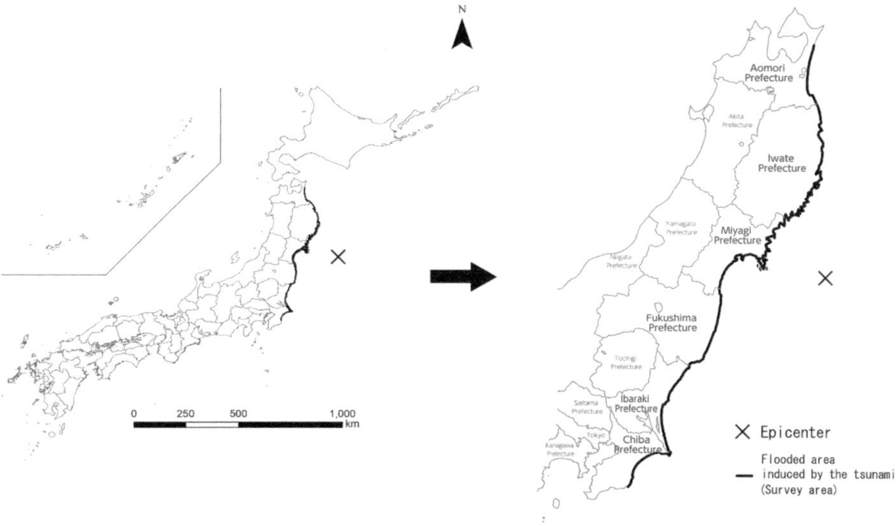

Fig. 23.1 Survey area

23.2 Survey Methods

The series of surveys includes mainly five categories as follows:

- Terrestrial vegetation
  - Terrestrial vegetation in the whole tsunami-flooded areas in 2000s, 2012, 2013, and 2014 were surveyed. Vegetation maps and the vegetation change maps were created.
  - Topographic maps from 1903 to 1917 were interpreted to detect the old topography before intensive development during 1945–2000.
- Fauna and flora in key sites
  - A total of 16 key sites in the tsunami-flooded areas were selected to record the fauna and flora in 2012, 2013, 2014, and 2015.
  - Belt transects were used to investigate the distribution of the living organisms in the key sites in 2012, 2013, 2014, and 2015 in order to reveal the secular change of the distributions.
- Shorelines and land cover in the coasts
  - Shorelines and the land cover statuses behind the shorelines on the tsunami-flooded areas were surveyed in 1970s, 2000s, 2011 (just after the earthquake), and 2014.
- Distribution of seagrass and seaweed beds

- Distribution of seagrass and seaweed beds before and after the earthquake in the tsunami-flooded areas was detected, and the distribution maps were created.

- Monitoring on specific sites where the pre-earthquake records are available
  - Tidal flats (a total of 18 sites), seagrass beds (a total of 5 sites), seaweed beds (a total of 4 sites), and seabirds breeding (a total of 4 sites) in the tsunami-flooded areas were surveyed in 2012, 2013, 2014, and 2015.

23.2.1 Terrestrial Vegetation

For the whole tsunami-flooded areas, we created and overlaid pre-earthquake vegetation maps in 2000s with post-earthquake vegetation maps made in 2012, 2013, and 2014 to detect the vegetation and land cover changes during the period. The vegetation maps in 2000s were created referring to the interpretation results of high-resolution satellite images (IKONOS and GeoEye-1) acquired from 2000 to 2010 and the 1/50,000 vegetation maps created based on the National Survey on the Natural Environment. The vegetation map of 2012 was created by interpreting the aerial photos taken by the Geospatial Information Authority of Japan (GIAJ) from March 2011 to April 2012. They were verified by the field surveys on detailed plant community composition and dominant species conducted at 448 sites and 2,537 sites, respectively, from September to November 2012. The survey in 2013 and 2014 extracted only the places with any changes (differences) of land cover through analysis of middle-resolution satellite images (RapidEye) and verified by the surveys on community composition and the dominant species conducted at 113 sites and 568 sites, respectively, from September to November 2013 and those conducted at 225 sites and 562 sites, respectively, from June to October 2014.

Furthermore, old topographic maps published from 1903 to 1917 were interpreted to detect old topography before intensive development during 1945–2000. It enabled us to know the distributions of former riverbeds, rivers, lakes, marshes, wetlands, sand dunes, and sandy beaches before intensive development, which were then compared with the conditions just before and after the earthquake.

23.2.2 Fauna and Flora in Key Sites

Some sites deemed especially important as the habitats of valuable organisms were selected among the tsunami-flooded areas (called "key sites"). On these sites, the fauna (mammals, birds, amphibians, reptiles, fish, terrestrial insects, and benthos) and flora (higher plants) were identified and recorded with remarks (e.g., rare species) if any. Besides, belt transects of 20 m width and 100–410 m length on each key site were used to investigate detailed distribution of vegetation types.

A total of the key sites were 16 sites: 3 sites were surveyed in 4 years (2012, 2013, 2014, and 2015), 2 sites were surveyed in 3 years (2013, 2014, and 2015), 7 sites were surveyed in 2 years (2013 and 2015), and 3 sites were surveyed in a year of 2015.

23.2.3 Shorelines and Land Cover Change of Coasts

The locations and status of shorelines of the whole tsunami-struck coasts (2,284 km in 2011 and 2,273 km in 2014) were identified from Aomori to Chiba Prefectures. The land cover statuses of the tsunami-struck beaches in the natural coast (680.5 km) were also identified four times in total, i.e., the 1970s (high economic growth period), pre-earthquake (2000s), and two post-earthquake periods (2011 and 2014) through interpretation of satellite and aerial photo images.

The images used for interpretation were color aerial photos taken in 1974–1977 for the 1970s, high-resolution satellite images (IKONOS and GeoEye-1) acquired in 2000–2009 for the pre-earthquake period (2000s). Color aerial photos taken during March 2011 to April 2012 were regarded as the images just after the earthquake. The middle-resolution satellite images (RapidEye) acquired in August 2014 was for the 2014 period.

The land cover statuses were classified into five environment types, i.e., sand beach, sand dune vegetation, coastal forest, coastal artificial structure (e.g. a port, tide embankment, and reclamation), and others (e.g. cultivated land, road, and house), and the statuses of land cover changes were identified.

23.2.4 Distribution of Seagrass and Seaweed Beds

Satellite or aerial photo images were utilized to identify the areas of seagrass and seaweed beds in both before and after the earthquake, and the distribution maps of them were created.

The pre-earthquake distribution maps covering the bordering coast from Iwate Prefecture to northern Fukushima Prefecture were created using high-resolution satellite images (WorldView-2, QuickBird, and GeoEye-1) acquired in December 2003 to February 2011. The post-earthquake distribution maps covering the coast of the whole tsunami-flooded areas from Aomori to Chiba Prefectures were created using color aerial photo images taken in October 2012 to January 2013 and high-resolution satellite images (WorldView-2, GeoEye-1, and QuickBird) acquired in July 2011 to February 2015.

The satellite images above were processed by following geometric correction, sun glint correction, atmospheric correction, water-depth correction, and object-oriented classification, and then seagrass and seaweed beds were automatically extracted from the processed satellite images. After that, the automated extracted seagrass and seaweed beds were verified by field survey conducted during July to August 2015.

23.2.5 Follow-Up of Monitoring Site 1000

"Monitoring Sites 1000" (Moni1000), which is a monitoring activity for various ecosystems since 2003, and "National Survey on the Natural Environment" (NSNE), which is one of the nationwide surveys on wildlife since the 1970s, have been conducted by BCMoE with a formulized instruction manual.[2] Thus, the pre-earthquake information on both organisms and their habitats are available at the specific monitoring sites.

This survey covered the almost same monitoring sites and methods of Moni1000 and NSNE in the tsunami-flooded area in order to follow up the information after the earthquake. These sites were tidal flats (18 sites, from Aomori to Chiba Prefectures), seagrass beds (5 sites, from Iwate to Chiba Prefecture), seaweed beds (4 sites, from Iwate to Ibaraki Prefectures), and seabirds breeding sites (4 sites, isolated islands of Aomori to Miyagi Prefectures).

In the tidal flats, the numbers of species and individuals of benthos were recorded. In the seagrass beds and the seaweed beds, dominant seagrass or seaweed species and their coverage were recorded. In the seabirds breeding sites, densities of nesting holes were recorded.

23.3 Results: Impacts of Earthquake and Tsunami and Changes in 5 Years After the Earthquake

23.3.1 Terrestrial Vegetation

Figure 23.2 shows an example of vegetation changes before and after the earthquake in the Sendai Plain, Miyagi Prefecture. The areas of paddy fields, sand dune vegetation, and coastal forests were the most greatly reduced due to the direct impacts of tsunami and ground subsidence. Most of them were turned into such as "uncultivated paddy fields" and "damaged afforestation sites" just after the tsunami. Some parts of them were turned into paddy fields again or artificial bare lands for new constructions afterward.

Some hygrophyte communities, which were not seen before the earthquake occurred at some newly created wetlands. They probably generated from buried seeds, and the dormant seeds might have germinated to be exposed due to the

[2] http://www.biodic.go.jp/moni1000/manual/index.html

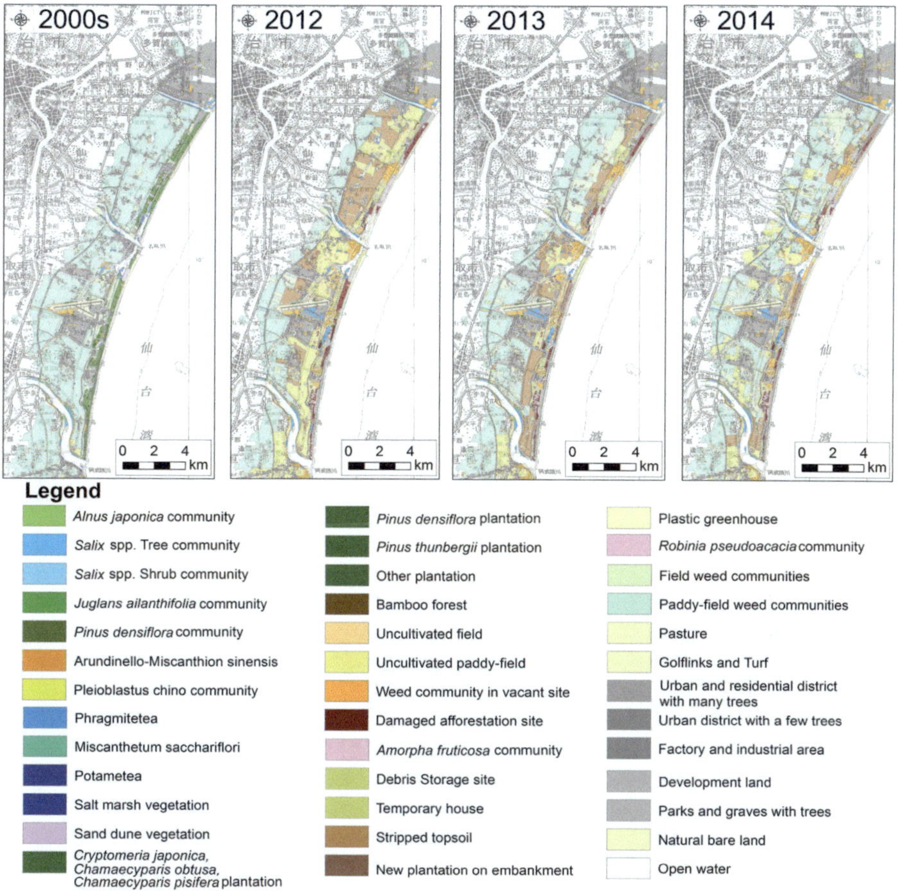

Fig. 23.2 Vegetation changes before (2000s) and after (2012, 2013, and 2014) the earthquake in the Sendai Plain, Miyagi Prefecture *Geographical maps: 1/200,000 map of GIAJ (After BCMoE forthcoming)

disturbance. The hygrophyte communities remained for 1 or 2 years after the earthquake, though gradually decreased afterward.

Natural recoveries were also observed at some wet grasslands or sand dunes (Fig. 23.3). However, artificial impacts such as reconstruction projects have been reducing these communities gradually.

Comparison of topographic maps published in 1903 to 1917 with the images taken just before and after the earthquake gave us interesting facts of the changes. Many examples in which the subsidence and tsunami-caused flood had turned the coastal lands back to their original state. In the case of Itagawaura (Fig. 23.4), a wet environment was reclaimed into paddy in 1929, though it reverted to wetlands due to the impact of the GEJE. Four years after the earthquake, the flood water was drained again. Such information gained from old maps is useful for under-

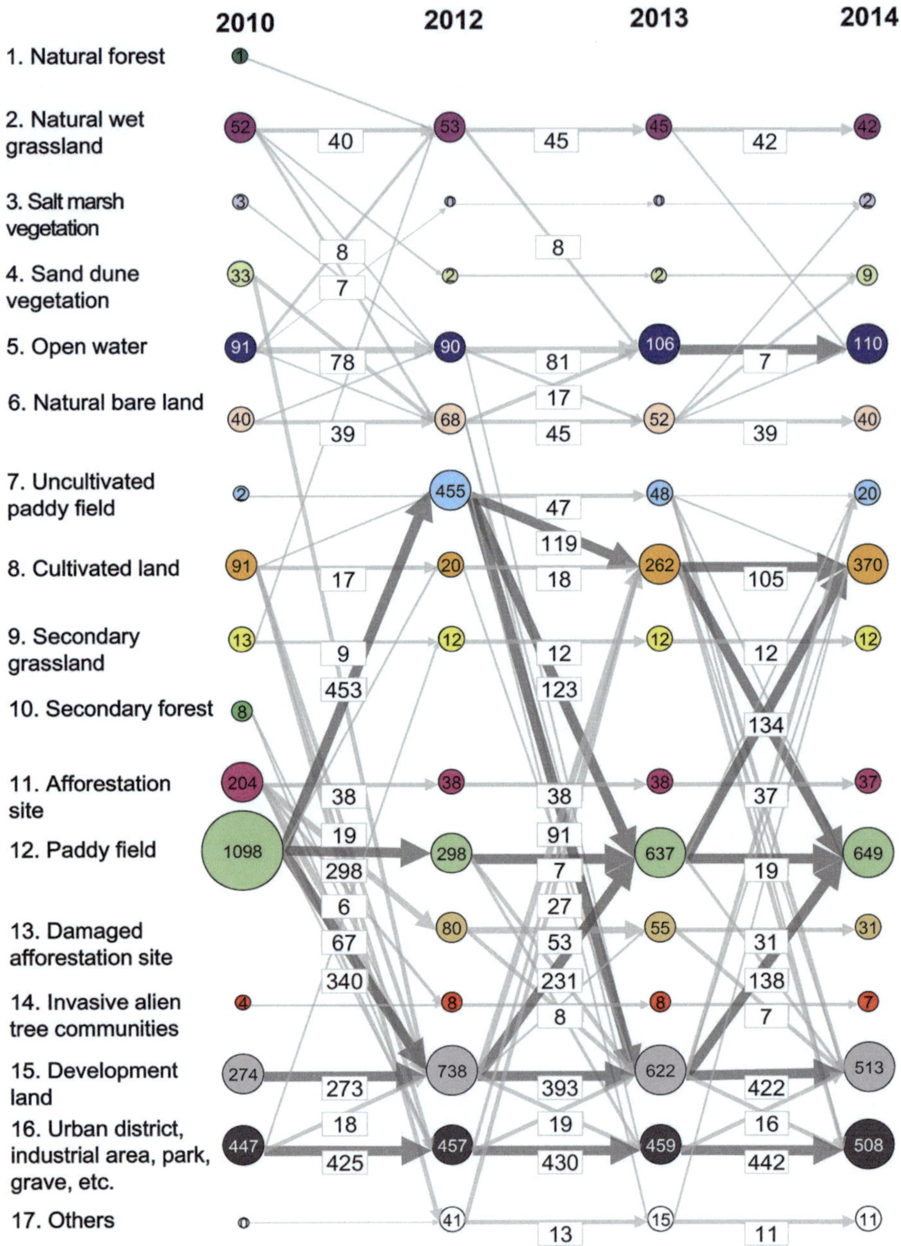

Fig. 23.3 Diagram of vegetation changes before and after the earthquake in Nanakita River mouth area in the Sendai Plain, Miyagi Prefecture. The legends of vegetation maps were categorized into 17 categories above. The up to five largest changes in each category of each year were charted as *arrows*. Less than 1 ha changes were not charted as *arrows* except for important changes (e.g., open water to salt marsh vegetation) (After BCMoE forthcoming)

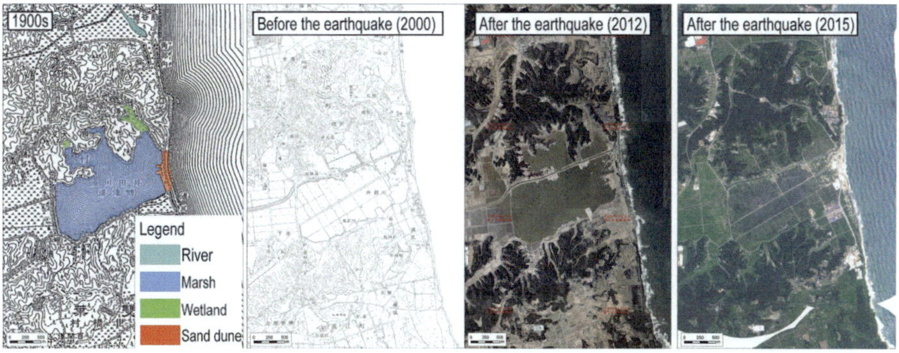

Fig. 23.4 Comparison between the topographic features in 1900s and the statuses before and after the earthquake in Itagawaura, Fukushima Prefecture *Geographical map: 1/25,000 map of GIAJ **Aerial Photos: Take by GIAJ ***Satellite imagery: "Includes material (c) (2015) BlackBridge AG. All rights reserved" (After BCMoE 2013)

standing the original natural environment where we do not know from the present condition.

23.3.2 Changes at Key Sites

23.3.2.1 Fauna and Flora

In a total of 16 key sites, the surveys on fauna and flora identified 1,893 species (334 animal and 947 plant species) including 128 rare species[3] (73 animal and 55 plant species). The results of these surveys were organized and recorded as a database of approx. 20,000 records of information on fauna and flora. Many of these organisms seemed to have adapted to the natural disturbance, including plants that regenerated from dormant seeds, which were buried before human turned the areas into rice paddies. Also, some animals that made use of such an environment appeared in the area after the earthquake.

23.3.2.2 Belt Transects Survey

Figure 23.5 shows an example of the result of belt transects survey in Idoura, Miyagi Prefecture. This example shows that, in 2 years after the earthquake, a variety of small-sized communities were distributed in a mosaic pattern, and later these communities merged into a single large community.

[3] Rare species are the species listed in National and Prefectural Red Lists.

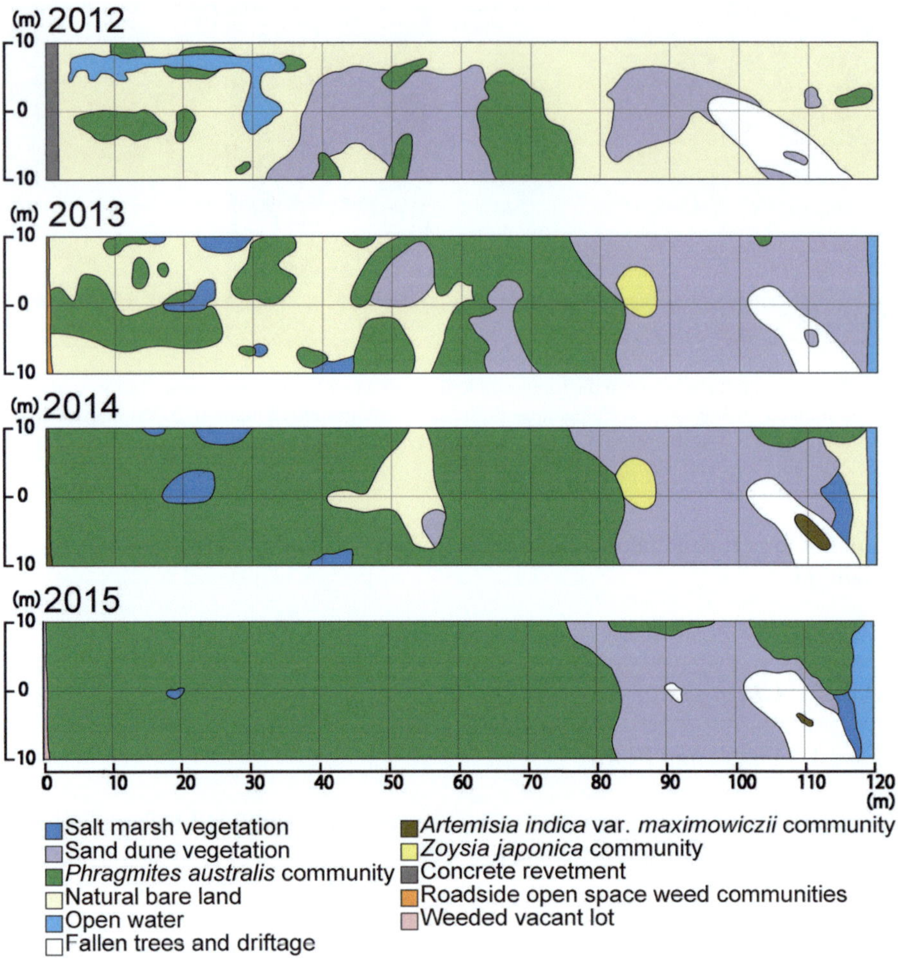

Fig. 23.5 The result of belt transects survey in Idoura, Miyagi Prefecture (After BCMoE forthcoming)

23.3.3 Shorelines and Land Cover Change of Coasts

The earthquake and tsunami had a great impact on the coast as shown by the recession of many shorelines due to the tsunami and the loss of sandy beaches and coastal forests behind them. Much of the shoreline which had receded showed some signs of recovery, while the recovery was slow at other coasts such as river mouth bars and pocket beaches.

Figure 23.6 shows the land cover areas of sand dunes, sand dune vegetation, coastal forests, and coastal artificial structures in the beaches (680.5 km) in 2000s (before the earthquake), 2011 (just after the earthquake), and 2014. Compared with in 2011 and in 2014, the land cover statuses of sand dunes, sand dune vegetations,

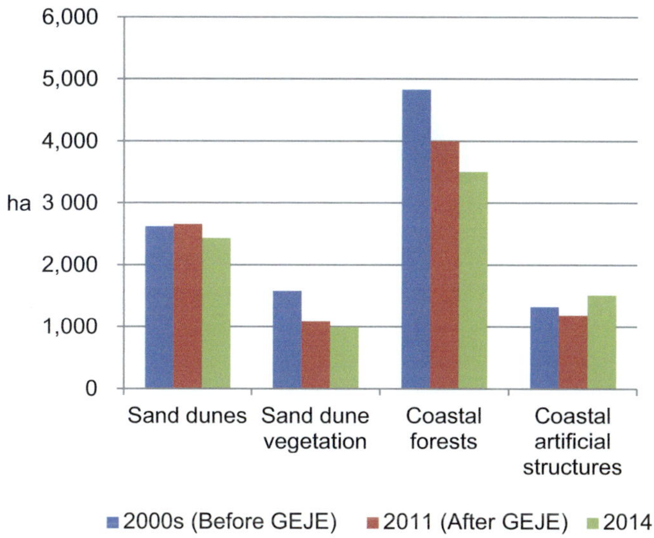

Fig. 23.6 Change of land cover status behind the shorelines of the beaches in the tsunami-struck natural coast (After BCMoE 2013, 2014, 2015, forthcoming)

and coastal forests were reduced respectively by 225 ha, 92 ha, and 496 ha. On the other hand, the areas of coastal artificial structures (e.g., tide embankments) were increased by 331 ha during the same period.

23.3.4 Seagrass and Seaweed Beds

Heavy impact due to the GEJE was given to the sandy and/or muddy seabed, and the majority of seagrass beds were lost because seagrass generally grow on the sediment of sand and silt (Fig. 23.7). Despite this, by 2015, some recoveries of seagrass beds were observed in various parts of the coasts with rather dense cover of seagrass. In contrast, seaweed beds, which generally grow on rocks and stones, were little affected by the tsunami and other events (Fig. 23.7).

23.3.5 Changes at Specific Sites of Monitoring Site 1000 Sites

23.3.5.1 Tidal Flats in the 18 Sites

The tsunami caused by the GEJE carried away a large number of living organisms from tidal flats, disrupted or destroyed reed beds, and reduced the areas of tidal flats. As a result of this survey, the number of species found increased in many sites in

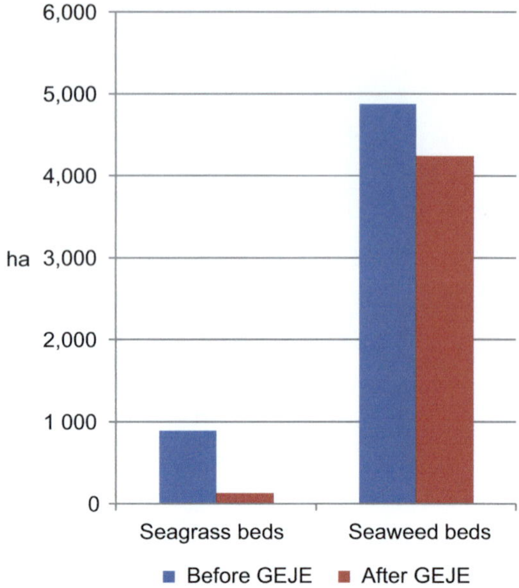

Fig. 23.7 Accumulated distribution areas of seagrass (*Zostera* spp.) and seaweed (Phaeophyceae spp.) beds in Iwate Prefecture to northern Fukushima Prefecture before the GEJE, the satellite images and aerial photographs were acquired in 2003 to 2011, and after GEJE, the satellite images and aerial photographs were acquired in 2011 to 2015 (After BCMoE forthcoming)

2015, though the community compositions at some sites were very different from those before the earthquake (Fig. 23.8). Furthermore, in contrast with the recovery of the number of species, the population densities have not recovered since 2012 in many sites. It seemed to take some more time until the populations recover to the level as before the earthquake.

23.3.5.2 Seagrass Beds in the Five Sites

The tsunami had carried seagrass away with the sediments of sand and silt where seagrass generally grow on with rootstocks spread on the seabed. Moreover, the change of light conditions due to the ground subsidence also influenced the growth of the remained seagrass. Thus, even after the earthquake, the coverage of the seagrass has not generally recovered in the five sites although the coverage has increased in some other sites. It revealed that large variation of recovery processes exists among the seagrass beds in the tsunami-flooded area.

23.3.5.3 Seaweed Beds in the Four Sites

Unlike seagrass, seaweed grows firmly fixed to rocks and stones with rhizoid, and the tsunami rarely carried them away in the four sites. Even if their phyllodes are torn away, they are able to regenerate soon. However, it was observed that the numbers of species and species compositions were changed in some sites after the

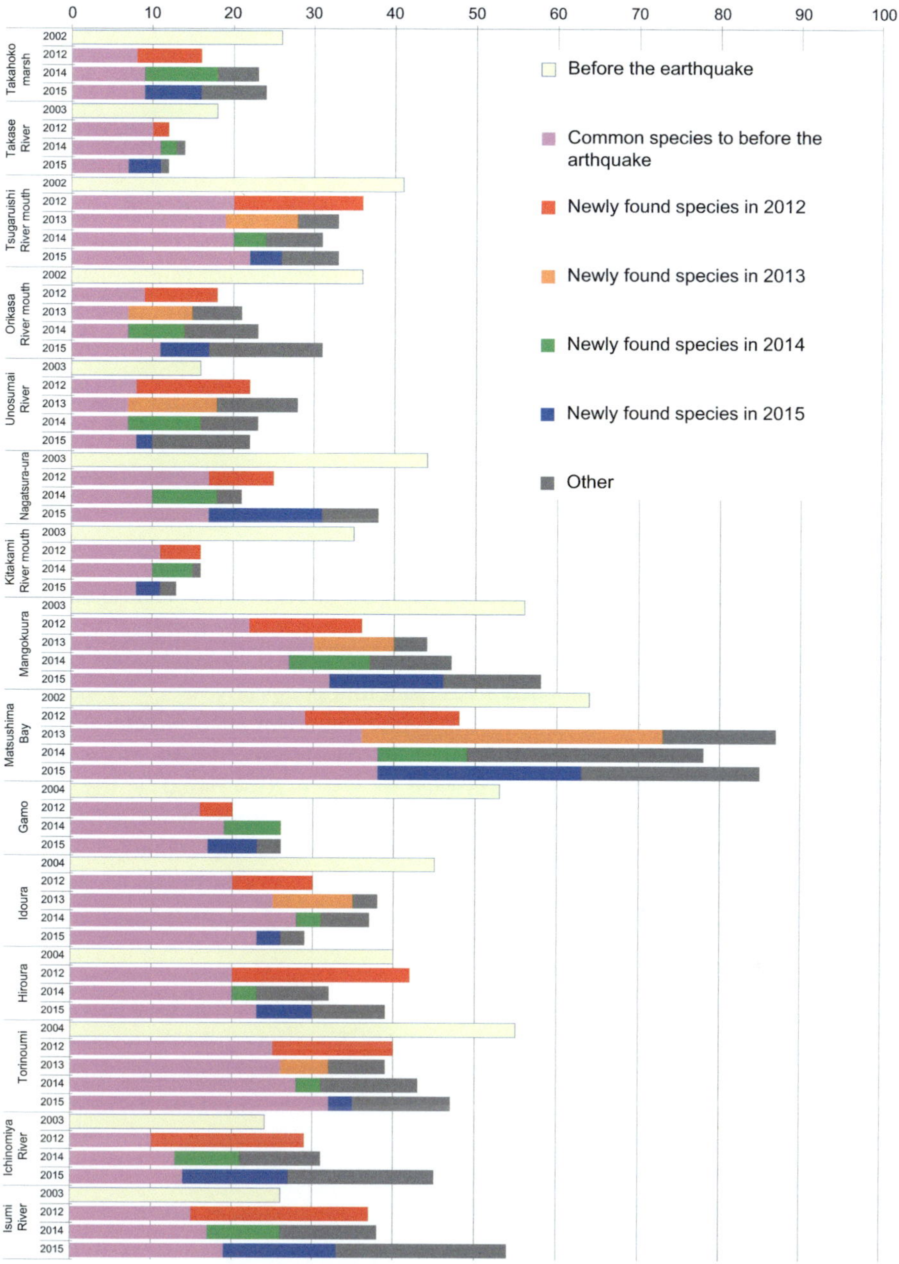

Fig. 23.8 Comparison of the number of species of benthos at various tidal flat sites before and after the earthquake. The data acquired in 7th National Surveys on the Natural Environment (BCMoE 2007) was used as the pre-earthquake data and the data acquired in this monitoring survey (BCMoE 2013, 2014, 2015, forthcoming) was used as the post-earthquake data (the number of species in qualitative research)

earthquake. Overall, the coverage of seaweed in 2015 was similar to or larger than before the earthquake in the surveyed sites.

23.3.5.4 Seabird Breeding

The GEJE occurred in March, not in a breeding season for seabirds in the four surveyed sites. Therefore, the earthquake and tsunami seemingly had no direct impact on seabirds although soil runoff and vegetation changes, which might disturb the seabird by changing the environments for breeding, were observed in some parts of the surveyed islands. The indirect impacts from these changes (i.e., soil runoff and vegetation changes) on seabirds are still unclear. Thus, further monitoring is needed.

23.4 Important Habitat Maps

Based on the results of the conducted surveys and experts interviews, "Important Habitat Maps" (Fig. 23.9) were created to show the statuses of the natural environment after the earthquake focusing on the important habitats for living organisms in order to conserve the natural environment. The maps define 11 habitats[4] as "important habitats" in the tsunami-flooded areas from four perspectives.[5]

Since it was urgently required to identify the important environment that need consideration and utilize this information in the reconstruction projects, the "Important Habitat Maps" were first created in March 2014 based on the survey results obtained in 2012. This information was updated as "Important Habitat Maps 2015" in March 2016 with in enlarged area (i.e., Fukushima Prefecture).

23.5 Information Platform

All of the data sets of the series of surveys (Table 23.1) are opened to access at the web site, "Shiokaze Natural Environment Log" <http://www.shiokaze.biodic.go.jp/>.

These data sets suggested that organisms and ecosystems were seemed to be rather adapted to the natural disturbance, such as earthquake and tsunami, than devastated, showing quite rapid recovery, although they are still in the middle of recovery process.

---

[4] Eleven important habitat: (1) seaweed beds, (2) seagrass beds, (3) tidal flats, (4) sand dune (sandy beach and sand dune vegetations), (5) natural vegetation in coastal cliff areas, (6) remind forest areas, (7) formerly forested area (mosaic-like varied disrupted environment), (8) marsh (salty and freshwater marsh vegetation), (9) grassland (seminatural grassland), (10) uncultivated paddy field, and (11) open waters (e.g., rivers and lakes).

[5] Four perspectives: (1) important places for inhabitation and growth of rare species, (2) places with high biodiversity, (3) places with high potentials of nature, and (4) important places to interact with nature.

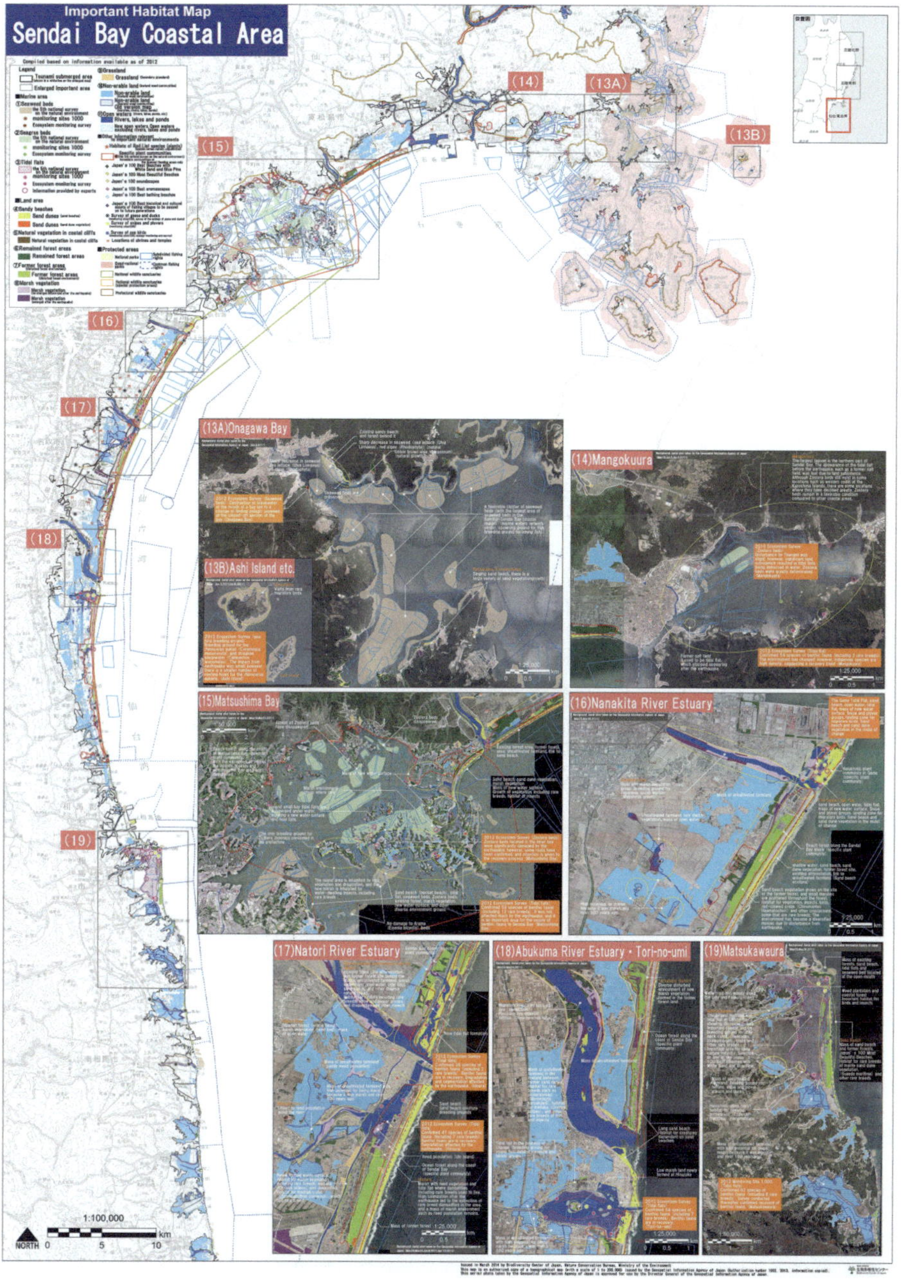

Fig. 23.9 Important habitat map (first edition of Sendai Bay Coastal Area) (After BCMoE 2014)

Table 23.1 List of main output

Output (format)	Details
Vegetation map (GIS data)	The tsunami-flooded areas (578 km^2) in 2000s, 2012, 2013, and 2014
Vegetation change map (GIS data)	Comparisons of the vegetation maps (578 km^2) of between 2000s and 2012, 2012 and 2013, as well as 2013 and 2014
Vegetation survey (data sheet)	2,985 sites in 2012, 681 sites in 2013, 787 sites in 2014, and 117 sites in 2015
Shoreline and land cover status of the coast survey (GIS data)	Land cover status of the beaches in the tsunami-flooded areas; 680.5 km in 1970s, 2000s, 2011, and 2014
	Shoreline status of the whole coast in the tsunami-flooded areas: 2,284 km in 2011, 2,273 km in 2014
Interpretation of the old edition Map (GIS data)	51 sheets of the Old Edition Map published by GIAJ from 1903 to 1917
Fauna and Flora in key sites (Schematic profile, database)	Belt transects survey and fauna and flora survey: 4 lines in 3 sites in 2012, 19 lines in 13 sites in 2013, 9 lines in 6 sites in 2014, and 23 lines in 16 sites in 2015
Distribution map of seagrass and seaweed beds (GIS data)	Iwate Prefecture to the northern part of Fukushima Prefecture before the GEJE and Aomori Prefecture to Chiba Prefecture after the GEJE
Distribution map of tidal flats (GIS data)	Iwate Prefecture to Fukushima Prefecture in 2014
Follow-up of monitoring site 1,000 (data sheet)	Tidal flats with benthos: 15 sites in 2012, 9 sites in 2013, 16 sites in 2014, 18 sites in 2015
	Seagrass beds: 5 sites in 2012 and 2013, 4 sites in 2014 and 2015
	Seaweed beds: 4 sites in 2012, 2013, 2014, and 2015
	Seabird breeding sites: 4sites in 2012, 2013, 2014, and 2015
Important habitat map (print, PDF)	2013 edition, 2015 edition
Brochure (print, PDF)	2013 edition, 2015 edition

References

Biodiversity Center of Japan, Ministry of the Environment (2007) 7th national surveys on the natural environment, fiscal year 2006 survey information. http://www.biodic.go.jp/kiso/fnd_list_h.html. Accessed 31 Mar 2016

Biodiversity Center of Japan, Ministry of the Environment (2013) Shiokaze natural environment, fiscal year 2012 survey information. http://www.shiokaze.biodic.go.jp. Accessed 31 Mar 2016

Biodiversity Center of Japan, Ministry of the Environment (2014) Shiokaze natural environment, fiscal year 2013 survey information. http://www.shiokaze.biodic.go.jp. Accessed 31 Mar 2016

Biodiversity Center of Japan, Ministry of the Environment (2015) Shiokaze natural environment, fiscal year 2014 survey information. http://www.shiokaze.biodic.go.jp. Accessed 31 Mar 2016

Biodiversity Center of Japan, Ministry of the Environment (forthcoming) Shiokaze natural environment, fiscal year 2015 survey information. http://www.shiokaze.biodic.go.jp.

Geospatial Information Authority of Japan of Land, Infrastructure and Transportation Ministry (2011) Press release. http://www.gsi.go.jp/chibankansi/chikakukansi_tohoku2.html. Accessed 31 Mar 2016

ERRATUM

Chapter 21
Flora of Freshwater Wetlands in the Tsunami-Affected Zone of the Tohoku Region

Mahoro Suzuki

J. Urabe, T. Nakashizuka (eds.), *Ecological Impacts of Tsunamis on Coastal Ecosystems*, Ecological Research Monographs,
DOI 10.1007/978-4-431-56448-5_21

DOI 10.1007/978-4-431-56448-5_24

In Chapter 21, titled "Flora of Freshwater Wetlands in the Tsunami-Affected Zone of the Tohoku Region," the author name should read as Suzuki Mahoro and this has been updated.

In the front matter, author name has been updated as Suzuki Mahoro in the table of contents.

The updated original online version for this chapter can be found at
DOI 10.1007/978-4-431-56448-5_21

J. Urabe, T. Nakashizuka (eds.), *Ecological Impacts of Tsunamis on Coastal Ecosystems*, Ecological Research Monographs,
DOI 10.1007/978-4-431-56448-5_24

MIX
Papier aus verantwortungsvollen Quellen
Paper from responsible sources
FSC® C105338

If you have any concerns about our products,
you can contact us on
ProductSafety@springernature.com

In case Publisher is established outside the EU,
the EU authorized representative is:
Springer Nature Customer Service Center GmbH
Europaplatz 3, 69115 Heidelberg, Germany

Printed by Libri Plureos GmbH
in Hamburg, Germany